Metal Ions in Fungi

MYCOLOGY SERIES

Editor

Paul A. Lemke

Department of Botany and Microbiology
Auburn University
Auburn, Alabama

1. *Viruses and Plasmids in Fungi*, edited by Paul A. Lemke
2. *The Fungal Community: Its Organization and Role in the Ecosystem*, edited by Donald T. Wicklow and George C. Carroll
3. *Fungi Pathogenic for Humans and Animals* (in three parts), edited by Dexter H. Howard
4. *Fungal Differentiation: A Contemporary Synthesis*, edited by John E. Smith
5. *Secondary Metabolism and Differentiation in Fungi*, edited by Joan W. Bennett and Alex Ciegler
6. *Fungal Protoplasts*, edited by John F. Peberdy and Lajos Ferenczy
7. *Viruses of Fungi and Simple Eukaryotes*, edited by Yigal Koltin and Michael J. Leibowitz
8. *Molecular Industrial Mycology: Systems and Applications for Filamentous Fungi*, edited by Sally A. Leong and Randy M. Berka
9. *The Fungal Community: Its Organization and Role in the Ecosystem, Second Edition*, edited by George C. Carroll and Donald T. Wicklow
10. *Stress Tolerance of Fungi*, edited by D. H. Jennings
11. *Metal Ions in Fungi*, edited by Günther Winkelmann and Dennis R. Winge

Additional Volumes in Preparation

Anaerobic Fungi: Biology, Ecology, and Function, edited by Douglas O. Mountfort and Colin G. Orpin

Metal Ions in Fungi

edited by

Günther Winkelmann
University of Tübingen
Tübingen, Germany

Dennis R. Winge
University of Utah
Salt Lake City, Utah

CRC Press
Taylor & Francis Group
Boca Raton London New York

CRC Press is an imprint of the
Taylor & Francis Group, an **informa** business

First published 1994 by Marcel Dekker, Inc.

Published 2019 by CRC Press
Taylor & Francis Group
6000 Broken Sound Parkway NW, Suite 300
Boca Raton, FL 33487-2742

First issued in paperback 2019

ISBN 13: 978-0-367-44957-5 (pbk)
ISBN 13: 978-0-8247-9172-8 (hbk)

Visit the Taylor & Francis Web site at
http://www.taylorandfrancis.com

and the CRC Press Web site at
http://www.crcpress.com

Library of Congress Cataloging-in-Publication Data

Metal ions in fungi / edited by Günther Winkelmann, Dennis R. Winge.
p. cm. -- (Mycology series; v. 11)
Includes bibliographical references and index.
ISBN 0-8247-9172-X
1. Fungi--Physiology. 2. Metal ions--Metabolism. I. Winkelmann, Günther. I. Winge, Dennis R. III. Series.
QK601.M48 1994
589.2'0419214-dc20 93-46026
CIP

Series Introduction

Mycology is the study of fungi, that vast assemblage of microorganisms which includes such things as molds, yeasts, and mushrooms. All of us in one way or another are influenced by fungi. Think of it for a moment—the good life without penicillin or a fine wine. Consider further the importance of fungi in the decomposition of wastes and the potential hazards of fungi as pathogens to plants and to humans. Yes, fungi are ubiquitous and important.

Mycologists study fungi either in nature or in the laboratory and at different experimental levels ranging from descriptive to molecular and from basic to applied. Since there are so many fungi and so many ways to study them, mycologists often find it difficult to communicate their results even to other mycologists, much less to other scientists or to society in general.

This series establishes a niche for publication of works dealing with all aspects of mycology. It is not intended to set the fungi apart, but rather to emphasize the study of fungi and of fungal processes as they relate to mankind and to science in general. Such a series of books is long overdue. It is broadly conceived as to scope, and should include textbooks and manuals as well as original and scholarly research works and monographs.

The scope of the series will be defined by, and hopefully will help define, progress in mycology.

Paul A. Lemke

Preface

This book was compiled as an overview of recent developments in the field of metal ions in fungi. Special emphasis has been given to mechanisms by which the cellular concentration of metal ions is regulated. Metal ion transport and intracellular metal ion sequestration are two prominent mechanisms discussed. In addition, the catalytic and/or structural role of metal ions as prosthetic groups in fungal enzymes is presented. The focus is on certain essential metal ions, such as iron, copper, zinc, and manganese. This book was not intended to provide a comprehensive review of all metal ions. Important metal ions like sodium, potassium, magnesium, and calcium, mainly involved in electrophysiological and signal-transducing cellular functions, could not be considered in this volume.

This book contains information on how and to what extent certain trace elements are recognized and inserted into cellular proteins. Rapid advances are taking place in bioinorganic chemistry and molecular biology, creating a rich interface between these disciplines. These advances have attracted an increasing number of scientists from different fields. We hope that this book will provide a detailed overview of many intriguing topics in the biology of metals to spur further research into this important field. Many questions still remain unanswered concerning mechanisms of homeostasis, transport, storage, and physiological functions of metal ions.

Our profound thanks go to all contributors to this book. We are grateful to the series editor, Paul Lemke, for inviting us to create and edit this book on metal ions as part of the Mycology Series. We also wish to express our appreciation to Elyce Misher and the staff at Marcel Dekker, Inc., for expert assistance in the preparation of this volume.

Günther Winkelmann
Dennis R. Winge

Contents

Contributors

Georg Auling, Ph.D. Professor, Institute of Microbiology, University of Hannover, Hannover, Germany

Michael Berreck Department of Microbiology, University of Innsbruck, Innsbruck, Austria

Jean-Michel Camadro, Ph.D. Laboratoire de Biochimie des Porphyrines, Institut Jacques Monod, University of Paris VII, Paris, France

Maria Teresa Carrì Department of Biology, University of Rome "Tor Vergata," Rome, Italy

Maria Rosa Ciriolo Department of Biology, University of Rome "Tor Vergata," Rome, Italy

Hans Diekmann, Ph.D. Professor, Institute of Microbiology, University of Hannover, Hannover, Germany

William H. Dvorachek, Jr. Department of Biology, University of New Mexico, Albuquerque, New Mexico

Francesca Galiazzo Department of Biology, University of Rome "Tor Vergata," Rome, Italy

Kurt Haselwandter Professor, Department of Microbiology, University of Innsbruck, Innsbruck, Austria

Yukimasa Hayashi, Ph.D. Department of Genetics, Institute for Developmental Research, Aichi Colony, Kasugai, Aichi, Japan

Teresa Keng, Ph.D. Assistant Professor, Department of Microbiology and Immunology, McGill University, Montreal, Quebec, Canada

Daniel J. Kosman, Ph.D. Professor, Department of Biochemistry, School of Medicine and Biomedical Sciences, State University of New York at Buffalo, Buffalo, New York

Pierre Labbe, Ph.D. Professor of Biochemistry, Laboratoire de Biochemie des Porphyrines, Institut Jacques Monod, University of Paris VII, Paris, France

Rosine Labbe-Bois, Ph.D. Laboratoire de Biochemie des Porphyrines, Institut Jacques Monod, University of Paris VII, Paris, France

Sally Ann Leong, Ph.D. Research Chemist, United States Department of Agriculture and Associate Professor, Department of Plant Pathology, University of Wisconsin, Madison, Wisconsin

Emmanuel Lesuisse, Ph.D. Chargé de Recherches au CNRS, Department of Microbiology Laboratoire de Biochemie des Porphyrines, Institut Jacques Monod, University of Paris VII, Paris, France

Ian G. Macreadie, B.Sc.(Hons.), Ph.D. Senior Research Scientist, Biomolecular Research Institute, Parkville, Victoria, Australia

Berthold F. Matzanke, Ph.D., Dipl.-Chem. Department of Microbiology/Biotechnology, University of Tübingen, Tübingen, Germany

Kent F. McCue Plant Gene Expression Center, United States Department of Agriculture and Department of Plant Biology, University of California—Berkeley, Albany, California

Baigen Mei, Ph.D. Department of Plant Pathology, University of Wisconsin, Madison, Wisconsin

Norihiro Mutoh Institute for Developmental Research, Aichi Colony, Kasugai, Aichi, Japan

Donald O. Natvig, Ph.D. Associate Professor, Department of Biology, University of New Mexico, Albuquerque, New Mexico

Daniel F. Ortiz Plant Gene Expression Center, United States Department of Agriculture and Department of Plant Biology, University of California—Berkeley, Albany, California

David W. Ow, Ph.D. Senior Scientist, Plant Gene Expression Center, United States Department of Agriculture and Department of Plant Biology, University of California—Berkeley, Albany, California

Jennifer L. Pinkham, Ph.D. Assistant Professor, Department of Biochemistry and Molecular Biology, University of Massachusetts, Amherst, Massachusetts

Hans Jürgen Plattner, Ph.D. Institute of Microbiology, University of Hannover, Hannover, Germany

I. S. Ross, B.Sc., Ph.D. Department of Biological Sciences, University of Keele, Keele, Staffordshire, England

Giuseppe Rotilio, M.D., Ph.D. Professor of Biochemistry, Department of Biology, University of Rome "Tor Vergata," Rome, Italy

Andrew K. Sewell, Ph.D. Departments of Medicine and Biochemistry, University of Utah Medical Center, Salt Lake City, Utah

David M. Speiser Plant Gene Expression Center, United States Department of Agriculture and Department of Plant Biology, University of California—Berkeley, Albany, California

Kenneth Sylvester Department of Biology, University of New Mexico, Albuquerque, New Mexico

Dick van der Helm, Ph.D. Professor, Department of Chemistry, University of Oklahoma, Norman, Oklahoma

Dennis R. Winge, Ph.D. Professor, Departments of Medicine and Biochemistry, University of Utah Medical Center, Salt Lake City, Utah

Günther Winkelmann, Ph.D. Professor, Department of Microbiology and Biotechnology, University of Tübingen, Tübingen, Germany

Metal Ions in Fungi

1

Transition Metal Ion Uptake in Yeasts and Filamentous Fungi

Daniel J. Kosman
School of Medicine and Biomedical Sciences, State University of New York at Buffalo, Buffalo, New York

I. INTRODUCTION

Free-living organisms are scavengers. They evolve in environments that are limited in many of the essential elemental and molecular chemical components required to sustain their growth and successful replication. Selective advantage is conferred to those species capable of accumulating and retaining those components present in limiting, trace amounts. By definition, the metals of the first transition series in Groups Ib, IIb, Va-VIIa, and VIII are among these trace components. The trace elements, Mn, Fe, Co, Ni, Cu, and Zn, as well as V and Cr, have all been demonstrated to be essential for life in all organisms. Yet, as illustrated in Table 1, with the exception of Fe, none represents a significant fraction of the earth's crust. The apparent bioavailability of these metals, particularly Fe, is restricted further by their limited solubility in the O_2-containing, commonly pH-neutral (aqueous) environments that many yeasts and filamentous fungi inhabit. This condition is due to the formation of essentially insoluble metal hydroxides of the higher valence ionic states of these metals, as illustrated also in Table 1. The common carbonates and phosphates of these metal ions are even less soluble. Thus even the apparent abundance of Fe (Table 1) is misleading.

Table 1 Composition and Solubility Data for Selected Transition Metals

Metal	Crust ppm	Sea Water ppm	Sea Water nM	YNB[a] nM	Solubility of metal hydroxides (Ksp)[b]
Mn	950	2×10^{-3}	40	6000	$<10^{-20}$
Fe	5×10^{4}	1×10^{-2}	200	3000	1.6×10^{-14} [Fe(II)] 1.1×10^{-36} [Fe(III)]
Co	25	1×10^{-4}	2	n.i.[c]	$<10^{-20}$
Ni	75	2×10^{-3}	40	n.i.[c]	$<10^{-20}$
Cu	55	3×10^{-3}	60	600	2×10^{-15}
Zn	70	1×10^{-2}	200	6000	1.8×10^{-14}

[a]Yeast nitrogen base (Difco).
[b]For divalent ion unless noted otherwise.
[c]Not included in medium.
Source: Adapted from Ochiai (1977).

This brief review of transition metal ion uptake in yeasts and filamentous fungi is written with this condition specifically in mind; it will focus on the first six of the metals listed above. This article will not review the biophysical aspects of transport (Borst-Pauwels, 1981) nor will it cover uptake of Group Ia or IIa cations, the alkali and alkaline earth metals. Its discussion of Mn(II) and Fe(II/III) uptake will also be limited, since these ions are the subject of other chapters in this volume. The purpose of this chapter will be to consider these various paradigms for transition metal ion uptake: (1) the nature of the uptake mechanism; (2) the specificity for the metal; (3) the role(s) of energy, and the types of energy, involved in the uptake; (4) the requirement for reduction of an ion to a lower valence state to increase bioavailability and/or uptake; and (5) the question of efflux, or, conversely, what drives accumulation. Since no other chapter in this volume will cover Cu uptake in detail, this metal will be a focus of this chapter. In this review, the term "uptake" will be confined to the actual translocation of a metal ion across the plasmalemma. "Accumulation" can be ambiguous in this regard, since organisms that have cell walls are notorious for binding ("entrapping" may be a more appropriate word) ionic species external to this membrane. Whether this cell wall binding is functional in sustaining normal cell growth, or is gratuitous, will also be discussed.

A chapter of this length cannot be all-inclusive. The budding yeast *Saccharomyces cerevisiae* will be discussed the most in the chapter in part because of its dominance in the literature. In selecting the literature examples, I mean no slight to the many others, just as important, that I have not included. These are listed in the bibliography, however. *Saccharomyces* figures strongly in the chapter for another reason: its ascendence as the classic eukaryotic (molecular) genetic system. I hope to indicate clearly that the genetic tools and reagents needed to reveal the many details about the components and mechanisms of metal uptake that we do not yet know are becoming available. The insight into details that this chapter lacks should be ours in the near future in *Saccharomyces,* and in other yeasts and filamentous fungi whose genetics are only now being made experimentally tractable.

II. MECHANISM OF TRANSITION METAL ION UPTAKE: AN OVERVIEW

In this section, the five aspects of metal uptake enumerated above are discussed in brief to provide a context for the more detailed sections that follow. These latter sections will include all relevant citations.

A. Transition Metal Ion Uptake Is Facilitated

Uptake of those metal ions noted above is facilitated in yeasts and filamentous fungi. That is, all reported studies have demonstrated that uptake is saturable with respect to [metal ion] in the surrounding medium. The data in Fig. 1 (open circles) illustrate the saturation behavior of copper uptake into a wild type strain of *Saccharomyces cerevisiae* grown in the absence of sulfate. The relevance of this growth condition will be discussed in more detail below. Data like these for copper, here added as $Cu{\cdot}(H_2O)_6{}^{2+}$, yield kinetic constants for uptake that are given in Table 2; included in the table also are kinetic constants for uptake for a number of other metal ions for this and other yeasts and filamentous fungi.

Another characteristic of metal ion transport that indicates that it is facilitated is the fact that the metal ion, although present in the surroundings in coordination with some ligand(s), passes through the plasmalemma as the "free" ion. This requires that a ligand displacement reaction must occur periplasmically. This ligand displacement presumably involves some cellular component, perhaps the transporter itself. This aspect of Cu^{2+} uptake has been studied in some detail.

These data show that K_M values, in particular, do not differ greatly among metals and among genera and species. Not included in this table are additional and higher K_M values reported for some of these ions. While it is apparent that such kinetic behavior has been observed, the physiologic significance of such uptake is not clear; the exclusion of environmental metal when in excess would appear to be more advantageous than having a second, lower affinity uptake system. As Gadd and White (1989) have

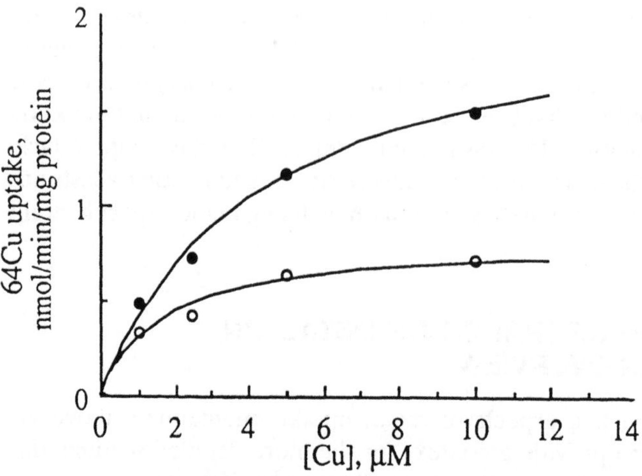

Figure 1 **Cu^{2+} concentration dependence of ^{64}Cu uptake in a wild type strain of *Saccharomyces cerevisiae* grown in sulfate-depleted (+methionine, open circles) or sulfate-supplemented media (closed circles). The lines represent the best fit to the Michealis-Menten equation with K_M=4 μM for both conditions, and V_{max}=0.10 nmol Cu/min/mg protein (sulfate-depleted) and 0.21 nmol Cu/min/mg protein (sulfate-supplemented). (Data adapted from Lin et al., 1993b.)**

pointed out, "where the concentration range [of the metal ion] used is high the affinity obtained is low." While the similarity in uptake kinetics for these several ions might be taken as evidence that the metals share a common "transporter," this fact has not been established, and, in fact, is not true for at least some of these ions (see below). In particular, there is a dearth of genetic evidence with respect to the transport of most of these metal ions; uptake mutants have, with the exception of Fe(III) uptake, gone unreported in the literature. On the other hand, this kinetic similarity presumably reflects the fact that the common laboratory strains of these microorganisms have been derived from similar metal-containing environments. For example, in the wine and brewing industries, strains of *S. cerevisiae* have been selected on the basis of their Cu resistance. Although this resistance certainly is a reflection of the tandem repeats present at the *CUP1* locus, the gene encoding the yeast Cu-thionein, it may also result from a reduced level of Cu uptake activity. The experiment has not been done to train a yeast or filamentous fungus on *limiting* Cu (or any other metal ion) to select for gain-of-function uptake "mutants." However, Cu uptake can be enhanced in *S. cerevisiae,* as indicated by the phenotype of a mutant in this yeast described in Sec. III.

Table 2 Kinetic Constants for Transition Metal Ion Uptake

Metal ion	Organism	K_M, μM	V_{max}[a]	Reference
Ni^{2+}	*N. crassa*	290	0.26 (mg dry wt)	Mohan et al. (1984)
Co^{2+}	*N. crassa*	1600	0.38 (mg dry wt)	Venkateswerlu and Sastry (1970)
	S. cerevisiae	77	5.4 (mg dry wt)	Norris and Kelly (1977)
Mn^{2+}	*C. utilis*	0.016	0.001 (mg dry wt)	Parkin and Ross (1986)
	A. niger	3	0.006 (mg dry wt)	Hockertz et al. (1987)
Zn^{2+}	*C. utilis*	1.3	0.21 (mg dry wt)	Failla et al. (1976)
	S. roseus	90	0.51 (mg dry wt)	Mowll and Gadd (1983)
	S. cerevisiae	3.7	1.6 (10^7 cells)	White and Gadd (1987)
Fe^{2+}	*S. cerevisiae*	5	0.0007 (10^7 cells)	Dancis et al. (1990)
Fe^{2+} or Fe^{3+}	*S. cerevisiae*	0.15	0.0026 (10^7 cells)	Eide et al. (1992)
Cu^{2+}	*P. ochro-chloron*	390	0.4 (10^7 cells)	Gadd and White (1985)
	S. cerevisiae	1.1	2.2 (mg dry wt)	De Rome and Gadd (1987)
	S. cerevisiae	3.8	0.1 (mg protein)	Lin et al. (1993b)

[a]V_{max} values given in nmol metal/min per unit of cell mass as indicated.

B. Specificity of Metal Ion Uptake

Early evidence suggested that the yeast (*S. cerevisiae*) plasma membrane contained a general divalent cation transport system, that, while apparently selective for Mn^+ (and Mg^{2+}), exhibited activity for Co^{2+}, Zn^{2+}, and Ni^{2+} among other cations. Unfortunately, this conclusion was based on data obtained at relatively high concentrations of the various metal ions, ≥ 10 μM, which were well above the range of K_M values now recognized for many of them (Table 2). More recent work has demonstrated high affinity and specific uptake systems for some of these ions in *S. cerevisiae* and other yeasts and fungi, e.g., Mn^{2+}, Cu^{2+}, Zn^{2+} and Co^{2+}. High affinity $Fe^{3+/2+}$ uptake is also specific.

Three aspects of metal metabolism bear on the issue of specificity: (1) a correspondence between environmental metal ion concentration and an experimental K_M value; (2) the aqueous chemistry of the various ions themselves; and (3) potential regulation of metal uptake. It is a biologic paradigm that affinity constants commonly reflect the range of solute concentration. Thus a K_M value of 1 mM, for example, does not make biologic sense for a solute whose concentration is 1–10 μM or less. Although there is some advantage to such a system in that the velocity of uptake is first order with respect to [solute], and thus maximally responsive to [solute], unless compensated for by a large V_{max} the net transport flux remains fairly low in such a low affinity system.

The second point relates to the quite different types of (coordination) chemistry for the members of this group of transition metals with respect to both geometry and ligand type. Mn^{2+}, for example, prefers oxygenous ligands and an octahedral coordination, while Cu^{2+} prefers nitrogenous ones and is most commonly four, not six, coordinate. This difference in affinity is reflected in the classic data on stabilities of the ethylene diamine complexes of these divalent transition metals (and Zn^{2+}), as illustrated in Fig. 2. Ligand displacement at these various metals differs as well, which is important because the exogenous ligand appears to be displaced, presumably by a cell component, prior to the uptake of the metal ion itself. For example, the water exchange rates of the aqueous complexes of the divalent states of these metal ions varies from 3×10^4 s^{-1} for Ni(II) to 8×10^9 s^{-1} for Cu(II). Just as significantly, these rates vary with the valence state of a given metal: for Fe(III) it is 3×10^3 s^{-1}, while for Fe(II) it is 3×10^6 s^{-1}. Lastly, the molecular functions of these metals are quite different. For example, Mn is a prosthetic group of pyruvate carboxylase (although Mg^{2+} is found in *S. cerevisiae*), and Zn^{2+} is a prosthetic group of hydrolases like the carboxypeptidases. In contrast, Fe is found as a component of O_2-activating enzymes, and Cu is required for superoxide detoxification. These functional differences are due to the fact that some of these metals,

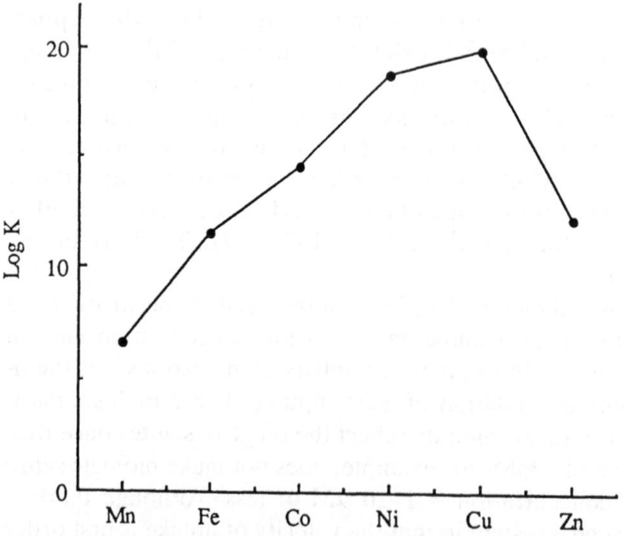

Figure 2 Stability constants of transition metal ion complexes with ethylenediamine. (Data adapted from Ochiai, 1977.)

like Mn^{2+} and Zn^{2+}, are considered "hard," and as such are good Lewis acids, while others, like Fe^{2+} and Cu^{2+}, are "softer" and more redox active.

Given the diverse functions for these metals, the cell's requirement for any one of them would reflect a specific environmental condition. Regulation of uptake of a particular metal would seem reasonable in order correctly to fulfill this need. To have a general transporter whose activity for all metals varied simultaneously would defeat the purpose of such a regulatory circuit. For all of these reasons, it is perhaps naive to consider *a priori* that there would be a transporter common to a group of metal ions with such different chemistries and biologic functions. The bulk of the evidence, elaborated on more fully below, confirms this suspicion.

C. Transition Metal Ion Uptake Is Energy Dependent

There is no question that the energy status of the cell modulates the uptake of all of these ions. What is not clear is the mechanism of this apparent energy coupling. Generally, there are four ways in which energy might be required to sustain the uptake of an extracellular solute. (1) Uptake is coupled directly to the hydrolysis of ATP. (2) Uptake is driven by a counterion concentration gradient established by the coupled hydrolysis of ATP. This uptake could involve either a symport or an antiport transport mechanism. With reference to known systems, the counterion could be H^+, K^+, Na^+, or some combination of these cations. With respect to H^+, uptake could be driven by a proton motive force established by the (a) plasma membrane ATPase. A true symport mechanism, however, which requires a concentration gradient of the cotransported ion, lower inside, is not reasonable for a free-living organism (unlike a cell within a multicellular organism). On the other hand, a symport with a counteranion, like Cl^- or PO_4^{3-}, is possible. (3) Concentration (uptake) of the ion within the cell requires or is linked to posttranslocation processing. Examples of this are synthesis of a metal-binding protein (or ligand, like glutathione); vesicular transport, e.g., uptake into a/the vacuole; endocytosis and vesicular trafficking. Of these, the last is least likely, and there is no evidence that metal uptake in yeasts and fungi involves endocytosis, since these organisms do not secrete metal-scavenging molecules. (4) Metal uptake requires reduction of the metal ion from a higher to lower valence state, e.g., Cu(II) → Cu(I) or Fe(III) → Fe(II). Although this reduction most likely would be dependent on a reduced pyridine nucleotide (NADH or NADPH), not ATP, the energy charge in the cell would reasonably be reflected in the redox balance (state) as well.

The first two of these mechanistic motifs would underlie what is commonly referred to as active transport, in that the energy released from ATP hydrolysis is used directly, or nearly so, to concentrate a solute across a membrane. The second two are no less "active," in the sense that cellular energy is used to effect this concentration, but the actual translocation across the plasmalemma is essentially diffusive, albeit facilitated. Furthermore, some combination of any of these four mechanisms is possible. As discussed below and in Chapter 20, reduction of Fe(III) to Fe(II) does occur prior to or during Fe uptake; Fe(II) appears to be the substrate for a putative Fe transport element. However, the transport itself may be energy dependent as well, and the concentration of Fe within the cell may be linked strongly to its binding to intracellular components, including ferritin. Also, transport of many ions into the vacuole in yeast is suggested, which might represent part of the energy dependence of cell accumulation of such metals.

D. Exogenous Metal Ion Reduction and Uptake

Practically, this potential feature of transition metal ion uptake only pertains to Fe(III) and Cu(II), although Co(III) might be considered as well. This limitation is due to the fact that only these metals, in possible biologic coordination complexes, have redox potentials within a few hundred millivolts of zero. This boundary condition ensures that this ionic form is neither so oxidizing that it is reduced nonspecifically by any one of a number of organic compounds, nor so reducing in the lower valence state that its autooxidation cannot be controlled. Thus Co(III) in N coordination can be expected to have $E^{\circ\prime}$ values around 100 mV, Cu(II) complexes around 300 mV, and Fe(III) complexes with oxygenous ligands around 500 mV. Why consider Cu(II) and Fe(III) reduction in the context of uptake? First, these metals are in these valence states in an aerobic environment; on the other hand, within the reducing milieu of the cell, unless stabilized in the higher valence state, they will be present as Cu(I) and Fe(II). Cu(I) and Fe(II), in comparison to Cu(II) and Fe(III), are also the most stable to hydrolysis to insoluble hydroxides (and phosphates).

However, the most compelling rationale for a reduction step prior to uptake is that reduction would strongly catalyze the displacement of the metal from the chelated form in which it is presented to the plasmalemma. As noted above, most evidence suggests that the metal is taken up as the free ion. The data for H_2O exchange of Fe(II) in comparison to Fe(III), the former being 10^3 times faster than the latter, illustrate how reduction promotes ligand exchange. Reduction of typically square planar Cu(II) to tetrahedral Cu(I) labilizes the metal to ligand displacement in a similar fashion. However, only Fe(III) uptake is demonstrably dependent on prior reduction. This was established by the isolation of a mutant unable to transport Fe(III), which, however, exhibited wild type Fe(II) uptake. This mutant lacks a plasma membrane Fe(III) reductase activity. Because the reduction is not rate limiting for Fe(III) uptake, it was not possible to prove a role for reduction in uptake by biochemical kinetic means. To do so required construction of a null allele at the Fe reductase gene locus. This demonstrates the importance of uptake mutant isolation and characterization to our understanding of metal uptake in yeasts and filamentous fungi.

E. Metal Ion Accumulation and Efflux

Aside from the periplasmic binding of metal ions, which is generally nonspecific (see below), there are three (intra)cellular fractions that can be demonstrated to accumulate newly arrived metals: (1) a particulate or "bound," fraction, which can include intact organelles (typically mitochondria or mitoplasts) as well as membranes; (2) a soluble fraction, which, depending on the the isolation procedure, can include all soluble components or just those in the cytosol; and (3) a vacuolar fraction. As indicated, the last fraction could be lost and separated as parts of the first two. Although data for all of these metal ions is lacking, much evidence indicates that none of them is "free" within the cell in the sense that they are not likely to be bound in simple, low M_r ionic complexes, "salts" like $CuSO_4$ or $FeCl_3$. Consequently, the liganded forms of these metal ions inside the cell are quite different thermodynamically from their forms in the external environment. The driving force for retention within the cell against a $[metal]_{total}$ gradient is due at least in part to this ligation difference. Examples of essentially *thermodynamically* irreversible intracellular ligation are many: Cu binding to thionein, Fe binding to ferritin, several metals binding to polyphosphate. It is important to recognize the difference between thermodynamic versus kinetic stability in this context: Cu in thionein and Fe in ferritin can be exchanged or mobilized, but these are kinetic, catalyzed processes. In a similar way, the uptake of a metal ion into the vacuole is an energy-dependent, catalyzed process, which viewed from the outside of the cell makes metal retention in the cell irreversible.

It is not surprising, therefore, that there is little evidence that any of these metals exhibits efflux from any yeast or filamentous fungus. A review of the literature indicates that selective advantage in fungi (at least in the industrialized world) is associated with efficient uptake of a metal, irreversible sequestration of the metal within the cell, and, closely related to the latter behavior, metal tolerance or resistance. That is, microorganisms appear more concerned about ensuring accumulation of sufficient metal in metal-limited environments than about getting rid of (excreting) excess metal in metal-rich ones. This point bears on the question of regulation of uptake, as well: its up-regulation in the former condition, and *not* down in the latter, has been demonstrated. The few examples of efflux are consistent with this conclusion in that they appear to be active, energy-requiring processes. However, the importance of this "pump" to the metabolism of the six metals of this review is not clear and, except for Mn^{2+}, is probably extremely limited.

III. UPTAKE AND RETENTION OF TRANSITION METALS IN YEASTS AND FILAMENTOUS FUNGI

In this section, the features of uptake and retention of these six transition metals are presented with the emphasis outlined above. An effort is made to provide a perspective on what questions or systems might be most fruitfully exploited at this time to delineate further the cellular handling of each of these metal ions. The metals are not taken in order of atomic number but in pairs in order of increasing knowledge at present: Ni and Co, Mn and Zn, and lastly Fe and Cu.

A. Nickel and Cobalt

1. *Nickel*

There is a relative dearth of explicit studies on the uptake of these two ions in both past and recent literature. They most often appear as potential competitive inhibitors of the uptake of another metal ion, such as Mn^{2+}, Cu^{2+}, or Zn^{2+}. Early work by Rothstein and coworkers showed that Ni^{2+} uptake in *S. cerevisiae* was both saturable and energy dependent (Fuhrmann and Rothstein, 1968). The energy dependence was demonstrated in that glucose-starved cells had 20% the uptake velocity of glucose-replete ones. A phosphate dependence was also noted and was suggested to be due to a specific role of phosphate in uptake. More likely is the PO_4^{3-} requirement for sustaining ATP production, or perhaps for the synthesis of polyphosphate (see below). Ni^{2+} uptake in *Neurospora crassa* is also saturable and energy dependent, although there was only a 50% decrease in glucose-starved cells (Mohan et al., 1984). On the other hand, N_3^- strongly inhibited uptake, by >90%. In both of these studies K_M values were found to be 200–500 μM, which, as noted in previous sections, would appear high for a trace metal that is not even included in standard minimal media.

In comparing these K_M values for uptake with the apparent activity of Ni^{2+} as inhibitor of uptake of another metal ion, one is struck by various inconsistencies. For example, 1 mM Ni^{2+} acts as a noncompetitive inhibitor of Co^{2+} uptake in *S. cerevisiae* with an estimated K_I of 65 μM, appreciably below the

K_M for Ni^{2+} uptake determined by the same workers (estimate from Fig. 6, Fuhrmann and Rothstein, 1968). On the other hand, in the same yeast, a K_I for inhibition of Co^{2+} uptake by Ni^{2+} can be estimated to be 200 μM (estimate from Fig. 5, Norris and Kelly, 1977). In the same experiment, a quantitatively similar inhibition of Ni^{2+} uptake by Co^{2+} was observed, suggesting the possibility that these two metal ions do share a common transporter.

Potentially, isolation and characterization of Ni uptake mutants would help to resolve this question. In fact, nickel-resistant strains of *N. crassa* have been described (Mohen et al., 1984). These mutants did show non–wild type Ni^{2+} uptake kinetics. The K_M values for uptake in the mutants were not strongly different from wild type; the difference in uptake velocity was due to differences in V_{max}. Two mutants had V_{max} values 50% larger than wild type, while one had a V_{max} that was only 25% of the wild type value. There was limited insight provided by these mutants, however, since they demonstrated no direct correlation between their kinetic behavior of Ni uptake and their Ni resistance. Nonetheless, it would be useful to know whether the differences in Ni^{2+} uptake among these isolates were reflected in differences in Co^{2+} uptake as well. These measurements have not been reported. In any event, the resistance of these strains to Ni^{2+} is not obviously due to differences in uptake; possibly, it is due to an alteration(s) in how the metal is handled intracellularly. This appears to be true of Ni-resistant isolates in *S. cerevisiae* as well (Joho et al., 1987). Unfortunately, there is essentially no information about the cellular distribution of Ni^{2+} in yeasts and filamentous fungi, or about cellular Ni-binding proteins. Whether Ni accumulates in the yeast vacuole is of interest, particularly with respect to the suggestion that PO_4^{3-} potentiates uptake. Polyphosphates are known to bind some metals in the vacuole: Mn^{2+} (Okorokov et al., 1977; Chang and Kosman, 1989) and Zn^{2+} (Doonan et al., 1979).

2. *Cobalt*

The picture for Co^{2+} uptake and retention is somewhat more complete. The kinetics of uptake of this metal ion by both *S. cerevisiae* (Norris and Kelly, 1977) and *N. crassa* (Venkateswerlu and Sastry, 1970) have been investigated in some detail. In these experiments care was taken to distinguish between cellular uptake and cell-wall binding. This was done by measuring cell-association at short times; at different [Co^{2+}]; and as a function of temperature. These measurements showed that an initial and rapid binding was relatively independent of metal concentration and of temperature, indicating that this binding was external to the plasmalemma and nonspecific. The K_M and V_{max} values for uptake measured in these two fungi were quite different, however: 1.6 mM and 0.4 nmol Co/min/mg dry wt for *N. crassa,* 77 μM and 5.4 nmol Co/min/mg dry wt for *S. cerevisiae.* What is striking about these values is that they show the fungus to be extremely inefficient at Co^{2+} uptake; at 1 μM Co^{2+}, for example, *Saccharomyces* would take up Co^{2+} >100 times faster than *Neurospora.*

The energy dependence of Co^{2+} uptake in these two organisms also differs. In yeast, uptake is strongly inhibited by energy depletion (Norris and Kelly, 1977) while in the filamentous fungus it is inhibited maximally by only 50% (Venkateswerlu and Sastry, 1970). No or limited efflux of metal could be demonstrated in either organism, however. The distribution of metal in the two microorganisms was also comparable as shown in Table 3. The distribution of the soluble Co in yeast shows that of the soluble cobalt, ⅓ was released from DEAE dextran-permeabilized cells (cytosolic), while ⅔ was apparently vacuolar (White and Gadd, 1986). Nonetheless, these several data suggest either that Co uptake (and trafficking) in these two organisms is fundamentally different or that the details of Co metabolism in the fungus need to be reinvestigated.

Some recent progress in the use of mutants in Co metabolism in *S. cerevisiae* has been reported. Conklin et al. (1992) transformed a wild type strain (sensitive to 2 mM Co^{2+} in solid, rich media) with a genomic DNA library in YEp24, a high-copy yeast vector, and selected transformants capable of growing in 10 mM $CoCl_2$. From this selection, they cloned a genetic locus designated *COT1,* which confers resistance to 20 mM Co^{2+} (and Rh^{2+}) but not to other divalent ions such as Cu^{2+}, Mn^{2+}, Ni^{2+}, and Zn^{2+}. *COT1* is not essential for Co uptake in that a null allele exhibits 55% of wild type uptake. Although the function of the *COT1* gene product is not known, Cot1 can be localized to mitochondria and is homologous to the *ZRC1* gene product. The ZRC1 protein is thought to bind Zn^{2+} and Cd^{2+} specifically and provide resistance to these metal ions by some uncharacterized mechanism (Kamizono et al., 1989).

Table 3 Distribution of Co in *S. cerevisiae* and *N. crassa* (% total accumulated Co)

Organism	Soluble (%)[a]	Bound (%)[b]
S. cerevisiae	70	30
N. crassa	88	12

[a]For *S. cerevisiae,* sum of soluble cytosolic and vacuolar Co; for *N. crassa,* H_2O-extractable Co.
[b]For *S. cerevisiae,* particulate Co; for *N. crassa,* TCA-extractable.
Source: Data from White and Gadd (1986) and Venkateswerlu and Sastry (1970).

Cot1 is not obviously required for normal mitochondrial function and could therefore be considered a candidate for a Co-binding protein involved in Co storage or detoxification. How its metal binding is linked kinetically to uptake is not clear; by comparison, deletion of the gene for yeast Cu-thionein has no effect on Cu uptake kinetics (Lin and Kosman, 1990). Also, the fact that deletion of *COT1* reduces Co uptake by only 50% suggests that this uptake may involve more than one pathway. Another question is raised by the apparent contradiction between the Co distribution studies summarized above in which a major function was localized to the yeast vacuole (White and Gadd, 1986) and the fact that Cot1 appears to be a mitochondrial protein. Did those former experiments fail to resolve a contribution from mitochondrial Co? They were not designed to distinguish between these two organelles. Lastly, how deletion of *COT1* effects the uptake of other divalent metal ions remains to be investigated. Can this or related mutants help to resolve the question of a general divalent cation transport system? It can be noted in this regard that transcription from *COT1* is stimulated 2- to 3-fold by Co (Conklin et al., 1992), suggesting the kind of specificity in regulation that would be expected of a Co-binding system. Whether *COT1* is stimulated by other metals has not been thoroughly investigated, however.

B. Manganese and Zinc

Manganese and zinc are chemically quite similar in that they are both strong Lewis acids and consequently serve a similar chemical function as an electrophilic prosthetic group in a variety of enzymes. On the other hand, their coordination chemistry tends to be different in that Mn^{2+} prefers octahedral coordination while Zn^{2+} is more commonly tetrahedral. Thus these two metal ions are not likely to share a common transporter, and in fact there is no strong evidence that they do.

1. *Manganese*

Mn^{2+} uptake and distribution has been described in some detail in a variety of yeasts and filamentous fungi. In particular, early studies were among the first that suggested a link between K^+ and PO_4^{3-} and divalent metal ion uptake in yeasts (Rothstein et al., 1958; Jennings et al., 1958). Glucose-, K^+-, and PO_4^{3-}-starved cultures of *S. cerevisiae* exhibited only a periplasmic binding of Mn^{2+}; uptake required energy repletion and the presence of the other two ions (Rothstein et al., 1958). The time dependence of Mn^{2+} uptake by K^+- and PO_4^{3-}-deficient cells indicated that it lagged behind the uptake of the latter two ions; Mn^{2+} uptake also depended on the amount of phosphate and potassium preequilibrated with the cells, although higher concentrations of K^+ (>3 mM) inhibited Mn^{2+} uptake. The inhibitory effect of arsenate on Mn^{2+} uptake was suggested to be related to arsenate inhibition of glycolysis at the level of triose phosphate isomerase, which blocked the synthesis of a phosphate carrier involved in the active accumulation of the PO_4^{3-} necessary for the synthesis of polyphosphate, that putative intracellular Mn^{2+} binding species.

This model was confirmed in some ways by work in *Saccharomyces carlsbergensis* and *S. cerevisiae* (Okorokov, et al., 1977). Mn^{2+} uptake was energy dependent, as demonstrated by its inhibition by 2-deoxyglucose, 2,4-dinitrophenol (DNP), and oligomycin. These workers also investigated the intracellular distribution of Mn^{2+} in *S. carlsbergensis*. The results are given in Table 4 and show that in this yeast approximately 25% of the metal is dissociable from cells permeabilized under hypotonic conditions to osmotically lyse (the) vacuole(s). The bulk of the Mn^{2+} is "bound," that is, it is not dissociable in water at neutral pH; of this fraction, approximately ⅓ is extracted by acetate (cell wall–bound metal) and ⅓ by perchlorate (intracellular compartments). The accumulation of Mn^{2+} in the tonoplast suggested that there was an energy-dependent transport associated with the vacuolar membrane and provided further evidence that polyphosphate was directly involved in Mn^{2+} retention in yeast.

Okorokov et al. (1983) subsequently demonstrated the stoichiometry of Mn^{2+} uptake into and K^+ efflux from *S. carlsbergensis,* and Mn^{2+} accumulation in the tonoplast and polyphosphate synthesis. These data suggested a model for Mn^{2+} (see Fig. 3) that includes the following features:

- Plasma and vacuolar membrane Mn^{2+}/K^+ antiport
- Vacuolar membrane Mn^{2+}/H^+ antiport

Table 4 Manganese Distribution in *Saccharomyces carlsbergensis*

Compartment	Fractionation step	Method	Mn content (% total)
Cell wall	1	pH 4.8 acetate wash	26.1
Cytosol	2	Cytochrome *c* permeabilization + pH 4.8 acetate wash of cell sample from Step 1	6.5
Vacuole	3	Osmotic shock of cell sample from Step 2	15.4
Osmotically stable intracellular compartments	4	0.5 *N* $HClO_4$, 0°C extraction of cell sample from Step 3	23.2
Residue	5	$HClO_4$, 180°C extraction of cell sample from Step 4	28.9

Source: Data from Okorokov et al. (1977).

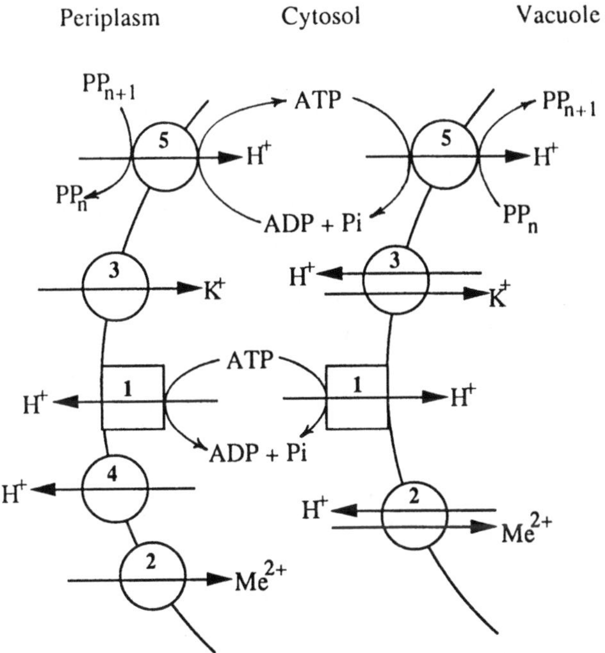

Figure 3 Model for cellular uptake and trafficking of transition metal ions with respect to the sources of energy required. (1) H^+ ATPase; (2) metal ion transporter coupled directly or indirectly to H^+ and/or K^+ gradients; (3) K^+ transporter or K^+/H^+ antiporter; (4) H^+ efflux perhaps coupled to Me^{2+} influx; (5) polyphosphate kinase acting as a H^+ pump and for retaining metals in the vacuole. (Adapted from Okorokov et al., 1973 and Gadd, 1990.)

- H^+ gradients generated by glycolysis-driven plasma and vacuolar membrane ATPases
- Glycolysis-driven plasma and vacuolar membrane polyphosphate kinases also driven by the K^+ efflux
- Accumulation of Mn^{2+} in the vacuole as a polyphosphate

Data supporting one aspect of this model are not available: there is no strong and direct evidence that the plasma membrane ATPase is required for Mn^{2+} uptake. Systematic efforts to correlate ATPase activity with Mn^{2+} uptake have not been reported, e.g., use of inhibitors [diethylstilbestrol (DES), dicyclohexylcarbodiimide (DCCD)] or ethanol (Cartwright et al., 1986, 1987), or cAMP manipulation (Ulaszewaski et al., 1989); nor have the effects of glucose on ATPase activity been considered in assessing the relationship between uptake and the energy charge in the cell (Serrano,

1983; Portillo and Mazon, 1986). On the other hand, two facts suggest that Mn^{2+} is bound in polyphosphates. First, in *S. cerevisiae,* 80% of the intracellular Mn^{2+} is not detectable by electron spin resonance, which is consistent with its chelation in an H_2O-accessible phosphate matrix (Chang and Kosman, 1989; Galiazzo et al., 1989). Second, this Mn^{2+} is dialyzable, also consistent with this kind of coordination; the coaccumulation of PO_4^{3-} and Mn^{2+} should also be noted (Chang and Kosman, 1989). This type of binding does not appear to occur in *Aspergillus niger,* however (Hockertz et al., 1987; Chap. 7 of this book).

Only two kinetic studies of Mn^{2+} uptake have been reported; Parkin and Ross (1986) and Hockertz et al. (1987) derived kinetic constants for uptake in *Candida utilis* and *A. niger,* respectively (Table 2). In both organisms, inhibition of Mn^{2+} uptake by other divalent metal ions was observed. For example, Mn^{2+} uptake ($[Mn^{2+}]$ = 2 nM) by *A. niger* in methane ethanesulfonic acid (MES) buffer was inhibited >90% by 10 μM Cu^{2+} and Zn^{2+} and >50% by 10 μM Fe^{2+} or Ni^{2+}. Although one is tempted to view these results as demonstrating competition for a general divalent transporter, another possibility cannot be excluded. In *A. niger,* the K_M for Mn^{2+} is 3 μM (Hockertz el al., 1987). At 2 nM Mn^{2+}, uptake is operating at only 0.1% of V_{max}; since the K_M values for uptake of the competing ions (for their transporter) are ≈3 μM also (Table 2), these transporters will be operating at approximately 75% V_{max} at a $[Me^{2+}]$ = 10 μM. If, for example, uptake of any metal ion uses the energy sources illustrated in Fig. 3, it is reasonable that inhibition by one metal of the uptake of another is indirect and due to depletion of the available energy gradient. One cannot impute mechanism simply by observing inhibition; true competition can only be demonstrated by the concentration dependence of inhibition.

There has been limited use of mutants in the study of Mn^{2+} uptake in yeasts and filamentous fungi. The only reported mutant, one containing a locus designated *mnr1,* has not been exploited biochemically or by molecular approaches, so it has provided limited information about the mechanism of Mn^{2+} handling by *S. cerevisiae* (Bianchi et al., 1981).

2. *Zinc*

Zinc uptake has been the subject of somewhat more definitive kinetic analysis, but, as for Mn^{2+} metabolism, little use of mutants has been reported. There is also limited information about the intracellular distribution of this metal ion. In a series of experiments, Failla and coworkers (Failla et al., 1976; Failla and Weinberg, 1977) characterized the uptake of Zn^{2+} by *Candida utilis* in some detail. Zn^{2+} uptake is energy dependent and clearly is specific for this metal. The only competitor was Cd^{2+}, which is interesting in light of the identification of the *ZRC1* locus in *Saccharomyces cerevisiae,* an allele which can confer both Zn and Cd resistance to that yeast (Kamizono et al., 1989). Although unproven, this phenotype and the sequence of the *ZRC1* gene product (with a putative metal-binding H×H×H motif) suggest that this protein could serve as a defense against an excess of these metal ions comparable to the function provided by the yeast thioneins with respect to Cu^{2+} toxicity (see below).

The only way that *C. utilis* reduces intracellular [Zn^{2+}] is by dilution due to cell growth and division; no efflux is observed, nor is there any detectable exchange of newly accumulated ^{65}Zn with an excess of nonradioactive Zn^{2+} added to the medium (Failla et al., 1976; Failla and Weinberg, 1977). Gel exclusion chromatography of soluble extracts from cells prelabeled with ^{65}Zn demonstrated that the major fraction of the label was bound to a protein that eluted as a species with M_r similar to that of the eukaryotic thioneins (6–7 kDa). Failla et al. (1976) demonstrated that *C. utilis* did not secrete any metal binding species; thus they concluded that the protein synthesis required for Zn uptake reflected the synthesis of this Zn-binding protein. Note that the ZRC1 protein is a 442 residue polypeptide (Kamizono et al., 1989), while thioneins commonly contain about 65 amino acids (Karin, 1985; Hamer, 1986). Also, there is no evidence that in *S. cerevisiae,* at any rate, thionein binds Zn (Winge et al., 1985). Consequently, the identity of the Zn-binding component separated chromatographically remains an open question. Nonetheless, this work reflects many of the features of Cu uptake (see below) with respect to cycloheximide inhibition and binding to protein species that are easily extractable from the cell. Whether this Zn- (or Cu-) binding protein is in the cytosol or the vacuole is not known. This work also indicated that Zn uptake by *C. utilis* was cyclic in batch culture, suggesting that some regulation of uptake might be occurring in this yeast (Failla and Weinberg, 1977). This suggestion has not been tested by molecular approaches, however.

Zn^{2+} uptake in *Sporobolomyces roseus* (Mowll and Gadd, 1983) and *S. cerevisiae* (Mowll and Gadd, 1983; White and Gadd, 1987) has been reported also. In the latter work, evidence for the involvement of the plasma membrane ATPase in Zn uptake was given by the inhibition of uptake by DCCD and DES. The role of K^+ efflux as a driving force for Zn uptake was evaluated by assessing the stoichiometry of Zn^{2+}(in)/K^+(out); no simple stoichiometry was observed, suggesting that the major fraction of the K^+ efflux was unrelated to Zn uptake (White and Gadd, 1987). On the other hand, good evidence was provided that the accumulation of Zn^{2+} into the vacuole (56% of the cell total) was due to a proton antiport associated with the vacuolar membrane (see Fig. 3). Little Zn was found in the cytosol, indicating that the chromatographically separable components described in *C. utilis* might have originated in the vacuole also. Chromatographic analysis of vacuolar vs. cytosolic metal binding species in yeasts and filamentous fungi has not been reported, so the precise cellular locale of such species is unclear. Zn does not efflux from *S. cerevisiae.*

No real mutants in Zn metabolism have been reported in a fungus. The only genetic information is provided by the isolation of the *ZRC1* locus, which when overexpressed episomally confers Zn^{2+} (and Cd^{2+}) resistance to the transformant (Kamizono, 1989). As noted above, *ZRC1*, like *COT1*, appears to encode an integral membrane protein; the cellular localization of this gene product has not been determined, however. Specific Zrc1 sequences include potential *N*-linked glycosylation sites and metal-binding (H-S-H-S-H) motifs. A deletion in *ZRC1* makes cells hypersensitive to both Zn^{2+} and Cd^{2+}; the biochemical basis of this sensitivity was not determined, however, e.g., alteration in uptake, retention, efflux, or localization of either metal. Surprisingly, expression of *ZRC1* mRNA is not stimulated by Zn^{2+}. Kamizono et al. (1989) concluded that despite the putative metal-binding potential of the *ZRC1* gene product it is more likely to be involved in the biosynthesis of cysteine than in metal detoxification. This conclusion is based on the homology between *ZRC1* and the *CysB* gene in *E. coli*; the latter gene encodes a protein that regulates cysteine biosynthesis in that microorganism (Ostrowski et al., 1987). Cysteine is essential to the detoxification of many metals, either as a dominant residue in the thioneins or as a component of the γ-glutamyl peptides found in some yeasts (Reese et al., 1988; Mehra and Winge, 1991). This could explain why Zn^{2+} does not transcriptionally activate the *ZRC1* locus, which would be expected if the Zrc1 protein were directly involved in Zn^{2+} uptake or storage. Another possibility is that this protein might be involved in pumping Zn^{2+} or Cd^{2+} out of the cell. This suggestion is not supported by evidence that such efflux occurs for either of these two metals, however. In summary, the *ZRC1* locus needs to be studied further, both genetically and biochemically.

C. Iron and Copper

More facts are known about the uptake of iron and copper in comparison to the others discussed above. There is also more information about the intracellular distribution of Fe and Cu and how their intracellular handling may relate to their transport. The distinguishing feature of the cellular metabolism of these two metal ions is the fact that they exist in the aerobic environment of a free-living organism in a higher valence state that is less stable and less soluble than the reduced form that may predominate in the cell. Therefore, Fe^{3+} and Cu^{2+} uptake may involve a reduction step. This step has been strongly implicated for the uptake of periplasmic Fe(III).

There is circumstantial evidence that reduction of Cu(II) to Cu(I) also occurs as a part of the cellular accumulation of this metal.

1. Iron

Since iron's handling by yeasts and fungi is discussed in Chapters 2, 4, and 5, it will only by covered here briefly with respect to the plasma membrane reductase activity associated with the gene product of the *FRE1* locus and its regulation by Fe, and the kinetics of Fe(III) uptake as a function of that activity. This focus is relevant to the question of whether Cu(II) uptake requires a prior reduction to Cu(I).

Yeasts exhibit nonspecific reductase activities that are found both in the plasma membrane and intracellularly (Crane et al., 1992; Lesuisse et al., 1990). In *S. cerevisiae,* the plasma membrane activity is associated with a number of complementation groups; two mutant alleles have been designated *fre1* and *fre2* (Dancis et al., 1990, 1992; A. Dancis, personal communication). Both alleles confer an Fe(II) dependence on growth and Fe uptake. One of these genes, *FRE1*, has been cloned; it codes for a protein containing 686 residues including a potential leader peptide characteristic of a membrane or secreted protein. Based on its sequence, the Fre1 protein may be an integral membrane protein, since it has up to seven transmembrane domains. It also has homology to the transmembrane cytochrome b_{558} component (the X-CGD protein) of the respiratory burst NADPH:O_2 oxidoreductase found in neutrophils (Clark, 1990). The number of complementation groups and the current picture of the structure of this latter enzyme suggest that the Fe(III) reductase in yeast is also multicomponent. This reductase can reduce a variety of species including Fe(III), ferricyanide (Crane et al., 1982), Cu(II), and formazans such as triphenyltetrazolium chloride (TTC) (Crawford et al., 1993). This latter reagent has long been used to test the respiration competence of yeasts due to its reduction by the mitochondrial cytochrome *bc* complex.

Fe(III) uptake requires some activity of the Fre1-dependent reductase, that is, null alleles at *FRE1* exhibit 5% the Fe(III) uptake of wild type. Fe(II) uptake by such strains is normal (Dancis et al., 1992). However, the rate of reduction of Fe(III) to Fe(II) by wild type cells is 10^3 times faster than uptake, so that reduction is not rate limiting even when strongly repressed physiologically (Dancis et al., 1990; Eide et al., 1992). Uptake is regulated (by Fe, for example), however. This appears due to the regulation of the transport(er) (Eide et al., 1992). The kinetics of uptake also indicate that Fe, as Fe(II), is transported into the cells as the free ion; the ligand to which the Fe(III) is bound periplasmically is probably displaced at the plasma membrane (Lesuisse et al., 1987). A reasonable assumption is that this displacement is catalyzed, in part, by the reduction of the Fe(III).

The fact that the K_M values for Fe^{2+} and Fe^{3+} uptake are the same (Table 2 and Eide et al., 1992) is consistent with transport, not reduction, being the kinetically detected step in "Fe^{3+}" uptake, that is, reduction is not rate limiting. It is also consistent with the hypothesis that reduction releases the Fe from the chelate in which it was bound extracellularly as Fe(III). Lastly, V_{max} of transport of Fe(II) is low in comparison to the V_{max} of uptake of other divalent metal ions (Table 2). It is not clear what the selective advantage is of a pathway (iron reduction and transport) that has the first step (reduction) generating a 10^3-fold excess of substrate per unit time in relation to the flux through the second step (transport).

Like other metals, Fe exhibits no efflux from *S. cerevisiae.* Within the cell, the metal is probably bound to a yeast ferritin, although this has not been confirmed (Raguzzi et al., 1988). Whether this binding occurs in the vacuole is also not definintely known, although the vacuole does accumulate the major fraction of intracellular Fe. This Fe may be bound to polyphosphate, as has been suggested for some of the other transition metals (Raguzzi et al., 1989). With respect to the energy status of the cells, Fe uptake does not require glucose, only some carbon source, e.g., ethanol or pyruvate (Lesuisse et al., 1987). Thus there appears to be no relationship between glycolysis and Fe uptake and accumulation as has been suggested for some transition metal ions (Rothstein et al., 1958).

FRE1 transcription is regulated strongly by Fe (Dancis et al., 1990, 1992); plasma membrane reductase activity is also regulated by heme (Lesuisse and Labbe, 1989; Anderson et al., 1992) and cAMP (Lesuisse et al., 1991). These latter effects have not been linked directly to the *FRE1* locus by northern analysis, for example. An important fact is that *fre1Δ*-containing strains still exhibit 10–15% plasma membrane reductase activity of wild type (Dancis et al., 1992; Hassett et al., 1993). Whether this activity is associated with another metal transport system is not known; as discussed below, it is possible that it is associated with Cu(II) reduction and uptake. It is also true that *S. cerevisiae* exhibits a nonreductive Fe(III) uptake as indicated by uptake of ferrioxamine B by heme-deficient strains that lacked an inducible (in iron-depleted medium) Fe(III) reductase (Lesuisse and Labbe, 1989). Whether this process functions outside of the laboratory is not known inasmuch as *S. cerevisiae* does not secrete any siderophorelike molecules.

Transcription from the *FRE1* locus is also dependent, apparently in *trans,* on the gene product of another locus, designated here as *MAC1*- (**M**etal **AC**tivated transcription factor). This gene is adjacent to *UBC7* on chromosome 13; it encodes a protein of 417 amino acids (Jungmann et al., 1993). It has strong homology to *ACE1* (*S. cerevisiae*) (Thiele, 1988; Furst et al., 1988; Buchman et al., 1989; Szczypka and Thiele, 1989) and *MATF*

(*Candida glabrata*) (Zhou and Thiele, 1991; Zhou et al., 1992). These two proteins can be termed "metal fists" in that they possess CysXCysXCys Cu-binding motifs. Both are Cu-activated positive effectors of transcription of Cu-inducible genes in the two organisms, such as *CUP1* (Cu-thionein) (Thiele, 1988; Furst et al., 1988; Buchman et al., 1989) and *SOD1* (Cu,Zn superoxide dismutase) (Gralla et al., 1991; Carri et al., 1991) in *S. cerevisiae.* Strains mutant at the *MAC1* locus exhibit no *FRE1* mRNA in iron-limited media (Jungmann et al., 1993; Romeo et al., 1993); wild type strains show strong induction from this locus in this nutrient condition. The plasma membrane reductase activity in *mac1*-carrying mutants is also reduced; quantitatively, it is similar to the activity observed in *fre1*Δ-carrying strains (Jungmann et al., 1993). Fe(III) uptake is also reduced (Hassett and Kosman, 1993). A dominant mutant allele of this locus, $MAC1^{up1}$, confers Cu-sensitivity. It also leads to high and unregulated *FRE1* transcription but has only a small effect on Fe(III) uptake (Jungmann et al., 1993). This is consistent with the fact that Fe(III) reduction is *not* rate limiting for uptake and the fact that reductase and transporter activities are differentially regulated (Eide et al., 1992). Some of these observations are discussed in more detail below.

In summary, the study of Fe uptake in yeast has provided important new information about how eukaryotes handle this metal. A number of genetic tools have been established and a variety of information has been gathered on the regulation of the protein species encoded by some of these loci. While most of this work has been done in *S. cerevisiae,* there is good reason to expect that the genetic reagents developed in this organism will be useful in isolating homologues in other fungi.

2. *Copper*

Copper uptake has been the most definitively studied with respect to mechanism. This is true in part because of the large literature on Cu resistance in yeasts, although this resistance is in general associated with the way the metal is sequestered on or in the cell. Nonetheless, given the number of mutants in *S. cerevisiae* in particular, which exhibit various degrees of Cu resistance, it has been of interest to assess whether such resistance is due to reduced uptake. On the whole, however, these various mutants have not yet been exploited in molecular terms and so represent potential for providing additional insight into Cu handling by this yeast.

There is a family of *cur* (Cu resistance) mutants that exhibit increased resistance to Cu (3 mM; compare to 0.8 mM Cu for wild type, see Gadd et al., 1984). Cu uptake by *cur1* cells and protoplasts has been described by Gadd et al. (1984). While the uptake by protoplasts was complicated by apparent lysis of wild type cells at the [Cu^{2+}] used (5 and 10 μM), *cur1*-

containing cells (and protoplasts) exhibited approximately 50% the initial uptake rate of wild type at both concentrations: 0.16 (wild type) versus 0.09 nmolCu/min/10^7 cells at 10 μM, for example. The amount of Cu accumulated at 2 h (an apparent limit of total uptake) was similarly different. The question this result raises is whether the reduced total accumulation on a per-cell basis can be directly correlated to the difference in uptake rate. That is, if uptake is reduced sufficiently, then it would be comparable to cell division. This would result in a constant dilution of the newly accumulated Cu by cell doubling. This type of analysis has not been provided for any of these putative uptake mutants. However, the fact that protoplasts derived from the *cur1*-containing mutant showed uptake and retention similar to cells from the same strain indicated that the uptake observed was due to a plasma membrane–dependent and not cell-wall–mediated accumulation.

This distinction is critical for the elucidation of Cu (and Zn) uptake by yeasts. Most yeasts are sulfate assimilators in which they take up and reduce medium sulfate to the level of sulfide; the sulfide is then incorporated into the carbon skeletons provided by acetylserine or homoserine (Fig. 4). Although this pathway is regulated by its sulfur-containing end products—methionine and, in particular, *S*-adenosylmethionine (SAM, Thomas et al., 1989)—there is a leakage of S^{2-} from the cell (Kikuchi, 1965). This sulfide generation has been thought to provide a Cu detoxification mechanism in that CuS has been observed to accumulate periplasmically (Ashida et al., 1963). This "uptake" can be confused with true transmembrane transport into the cell. The data above with protoplasts show that this is not the underlying difference in the *cur1*-containing mutant; that is, it is not a hyperproducer of H_2S, a phenotype that has been described (Ashida, 1965;

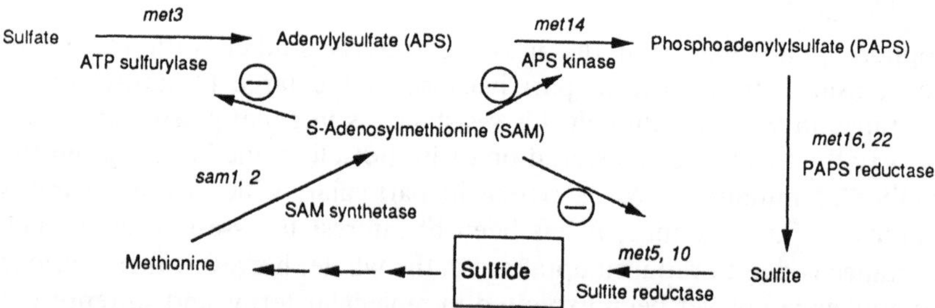

Figure 4 Sulfide formation by *Saccharomyces cerevisiae* and its regulation (Jones and Fink, 1982; Thomas et al., 1989). The steps in the reductive assimilation of SO_4^{2-} up to the generation of S^{2-} and the corresponding enzymes and genes (as mutant alleles) are illustrated. The central role of the end product S-adenosylmethionine (SAM) in the repression of expression of three of the four structural genes involved in the generation of S^{2-} is indicated also.

Kikuchi, 1965). Nonetheless, these *cur* mutants remain uncharacterized genetically; their study by molecular techniques should be encouraged.

The question of the energy source for the active transport of Cu was addressed by Gadd et al. (1984) and DeRome and Gadd (1987). The inhibition of Cu uptake by glucose depletion, cyanide, 2-deoxyglucose, DNP, or 4-(trifluoromethoxy)phenylhydrazone (FCCP) indicated that sustained ATP production was required for uptake (and not surface binding). Treatment of the cells with DCCD also inhibited uptake, suggesting that the plasma membrane ATPase was also involved. All of these inhibitions were strong, although the data were not quantitated. The stoichiometry of Cu uptake (1.17 nmol Cu/min/mg dry wt) and K^+ efflux (2.28 nmol/min/mg dry wt)—or 1:2—was also determined, as was the effect of the K^+ specific ionophore valinomycin, which increased K^+ efflux without altering Cu uptake. This indicates a close coupling between Cu_{in} and $K^+{}_{out}$, rather than a ψ due to a K^+ gradient. The Cu uptake rate did apparently depend on the initial $[K^+]$ in the cell (which varied from 82 to 139 mM); again the data were not quantitated, but the Cu uptake in the K^+-replete cells can be estimated to be approximately 5-fold higher than in the K^+-depleted ones (estimate from Fig. 4, DeRome and Gadd, 1987). Taken together, these data support a model of Cu uptake that includes the general features included in Fig. 3.

Other data are not entirely in agreement with this model, however. On the one hand, the fact that ^{64}Cu uptake is qualitatively independent of the carbon source indicates that glycolysis is not required for uptake. ^{64}Cu transport velocity in gylcerol/ethanol grown cells is 74% the velocity in glucose grown ones (Lin et al., 1993b; see also Table 5). Thus some specific role for glycolysis-generated ATP (or P_i) is not likely for Cu uptake or intracellular trafficking (Lin and Kosman, 1990; Lin et al., 1993a,b). Also, Hassett and Kosman (1993) failed to establish a consistent difference in ^{64}Cu uptake between cells grown in the absence and presence of K^+. Atomic absorption analysis demonstrated the K-depletion of the former cultures. Similarly, treatment of cells with ethanol up to 2 M failed to produce a systematic effect on ^{64}Cu uptake as would be predicted on the basis of a change in the permeability of the plasma membrane (Hassett and Kosman, 1993). Cartwright et al. (1986, 1987) have shown that ethanol causes a depolarization of this membrane leading to a reduced uptake of glycine. This depolarization can be correlated to a decreased activity of the plasma membrane ATPase (Cartwright et al., 1987). That this ATPase might be involved in Cu uptake is suggested by the strong inhibition of uptake by DCCD and DES as noted above. Conversely, vanadate, which inhibits this ATPase by 85% (Willsky, 1979), inhibits ^{64}Cu uptake by only 50% (Hassett and Kosman, 1993), suggesting that the inhibition of uptake

Table 5 Kinetic Dissection of ^{64}Cu Uptake into the Intracellular and Periplasmic Pools

Specific medium condition or Uptake addition	Intracellular uptake, % glucose control (% inhibition)[a]	Total uptake, % glucose control (% inhibition, periplasmic uptake)[b]
+glucose	100	100
−glucose	16 (84)	3 (100)
+glycerol/ethanol	74 (26)	56 (51)
+DNP	26 (74)	23 (78)
$+N_3$	13 (87)	20 (77)
+cycloheximide	50 (50)	18 (96)
$+Zn^{2+}$ or Ni^{2+} (50 μM)	100 (0)	50 (71)
Kinetic constants		
V_{max} (nmol Cu/min/mg)	0.10	0.21
K_M (μM)	3.8	4.4

[a]Values in parentheses are % inhibition of accumulation into the intracellular pool only.
[b]Values in parentheses are % inhibition into the periplasmic pool only, calculated on the basis that approximately 30% of the total ^{64}Cu accumulation is into the intracellular pool (at 10 μM ^{64}Cu), corrected for the % inhibition into that pool as indicated by the data in the first column.
Source: Data for wild type *S. cerevisiae* (Met prototroph) grown in sulfate-supplemented media; from Lin and Kosman (1990) and Lin et al. (1993b).

by DES or DCCD is not altogether specific (Borst-Pauwels et al., 1983). These several results indicate that a clear understanding of the mechanism of Cu uptake by yeast remains a goal and not a reality.

Although Cu accumulation has been described for other fungi (Gadd and Griffiths, 1980; Townsley and Ross, 1985; Germann and Lerch, 1987; Phelan et al., 1990), only in *Penicillium ochro-chloron* have uptake kinetics and mechanism been addressed (Gadd and White, 1985). The significance of this work is due, in part, to the fact that it employed protoplasts derived from this filamentous fungus, thus obviating the problems associated with using hyphae in transport studies. The Cu resistance exhibited by this strain could be correlated to its rather high K_M for uptake (390 μM) but not to its V_{max} (0.37 nmol Cu/min/10^7 cells), which is not substantially different from V_{max} values for wild type strains of *S. cerevisiae,* for example (Gadd et al., 1984; De Rome and Gadd, 1987; Lin and Kosman, 1990; Lin et al., 1993b). Cu uptake by *P. ochro-chloron* is stimulated by glucose and inhibited by KCN, indicating that it is active as well as facilitated.

The specificity of Cu uptake in *S. cerevisiae* can be obscured by the trapping of metal sulfides in the periplasm. Lin et al. (1993a,b) have dissected the cell accumulation of Cu by this yeast into two pools, one extracellular and dependent on the concurrent generation of H_2S, the other intracellular, resulting from the active transport of Cu across the plasmalemma. While Zn^{2+} strongly inhibits uptake into the former pool, apparently by competing with Cu for the S^{2-}, it has little effect on the intracellular uptake of Cu. The decrease in accumulation of ^{64}Cu into the periplasmic pool due to Zn^{2+} can be quantitatively accounted for by the periplasmic accumulation of ^{65}Zn. The partitioning of newly accumulated ^{64}Cu in *S. cerevisiae* can be manipulated by altering the H_2S-generating capacity of the cell. This can be done by repressing sulfate reduction to sulfide by addition of methionine or SAM, or by using a strain that carries a mutation in one of the genes encoding a component upstream from or at the sulfite reductase step in the reduction pathway (Fig. 4). As illustrated in Table 5, Cu uptake into the two pools can be distinguished using such strategies.

The data in Table 5 show that the V_{max} for uptake is, in part, directly correlated to the sulfide-generating capacity of a given strain. The kinetics of uptake (into both pools) are different only with regard to this V_{max}, however; the K_M value is essentially insensitive to this capacity. This behavior is illustrated in Fig. 1 above also [compare the data for the sulfate-depleted cells (open circles) with the sulfate-replete ones (closed circles)]. Both pathways of accumulation require energy, and both are inhibited by cycloheximide, although as indicated in Table 5 extracellular accumulation is somewhat more sensitive to cycloheximide. The fact that the K_M for uptake is independent of where the Cu accumulates suggests but does not require that the two pathways share a common saturable kinetic intermediate. The time dependence of ^{64}Cu accumulation, however, also suggests that a kinetic intermediate exists as indicated by the appearance at short times of ^{64}Cu in the periplasmic pool; the ^{64}Cu in this pool is exchangeable with added Cu and can be quantitated in this way (Fig. 5). These kinetic and fractionation experiments can be fitted to a model of Cu uptake by *S. cerevisiae* shown in Fig. 6 (Lin et al., 1993b).

The intracellular locale of Cu is not entirely clear. White and Gadd (1986) presented data showing that the majority of the Cu was "bound" (particulate) with minor, soluble fractions released from the cytosol and the vacuole. Lin et al. (1993a) fractionated newly arrived ^{64}Cu chromatographically; the soluble ^{64}Cu (cytosol and vacuole; these two compartments were not distinguished in this work, unfortunately) could be recovered quantitatively from Superose 12 fast protein liquid chromatography (FPLC) in up to five ^{64}Cu-binding fractions. Two fractions were identified as Cu-thionein and Cu, Zn

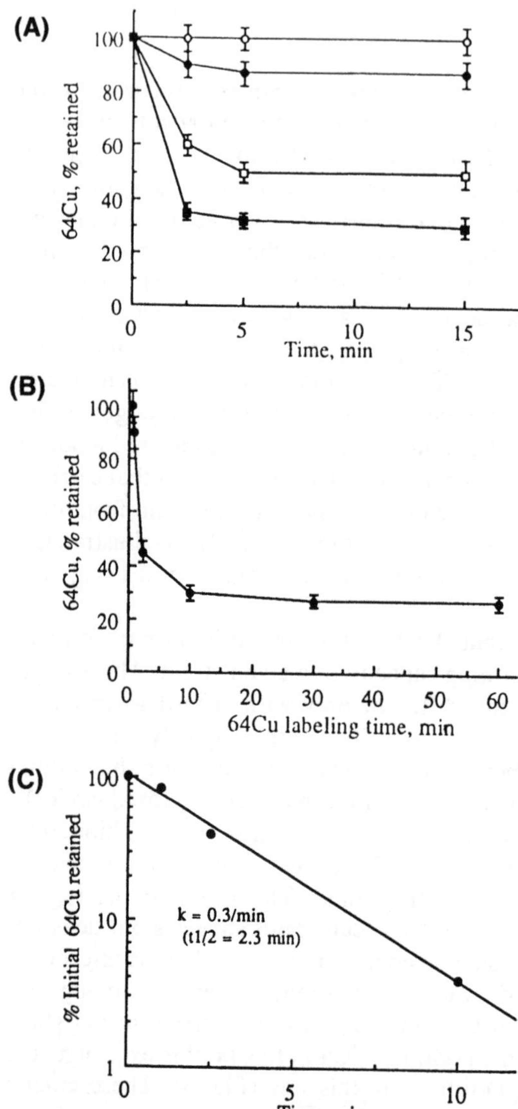

Figure 5 Time dependence of the appearance of the exchangeable, periplasmic ^{64}Cu-binding pool in a wild type strain of *S. cerevisiae*. Cell samples were labeled for varying lengths of time with ^{64}Cu and then challenged with cold Cu (10 μM). The fraction of the initial ^{64}Cu retained was determined as a function of duration of the chase. In panel A, the labeling times were 30 s (open circles), 1 min (closed circles), 2.5 min (open squares), 10 min (closed squares). In panel B, the fraction of the initial ^{64}Cu retained after a 30 min chase is plotted against the labeling time. In panel C, the data from panel B are replotted on a semilog scale. The line represents a nonlinear fit of the data to a single exponential using the computer program Enzfitter with $k = 0.3 \pm 0.04\ \text{min}^{-1}$. (Data from Lin et al., 1993b.)

superoxide dismutase, respectively, while a third was shown to contain, at least, Cu-glutathione. Cu, Zn superoxide dismutase is a cytosolic protein (*not* peroxisomal, see Crapo et al., 1992). Cu-thionein is probably in both compartments (Wright et al., 1987), although more recent study may localize it predominantly to the vacuole (D. Thiele, personal communication). How this distribution bears on the question of uptake kinetics and mechanism also needs to be established. However, Cu-thionein is *not* required for uptake in that strains carrying a null allele at the *CUP1* locus have wild type Cu uptake kinetics (Hamer et al., 1985; Lin and Kosman, 1990). However, if Cu is localized to the vacuole (and bound to thionein there), then the transport of Cu into the vacuole (Fig. 3) might be reflected in the kinetics of cellular uptake. While this part of the model for Cu uptake and trafficking remains to be fully elucidated, a mutant in the vacuolar H^+ ATPase does exhibit altered Cu accumulation (D. Thiele, personal communication).

The glutathione content of the cell effects the uptake of ^{64}Cu (Hassett et al., 1993). GSH/GSSG levels in *S. cerevisiae* can be manipulated by treatment of cells with *S*-buthionine sulfoximine (BTS), a potent inhibitor of γ-glutamylcysteine synthetase that catalyzes the first step in the biosynthesis of glutathione (Griffith and Meister, 1979). Cells treated with 10 mM BTS exhibit no growth deficit. In 4 h, the total GSH/GSSG content of the cells decreases by $95\pm3\%$; this is accompanied by a 2.5-fold *increase* in ^{64}Cu uptake velocity by the treated cells. The decrease in cellular glutathione is reflected in a 90% reduction in the ^{64}Cu which accumulates in the chromatographic fraction shown to contain Cu·glutathione (above and Hassett and Kosman, 1993). These results suggest that the intracellular handling of Cu can effect uptake. However, note that while eliminating Cu-thionein has no effect on uptake (above and Lin and Kosman, 1990), reducing glutathione levels does. This difference would indicate that glutathione is more directly involved *kinetically* with uptake in comparison to thionein as is illustrated in the model of Cu-handling shown in Fig. 6, although a nonspecific effect of glutathione depletion on Cu uptake cannot be ruled out at this time.

Cu uptake in *S. cerevisiae* can be regulated. The strongest evidence for this comes from studies on a Cu-uptake mutant, originally designated UPC1 (Uptake of Cu; Crawford et al., 1993). This mutant strain exhibits several phenotypes. Chief among these phenotypes are 7-fold larger V_{max} and a K_M value of 0.1 that of wild type; a 4-fold enhancement of plasma membrane reductase activity and *FRE1* mRNA; and the unregulated expression of *CUP1*. Neither *FRE1* nor *CUP1* is associated with the elevated Cu uptake, however, since placing the *UPC1* allele, which is dominant, into backgrounds carrying deletions of either these genes has no effect on the UPC1 Cu uptake phenotype. As indicated above, the *UPC1* allele has been designated *MAC1*^{up1}.

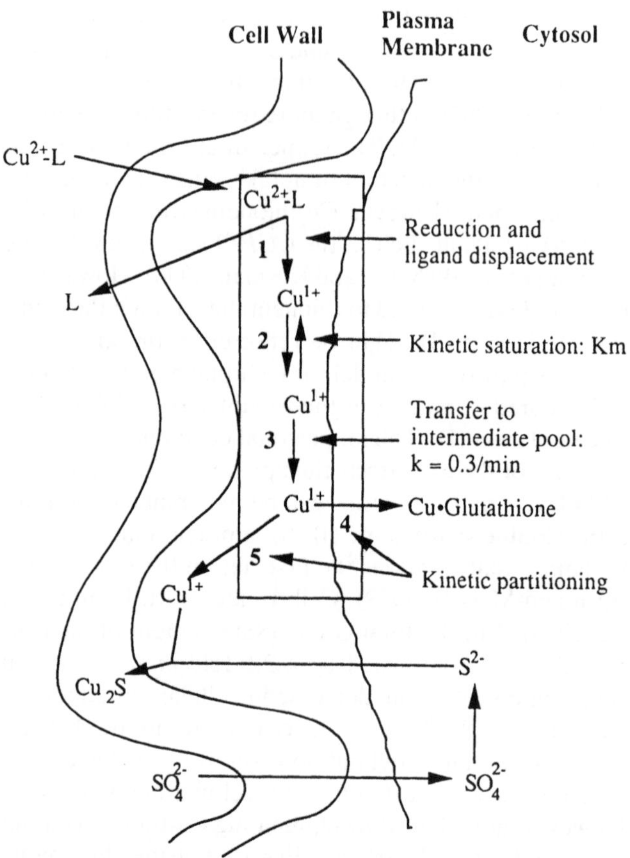

Figure 6 Model for the processing of newly arrived ^{64}Cu in the yeast *Saccharomyces cerevisiae* (Lin et al., 1993b). The numbers in the model refer to the following steps in this pathway. (1) Reduction of Cu(II) to Cu(I) by a membrane metal reductase; the change in redox state promotes the release of the ^{64}Cu from the presenting ligand. (2) The metal associates with a membrane component in a kinetically detectable process (K_M). (3) The Cu(I) is transferred to an intermediate Cu-binding pool, $k = 0.3\ min^{-1}$. The ^{64}Cu in this pool is subsequently partitioned between (4) translocation (limiting $V_{max} = 0.1$ nmol/min/mg protein) and (5) simple diffusion away from the cell surface into the periplasm where it can encounter H_2S and be trapped in the cell wall as copper sulfide. The Cu translocated into the cytosol is suggested to bind initially to glutathione (GSH). This binding may influence the measured uptake velocity. (Hassett et al., 1993.)

MAC1^{wt} can be localized to the nucleus (Jungmann et al., 1993). It is apparently a Cu- and DNA-binding protein that acts a positive transcription factor for the expression of several genes, e.g., *FRE1, CCP1* (cytochrome *c* peroxidase), *CTT1* (catalase T), and *SOD2* (mitochondrial Mn superoxide dismutase) (Lee et al., 1993). Based on the Cu uptake phenotype of strains carrying *MAC1*^{up1}, it may also regulate expression of some gene(s) associated with that cellular activity. This remains only a hypothesis, however, but one which deserves further investigation.

A major question that the phenotype associated with *MAC1*^{up1} raises is the potential role of Cu(II) reduction in Cu^{2+} uptake. Cu, like the other metals, is taken into the cell as the free ion. This is indicated by the independence of the K_M (and V_{max}) for ^{64}Cu uptake from the ligation state of the ^{64}Cu in the uptake medium (Table 6). As suggested in Sec. II, reduction can catalyze ligand dissociation due to the coordination preferences of the higher vs. lower valence states of a metal ion, e.g., Cu(II) (tetragonally disorted square planar) and Cu(I) (tetrahedral). This reduction step is included in the model of Cu uptake illustrated in Fig. 6. There are no data that directly implicate Cu(II) reduction in Cu uptake, however. Studies on a Cu reductase activity in the cell-wall materials of the yeast *Debaryomyces hansenii* (Wakatsuki et al., 1988, 1991a,b) do not correspond with the localization of such activities in *S. cerevisiae* (plasma membrane, intracellular) (Lesuisse et al., 1990). Possibly, the former yeast handles Cu similarly but has a cell anatomy that is somewhat different. In any event, even in that yeast, there is no direct evidence that the reductase activity detected is required for or associated with Cu uptake. As noted above, only the isolation of mutants in *fre1* allowed for the unequivocal result that that reductase activity was required for Fe(III) reduction (and uptake).

Table 6 Kinetic Constants for ^{64}Cu Uptake by *S. cerevisiae:* Effect of ^{64}Cu Ligation State

Conditions, substrate	V_{max} (nmol ^{64}Cu/min/mg protein)	K_M (μM)
Sulfate-depeleted (+Met)		
$^{64}Cu(H_2O)_6^{2+}$	0.103±0.002	4.0±0.4
$^{64}Cu{\cdot}His_2$	0.100±0.005	3.8±0.3
Sulfate-supplemented		
$^{64}Cu{\cdot}His_2$	0.210±0.007	4.4±0.4
$^{64}Cu(H_2O)_6^{2+}$	0.243±0.007	5.1±0.4

The kinetic constants were derived from a direct nonlinear fit of initial velocities of uptake *versus* [^{64}Cu] using the computer program Enzfitter.
Source: Values from Lin and Kosman (1990) and Lin et al. (1993b).

Potentially, the reductase activity associated with Cu uptake, which the results outlined above show is not provided by the *FRE1* gene product, could be represented by the residual reductase activity in *fre1Δ*-containing strains. These strains thus provide the appropriate background in which to select for specific Cu reductase/uptake mutants. A strategy to do so is outlined in Fig. 7 and is currently being employed to identify useful mutants in Cu-specific genetic loci in *S. cerevisiae* that can then be cloned (Kosman et al., 1993). The apparent up-regulation of a variety of genes in *MAC1*^{up1}-containing strains may provide another source of Cu-specific genes. One can assume that a collection of transcripts, in addition to *FRE1* and *CUP1* mRNAs, are elevated in this background. The difference in mRNA populations between wild type and mutant could be used to generate a subtracted cDNA library that would include other target genes for the *MAC1* gene product, some of which could be Cu-specific. The construction of this library is also in progress (Kosman et al., 1993).

IV. CONCLUSION

The picture that emerges in assessing the literature on transition metal ion uptake by yeasts and filamentous fungi is both tantalizing and frustrating. With the exception of the work on Fe and Cu metabolism in *S. cerevisiae,* and a number of other useful and well-designed studies, there is little sustained and completely thorough work on any other metal ion in any other fungus. The isolation of mutants and their use in cloning complementing wild type alleles associated with metal metabolism in *S. cerevisiae,* for example, has been a neglected strategy. Other emerging genetic systems such as *Schizosaccharomyces pombe* have also been ignored as organisms in which to identify the components needed for specific metal metabolism and to characterize these components in molecular terms. This is disappointing because metals *are* essential to life; their study in higher eukaryotes is difficult, expensive, and complicated by ethical concerns; and their chemistry is such that much can be deduced *a priori* about *how* one metal ion might be handled in comparison to another. I have attempted to remind the reader of the need to keep the known coordination and redox chemistry of a given metal in mind; this chemistry will dictate how a metal ion is likely to be handled by the cell. While the details of metal metabolism in higher eukaryotes is likely to be somewhat different than those in yeasts and other fungi, all organisms have to deal with this chemistry and will likely do so in functionally homologous ways. While this book describes various aspects of metal ions in fungi, the reader should appreciate that the work discussed here is a paradigm of metal handling by cells in general.

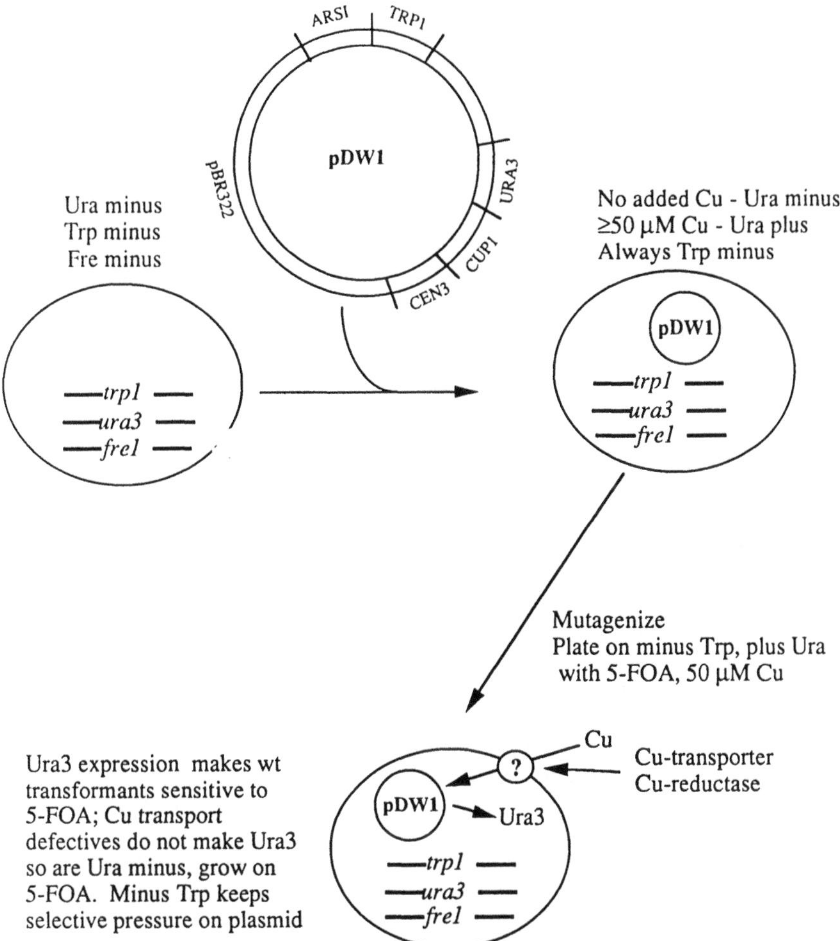

Figure 7 Strategy for selecting Cu uptake mutants in *S. cerevisiae.* Plasmid pDW1 expresses the *URA3* gene under control of the *CUP1* promoter. Cells that take up Cu are Ura$^+$ and growth-inhibited by 5-fluoroorotic acid. Cells that are Cu uptake defective will be Ura$^-$ and growth-insensitive to 5-FOA. This provides a positive screen for useful mutants. Uptake mutants can occur in either a transporter or another component such as the putative Cu-reductase. The mutagenesis is carried out in a *fre1Δ*-containing strain to eliminate the large background reductase activity due to the locus that supports Fe(III) but not Cu(II) uptake (Kosman et al., 1993).

BIBLIOGRAPHY

This is a bibliography of citations to work relevant to the metabolism of transition metals in yeasts and fungi. Included are several publications not cited in the text. They are included to provide the reader with a more complete bibliography than is reflected in the material selected for discussion in the article itself.

Anderson, G. J., Lesuisse, E., Dancis, A., Roman, D. G., Labbe, P., and Klausner, R. D. (1992). Ferric ion reduction and iron assimilation in *Saccharomyces cerevisiae, J. Inorgan. Biochem., 47:* 249–255.

Ashida, J. (1965). Adaptation of fungi to metal toxicants, *Annu. Rev. Plant Physiol., 3:* 153–174.

Ashida, J., Higashi, N., and Kikuchi, T. (1963). An electronmicroscopic study of copper precipitation by copper-resistant yeast cells, *Protoplasma, 57:* 27–32.

Auling, G. (1989). Bacterial and fungal high-affinity transport systems specific for manganese, *Metal Ion Homeostasis: Molecular Biology and Chemistry,* Alan Liss, New York, pp. 459–468.

Baldry, M. G. C., and Dean, A. C. R. (1980). Copper accumulation by bacteria, moulds and yeasts, *Microbios, 29:* 7–14.

Bérczi, A., Sizensky, J. A., Crane, F. L., and Faulk, W. P. (1991). Diferric transferrin reduction by K562 cells. A critical study, *Biochim. Biophys. Acta, 1073:* 562–670.

Beswick, P. H., Hall, G. H., Hook, A. J., Little, K., McBrien, D. C. H., and Lott, K. A. K. (1976). Copper toxicity: Evidence for the conversion of cupric to cuprous copper *in vivo* under anaerobic conditions, *Chem.-Biol. Interactions, 14:* 347–356.

Bianchi, M. E., Carbone, M. L., Lucchini, G., and Magni, G. E. (1981). Mutants resistant to managanese in *Saccharomyces cerevisiae, Curr. Genet., 4:* 215–220.

Borst-Pauwels, G. W. F. H. (1981). Ion transport in yeast, *Biochim. Biophys. Acta, 650:* 88–127.

Borst-Pauwels, G. W. F. H., Theuvenet, A. P. R., and Stols, A. L. H. (1983). All-or-none interactions of inhibitors of the plasma membrane ATPase with *Saccharomyces cerevisiae, Biochim. Biophys. Acta, 732:* 186–193.

Buchman, D., Skroch, P. Welch, J., Fogel, S., and Karin, M. (1989). The *CUP2* gene product, regulator of yeast metallothionein expression, is a copper-activated DNA-binding protein, *Mol. Cell. Biol., 9:* 4091–4095.

Carri, M. T., Galiazzo, F., Ciriolo, M. R., and Rotilio, G. (1991). Evidence for co-regulation of Cu,Zn superoxide dismutase and metallothionein gene expression in yeast through transcriptional control by copper via the ACE1 factor, *FEBS Lett., 2278:* 263–266.

Cartwright, C. P., Juroszek, J.-R., Beavan, M. J., Ruby, F. M. S., De Orais, S. M. F., and Rose, A. H. (1986). Ethanol dissipates the proton-motive force across the plasma membrane of *Saccharomyces cerevisiae, J. Gen. Microbiol., 132:* 369–377.

Cartwright, C. P., Veazey, F. J., and Rose, A. H. (1987). Effect of ethanol on activity of plasma-membrane ATPase in, and accumulation of glycine by, *Saccharomyces cerevisiae, J. Gen. Microbiol., 133:* 857–865.

Chang, E. and Kosman, D. J. (1989). Intracellular Mn (II)-associated superoxide scavenging activity protects Cu,Zn superoxide dismutase-deficient *Saccharomyces cerevisiae* against dioxygen stress, *J. Biol. Chem. 364:* 12172–12178.

Clark, R. A. (1990). The human neutrophil respiratory burst oxidase, *J. Infect. Dis., 161:* 1140–1147.

Conklin, D. S., McMaster, J. A., Culbertson, M. R., and Kung, C. (1992). *COT1,* a gene involved in cobalt accumulation in *Saccharomyces cerevisiae, Mol. Cell. Biol., 12:* 3678–3688.

Crane, F. L., Roberts, H., Linnane, A. W., and Low, H. (1982). Transmembrane ferricyanide reduction by cells of the yeast *Saccharomyces cerevisiae, J. Bioener. Biomem., 14:* 191–204.

Crapo, J. D., Oury, T., Rabouille, C., Slot, J. W., and Chang. L. Y. (1992). Copper, zinc superoxide dismutase is primarily a cytosolic protein in human cells, *Proc. Natl. Acad. Sci. USA, 89:* 10405–10409.

Crawford, B. S., Romeo, A., Hassett, R., and Kosman, D. J. (1993). Unpublished.

Dameron, C. T., Smith, B. R., and Winge D. R. (1989). Glutathione-coated cadmium-sulfide crystallites in *Candida glabrata, J. Biol. Chem., 264:* 17355–17360.

Dancis, A., Klausner, R. D., Hinnebusch, A. G., and Barriocanal, J. G. (1990). Genetic evidence that ferric reductase is required for iron uptake in *Saccharomyces cerevisiae, Mol. Cell. Biol., 10:* 2294–2301.

Dancis, A., Roman, D. G., Anderson, G. J., Hinnebusch, A. G., and Klausner, R. D. (1992). Ferric reductase of *Saccharomyces cerevisiae:* Molecular characterization, role in iron uptake, and transcriptional control by iron, *Proc. Natl. Acad. Sci. USA, 89:* 3869–3873.

De Nobel, J. G., and Barnett, J. A. (1991). Passage of molecules through yeast cell walls: A brief essay-review, *Yeast, 7:* 313–323.

De Rome, L. and Gadd, G. M. (1987). Measurement of copper uptake in *Saccharomyces cerevisiae* using a Cu^{2+}-sensitive electrode, *FEMS Microbiol. Lett., 43:* 283–287.

Doonan, B. B., Crang, R. E., Jensen, T. E., and Baxter, M. (1979). *In situ* x-ray dispersive microanalysis of polyphosphate bodies in *Aureobasidium pullulans, J. Ultrastruct. Res., 69:* 232–238.

Eide, D., and Guarente, L. (1992). Increased dosage of a transcriptional activator gene enchances iron-limited growth of *Saccharomyces cerevisiae, J. Gen. Microbiol., 138:* 347–354.

Eide, D., Davis-Kaplan, S., Jordan, I., Sipe, D., and Kaplan, J. (1992). Regulation of iron uptake in *Saccharomyces cerevisiae.* The ferrireductase and Fe(II) transporter are regulated independently, *J. Biol. Chem., 267:* 20774–20781.

Failla, M. L., and Weinberg, E. D. (1977). Cyclic accumulation of zinc by *Candida utilis* during growth in batch culture, *J. Gen. Microbiol., 99:* 85–97.

Failla, M. L., Benedict, C. D., and Weinberg, E. D. (1976). Accumulation and storage of Zn^{2+} by *Candida utilis, J. Gen. Microbiol., 94:* 23–36.

Fogel, S., Welch, J. W., and Maloney, D. H. (1988). The molecular genetics of copper resistance in *Saccharomyces cerevisiae*—a paradigm for non-conventional yeasts, *Basic Microbiol., 28:* 147–160.

Freedman, J. H., and Peisach, J. (1989). Intracellular copper transport in cultured hepatoma cells, *Biochem. Biophys. Research Commun., 164:* 134–140.

Fuhrmann, G.-F., and Rothstein, A. (1968). The transport of Zn^{2+}, Co^{2+} and Ni^{2+} into yeast cells, *Biochim. Biophys. Acta, 163:* 325–338.

Furst, P., Ju, S., Hackett, R., and Hamer, D. (1988). Copper activates metallothionein gene transcription by altering the conformation of a specific DNA binding protein, *Cell, 55:* 705–717.

Gaber, R. R., Styles, C. A., and Fink, G. R. (1988). *TRK1* encodes a plasma membrane protein required for high-affinity potassium transport in *Saccharomyces cerevisiae, Mol. Cell. Biol., 8:* 2848–2859.

Gadd, G. M. (1990). Metal tolerance, *Microbiology of Extreme Environments* (C. Edwards, ed.), Open University Press (Milton Keynes), Baltimore, pp. 178–210.

Gadd, G. M., and Griffiths, A. J. (1980). Influence of pH on toxicity and uptake of copper in *Aureobasidium pullulans, Trans. Br. Mycol. Soc., 75:* 91–96.

Gadd, G. M., and Mowll, J. L. (1983). The relationship between cadmium uptake, potassium release and viability in *Saccharomyces cerevisiae, FEMS Microbiol. Lett., 16:* 45–48.

Gadd, G. M., and White, C. (1985). Copper uptake by *Penicillium ochro-chloron:* Influence of pH on toxicity and demonstration of energy-dependent copper influx using protoplasts, *J. Gen. Microbiol., 131:* 1875–1879.

Gadd, G. M., and White, C. (1989). Heavy metal and radionuclide accumulation and toxicity in fungi and yeasts, *Metal-Microbe Interactions* (G. M. Gadd, ed.), IRL Press, Oxford, pp. 19–38.

Gadd, G. M., Stewart, A., White, C., and Mowll, J. L. (1984). Copper uptake by whole cells and protoplasts of a wild-type and copper-resistant strain of *Saccharomyces cerevisiae, FEMS Microbiol. Lett., 24:* 231–234.

Galiazzo, F., Pederson, J. Z., Civitarealle, P., Schiesser, A., and Rotilio, G. (1989). Manganese accumulation in yeast cells, *Biol. Metals, 2:* 6–10.

Germann, U. A., and Lerch, K. (1987). Copper accumulation in the cell-wall-deficient slime variant of *Neurospora crassa, Biochem. J., 245:* 479–484.

Gralla, E. B., Thiele, D. J., Silar, P., and Valentine, J. S. (1991). ACE1, a copper-dependent transcription factor, activates expression of the yeast copper, zinc superoxide dismutase gene, *Proc. Natl. Acad. Sci. USA, 88:* 8558–8562.

Griffith, O. W., and Meister, A. (1979). Potent and specific inhibition of glutathione synthesis by buthionine sulfoximine (*S-n*-butyl homocysteine sulfoximine), *J. Biol. Chem., 254:* 7558–7560.

Hamer, D. (1986). Metallothioneins, *Ann. Rev. Biochem., 55:* 913–951.

Hamer, D. H., Thiele, D. J., and Lemontt, J. F. (1985). Function and autoregulation of yeast copperthionein, *Science, 228:* 685–680.

Hassett, R., and Kosman, D. J. (1993). Unpublished.

Hassett, R., Pardee, T., and Kosman, D. J. (1993). Unpublished.

Hockertz, S., Schmid, J., and Auling, G. (1987). A specific transport system for manganese in the filamentous fungus *Asperigillus niger, J. Gen. Micro., 133:* 3513–3519.

Jennings, D. H., Hooper, D. C., and Rothstein, A. (1958). The participation of phosphate in the formation of a "carrier" for the transport of Mg^{++} and Mn^{++} ions into yeast cells, *J. Gen. Physiol., 41:* 1019–1026.

Joho, M., Imada, Y., and Murayama, T. (1987). The isolation and characterization of Ni^{2+} resistant mutants of *Saccharomyces cerevisiae, Microbios, 51:* 183–190.

Jones, E. W., and Fink, G. R. (1982). Regulation of amino acid and nucleotide biosynthesis in yeast, *The Molecular Biology of the Yeast Saccharomyces—Metabolism and Gene Expression* (J. N. Strathern, E. W. Jones, and J. R. Broach, eds.), Cold Spring Harbor, New York, pp. 181–299.

Jungmann, J., Lee, J., Romeo, A., Hassett, R., Reins, H.-A., Kosman, D., and Jentsch, S. (1993). MAC1, a homolog of Cu-dependent transcription factors linked to Cu/Fe metabolism and stress resistance in yeast, *EMBO J.,* in press.

Kamizono, A., Nishizawa, M., Teranishi, Y., Murata, K., and Kimura, A. (1989). Identification of a gene conferring resistance to zinc and cadmium ions in the yeast *Saccharomyces cerevisiae, Mol. Gen. Genet., 219:* 161–167.

Karin, M. (1985). Metallothioneins: Proteins in search of function, *Cell, 41:* 9–10.

Kihn, J. C., Mestagh, M. M., and Rouxhet, P. G. (1987). ESR study of copper(II) retention by entire cell, cell walls, and protoplasts of *Saccharomyces cerevisiae, Can. J. Microbiol., 33:* 777–782.

Kikuchi, T. (1965). Studies on the pathway of sulfide production in a copper-adapted yeast, *Plant Cell Physiol., 6:* 195–210.

Kosman, D. J., Romeo, A., Lee, J., Hassett, R., and Plotkin, J. (1993). Unpublished.

Kremer, S. M., and Wood, P. M. (1992). Evidence that cellobiose oxidase from *Phanerochaete chrysosporium* is primarily an Fe(III) reductase, *Eur. J. Biochem., 205:* 133–138.

Lee, J., Romeo, A., and Kosman, D. J. (1993). Unpublished.

Lesuisse, E., and Labbe, P. (1989). Reductive and non-reductive mechanisms of iron accumulation by the yeast *Saccharomyces cerevisiae, J. Gen. Microbiol., 135:* 257–263.

Lesuisse, E., Raguzzi, F., and Crichton, R. R. (1987). Iron uptake by the yeast *Saccharomyces cerevisiae:* Involvement of a reduction step, *J. Gen. Microbiol., 133:* 3229–3236.

Lesuisse, E., Crichton, R. R., and Labbe, P. (1990). Iron-reductases in the yeast *Saccharomyces cerevisiae, Biochim. Biophys. Acta, 1038:* 253–259.

Lesuisse, E., Horion, B., Labbe, P., and Hilger, F. (1991). The plasma membrane ferrireductase activity of *Saccharomyces cerevisiae* is partially controlled by cyclic AMP, *Biochem. J., 280:* 545–548.

Lin, C.-M., and Kosman, D. J. (1990). Copper uptake in wild type and copper metallothionein-deficient *Saccharomyces cerevisiae, J. Biol. Chem., 265:* 9194–9200.

Lin, C.-M., Crawford, B. S., and Kosman, D. J. (1993a). Distribution of ^{64}Cu in

Saccharomyces cerevisiae: Cellular locale and metabolism, *J. Gen. Microbiol., 139:* 1605–1615.

Lin, C.-M., Crawford, B. S., and Kosman, D. J. (1993b). Distribution of ^{64}Cu in *Saccharomyces cerevisiae:* Kinetic analysis of partitioning, *J. Gen. Microbiol., 139:* 1617–1626.

Mehra, R. K., and Winge, D. R. (1991). Metal ion resistance in fungi: Molecular mechanisms and their regulated expression, *J. Cell. Biochem., 45:* 30–40.

Meikle, A. J., Gadd, G. M., and Reed, R. H. (1990). Manipulation of yeast for transport studies: Critical assessment of cultural and experimental procedures, *Enzyme Microbiol. Technol., 12:* 865–872.

Mohan, P. M., Rudra, M. P. P., and Sastry, K. S. (1984). Nickel transport in nickel-resistant strains of *Neurospora crassa, Curr. Microbiol., 10:* 125–128.

Mowll, J. L., and Gadd, G. M. (1983). Zinc uptake and toxicity in the yeasts *Sporobolomyces roseus* and *Saccharomyces cerevisiae, J. Gen. Microbiol., 129:* 3421–3425.

Nieuwenhuis, B. J. W. M., Weigers, C. A. G. M., and Borst-Pauwels, G. W. F. H. (1981). Uptake and accumulation of Mn^{2+} and Sr^{2+} in *Saccharomyces cerevisiae, Biochim. Biophys. Acta, 649:* 83–88.

Norris, P. R., and Kelly, D. P. (1977). Accumulation of cadmium and cobalt by *Saccharomyces cerevisiae, J. Gen. Microbiol., 99:* 317–324.

Ochiai, E.-I. (1977). *Bioinorganic Chemistry,* Allyn and Bacon, Boston, Mass.

Ohsumi, Y., Kitamoto, K., and Anraku, Y. (1988). Changes induced in the permeability barrier of the yeast plasma membrane by cupric ion, *J. Bacteriol., 170:* 2676–2682.

Okorokov, L. A., Lichko, L. P., Kadomtseva, V. M., Kholodenko, V. P., Titovsky, V. T., and Kulaev, I. S. (1977). Energy-dependent transport of manganese into yeast cells and distribution of accumulated ions, *Eur. J. Biochem., 75:* 373–377.

Okorokov, L. A., Lichko, L. P., and Andreeva, N. A. (1983). Changes of ATP, polyphosphate, and K^+ contents in *Saccharomyces carlsbergensis* during uptake of Mn^{2+} and glucose, *Biochem. Internat., 6:* 481–488.

Ostrowski, J., Burdzy, G. J., and Kredich, N. M. (1987). DNA sequence of the *cysB* regions of *Salmonella typhimurium* and *E. coli, J. Biol. Chem. 262:* 5999–6005.

Parkin, M. and Ross, I. S. (1986). Specific uptake of manganese in the yeast *Candida utilis, J. Gen. Microbiol., 132:* 2155–2160.

Phelan, A., Thurman, D. A., and Tomsett, A. B. (1990). The isolation and characterization of copper-resistant mutants of *Aspergillus nidulans, Curr. Microbiol., 21:* 255–260.

Pilz, F., Auling, G., Stephan, D., Rau, O., and Wagner, F. (1991). A high affinity Zn^{2+} uptake system controls growth and biosynthesis of an extracellular, branched β-1,3-β-1,6-glucan in *Sclerotium rolfsii* ATCC 15205, *Exper. Mycol., 15:* 181–192.

Portillo, F., and Mazon, M. J. (1986). The *Saccharomyces cerevisiae* start mutant carrying the *cdc25* mutation is defective in activation of plasma membrane ATPase by glucose, *J. Bacteriol., 168:* 1254–1257.

Raguzzi, F., Lesuisse, F., and Crichton, R. R. (1988). Iron storage in *Saccharomyces cerevisiae, FEBS Lett., 231:* 253–258.

Reese, R. N., Mehra, R. K., Tarbe, E. B., and Winge, D. R. (1988). Studies on the γ-glutamyl Cu-binding peptide from *Schizosaccharomyces pombe, J. Biol. Chem., 263:* 4186–4192.

Romeo, A., Lee, J., Kosman, D. J., Jungmann, J., and Jentsch, S. (1993). Unpublished.

Ross, I. S. (1977). Effect of glucose on copper uptake and toxicity in *Saccharomyces cerevisiae, Trans. Br. Mycol. Soc., 69:* 77–81.

Ross, I. S., and Walsh, A. L. (1981). Resistance to copper in *Saccharomyces cerevisiae, Trans. Br. Mycol. Soc., 77:* 27–32.

Rothstein, A., Hayes, A., Jennings, D., and Hooper, D. (1958). The active transport of Mg^{++} and Mn^{++} into the yeast cell, *J. Gen. Physiol., 41:* 585–594.

Sabie, F. T., and Gadd, G. M. (1990). Effect of zinc on the yeast-mycelium transition of *Candida albicans* and examination of zinc uptake at different stages of growth, *Mycol. Res., 94:* 952–958.

Serrano, R. (1983). In vivo glucose activation of the yeast plasma membrane ATPase, *FEBS Lett., 156:* 11–14.

Serrano, R., Kielland-Brandt, M. C., and Fink, G. R. (1986). Yeast plasma membrane ATPase is essential for growth and has homology with (Na^+ + K^+), K^+ and Ca^{2+} ATPases, *Nature, 319:* 689–693.

Szczypka, M. S., and Thiele, D. J. (1989). A cysteine-rich nuclear protein activates yeast metallothionein gene transcription, *Mol. Cell. Biol., 9:* 421–429.

Theuvenet, A. P. R., and Bindels, R. J. M. (1980). An investigation into the feasibility of using yeast protoplasts to study the ion transport properties of the plasma membrane, *Biochim. Biophys. Acta, 599:* 587–595.

Theuvenet, A. P. R., Nieuwenhuis, B. J. W. M., van de Mortel, J., and Borst-Pauwels, G. W. F. H. (1986). Effect of ethidium bromide and DEAE-dextran on divalent cation accumulation in yeast. Evidence for an ion-selective extrusion pump for divalent cations, *Biochim. Biophys. Acta, 855:* 383–390.

Thiele, D. J. (1988). *ACE1* regulates expression of the *Saccharomyces cerevisiae* metallothionein gene, *Mol. Cell. Biol., 8:* 2745–2752.

Thomas, D., Cherest, H., and Surdin-Kerjan, Y. (1989). Elements involved in *S*-adenosylmethionine-mediated regulation of the *Saccharomyces cerevisiae MET25* gene, *Mol. Cell. Biol., 9:* 3292–3298.

Townsley, C. C., and Ross, I. S. (1985). Copper uptake by *Penicillium spinulosum, Microbios., 44:* 125–132.

Ulaszewski, S., Hilger, F., and Goffeau, A. (1989). Cyclic AMP controls the plasma membrane H^+-ATPase activity from *Saccharomyces cerevisiae, FEBS Lett., 245:* 131–136.

van der Pal, R. H. M., Belde, P. J. M., Theuvenet, A. P. R., Peters, P. H. J., and Borst-Pauwels, G. W. F. H. (1987). Effect of ruthenium red upon Ca^{2+} and Mn^{2+} uptake in *Saccharomyces cerevisiae.* Comparison with the effect of La^{2+}, *Biochim. Biophys. Acta, 902:* 19–23.

Venkateswerlu, G., and Sastry, K. S. (1970). The mechanism of uptake of cobalt ions by *Neurospora crassa, Biochem. J., 118:* 497–503.

Wakatsuki, T., Imahara, H., Kitamura, T., and Tanaka, H. (1979). On the absorption of copper into yeast cell, *Agric. Biol. Chem., 43:* 1687–1692.

Wakatsuki, T., Iba, M., and Imahara, H. (1988). Copper reduction by yeast cell wall materials and its role on copper uptake in *Debaryomyces hansenii, J. Ferment. Technol., 66:* 257–265.

Wakatsuki, T., Hayakawa, S., Hatayama, T., Kitamura, T., and Imahara, H. (1991a). Solubilization and properties of copper reducing enzyme systems from the yeast cell surface in *Debaryomyces hansenii, J. Fermen. Bioengin., 72:* 79–86.

Wakatsuki T., Hayakawa, S., Hatayama, T., Kitamura, T., and Imahara, H. (1991b). Purification and some properties of copper reductase from cell surface of *Debaryomyces hansenii, J. Fermen. Bioengin., 72:* 158–161.

White, C., and Gadd, G. M. (1986). Uptake and cellular distribution of copper, cobalt and cadmium in strains of *Saccharomyces cerevisiae* cultured on elevated concentrations of these metals, *FEMS Microbiol. Ecol. 38:* 277–283.

White, C., and Gadd, G. H. (1987). The uptake and cellular distribution of zinc in *Saccharomyces cerevisiae, J. Gen. Microbiol., 133:* 727–737.

Willsky, G. R. (1979). Characterization of the plasma membrane Mg^{2+}-ATPase from the yeast, *Saccharomyces cerevisiae, J. Biol. Chem., 254:* 3326–3332.

Winge, D. R., Nielson, K. B., Gray, W. R., and Hamer, D. H. (1985). Yeast metallothionein. Sequence and metal-binding properties, *J. Biol. Chem., 260:* 14464–14470.

Wright, C. F., McKenney, K., Hamer, D. H., Byrd, J., and Winge, D. R. (1987). Structural and functional studies of the amino terminus of yeast metallothionein, *J. Biol. Chem., 262:* 12912–12919.

Zhou, P., and Thiele, D. J. (1991). Isolation of a metal-activated transcription factor gene in *Candida glabrata* by complemenation in *Saccharomyces cerevisiae, Proc. Natl. Acad. Sci. USA, 88:* 6112–6116.

Zhou, P., Szczypka, M. S., Sosinowski, T., and Thiele, D. J. (1992). Expression of a yeast metallothionein gene family is activated by a single metallothionein transcription factor, *Mol. Cell. Biol., 12:* 3766–3775.

2

Hydroxamates and Polycarboxylates as Iron Transport Agents (Siderophores) in Fungi

Dick van der Helm
University of Oklahoma, Norman, Oklahoma

Günther Winkelmann
University of Tübingen, Tübingen, Germany

I. INTRODUCTION

In oxidation reduction enzymes, metal ions with two accessible ionization states are often used as cofactors. Similarly such ions are used in transport proteins. Many different metals such as copper (Cu^{+}/Cu^{++}) or cobalt (Co^{2+}/Co^{3+}) can be used for this purpose, but more often than not iron (Fe^{2+}/Fe^{3+}) was selected by nature. This is also the case in microorganisms. However, in an aerobic environment the acquisition of iron poses a problem, because iron exists as Fe^{3+} and precipitates as an oxide-hydroxide, which has a very small solubility constant ($\sim 10^{-38}$). This means that the concentration of the free Fe^{3+} is extremely small, of the order of 10^{-18} M. Therefore, soluble iron is not available to microorganisms without a solubilization system. Many aerobic and faculative anaerobic microorganisms therefore produce, under iron-deficient circumstances, chelating agents, siderophores that form complexes with Fe^{3+} with very high complexation constants, thus solubilizing Fe^{3+}.

The siderophores are biosynthesized by the organisms under negative iron control. They are released to the environment where the ferrisiderophore complexes are formed. The complexes are subsequently taken up by the microorganisms. In bacteria it is known that this uptake mechanism is often specific and that it involves a number of proteins, which are also produced only under negative iron control. In fungi this is presumed to be the case as well, but no transport proteins have been identified so far.

Many different hydroxymate siderophores have been isolated and identified, and many more may be expected to be discovered. There is a simple reason for this variety in siderophores that are found in nature. There is a fierce competition for iron by microorganisms in nature. It is obvious that for competitive reasons, ideally, each microorganism would make a specific siderophore that only the particular organism would recognize and assimilate. This idealized situation has not been achieved in nature, but it is approached to some extent, and this is the reason for the large variety of siderophores that have been discovered.

There are, in fact, two other aspects to this same attempt by nature for competitiveness and conservation of metabolic energy. One is that a microorganism may synthesize different siderophores for different environments, and this is a possible reason for the experimental fact that a microorganism quite often makes several siderophores, sometimes several within one family, but sometimes siderophores from different hydroxamate families. Still one other aspect is important, which is the uptake system used by the microorganism to take up or assimilate the siderophore. It is known that this is an active uptake system from the fact that it can be poisoned by metabolic inhibitors. It is obvious that when this uptake system is able to assimilate and transport not only the siderophore produced by the microorganism but also other similar ones, or, in an even more radical adaptation, has developed an additional uptake system for siderophores it does not produce, the microorganism has a definite competitive advantage. Both these adaptations have already been shown in some bacterial systems. Much less is known for fungal uptake systems, but it seems logical to assume that these do occur in those microorganisms as well. One could even take this argument further and expect the least specific uptake system to have the most advantage; but this does not seem to be the case, and most known uptake systems are quite specific.

The present chapter concentrates on siderophores produced by fungi. The siderophores that have been found thus far are categorized as hydroxamates and hydroxyacids, of which the latter type have only recently been discovered. The physical chemical properties, the solid and solution structures, will be described as well as other properties. A large section of the chapter covers the uptake processes of ferrisiderophores in fungi that have been studied.

A number of reviews, as well as several books, have been published in the last few years that can be consulted for more details (Neilands, 1981;

Hider, 1984; Winkelmann et al., 1987; Matzanke et al., 1989; and Winkelmann, 1991).

II. THE HYDROXAMATE GROUP AND METAL COMPLEXATION

The structures of most fungal siderophores are derived from acylated N^{δ}-hydroxy ornithine. The >N–OH group is deprotonated, and this group together with C=O of the acyl group forms a bidentate ligand: the hydroxamate group (Scheme 1): –N(O–)–C(=O)–, which forms thermodynamically stable chelates with spherically symmetric trivalent metal ions. They exhibit unusually high specificity for Fe^{3+}. The affinity constants for the complexes with Fe^{3+} for hydroxamates are higher than the affinities for other ions. The reason for this is the high charge and small size of the ferric ion that make it a hard acid. This in turn requires the ligand to be a hard base, and charged oxygen atoms as occur in the hydroxamate group behave as hard basic ligands. Siderophores, therefore, most often have all oxygen atoms as ligand atoms, although there are some exceptions, where one of six ligand atoms is a negatively charged nitrogen atom (Hancock and Martell, 1989).

The normal electronic configuration of Fe^{3+} in the complexes with siderophores is the one in which each of the 5d electrons occupies one of the 3d orbitals, thus giving a stable high spin electronic configuration (Oosterhuis, 1974). It can be observed with magnetic measurements. This electronic configuration has several consequences for the structure and properties of the complexes. High spin d^5 Fe^{3+} complexes have no crystal field stabilization energy, and most of the interaction energy is the result of the coulombic interactions between the positively charged metal ion and the negatively charged oxygen atoms. Another consequence is that the metal ion is spherically symmetric, and the Fe^{3+} in the siderophore complexes has no directional preferences for the Fe-O bonds. Therefore, although the Fe^{3+}

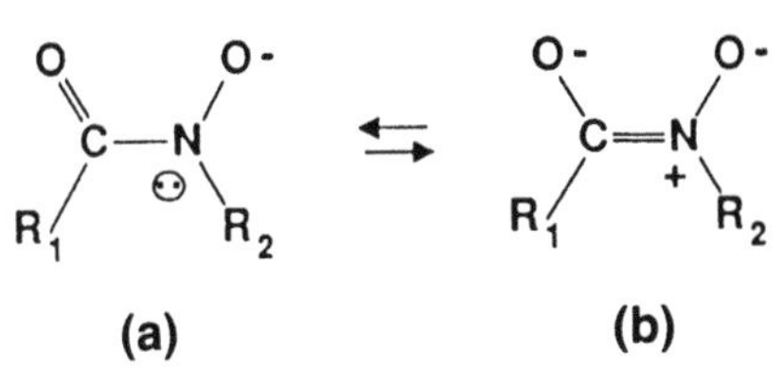

(a) (b)

Scheme I

ion is normally octahedrally surrounded by six oxygen atoms, this octahedron is always irregular (van der Helm et al., 1987) (see below). The Fe^{3+} ion seems to have the optimum size for this type of interaction. If the ion were larger, the charge density would be less and the attractive coulombic energy smaller. If the ion were smaller (as for instance in Al^{3+}), the smaller distance between the metal ion and the charged oxygen atoms would increase the repulsive coulombic interactions between the liganding atoms (Xiao et al., 1992). The high spin ferric ion seems, therefore, to have the optimum size, explaining the specificity of hydroxamates for this metal ion.

On the other hand, one can look at the suitability of the hydroxamate group as a ligand. The most important characteristic of the group is the resonance indicated in Scheme I (Monzyk and Crumbliss, 1979). Form (a) would put all the negative charge on only one of the oxygen atoms, whereas form (b) distributes this charge over both oxygen atoms and improves the stability of the resulting metal complex. The importance of resonance from (b) can be seen in structural and thermodynamic results (see below) of model complexes. Inductive effects of R_2 are effective, but more important in actual siderophores is the possible conjugation of R_1 with the hydroxamate group, for instance when the acyl group is derived from 3-methyl-5-hydroxy-pent-2-enoic acid (anhydro mevalonic acid) rather than acetic acid. Both groups are found in many hydroxamate siderophores. The normal pK_a values of a hydroxamic acid is about 9. A much higher pK_a value would make the resulting metal complexes relatively unstable at low pH values due to competition with H^+ ions. A lower pK_a, on the other hand, indicates that the ion is quite stable compared to the protonated (or chelated) form and indicates lower thermodynamic stability of the complex. Therefore the hydroxamate group seems to be very well suited to be a hard basic ligand, required by the ferric ion and thus able to make stable and specific complexes (Hancock and Martell, 1989). It should also be noticed that tris complexes of hydroxamates with Fe^{3+} are uncharged and that this property has obvious implications for membrane translocations. Actually the 3-hydroxy-4-pyridones have similar properties to hydroxamic acids. These, with one exception, are not found in nature but have recently been synthesized and tried clinically for removing excess iron (Hider and Hall, 1991).

A. Model Compounds and Structural Parameters

A number of the features described in the previous paragraphs have been explored by solid-state structure determinations of metal chelates of model hydroxamates (Dietrich et al., 1991). The relatively small molecular weights of these compounds allow quite accurate determinations of bond distances and angles, thus allowing comparisons.

There is a reciprocal influence of the metal ion and the ligand. For instance, when the chelate structure is compared with the protonated form of the ligand, hydroxamic acid, it is obvious that in the chelate, resonance form (b) is more important than it is in the hydroxamic acid. In hydroxamate the C–N bond decreases by approximately 0.04 Å while the C=O distance increases by about 0.03 Å when compared to the acid form of the group (Dietrich et al., 1990, 1991). These are significant differences for structures where the standard deviations are about 0.003 Å. One could, therefore, say that the proton is not a good metal ion.

Even in the chelates, however, there is not an equal distribution of charges over the two oxygen atoms. This is apparent from the fact that the Fe–O–(N) distance is found to be 0.06–0.10 Å smaller than the Fe–O–(C) distance in all hydroxamate structures, indicating that the charge on (N)–O is larger than on (C)–O in all chelates (van der Helm et al., 1987). It is also apparent that for the best hydroxamate ligand (i.e., the ligand that gives the most negative charge on the (C)–O oxygen atom) and forming the strongest complex, this difference in Fe–O distances is the least, proving the importance of resonance form (b) (Dietrich et al., 1991).

Other structural parameters should be considered. Hydroxamate is not a symmetric ligand, and if a tris complex forms one can anticipate both cis and trans complexes (Scheme II). For an unknown reason, the tris bidentate complexes prefer the cis configuration. One can imagine two triangles, one through (N)↺O oxygen atoms and the other through (C)↺O oxygen atoms (Scheme III). Because there is more charge on the

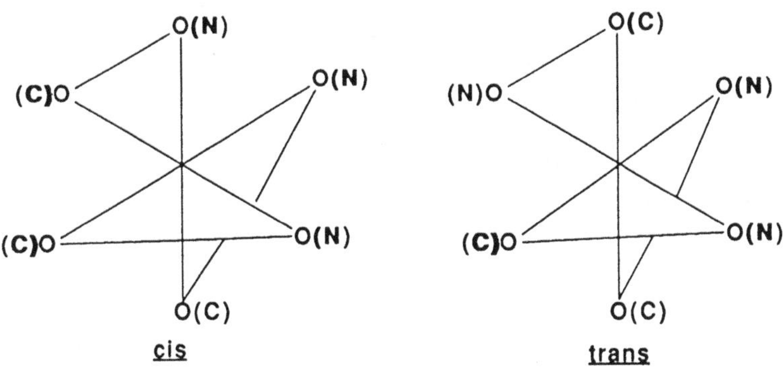

Scheme II

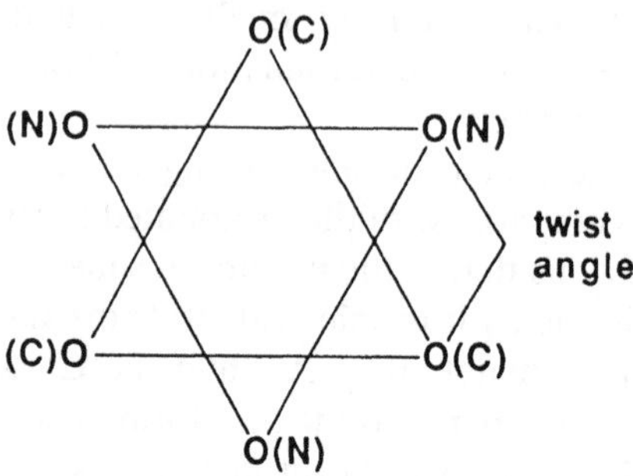

Scheme III

O(N) atoms than on the O(C) atoms, one always observes the distance of the Fe^{3+} ion to these planes to be different, and smaller to the plane of the O(N) atoms, by about 0.12 Å. For the most stable complexes this difference will be the smallest, and this is thus another structural parameter (Dietrich et al., 1991).

The O . . . O distance in the bidentate hydroxamate ligand remains constant at 2.55 Å. Also the radius of the Fe^{3+} ion is a particular value and constant. It is therefore, the angle at the metal ion, the (N)O–Fe–O(C) angle, that is truly a structural parameter despite a possible decrease in the Fe–O(C) distance with concomitant increase in the Fe–O(N) distance. This angle is 78° and constant in Fe^{3+} hydroxamates, and also the ligand bite, which is the O . . . O distance divided by the average Fe–O distance, is constant with a value of 1.26 (ligand bite) for all ferric hydroxamates (van der Helm et al., 1987). The value of the O–Fe–O angle of 78° is considerably different from 90°, the angle one expects in a regular octahedron, and this is, therefore, the main cause of the irregularity of the octahedral metal surrounding in ferric hydroxamate complexes.

Other structural parameters are used, but these are not necessarily constants for a particular metal ion. One of these is interesting and important; it is called the twist angle. Viewing Scheme III, one can see that in a regular octahedron the two triangles make an angle of 60°. However, in the trigonal prism this angle would be 0°. In the ferric hydroxamate complexes, the twist angles are between 0° and 60°, and they are observed to be as small as 36° and as large as 45°. This parameter is not a constant (Dietrich et al., 1991), as was shown by observing the structural results of the same complex in two crystal structures that contained different amounts of water of crystallization (Xiao et al., 1992).

B. Hydroxamate Complexes with Other Metal Ions

The binding constant of a hydroxamate with Fe^{3+} is normally larger than for any other metal ion. However, ions that have characteristics similar to those of Fe^{3+} will have large binding constants as well. Most important of these is Ga^{3+}, which has a radius slightly smaller than that of Fe^{3+}. Also Al^{3+} with a still smaller radius and In^{3+} with a larger ionic radius form hydroxamate complexes with high binding constants. Each of these three ions is also spherically symmetric, but in these cases this is caused by a closed (full) shell of electrons. Specifically Ga^{3+} complexes with hydroxamates have similar binding constants to those of the corresponding Fe^{3+} complexes. Thermodynamically a formation constant is determined by the difference of the standard free energy of formation of the product (complex) and reagents (metal ion and ligand). In Ga^{3+}, due to its slightly smaller radius, the angle O–Ga–O will be larger, and consequently the ligand bite will be larger. The absolute stability of the complex itself will therefore be greater due to larger coulombic interactions between the metal ion and the oxygen atoms. This is compensated, however, by a larger hydration energy of the aqueous Ga^{3+} ion compared to that of the Fe^{3+} ion, yielding therefore similar binding constants (Dietrich et al., 1991). The Al^{3+} ion is small enough so that if it is surrounded by six charged oxygen atoms the repulsive energy between these negative charges will become significant, and the ion will have a slight preference for a tetrahedral rather than an octahedral surrounding; and normally the formation constants for Al^{3+} are smaller than those for Fe^{3+}. Even so, the similarity of the Ga^{3+}, Al^{3+}, and In^{3+}, and Fe^{3+} complexes with hydroxamates provides an opportunity for a number of experiments:

(a) The acquisition mechanism of a microorganism often can be tricked to take up Ga^{3+}, Al^{3+}, or In^{3+} siderophore complexes.

(b) The substitution of either Al^{3+}, or Ga^{3+} ions for Fe^{3+} in siderophore complexes allows the use of NMR experiments to study their structures, because neither Al^{3+} or Ga^{3+} ions have electronic magnetic moments that broaden the lines in NMR spectra. This line broadening does occur in Fe^{3+} siderophore complexes due to the parallel alignment of the five 3d electrons in this ion, making it impossible to interpret the spectra (Llinás et al., 1970; Jalal et al., 1986; Jalal et al., 1985).

(c) None of the three ions plays a physiological role. However, excess Al^{3+} is strongly implicated in impairing renal function. Excess Al^{3+} can be removed by means of hydroxamic acids, siderophores (Desferal), or other similar ligands (hydroxypridones).

(d) Ga^{67} is used as a radiolabel for diagnostic purposes. It is important that the metal, when infused, be incorporated in a complex that is very stable, and siderophore complexes or their analogs can be used for this purpose.

There is a one-step reductive reaction known to make Al^{3+} and Ga^{3+} complexes directly from Fe^{3+} complexes of hydroxamates (Emery, 1986). The reaction is based on the fact that neither Al^{3+} nor Ga^{3+} can be reduced but that Fe^{3+} can be reduced.

Of the divalent ions, only Cu^{2+} forms relatively strong complexes with hydroxamates. One can, however, expect a very different structure for these Cu^{2+} complexes from the ones observed for Fe^{3+}. Even weaker than the Cu^{2+} complexes are the hydroxamate complexes with Fe^{2+}. This ion can be considered a soft acid with a distinct preference for soft basic nitrogen ligands (Hancock and Martell, 1989). Reduction of a ferric siderophore complex, therefore, allows the release of the iron in the form of Fe^{2+}, as for instance in the Ga^{3+} and Al^{3+} substitution reactions mentioned in the previous paragraph, or in the release of iron within a fungal cell after reduction, once the ferric siderophore complex is taken up by the cell.

Even though the ferric hydroxamates are thermodynamically very stable, the spherical symmetry of the ion and the lack of crystal field stabilization make the complexes kinetically labile. The ferric complexes have the possibility of both geometric isomerism (e.g., Scheme III and below) and optical isomerism (if the ligand is chiral, as most natural hydroxamate siderophores are). Due to the kinetic lability it is not generally possible to isolate such isomers for the ferric complexes in the liquid state. The uptake of ferric siderophores is considered to be sensitive and specific for a particular isomer. In order to study this, investigators have resorted to d^3 Cr(III) and d^6 Rh(III) complexes (Leong and Raymond, 1974; Raymond et al., 1984). These complexes are kinetically stable due to crystal field stabilization and can therefore be separated, to some extent, in geometrical and optical isomers. The disadvantage is that these complexes are regular octahedral (due to crystal field stabilization) and are thus only roughly comparable to the Fe(III) complexes of siderophores.

C. Formal Structures of Natural Hydroxamate Siderophores

Natural hydroxamate siderophores are sometimes tetradentate but mostly hexadentate, that is, the molecule contains either two or three hydroxamate groups. The hydroxamate groups are derived from ornithine as indicated in Scheme IV. The various N^{δ}-acyl groups found in natural siderophores are shown in Scheme V and are thus derived from various

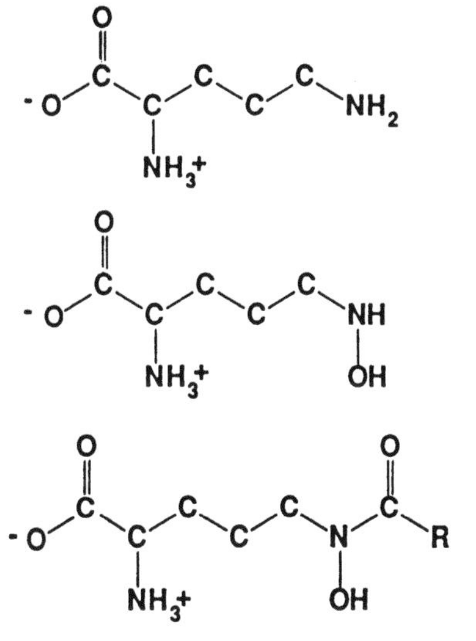

Scheme IV

acids. Actually only (a), (b), and (c) in Scheme V are common, and these groups are derived from acetic acid and cis and trans anhydromevalonic acid. All natural hydroxamate siderophores, with one exception, are derived from *L*-ornithine.

Although ornithine based hydroxamates are usual for fungal systems, a cyclized N^{δ}-acyl hydroxylated ornithine is quite often found in the pseudobactin siderophores from fluorescent Pseudomonas bacterial species.

The free hydroxamic acids have normally the trans configuration for the N–O–H and C=O group, as indicated in Scheme IV (Dietrich et al., 1990). However, in the chelates this group is necessarily in the cis configuration.

One recognizes a number of families of ornithine derived hydroxamate siderophores, dependent upon the way the ornithines are connected with one another. All are natural siderophores produced by fungi.

1. *Rhodotorulic Acids* (Atkin and Neilands, 1968)

The α-amino group and the acid group of two amino acids can form a pair of peptide linkages and thus form a diketopiperazine ring. This is the structure for two known siderophores, rhodotorulic acid and dimerum acid (Diekmann, 1970) (Scheme VI), in which the acyl groups are, respectively, (a) and (b) in Scheme V. They are necessarily tetradentate ligands containing two hydroxamate groups.

(a) $-CH_3$

(b) 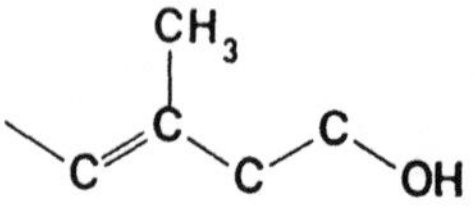 $-C{=}C(CH_3){-}C{-}C{-}OH$

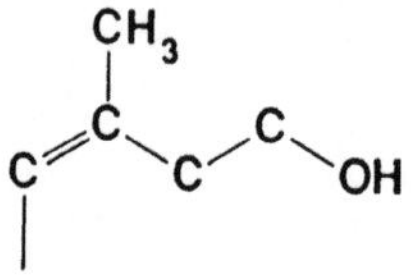

(c) $C{=}C(CH_3){-}C{-}C{-}OH$

(d) 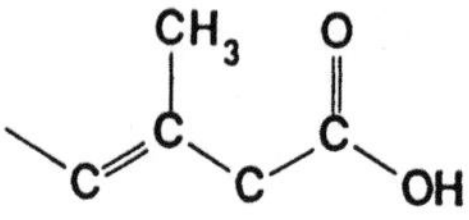 $-C{=}C(CH_3){-}C{-}C(=O){-}OH$

(e) 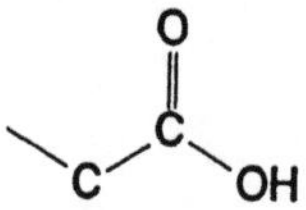 $-C{-}C(=O){-}OH$

(f) 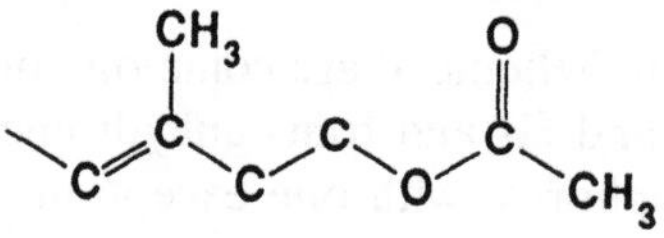 $-C{=}C(CH_3){-}C{-}C{-}O{-}C(=O){-}CH_3$

(g) 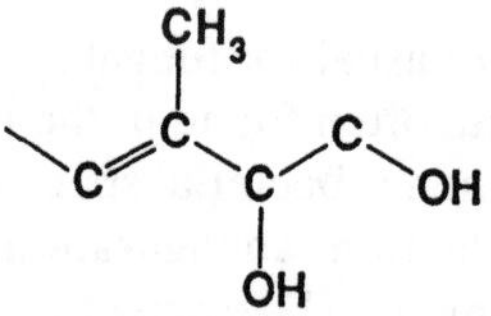 $-C{=}C(CH_3){-}C(OH){-}C{-}OH$

(h) 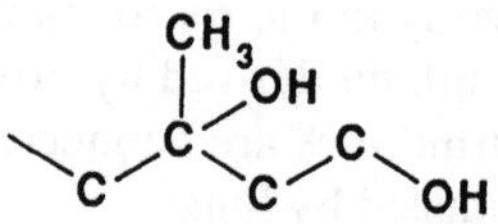 $-C{-}C(CH_3)(OH){-}C{-}C{-}OH$

(i) 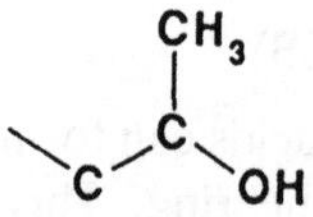 $-C{-}C(CH_3){-}OH$

Scheme V

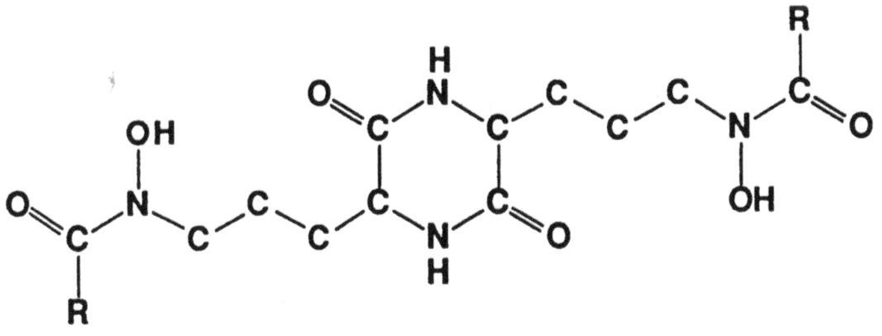

R = CH_3 : **Rhodotorulic Acid**

R = b : **Dimerum Acid**

Scheme VI

2. *Coprogens* (Hesseltine et al., 1952)

In a few cases fungi produce a single type of siderophore, e.g., *Rhodotorula rubra* produces only rhodotorulic acid. A number of fungi, however, produce a number of different siderophores, sometimes within one family (Jalal et al., 1984) and sometimes from different families (Jalal et al., 1986), to adapt, in the best possible way, to all possible environments. Even when the literature reports only one siderophore from a particular organism, one may anticipate that a further analysis would disclose that more than one siderophore is produced. This appears to be the case for the coprogens as well as the fusaranines and ferrichromes.

The coprogens have as a central unit dimerum acid, although in a few cases the N^{δ}-acyl group of one of the hydroxamate groups is either (a) or (g) instead of (b) (Jalal et al., 1988, 1989; Frederick et al., 1981). In all coprogens one of the N^{δ}-acyl groups of the diketopiperazine unit is (b) in Scheme V, and the terminal OH functionality of this group is esterified with the α-COOH group of a third ornithine that is hydroxylated and acylated to form the third hydroxamate group of the siderophore. All coprogens are thus hexadentate ligands forming 1:1 complexes with Fe^{3+}. The coprogens can easily be hydrolyzed at pH values above 9, due to the ester group that is present in the structure, and solutions should therefore be buffered. The α-amino group of the third ornithine can be free, as in coprogen B (Diekmann, 1970), or acetylated, as in coprogen, or dimethylated (Jalal et al., 1988). Some of the deferri coprogens besides their siderophore growth factor and antibiotic activity have been found to possess cytotoxic activity as well (Frederick et al., 1981). Coprogen B is an important experimental tool because it can be easily acetylated with (C^{14}) acetic acid, forming

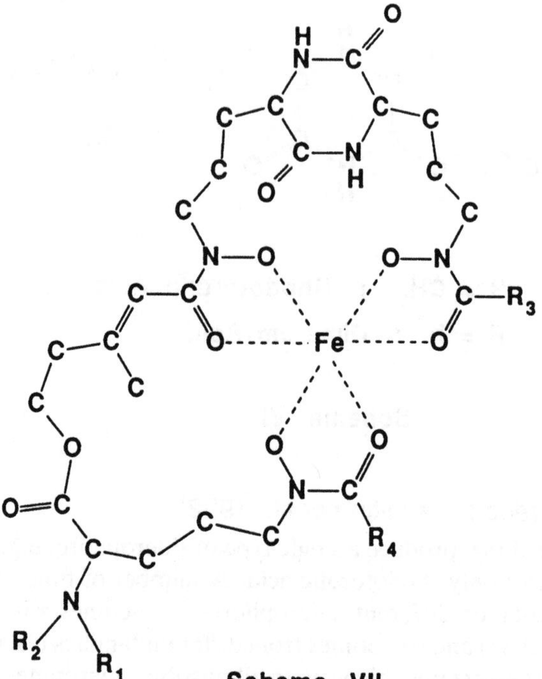

Scheme VII

	R_1	R_2	R_3	R_4
Coprogen	H	$-COCH_3$	b	b
Coprogen B	H	H	b	b
Neocoprogen I	H	$-COCH_3$	CH_3	b
Isoneocoprogen I	H	$-COCH_3$	b	CH_3
Neocoprogen II	H	$-COCH_3$	CH_3	CH_3
N^{α}-Dimethylcoprogen	CH_3	CH_3	b	b
N^{α}-Dimethylneocoprogen I	CH_3	CH_3	CH_3	b
N^{α}-Dimethylisoneocoprogen I	CH_3	CH_3	b	CH_3
Hydroxycoprogen	H	$-COCH_3$	b	g
Hydroxyneocoprogen I	H	$-COCH_3$	CH_3	g
Hydroxyisoneocoprogen I	H	$-COCH_3$	g	CH_3

labeled coprogen, and the fate of the ligand can be followed in uptake experiments (see below). All known coprogens are shown in Scheme VII. One may anticipate that other new coprogens produced by fungi will be found in the future.

3. *Fusarinines (Fusigens)* (Diekmann, 1967; Sayer and Emery, 1968)

In the fusarinines all three N^{δ}-acyl hydroxy ornithines are bonded to one another by means of ester linkages, in the same way as the third unit in the coprogens is attached to the dimeric diketopiperazine unit (Scheme VIII). However, the *N*-acyl group is derived from *cis*-anhydromevalonic acid rather than the trans form. The free amino groups adjacent to the ester linkages make the ester groups very labile, and any isolation procedure always will yield, beside the cyclic compound (fusarinine C), the linear trimer (fusarinine B), the linear dimer (fusarinine A), and the monomer (*cis*-fusarinine). It is, therefore, not clear if the linear forms are natural

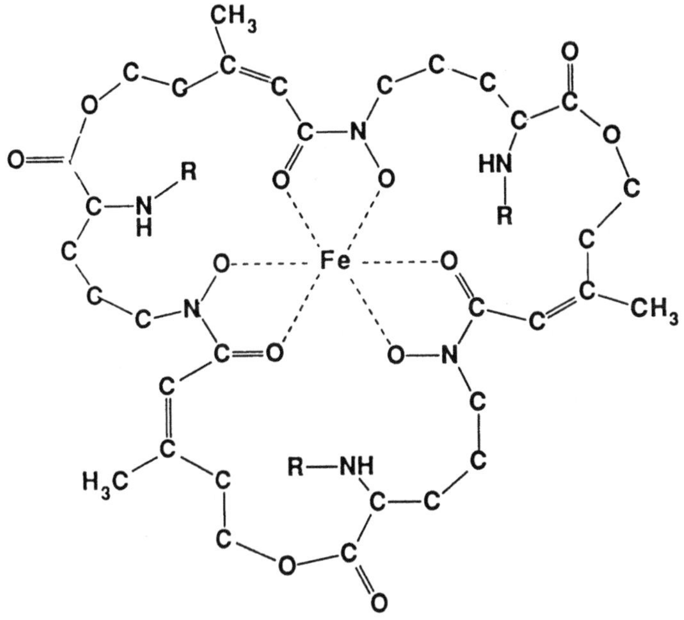

Scheme VIII

N^{α}-Triacetylfusarinine C R = $-COCH_3$

products biosynthesized by the organism or if they are hydrolysis products. The same is true for the monomer *trans*-fusarinine, which has been isolated together with dimerum acid, which is built from two *trans*-fusarinine units (see above). Another siderophore in this family is *N,N′,N″*-triacetylfusarinine C, in which the three free amino groups in fusarinine C are acetylated (Diekmann and Krezdorn, 1975; Moore and Emery, 1976). This siderophore is found in some *Aspergillus* and *Pencillium* species. The compound is more stable but will still instantly hydrolyze above a pH of 9.

The unusual siderophore neurosporin (Eng-Wilmot et al., 1984) belongs in this family. The N^{δ}-acyl group is not derived from anhydromevalonic acid but it is unit (i) in Scheme V while it is built from D-ornithine rather than L-ornithine.

The fusarinines are in general difficult to purify, and this may be the reason that relatively few members of this family have so far been found. On the other hand, the positive charge of the ferric chelates of the linear fusarinines might give these compounds a competitive advantage in certain circumstances. It has now been shown in several instances that the ferric chelates of the monomeric fusarinines are actively taken up by microorganisms (Jalal et al., 1987; Eng-Wilmot et al., 1992).

4. *Ferrichromes* (Emery and Neilands, 1961)

Historically the name ferrichrome refers to the ferric chelate of the siderophore. The siderophore itself, therefore, is identified as deferriferrichrome. There are more known members of this family than of any other. This is due partially to their chemical stability and partially to the relative ease of purification, especially with reverse phase column chromatography. Also in this family, the N^{δ}-acylated hydroxy L-ornithine is the basic building block, but here three ornithines are linked to one another by normal peptide linkages. This linear tripeptide with cis anhydromevalonic acid as acyl group is the siderophore found in *Aspergillus ochraceous* and more recently in *Agaricus bisporus* with the given name desdiserylglycyl desferriferrirhodin (DDF) (Jalal et al., 1985).

In all other deferriferrichromes one finds the triornithine peptide linked into a cyclic peptide with the use of glycine, serine, and alanine residues. The best known member is deferriferrichrome itself. It was the first hydroxamate siderophore reported in the literature, and it is produced by a number of fungal species. In deferriferrichrome, the cyclic hexapeptide is formed by the three ornithines and completed with threee glycine residues. It is not only the best known siderophore but also the most studied one. It is also one of the few siderophores on which biosynthetic studies have been carried out (Emery, 1971). Sometimes, simultaneously with deferriferrichrome, deferriferrichrome A is produced. In this compound, the acyl groups are acidic (residue (d) in Scheme 5), and two of the glycines are replaced by serine residues in the cyclic hexapeptide. There are four other members in this family that are quite common and are often isolated from microorganisms: the deferri forms of ferricrocin, ferrichrysin, ferrirubin, and ferrirhodin (Scheme IX) (Keller-Schierlein, 1963). In these compounds, one or two serine residues occur in the cyclic hexapeptide. The three acyl groups in each case for these four compounds are identical (Scheme IX): (a), (b), or (c) in Scheme V.

There are, however, many other members known in this family, but these compounds are less common and have been isolated so far only from one particular organism: *Aspergillus ochraceous* (Jalal et al., 1984). These are called asperchromes, and eleven have been identified. The hexapeptides in these compounds are orn-orn-orn-ser-ser-gly as in ferrichrysin, ferrirhodin, and ferrirubin. However, the three acyl group for each is different, mostly derived from a combination of residues (a) and (b) shown in Scheme V. One can, therefore, consider those to be compounds in between ferrichrysin and ferrirubin. Other new acyl groups are found as well. For instance, asperchrome C contains one esterified (b) group (this is (f) in Scheme V), and in

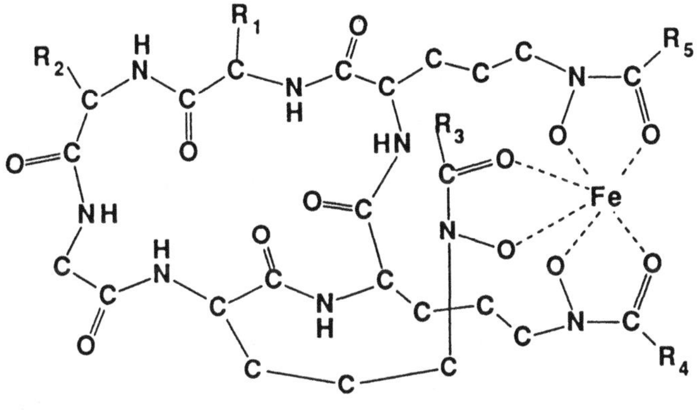

Scheme IX

	R_1	R_2	R_3	R_4	R_5
Ferrichrome	H	H	a	a	a
Ferrichrome A	CH_2OH	CH_2OH	d	d	d
Ferricrocin	H	CH_2OH	a	a	a
Ferrichrysin	CH_2OH	CH_2OH	a	a	a
Ferrirubin	CH_2OH	CH_2OH	b	b	b
Ferrirhodin	CH_2OH	CH_2OH	c	c	c

asperchrome A the acyl group (h) is found. In the asperchromes, many chemical isomers are possible because the three ornithines are not equivalent. Other members in this family that are not commonly found are malinochrome (Emery, 1980) with one alanine in the cyclic peptide and (e) (Scheme V) as the *N*-acyl groups, ferrichrome C also with one alanyl residue and (a) as the N^{δ}-acyl groups, asperchrome A with one alanyl and one serine and (b) as the acyl groups, and tetraglycyl ferrichrome (Deml et al., 1984) with four glycine residues forming a cyclic heptapeptide but with the same acyl groups (a) as ferrichrome. It is obvious that one can expect still other members in this family. It is interesting to note, however, that reside 6 in all ferrichromes is a glycine residue. This is a structural requirement for the conformation of the cyclic peptides (see below).

The albomycins, potent antibiotics against both gram-positive and gram-negative bacteria, belong to the ferrichrome family. They are hydroxamates derived from a linear ornithine peptide structure (Benz et al., 1982). The fourth residue in the peptide is L-serine, which is covalently (peptide) linked to a molecular system that has antibiotic activity. Presumably the ferrichrome uptake system is used to bring the compound into the cell. Other similar natural siderophore based antibiotics are known, and other compounds have been synthesized further to explore the effectiveness of this type of antibiotic activity.

Earlier the argument was made that a microorganism will have an advantage by producing several siderophores adapted to different environments. In the case of asperchromes, many of them are made by the same organism, and this argument does not hold, because most of the asperchromes have very similar physical properties. In this case, the production of these similar compounds seems to indicate a lack of specificity of the enzymes that biosynthesize the siderophores.

D. The Advantage of Hexadentate Ligands

The thermodynamics of formation constants for model and siderophore hydroxamates have been reviewed (Raymond et al., 1984; Crumbliss, 1991). Formation constants are determined by standard enthalpy and entropy changes; that is, a higher formation constant is found when in the process of forming the complex heat is given off and order is decreased, and both, therefore, need to be considered. For a bidentate ligand the formation of the tris complex proceeds as

$$Fe^{3+} + L^- \rightarrow FeL^{2+} \qquad K_1$$

$$FeL^{2+} + L^- \rightarrow FeL_2^+ \qquad K_2$$

$$FeL_2^+ + L^- \rightarrow FeL_3 \qquad K_3$$

or overall

$$\beta_3 = \frac{[FeL_3]}{[Fe^{3+}]\,[L^-]^3}$$

in which

$$\beta_3 = K_1K_2K_3$$

For the hexadentate ligand it is

$$Fe^{3+} + L^{3-} \rightarrow FeL$$

$$K = \frac{[FeL]}{[Fe^{3+}]\,[L^{3-}]}$$

Comparing the effectiveness of hexadentate ligands with bidentate ligands one needs to consider several processes (Hancock and Martell 1989; Crumbliss, 1991).

(a) The chelate effect favors a bidentate ligand over an unidentate one. This is an entropic effect that increases still more for a hexadentate ligand and therefore favors the formation of a complex with a hexadentate ligand.

(b) The atomic connection between different hydroxamate groups may impose a steric strain on the complex with the hexadentate ligand. This could be indicated by the observation that K for a hexadentate ligand is smaller than β_3 for the structurally similar bidentate ligand. This is difficult to assess experimentally but may occur in ferric coprogens and ferrichromes.

(c) Too long an atomic connection between hydroxamate groups may give a disadvantage to the hexadentate ligands for entropic reasons. This probably is an effect in ferric tri-N^{α}-acetyl fusarinine.

(d) The ligand may have to be rearranged before chelation occurs. Preordering of liganding atoms is attempted in the synthetic ligands. This may be an effect in ferrichrome.

All effects, however, either are small or compensate such that K is relatively close to β_3 for the tris complex of the similar bidentate hydroxamate ligand group.

Much more important in the comparison between hexadentate and bidentate ligands is the third-order power of $[L^-]$ in the equation for β_3 in comparison to the first-order power of $[L^{3-}]$ in K (Crumbliss, 1991). One should focus on the important concentration, which is the noncomplexed $[Fe^{3+}]$, which should be small. Even better is to evaluate the ratio of free and bound iron which is for the bidentate ligand:

$$\frac{[Fe^{3+}]}{[FeL_3]} = \frac{1}{\beta_3[L^{-1}]^3}$$

and for the hexadentate ligand:

$$\frac{[Fe^{3+}]}{[FeL]} = \frac{1}{K[L^{3-}]}$$

One can easily see that there is a very large advantage for the hexadentate ligand at low concentrations of the ligand, i.e., concentrations that occur in physiological conditions: 10^{-4} M and lower.

Common values of K (and β_3) are approximately 10^{30}, and if the ligand has a concentration of 10^{-5} M the ratio for uncomplexed to complexed iron for the bidentate ligand would be 10^{-15} and for the hexadentate ligand 10^{-25}, giving a huge advantage to the hexadentate ligand.

The large complexation constants of hexadentate hydroxamates are remarkable, but they are necessary for the ligands to complex Fe^{3+} in an aerobic environment. The concentration of Fe^{3+} in an aerobic environment can be estimated to be 10^{-18} M. One can use the formulas given above to calculate the concentrations in the presence of hydroxamate ligands. In the presence of siderophore at a concentration of 10^{-5} M, the concentration for free iron is only $\sim 10^{-30}$ M for a hexadentate ligand and 10^{-20} M for a bidentate ligand. This indicates that the hexadentate ligand can easily compete for Fe^{3+} in an aerobic environment but a bidentate ligand does so only with difficulty, while a tetradentate ligand, by implication, would do better than a bidentate ligand. It is an observed fact that natural siderophores easily can leach iron from glass and even from nails, although in the latter case they may dissolve the rust more than the metal.

E. Structure Determinations of Hydroxamate Siderophores

Ferrichrome was the first hydroxamic siderophore for which the structure was determined (Emery and Neilands, 1961). This was done primarily by classical chemical methods. These chemical methods developed specifically for hydroxamates are still appropriate and still are used with success. However, in recent decennia instrumental methods for structure determination have become more important, primarily single crystal x-ray diffraction and nuclear magnetic resonance (NMR) (van der Helm et al., 1987; Jalal and van der Helm, 1991). The advantage of x-ray diffraction is the large amount of objective information that is obtained from the structure determination. One observes the conformation, the configuration, and the geometry of the ligand and surrounding of iron, the bond distances and bond angles. There are of course disadvantages as well. For instance, it is often difficult or impossible to obtain single crystals. The conformation that is observed may be affected by the intermolecular interactions in the crystal, while the accuracy of the bond distances and angles is limited due to the large molecular weights of the siderophores and the instability of the crystals. NMR, on the other hand, can always be applied on the ligands. It is not possible to do NMR on the iron chelates due to line broadening (the reason for this was given earlier). On the other hand, NMR can be used for the free ligands, as well as the Al(III) and Ga(III) chelates of the siderophores. The disadvantage of NMR is that it is not a truly objective method and is, as all spectral techniques are, a comparison technique. Even so, when a new siderophore is found that is similar to a known one, NMR is the method of choice, because of its speed and convenience.

1. Crystal Structures of Ferrichromes

Ferrichrome A was the first ferric chelate of a hydroxamate siderophore for which a crystal structure was determined (Zalkin, Forrester, and Templeton, 1966). It showed the basic features found in all ferrichromes. This structure was the basis for the extensive NMR work and H^1 and C^{13} assignments made by Llinás and coworkers for various ferrichromes. Additional NMR assignments on asperchromes, coprogens, and fusarinines were tabulated more recently (Jalal and van der Helm, 1991).

Accurate structures were determined for both ferrichrome A and alumichrome A (van der Helm et al., 1981) that confirmed the general conclusion made earlier regarding the differences between Fe(III) and Al(III) chelates of hydroxamates using bidentate ligands.

The structures of a fairly large number of molecules in the ferrichrome family have been determined by single crystal x-ray diffraction (van der Helm et al., 1987). They crystallize quite easily as metal chelates, but it has not been possible to crystallize any of the unchelated ligands. A stereoview of the structure of asperchrome D_1 is shown (Fig. 1). It can be seen that ornithine 2 and 3 have *N*-acyl groups derived from acetic acid, while the acyl group of ornithine 1 is derived from *trans*-anhydromevalonic acid. If all *N*-acyl groups were of the first type, the molecule would be ferrichrysin. If they were all of the latter type, the molecule would be ferrirubin. The three ornithines are not equivalent, and one can therefore expect chemical isomers of this compound, and these are indeed found, i.e., asperchromes D_2 and D_3, which were isolated from the same organism, *Aspergillus ochraceous* (Jalal et al., 1988).

All structures determined in the ferrichrome family show similar features.

(a) The Fe(III) is on one side of the molecule.
(b) The coordination of the metal ion is Λ-*cis*.
(c) The configuration of the amino acids is L.
(d) The conformation of residue 5 and 6 show it to form a β(II) bend.

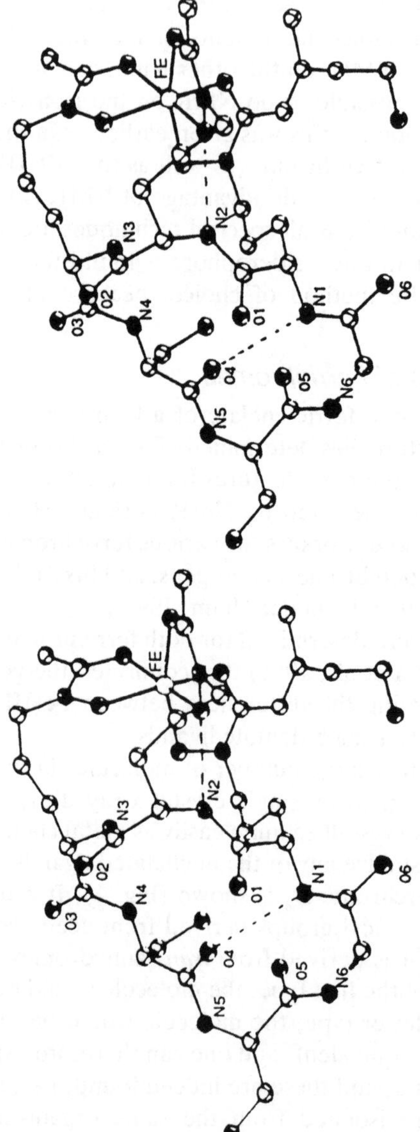

Figure 1 Stereoview of the structure of asperchrome D_1, a siderophore of the ferrichrome family. A three-dimensional view can be obtained by placing a piece of cardboard 10–12 inches long between the two views; staring at the two pictures will make them coalesce into one that shows depth. (From Jalal et al., 1988, with permission.)

(e) The result of this β(II) bend is a weak hydrogen bond between the C=O of the preceding residue (4) and the NH of the succeeding residue (1).

(f) The β(II) bend requires a D-amino acid in the second residue of the bend or a glycine. This is the reason why all compounds in the ferrichrome family of siderophores have a glycine in position 6.

(g) The conformational angles of ornithines 2 and 3 indicate that they form a β(I) bend, which requires two L-amino acids. Residues 2 and 3 are always L-ornithine.

(h) This β(I) bend does not result in a hydrogen bond between preceding and succeeding residues. The C(1)=O group is upward from the plane of the ring and not directed to N(4)–H; see Fig. 1.

(i) A hydrogen bond is formed between N(2)–H and the hydroxyl oxygen atom of the same ornithine residue.

(j) One can superpose the structures of various members of the ferrichrome family and obtain a root-mean-square deviation of the 49 atoms, which are common to all members, of approximately 0.30 Å (Barnes et al., 1985; Fidelis et al., 1990). Or one can compare the conformational angles around the cyclic hexapeptide ring and find maximal differences of 25°. This indicates that the structures and conformations are not the same but that they are similar with some conformational freedom. (This excludes one of two independently determined ferricrocin molecules, for which there is a clear indication that intermolecular forces influence the conformation.)

(k) There are several indications for strain in these molecules both in bond angles and in that the hydrogen bond between residues 1 and 4 due to the β(I) bend is not formed. It is also clear, although there is no crystal structure to prove it, that the orientations of the hydroxamate groups are not prearranged or preformed. Both these aspects lower the formation constant to some extent. However, the formation constant of deferriferrichrome ($\log K = 29.1$) is sufficiently large for efficient chelation of Fe^{3+} in an aerobic environment.

(l) Despite the structural similarities, the members in the ferrichrome family show distinct differences in uptake characteristics (see below). Although this can most easily be correlated with the immediate environment of the metal ion, other features of the molecules should not be discounted in playing a part in the uptake mechanism (Huschka et al., 1986).

2. *Ferric Coprogen*

The crystal structure of only one member in the coprogen family has been determined (Hossain et al., 1987). This is the structure of ferric neocoprogen I. Neocoprogen I differs from coprogen in that one N^{δ}-acyl group (part of the diketopiperazine ring) is derived from acetic acid rather than from *trans*-anhydromevalonic acid. Two independent molecules are observed in the crystal structure, and this allows a comparison between the two molecules. A stereoview of one of the molecules is shown in Fig. 2. Again the structure determination allows a number of observations.

(a) The metal ion is not in the center of the molecule and appears to be rather accessible as in the ferrichromes.

(b) The first two residues, containing the diketopiperazine ring, can be directly compared to dimerum acid. The structure of ferric neocoprogen I shows that also dimerum acid could chelate tetradentately to Fe^{3+} (but see below). There is no excessive strain in this chelation, because one of the atoms in one of the molecules is disordered, indicating different possible conformations.

(c) Two different conformations for the third ornithine are found in the two independent molecules. This is not unexpected, because the connection between the second and the third hydroxamate group involves a large number of atoms. This may have an adverse effect on the formation constant.

(d) The compound has 16 possible configurations for the iron surrounding (see below), but both molecules in the crystal structure have the same ΔC-trans-trans configuration.

(e) The trans configuration of the hydroxamate groups does not seem to have any effect on the geometry of the iron chelate structure. The Fe–O(N) and Fe–O(C) bond distances, the O–Fe–O chelate angle, the ligand bite,

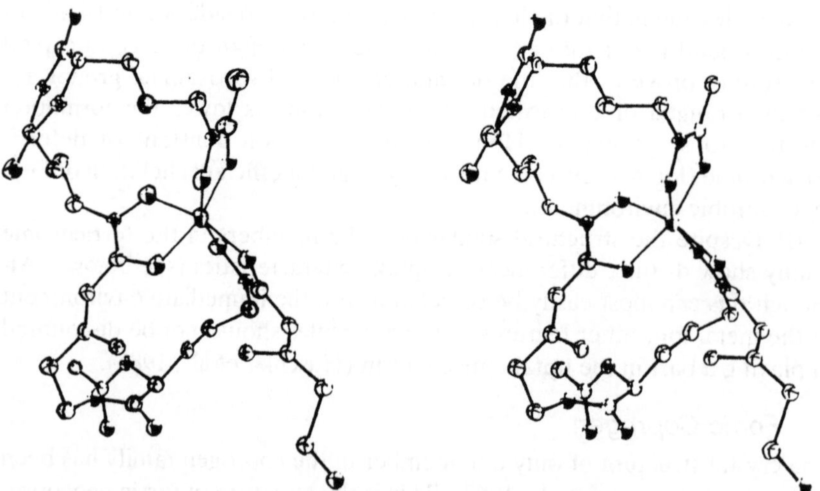

Figure 2 A stereoview of the structure of neocoprogen I, a siderophore in the coprogen family. (From Hossain et al., 1987, with permission.)

and the twist angle are all closely similar to those observed in other Fe(III) hydroxamate complexes that have a cis configuration.

The structure of the first known members in this family, coprogen and coprogen B, were determined by chemical methods. The structure of ferric neocoprogen I and the NMR assignments allowed the structure determination by NMR of a number of other members in this family (Jalal et al., 1988, 1989).

There are no crystal structures for the iron chelates of natural bidentate or tetradentate siderophores. There is, however, a crystal structure for the iron chelate of a tetradentate synthetic hydroxypridone (Fig. 3) (Scarrow et al., 1985), and the ligand in that respect can be seen as an analog of rhodotorulic acid. It shows a Fe_2L_3 structure in which each of the tetradentate L ligands bind both metal atoms. This indicates that all three ligands are equivalent in this structure. Research has been carried out on the chelation behavior in the liquid state on both the Fe(III) and Ga(III) chelates of rhodotorulic acid (RA) and dimerum acid (DA) (Jalal et al., 1986). In the latter case there is a clear indication for an $Fe_2(DA)_3$ complex with predominantly a Δ configuration around the metal ion. The NMR

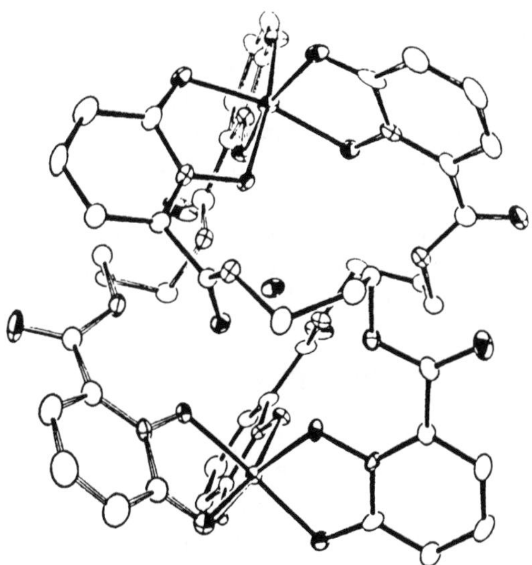

Figure 3 A perspective view of the 3:2 complex for *N*[1],*N*[3]-di(1-oxy-2(H)-pyridinonate-6-carboxyl)-1,3-propane diamine with Fe^{3+}. The structure for $Fe_2(RA)_3$ is proposed to be an analog of the structure shown in the figure. (From Scarrow, White, and Raymond, 1985, with permission.)

spectra for $Ga_2(DA)_3$, however, show two different conformations for the ligand in a 2:1 ratio. This is not in agreement with the structure shown in Fig. 3. This observation could indicate a structure in which each metal atom is coordinated tetradentally by one DA as shown in Fig. 2 for part of the neocoprogen structure while the third DA ligand connects the two metal atoms. This is a much less elegant structure than the one shown in Fig. 3, but it is certainly possible.

The difficulty in obtaining crystals for iron chelates of natural tetradentate and bidentate siderophores may be caused by a solution equilibrium of different coordination and structural isomers.

3. *Fusarinines*

There are two known crystal structures of ferric chelates of fusarinines (Hossain et al., 1980; Eng-Wilmot et al., 1984) and the one for ferric *N,N′,N″*-triacetylfusarinine C (FeTAFC) is shown in Fig. 4. Again a number of observations can be made.

(a) The iron atom is located in the center of the molecule. However, the metal atom is still accessible (from below in Fig. 4).

(b) The ligand is cyclic, made from three L-ornithines, while the *N*-acyl groups are derived from *cis*-anhydromevalonic acid.

(c) The different hydroxamate groups are connected by a long chain involving ten atoms. This may give entropic disadvantage to the formation constant. On the other hand, the double bond in the *N*-acyl group may give a hydroxamate group (Scheme 1) with stronger binding to the metal ion. The structural results are not sufficiently accurate to assess this for the compound.

(d) The configuration for the chelate in the crystal is *Λ-cis* (but see section on isomers for more details). The geometry of the iron hydroxamate chelation is very similar to that found in other structures.

F. Identification of Hydroxamate Siderophores

It is clear that a rather large number of different siderophores are known at this time. The need for iron and competition between species is the reason for this variety in structures. The search for siderophore(s) from an organism not previously investigated may, therefore, yield either new or known siderophores. Even for organisms that were previously investigated, more often than not only the siderophore produced in the largest quantity was investigated; thus an investigation of the other siderophores produced in minor amounts may yield new siderophores. In either case, it is important to be able to identify siderophores by comparison with known ones.

The isolation, purification, and identification of hydroxamate siderophores were reviewed just recently (Jalal and van der Helm, 1991). Siderophores have both hydrophilic and lipophilic characterisics, which is the basis for the extraction procedures used originally for isolation. This isolation method, however, has been replaced by XAD chromatography. The modern purification methods also involve chromatography, most often ion exchange and reverse phase. A fairly pure sample is necessary before an identification can be successful.

There are four methods that can be used for the identification of siderophores: nuclear magnetic resonance (NMR), mass spectrometry (MS), chromatography (HPLC), and thin layer chromatography (TLC). The easiest of these methods is TLC using observed and published RF values. However, RF values depend on the plates used and the technique of the investigator, and so TLC is not an accurate method. The only way it can be successful is by having, in the laboratory, samples of a number of siderophores for direct comparison.

Reverse phase HPLC is of course much more precise and also more accurate, but it still depends on the instrumentation used and definitely on the column used for the separation. The great advantage of the technique is that it is reproducible, and for a particular experimental set up the retention times can be used for a reliable identification (Konetschny-Rapp et al., 1988). Also here, in practice, it requires authentic samples of known siderophores, to be used in comparison, to make this a useful technique. If these are available, the method is accurate, rapid, and reliable for identification.

NMR is the most complete method of identifying known siderophores. It was already explained that only the ligands or the Al(III) or Ga(III) chelates of the siderophores can be used for NMR. There are known methods to deferriate Fe(III) chelates of siderophores (Jalal et al., 1985), and there is also a one-step method to produce either the Al(III) or Ga(III) chelates directly from the ferrisiderophore complex (Emery, 1986). The initial work by Llinás on ferrichromes and later the research on asperchromes and all the different coprogens has allowed rather complete assignments of all ^{1}H and ^{13}C resonances in the known hydroxamate siderophores (Jalal and van der Helm, 1991). These assignments have been tabulated, and especially the ^{13}C assignments are useful in the unique identification of a known siderophore. The method can of course also be used for the structure determination of a new siderophore, if the new compound has a similarity to compounds already known. The compound needs to be pure for the method to be useful. The advantage of the method is that observed resonances are not dependent on the instrument used. A possible disadvantage is that large quantities (mg) are required.

Purity is also a requirement for MS. The most common technique is to use FAB MS, which yields molecular weight peaks of MH^+ and MNa^+

separated, therefore, by 22 molecular weight units. The M.W. of a siderophore can be in itself a method to identify a siderophore. Chemical isomers, like asperchrome D_1, D_2, and D_3 will, of course, give the same M.W. It is also possible to do high accuracy FAB MS, which yields accurate masses of compounds (accuracy about 1 in 10^6 units). Such an accurate molecular weight allows the analysis of the number of atoms of each type in the compound. This analysis can often be more accurate than the classical chemical analysis. Also FAB MS can be used as one method to contribute to the structure determination of a new siderophore. The method is not dependent on the instrument, but the MS needs to be equipped with FAB or a similar device to vaporize the molecule.

It is possible that there are a large number of hydroxamate siderophores in nature that are not yet known. Similarly, there are so-called new siderophores reported in the literature, which on further investigation will prove to be known siderophores.

G. Geometric and Optical Isomers

Uptake of ferric chelates by microorganisms shows stereospecific preferences (Winkelmann and Braun, 1981). It may even be that the uptake is totally stereospecific, but this is both difficult to prove nor once proven would it necessarily be true for all uptake mechanisms. Still there is ample evidence that the stereochemistry of ferric siderophores is important.

There are three causes for isomerism: (a) cis and trans geometric isomers for the iron coordination (Scheme II), which is only important if the bidentate chelate group is asymmetric and of course the hydroxamate group is asymmetric; (b) there are two causes for optical isomerism: i. the configuration for the ligand, i.e., if the ligand is made from L or D amino acids; ii. the configuration around the metal ion, which can be Λ or Δ as indicated in Scheme X.

It is instructive to realize that the two causes for optical isomerism can be directly related to the optical isomerism that exists for a dipeptide made

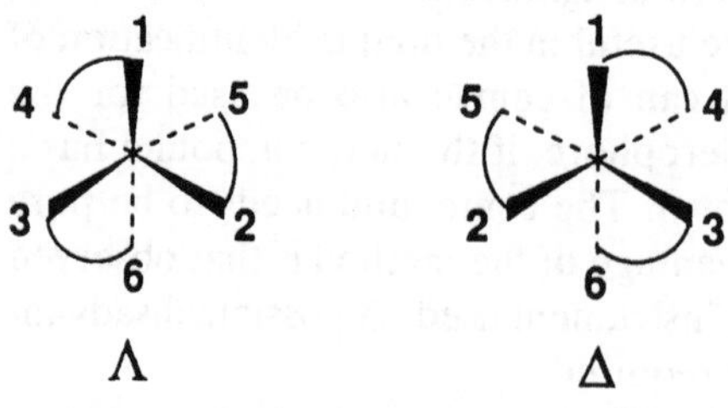

Scheme X

from two different amino acids. In case of a peptide with two residues. I and II, we have (a) I(L) II(L); (b) I(L) II(D); (c) I(D) II(L); and (d) I(D) II(D), in which (a) and (d), and (b) and (c), in pairs are enantiomers, mirror images of each other. Enantiomers have the same chemical properties in a nonchiral environment but will show opposite optical behavior (CD spectroscopy). Enantiomers are also energetically identical or in thermodynamic terms have identical standard free energy of formation. On the other hand, (a) and (b) are diastereomers that are not each other's mirror images and have different chemical properties, i.e., different melting points and different energies. However, if one of the residues (e.g., I) is glycine, which is not optically active, then (a) and (c) and (b) and (d) in pairs are identical, while (a) and (b) (and of course (c) and (d)) are enantiomers that cannot be separated by normal chemical methods and have the same standard free energy of formation.

Turning now to siderophore complexes and associating I with the optical activity of the ligand and II with that of the metal ion (Λ or Δ), one can recognize two situations.

(i) If the ligand is not optically active, one will find an equal amount of Λ and Δ chelate being made, because the standard free energy of formation is the same for both. Examples of this situation are found in the ferrioxamine family of siderophores. The two compounds are optically not identical but are enantiomers, and the chelate is a racemic mixture (50/50) of the two by thermodynamic necessity. Because they are not optically identical, an uptake mechanism may show a preference for either the Λ or the Δ isomer (enantiomer).

(ii) If the ligand (siderophore) is chiral, one may assume that only one isomer is biosynthesized, e.g., L. In this case we therefore expect the possibility of two complexes, LΛ and LΔ but not necessarily in equal amounts, because they have different standard free energies of formation. One may even expect only one of the two to be produced (as for instance in the case of the ferrichromes, where exclusively LΛ is observed), because of a large thermodynamic advantage. If one synthesizes the siderophore from D amino acids, as was done for ferrichrome (Winkelmann and Braun, 1981), one should expect in this case only the DΔ isomer (enantiomer of LΛ) to be observed. This was shown to be the case. However, uptake experiments in a fungal culture showed exclusive uptake of LΛ over DΔ ferrichrome. This may of course be due to the Λ stereochemistry of the iron coordination, as is normally assumed, but in principle it may equally well be due to the recognition of the chirality of the ligand by the uptake mechanism.

Enantiomers can, in principle, be separated on column chromatography with chiral support. The difficulty is that even though the formation constants of the Fe(III) complexes are very large, they still have large kinetic

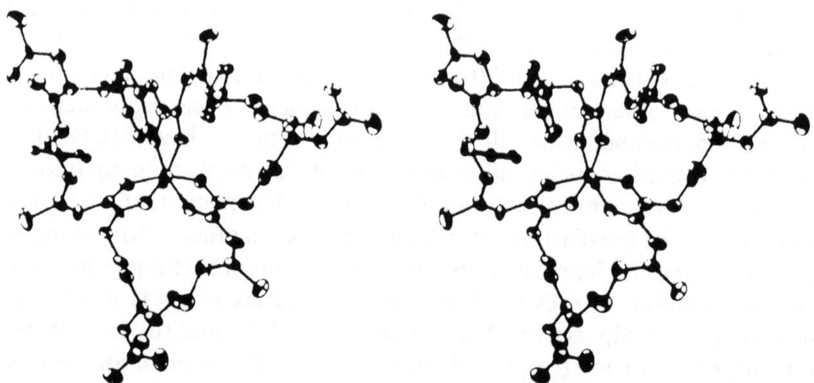

Figure 4 A stereoview of the structure of ferric *N,N′,N″*-triacetylfusarinine C. The structure is a benzene solvate. (From Hossain et al., 1980, with permission.)

lability, and one enantiomer, once isolated, will quickly racemize. Raymond and coworkers took the approach to replace the Fe(III) ion by metal ions that use CFSE for stabilization and therefore are kinetically stable (Leong and Raymond, 1974). The isolation, therefore, is in principle possible. These experiments were partially successful, and more modern techniques now available could improve the results.

We can now include geometric isomers. One should remember that the hydroxamate ligand is asymmetric (Scheme I). For a tris complex with a bidentate ligand, one has one cis and one trans complex. When they are prepared, a racemic *Δ-trans* and *Λ-trans* will be formed and a racemic mixture of *Δ-cis* and *Λ-cis* (Abu-Dari et al., 1979). In other words, there are four diastereomers, but pairs of these four are enantiomeric.

If one considers a tris complex of an optically active ligand, one can have eight diastereomers. However, if one starts with only one of the optically active ligands, this reduces to four: L-*Δ-trans,* L-*Λ-trans,* L-*Δ-cis,* and L-*Λ-cis,* none of which is enantiomeric to another. The partial separation was accomplished for the Cr(III) tris complexes of *N*-methyl-L-menthoxyacethydroxamate, and these were characterized (Leong and Raymond, 1974). The same stereoisomerism can be expected for hexadentate ligands that are cyclically symmetric, as for instance for ferric tri-*N*-acetyl fusarinine. One expects, for this compound, four stereoisomers if the ligand is built from L-ornithines, and the four can be identified as L-*Λ*-cis, L-*Δ-cis,* L-*Λ*-trans, and L-*Δ-trans.* The diastereomer observed for FeTFAC, in the crystal structure, is L-*Λ-cis* (Hossain et al., 1980).

If the hexadentate is linear, as in coprogen, or the ornithines are ar-

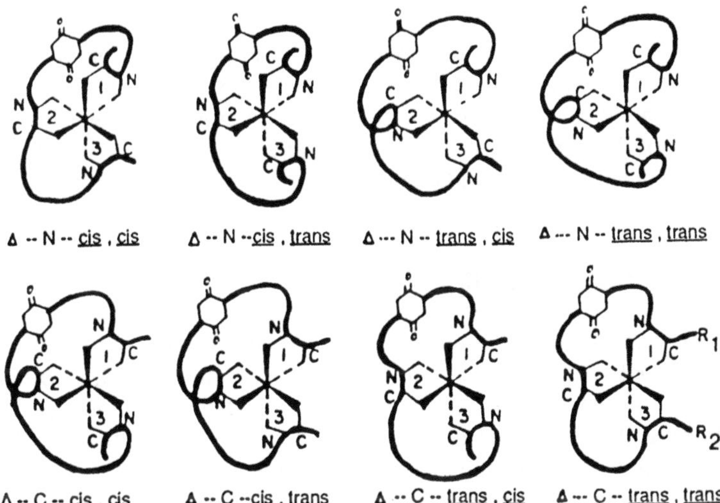

Figure 5 Eight diastereomeric isomers of coprogen with Δ configuration. There is an additional set of eight Λ diastereomers. See text for nomenclature. (From Hossain et al., 1987, with permission.)

ranged linearly, as in ferrichrome, the hydroxamate groups are no longer equivalent, and there will be $2^3=8$ *cis* and *trans* iosmers. The nomenclature should indicate the orientation of each hydroxamate group separately.[1] Each can have Λ and Δ metal environment, and when the siderophore is made from L-residues as they are in coprogen and ferrichrome there are 16 diastereomers, none of which is enantiomeric to another. Eight of the 16 are shown in Fig. 5 (for Fe coprogen); the other eight have thus the Λ configuration (Hossain et al., 1987).

It is remarkable that in all structures of members in the ferrichrome family only one of these 16 diastereomers is observed, and also in solution only one is observed. The one observed is properly identified as *L-Λ-N-cis-cis;* mostly this is shortened to *Λ-cis.* In the crystal structure of ferric neocoprogen I, also, only one diastereomer is observed, L-*Δ*-C-*trans-trans*

[1](i) Viewed down the C_3 axis, the chelate rings, 1, 2, and 3 are arranged clockwise for L isomers and counterclockwise for D isomers. (ii) If ring 1 has the carbon atom of the hydroxamate group below the nitrogen, it is denoted C. If the reverse is true it is called N. (iii) For rings 2 and 3, each is called cis or trans, depending upon whether it has the same or opposite relative orientation with respect to the C_3 axis as does ring 1. (iv) The ring nearest to the free amino terminus (in ferrioxamine B) is designated as ring 1. Because in the coprogens there is no unique N-terminus (as in ferrioxamine B or D_1), the chelate rings are designated so that the diketopiperazine ring is placed between rings 1 and 2.

(Hossain et al., 1987). However, solution studies indicate both trans and cis isomers to be in equilibrium (Wong et al., 1983).

The kinetic lability of ferric siderophores makes diastereoisomerism a difficult subject for study. Also, the occurrence in solution of different diastereomers might make it very difficult to obtain crystals. However, an understanding of diastereoisomerism is important for the understanding of the stereospecificity of the uptake mechanism of ferric siderophores in fungal systems.

H. Determination of Configuration

A normal structure determination by single crystal x-ray diffraction does not distinguish between a molecule and its mirror related image, in other words, it does not distinguish between enantiomers. There is, however, an additional physical phenomenon, anomalous x-ray dispersion, that can be applied to distinguish the structure of one enantiomer from another. It is an experimental technique used originally by Bijvoet to determine the absolute (real) configuration of (+)-tartaric (which happened to be identical to assumed configuration based on the Fischer convention).

Therefore, if it is known that the siderophore is built from L-ornithines, a normal x-ray structure determination would choose the enantiomer structure that shows the L-configuration for L-ornithine and thus, by implication, obtain the absolute configuration of the metal environment: Λ or Δ.

This can then be checked by doing a Bijvoet experiment, which necessarily would show once again, but now objectively, the L-configuration of the ornithine and the absolute configuration of the metal environment. Actually this Bijvoet experiment was done for most of the x-ray structures of ferric siderophores and has thus allowed a double check on the configuration determination of the particular diastereomer occurring in the crystal. In all studies, the particular interest is the cis and trans arrangement of the hydroxamate groups and in the Λ or Δ determination of the chelate rings around the Fe(III) ion.

The absorption spectrum of Fe(III) hydroxamates shows a charge-transfer band at around 450 nm that is quite well isolated from other absorption bands of the compounds, which occur at lower wavelengths. The absorbtion is not very strong, with a molar absorbtivity of abour 3×10^3 for a hexadentate ligand. This absorption gives the distinctive red-brown color to ferric hydroxamates. It does not occur for Ga(III) or Al(III) chelates of siderophores, which therefore are colorless. This absorption band can be used to distinguish tris bidentate complexes from hexadentate chelates, because upon acidification of the solution the former complexes show a shift of the absorption band to a higher wavelength (490 nm), giving the solution a purplish color, while the hexadentate complexes are not affected and therefore do not change their red-brown appearance in solution.

This same absorption band, because it is caused solely by the Fe(III) environment, can be used to explore the identity of an optical isomer for the Fe(III) environment and to establish it as either Λ or Δ. This can be done with CD spectroscopy, which in principle explores the chirality of the molecular system, causing the absorption observed in an optical spectrum, in this case the metal environment. The original experiment was done with the Cr(III) tris chelates of *N*-methyl-L-menthoxyacethydroxamate (Leong and Raymond, 1974). The four possible diastereomers were separated into *Λ-cis, Δ-cis,* and a mixture (*Λ,Δ-trans*). The Cr-complex gave an absorption band at about 610 nm centered on Cr(III). The Λ-optical isomer showed a CD spectrum with a negative band at low wavelength and a positive band at higher wavelength (low (−) high (+), while the Δ optical isomer showed just the opposite spectrum. This low (−) high (+) CD spectrum as an identification for the Λ-optical isomer was then applied to solution spectra of ferric siderophores, absorbing at 450 nm, to identify their Λ or Δ character.

A more direct method was used employing the crystals and crystal structures of ferrichrome (van der Helm et al., 1980), ferric neocoprogen I (Hossain et al., 1987), and ferric tri-*N*-acetyl fusarinine (Hossain et al., 1980). Powdered crystals of the compound were dispersed in a KBr pellet, and the CD spectrum was taken on this pellet. Because the objective absolute configuration was known from the crystal structure determination (and Bijvoet experiment), this experiment directly related the observed absolute configuration with the CD spectrum. The earlier solution results for the Cr(III) complexes were found to be correct, that is, low (−) high (+) shows the configuration around the Fe atom to be Λ, while a low (+) high (−) set of bands is characteristic for a Δ Fe environment. It is believed that these are the only experiments in the literature where solid state CD spectra have been employed for a useful purpose. The solid spectra firmly and objectively established the identification of Λ and Δ optical isomers from the CD spectrum.

CD spectra can, therefore, be used in solution to identify the optical isomer in solution. This poses a possible problem in that an equilibrium between diastereomers can occur in solution. If one predominates, the CD spectrum will indicate that particular diastereomer, but the band strengths ($\Delta\epsilon$) will be decreased. A very clear indication of an equilibrium of diastereomers was found in the case of ferric tri-*N*-acetyl fusarinine (Hossain et al., 1980). The x-ray structure determination showed it to be Λ, and so did the solid spectrum of the powdered crystals. However, in solution the CD spectrum showed the predominant form to be Δ. In other words, in solution an equilibrium between Λ and Δ exists in which the latter predominates, while the crystallization procedure selectively crystallizes the Λ isomer. It is not unexpected that an equilibrium between diastereomers can exist for this siderophore, because a large number of atoms (10) are involved connecting adjacent hydroxamate groups, giving the compound the possibility to form different diastereomers with similar formation energies. It also points out the possible problem with x-ray diffraction structure determinations. The structure and particular diastereomer that is determined in the solid state may not be the only one or the predominant one occurring in solution.

CD specctra in solid and solution, x-ray structure determination, and separation of kinetically stable metal chelates of hydroxamates are three methods of exploring experimentally the diastereomers of siderophores. It seems possible that also NMR could be used, but this has not been done so far.

I. Modeling

Another method that should be added to the study of diastereoisomerism of siderophore complexes is modeling. It is very informative but also very dangerous, because such studies need experiments to check the results, and such experiments are quite often difficult to design.

An earlier modeling experiment on ferrichrome cannot be considered successful (Sheridan et al., 1983). It showed, for instance, a hydrogen bond for the β(I) bend that is not observed in the crystal structure of any member in the ferrichrome family.

However, more has become known about force constants for bonds as well as bond and conformational angles. Modeling the iron environment properly is still a problem (see below), and the most vexing problem is the need to investigate all possible conformations, which becomes a very large enterprise for larger molecules.

A more sophisticated calculation was made for ferric neocoprogen I, for which 16 diastereomers are possible (Raymond et al., 1984; Hossain et al., 1987). This seemed to be a good example in being less restricted than ferrichrome but not as free to assume different diastereomers as FeTFAC. All 16 diastereomers were built, six of which required unreasonable bond angles and were not further considered. An idealized ferric hydroxamate coordination geometry derived from crystal structures was used and partially restrained. Molecular dynamics was applied on each of the 10 diastereomers. It is believed that such a simulation hovers over the minimum energy conformation of the molecule. For each diastereomer, the M.D. simulation was then annealed, i.e., energy minimized. The result of these calculations is that the diastereomer Δ-C-*trans-trans* is calculated to have the lowest energy (Fidelis, 1990). This is the diastereomer observed in the crystal structure. The result may of course be fortuitous. It should also be noted that the conformation of the ligand is somewhat different from the two molecules observed in the crystal structure. The next lowest energy diastereomer is calculated to be Δ-*N-cis-cis* (Fig. 5), and the next two are the two Λ-diastereomers having these geometric arrangements. This certainly is an encouraging result, but other and better calculations are required to make modeling a truly reliable method of exploring diastereomers in ferric siderophores.

III. THE FATE OF IRON AFTER UPTAKE

After the iron chelate of the siderophore is assimilated by the cell of the microorganism, one would like to explore the process by which the Fe^{3+} ion is released and in which form it is stored in the cell.

The redox potential of the Fe^{3+}/Fe^{2+} couple is positive against the normal hydrogen electrode (NHE), and one can expect it to become negative in a complex with Fe^{3+} with a large formation constant, or rather when a complex is formed that is specific for Fe^{3+} and not specific for Fe^{2+}. The redox potential for the ferric coprogen complex is, for instance, −0.450 V against NHE, corresponding to −0.691 V against the commonly used standard calomel electrode. Other hexadentate hydroxamate siderophore complexes have similar values (Matzanke et al., 1989). These redox potentials are within the range of physiological reductants such as NADH and $FADH_2$. Most often, therefore, a reductive enzymatic mechanism can be anticipated to occur. It will reduce ferric siderophore to the ferrous complex, after which the Fe^{2+} is released. Such reductive enzymatic mechanisms have indeed been found, but so far none has been determined to be specific. The Fe^{2+} will be released by the siderophore ligand after reduction because it has little affinity for Fe^{2+}. In one case, however, an esterase enzyme was found to be operative (Adjimani and Emery, 1988) This is in the case of ferric tri-*N*-acetyl fusarinine. The resulting bidentate *cis*-fusarinine, after hydrolysis, would more easily release Fe^{3+}, which can then be reduced.

Free Fe^{3+} or Fe^{2+} is definitely poisonous to the cell due to the production of free oxygen radicals. A storage mechanism for the iron ions is therefore required. The only feasible way to study this is by Mössbauer spectroscopy (Matzanke et al., 1987, 1989). Although the amount of iron within a cell limits the sensitivity of the method, it is possible to differentiate the spectrum of ferricrocin and ferric coprogen. Also ferritin and Fe(II) complexes can be recognized. It is possible to do the experiments *in vivo* and to follow the metabolization of the compounds. Both uptake and metabolism in *Neurospora crassa* and *Rhodotorule minuta* has been studied, and much more and important information can be expected in the future from using this spectroscopic technique.

IV. THE POLYCARBOXYLATE SIDEROPHORE RHIZOFERRIN

A new fungal siderophore, named rhizoferrin (Scheme XI) has recently been isolated from low-iron culture filtrates of *Rhizopus* strains and other members of the Mucorales (Zygomycetes) (Drechsel et al., 1991, Thieken and Winkelmann, 1992). Structure elucidation of this compound by gas chromatography/mass spectroscopy, electrospray mass specroscopy, and NMR techniques revealed that two critic acid residues are linked to diaminobutane (putrescine), resulting in N^1,N^4-bis(1-oxo-3-hydroxy-3,4-dicarboxybutyl)-diaminobutane as the structure of rhizoferrin. A stereochemical analysis of rhizoferrin by circular dichroism spectra revealed that the two quarternary carbon atoms of the citric acid residues possess idential R-configurations and that the ferric rhizoferrin complex adopts Λ-configuration about the iron center (Drechsel et al., 1992). In acid solution, rhizoferrin undergoes dehydration and cyclization to imido-rhizoferrin and bisimido-rhizoferrin. Rhizoferrin and its dehydration products can be separated and characterized using HPLC and capillary electrophoresis. Based on the absorption maximum of the metal/ligand charge transfer band at 335 nm, a molar extinction coefficient of 2300 M^{-1} cm^{-1} was determined (Drechsel et al., 1992). A bioassay for the detection of rhizoferrin in fungal culture filtrates has been developed by using a hydroxamate-transport defective strain of *Morganella morganii* (SBK3) (Thieken and Winkelmann, 1993). Strains of the *Proteus-Providencia-Morganella* group have been shown to recognize rhizoferrin and other carboxylate type siderophores (Drechsel et al., 1993).

V. CONDITIONS OF SIDEROPHORE PRODUCTION

Siderophores are defined by their ferric specific chelating property and their production and utilization by microorganisms. However, the definition of siderophores also includes that their biosynthesis is regulated by internal iron-sensing mechanisms (see Chap. 4 by Mei and Leong). Thus the production of siderophores in fungi is induced only under low-iron conditions or iron stress. In iron-sufficient media containing >1 mM Fe(III), the biosyn-

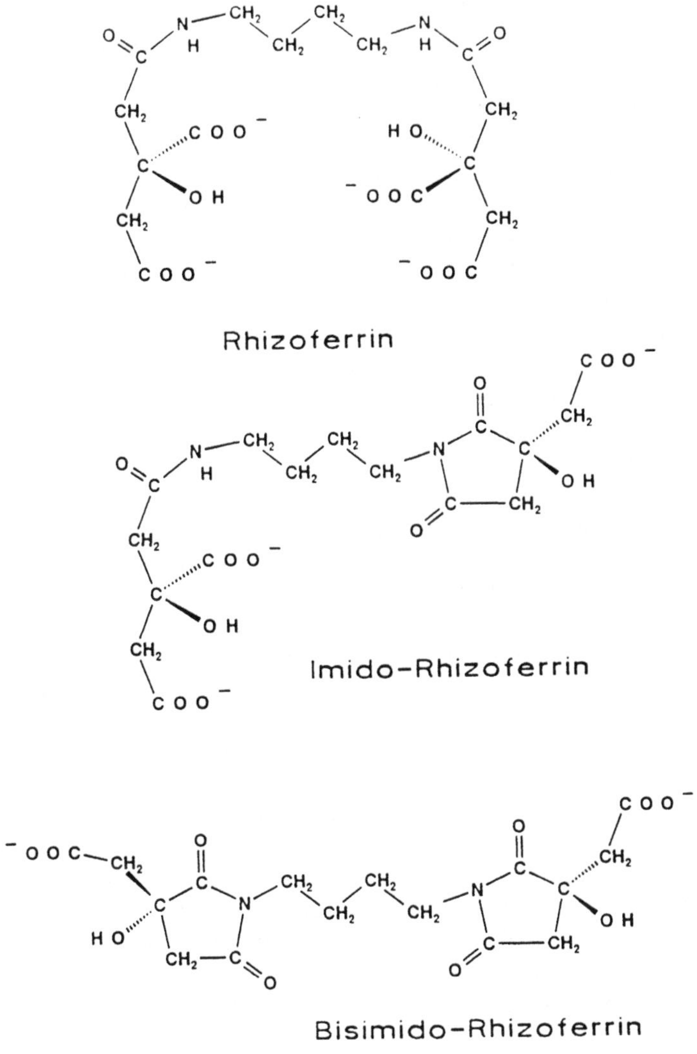

Rhizoferrin

Imido-Rhizoferrin

Bisimido-Rhizoferrin

Scheme XI

thesis of siderophores is repressed or greatly reduced; therefore overproduction of siderophores can only be detected when fungi are grown in low-iron media. Most chemically defined or complex fungal growth media, however, contain sufficient iron (>1 mM) so that iron deficiency has to be controlled by careful selection of low-iron chemicals or by special methods of deferration. The use of the so-called low-iron media (LIM) for *Saccharomyces cerevisiae* has been recently described (Eide and Guarante, 1991). The following methods are currently used to prepare low-iron media:

(a) Extraction of media with 8-hydroxyquinoline in chloroform (or dichloromethane). However, extraction of chemically defined media with 8-hydroxyquinoline seems to be appropriate only for small volumes and requires subsequent removal of the organic solvents.

(b) The most common procedure is to use Chelex-100 columns to deferrate salt media and glucose (separately). This is the method of choice for larger volumes of media.

(c) The addition of synthetic chelators, e.g., ethylenediamine di-(*o*-hydroxyphenyl acetic acid (abbreviated EDDHA or EDDA) is now often used to reduce the amount of bioavailable iron in bioassays but also in liquid media (Minnick et al., 1991; Jacobson and Vartivarian, 1993). When utilizable siderophores are excreted or added, interligand iron exchange occurs so that iron bound by EDDA can then be mobilized. Thus siderophore-dependent growth is the result of the reversal of growth inhibition by nonutilizable synthetic iron chelators.

(d) Rapid growth of fungi also leads to a continuous decrease of iron in the medium. Thus growing fungi gradually deplete their growth medium of iron by excreting siderophores with subsequent uptake of iron via siderophore transport systems. This in turn may increase siderophore biosynthesis. Various fungi are also able to acidify the medium, which ensures a better solubility of ferric ions and enables iron uptake via organic acids (Winkelmann, 1979a; Winkelmann, 1992). The question whether or not iron depletion is strong enough to initiate siderophore biosynthesis depends on the fungal system used and on the cultivation medium.

(e) A further method to achieve a kind of iron-deficinecy is the use of ions that interfere with the uptake or metabolism of iron. Thus the presence of large amounts of Al(III), Cr(III), or Co(III) in the incubation medium may induce iron-deficiency symptoms in fungi (Sivarama-Sastry et al., 1962). Laboratory conditions often do not reflect the natural ecological conditions. Thus several grams of siderophores per liter can be produced in shake cultures, whereas in a natural environment concentrations of only several micrograms per kilogram soil have been reported (Powell et al., 1980; Bossier et al., 1988). It may rather be assumed that the production of siderophores in ecological microenvironments is finely tuned by continuous repression and derepression of siderophore biosynthesis to avoid wasting of synthetic energy. Another aspect of siderophores, however, seems to be of even greater importance. Those organisms that are equipped with an efficient siderophore-mediated iron acquisition system are preferably able to inhabit certain low-iron ecological niches. This is of importance in alkaline environments or in various host tissues of plants and animals. Iron binding proteins like transferrin and lactoferrin can reduce the available iron so effectively that even siderophores may fail to extract iron. Recently six different dermatophytic fungi have been shown to produce ferrichrome type siderophores under iron-deficient conditions (Mor et al., 1992), which underlines the importance of siderophores in pathogenic fungi. Despite the presence of siderophores and their cognate transport systems, the dermatophytic fungi seem to be unable to invade deeper tissues because of the inhibitory action of the transferrins. Irrespective of the presence of siderophores, pathogenic and nonpathogenic fungi are both able to produce siderophores, which makes a simple distinction based on iron-acquisition systems between these groups impossible. Moreover, certain highly pathogenic *Candida* and *Cryptococcus* species seem to be unable to synthesize any hydroxamate type siderophores. This indicates that iron must be scavenged by alternative mechanisms and that additional virulence factors are involved. The mode of iron utilization in these fungi is still under investigation, and our knowledge on how and to what extent iron is metabolized is still limited. The general requirement of iron in all fungal pathogens, however, is evident, and blockage of iron transport or its further metabolism seems to be an effective method to inhibit growth in host tissues.

VI. MECHANISMS OF SIDEROPHORE-MEDIATED IRON TRANSPORT IN FUNGI

The crucial event leading to the rapid entrance of iron into the fungal cells is the interaction of siderophores with the membrane located siderophore transport system(s). Although the putative proteins of fungal siderophore iron transport systems have not been identified so far, there is considerable kinetic evidence that special siderophore transport system(s) are involved. Kinetic measurements are commonly performed in a time-dependent and concentration-dependent manner. Time-dependent measurements are required to determine the actual transport rates, while concentration-dependent measurements are designed to determine whether or not saturation of a transport system or diffusion is involved. The transport system may be a carrier protein or a channel. If concentration-dependent uptake yields a straight line, then simple diffusion is the underlying principle of siderophore iron transport. In this case laws of diffusion predict that the intracellular concentration is always proportional to the supplied outer concentration. Diffusion lines are seldom observed in siderophore iron transport but seem to occur in the presence of inhibitors (NaN_3, KCN) or when synthetic and enantiomeric analogs are used (Müller et al., 1985; Winkelmann and Braun, 1981). Most natural siderophores, however, show saturation kinetics enabling the determination of K_M and V_{max} values. However, there is only a formal analogy between enzymatic and transport kinetic data, as transport is often a multistep reaction involving a variety of interacting membrane proteins. Irrespective of the actual process of molecular interaction of siderophores with membrane proteins, the determined K_M and V_{max} values are important characteristics for the efficiency of siderophore transport systems. As an alternative, the designation K_T has been introduced for half-maximal saturation of transport systems, enabling a distinction from enzymatically determined K_M values. However, as long as the experimental procedure is in principle identical there is no need to change the commonly used terms. Greater problems arise when the saturation kinetics deviate considerably from the theoretically derived hyperbolic curves. In this case a mixture of diffusion and saturation has to be taken into consideration, which may be distinguished by subtraction procedures. Iron uptake via siderophores can be measured by isotopic labeling of the iron atom (^{55}Fe or ^{59}Fe) and/or of the ligand (^{3}H or ^{14}C). A detailed protocol of the currently used methods for transport measurement in fungal siderophore research has been described recently (Winkelmann, 1993). After uptake of the labeled siderophores, the cells are filtered off through membrane filters and washed; then the radioactivity is counted in a liquid scintillation counter. Precautions can be taken to avoid adsorption of labeled siderophores to the filter and also to the surface of the cells by washing with cold siderophore solutions. However, as the amount of adsorbed label remains constant for all samples in a kinetic assay, the contribution of radioactivity based on adsorbed material can be determined separately with dead cells or in the presence of inhibitors and may then subsequently be subtracted. Although only kinetic data are so far available, the observed differences during uptake of fungal siderophores suggest that the mechanisms of uptake may be different with regard to the fate of iron and ligand. Four distinct mechanisms have been described (Fig. 6):

(a) Uptake of the intact siderophore complex into the fungal cell with subsequent egress of the ligand molecule. This so-called "shuttle mechanism" is typical for all the siderophores of the ferrichrome and coprogen families (Emery, 1971; Winkelmann and Zähner, 1973).

(b) Uptake of the iron without entrance of the ligand molecule. This so-called "taxicab mechanism" has been observed with ferric rhodotorulate in *Rhodotorula* species (Carrano and Raymond, 1978). This iron transfer is

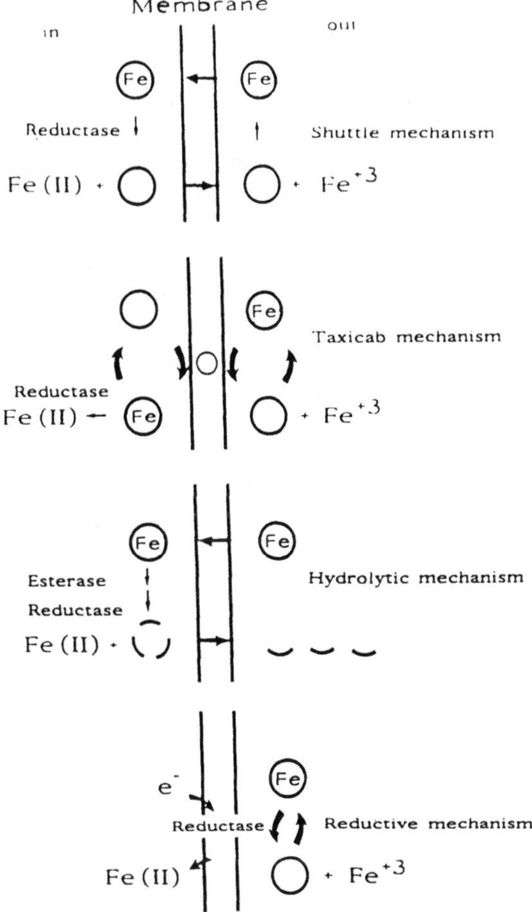

Figure 6 Mechanisms of siderophore iron uptake across the cytoplasmic membrane of fungi.

obviously not accompanied by a reduction step but rather a membrane-mediated exchange of the metal between the intra- and extracellular siderophores (Müller et al., 1985).

(c) A third mechanism has been shown to occur when the ester type ferric triacetylfusarinine C is taken up in *Mycelia sterilia* (Adjimani and Emery, 1987, 1988). After uptake of ferric triacetylfusarinine C, the ester bonds are split off by specific esterases, and the monomeric fusarinines are excreted. A simultaneous reduction of Fe(III) to Fe(II) seems to be also involved. This "degradative mechanism" is of importance for ridding of toxic metal atoms like Al(III), Ga(III), or Cr(III), which are nonreducible and therefore remain bound to the monomers and are excreted again.

(d) A further "reductive mechanism" seem to occur in certain cases where siderophores are bound to the transport systems without being internalized. Thus ferrichrome A was found to be unable to enter the cells of *Ustilago shaerogena.* EPR spectroscopic measurements later revealed that membrane reduction of Fe(III) to Fe(II) occurred and Fe(II) was subsequently taken up (Ecker et al., 1982). Thus the reductive iron uptake from stable ferric complexes implies a source of electrons at the cellular membrane and is assumed to be generally unspecific as to the chelate structure (Emery, 1987). As shown in the chapter on iron uptake in *Saccharomyces cerevisiae,* the reductive pathway is of special importance in certain nonhydroxamate-producing yeasts. Spectroscopic measurements (ESR, Mössbauer) give additional information on the oxidation state and ligand environment (see the chapter on the storage of iron). As there is now a great diversity of structures and functional specificities of siderophores in the fungal genera, no general description of the functional properties of siderophores will be presented here. We rather describe these aspects related to certain fungal species or groups that have received most attention in siderophore research for historical reasons or that have gained interest because of various chemosystematic reasons. Also, as long as the different fungal receptors, binding sites, or transport systems are not characterized in molecular terms, the uptake of siderophores will be dealt with in close connection with the corresponding fungi.

VII. IRON TRANSPORT IN SELECTED FUNGI

A. *Ustilago*

Emery (1971b) was the first to show that ferrichrome functions as an iron transport molecule in the smut fungus *Ustilago sphaerogena.* Transport of iron via ferrichrome in this fungus was shown to be an active transport process, which is inhibited by respiratory poisons like azide, cyanide, anaerobiosis (nitrogen) and to a lesser extent by uncouplers like 2,4-dinitrophenol. Using ^{14}C-labeled desferriferrichrome it was also shown that after intracellular removal of iron from ferrichrome the ligand was again excreted for another round of iron transport. Thus ferrichrome behaves as an iron shuttle, acting as a specific membrane iron carrier. Later the name "siderophore" was coined to designate a family of low molecular mass, virtually ferric specific ligands elaborated by microorganisms to combat iron deficiency (Lankford, 1973).

Surprisingly, concomitantly produced ferrichrome A was not a good iron transporting agent in *Ustilago,* although the structure was similar to ferrichrome (Emery, 1971). The mode of iron uptake mediated by ferrichrome A was shown to depend on a membrane-dependent reductive removal of iron with subsequent internalization of iron (Ecker et al., 1982; Ecker and Emery, 1983). Thus while ferrichrome was transported across the fungal membrane as an entity, ferrichrome A slowly delivered iron to membrane located acceptors. The functional differences in transport between ferrichrome and ferrichrome A were shown to reside in the iron-surrounding *N*-acyl moieties that consisted of bulky and charged methylglutaconic acid residues in the case of ferrichrome A, whereas ferrichrome contained short acetyl residues.

Several other phytopathogenic *Ustilago* species, like *U. hordei, U. nigra,* and *U. maydis,* have recently become of interest for genetic manipulation, which is referred to in the chapter dealing with genetic aspects of iron transport. The pattern of siderophore production by *U. maydis,* the cause of corn smut disease, has been found to be very similar if not identical with that of *U. sphaerogena* (Budde and Leong, 1989). In a taxonomic survey of the *Ustilaginaceae,* 7 genera (12 species) and some related genera, such as *Graphiola, Protomyces,* and *Tilletiaria,* it was shown that the parasitic members of the Ustilaginales produce ferrichrome A and/or ferrichrome, while the saprophytic members are characterized by the formation of rhodotorulic acid. Since the parasitic *Ustilago violacea* produced rhodotorulic acid and also because of other characteristics, this species was transferred to a separate genus, *Microbotryum* (Deml, 1985). Moreover, the smut fungi (*U. anomala, U. cordai, U. reticulosa, U. vinosa*) parasitizing dicotelydoneous host species (Polygonaceae) also did not produce ferrichromes but instead rhodotorulic acid, which suggests a natural relationship to *Microbotryum* and *Sphacelotheca* (Deml, 1985).

From a strain of *Neovosssia indica,* a smut fungus of the order Tilletiales, a novel ferrichrome type siderophore was isolated that consisted of a heptapeptide (Deml et al., 1984). While the usual ferrichrome contained a $(Gly)_3$ sequence within a hexapeptide ring system, the novel ferrichrome contained a $(Gly)_4$ sequence and was therefore named tetraglycylferrichrome. A comparison of transport between ferrichrome and tetraglycylferrichrome revealed identical transport rates, indicating that minor changes in the peptide moiety did not qualitatively alter the transport properties of ferrichrome type siderophores.

B. *Agaricus bisporus*

Another recently described siderophore producing basidiomycetous fungus is the edible common white mushroom, *Agaricus bisporus,* which has been shown to produce several structurally different siderophores (Eng-Wilmot et al., 1992). The chelates were identified as ferrichrome, ferric fusarinine

C, and des(diserylglycycl)ferrirhodin (DDF). The iron transport properties of the three compounds and some structurally similar exogenous compounds, like ferrichrome A, ferrirhodin, and triacetylfusarinine C, were studied using the ^{55}Fe-labeled chelates. As shown by the authors, the transport of these siderophores was via a high-affinity, energy-dependent process, and the transport effectiveness was in the order ferrichrome > DDF > ferric fusarinine C. The relative uptake of iron by Λ-*cis* ferrichromes was ferrichrome > ferrirhodin > ferrichrome A, and the corresponding transport activity of Δ-*cis* fusarinines was ferric fusarinine C > tris-*cis*- (and *trans*-) fusarinine iron (III) > ferric N^1-triacetylfusarinine C.

C. *Rhodotorula*

The basidiomycetous yeast *Rhodotorula pilimanae* has been shown to produce the siderophore rhodotorulic acid (RA) under iron deficient conditions (Atkin and Neilands, 1968). This chelator represents a dihydroxamate type siderophore that forms a 2:3 complex (Fe_2RA_3) with iron at neutral pH, in which the predominant configuration about the metal center is Δ (Carrano and Raymond, 1978). However, at lower pH values positively charged complexed species may prevail, such as $FeRA^{2+}$ and $Fe(RA)_2^+$, which may differ in their transport properties. Transport of iron via RA was sensitive to respiratory inhibitors, such as cyanide and azide, and also to 2,4-dinitrophenol, an uncoupler of oxidative phosphorylation, and idoacetamide, an alkylating agent of sulfhydryl groups. Therefore iron transport via RA is regarded as an active transport process requiring metabolic energy for iron accumulation. The transport model proposed by Carrano and Raymond (1978) suggested a taxi mechanism by which iron is delivered to the cell without entering of the ligand into the interior of the cell. This could be confirmed by using Cr^{+3}-substituted and ^{3}H-labeled $Fe_2(RA)_3$. Moreover, a possible redox catalysis during ligand exchange could be excluded by using Ga(III)-substituted RA complexes, which cannot be reduced by biological systems (Müller et al., 1985). As was shown by these authors, the trihydroxamate type siderophore, ferrioxamine B, was completely ineffective at delivering iron to *R. pilimanae,* confirming the view that this transport system is specific for Fe-rhodotorulate. Several synthetic analogs of rhodotorulic acid have been prepared in which the diketopiperazine ring was replaced by a simple chain of *n* methylene groups. It was found that *R. pilimanae* is able to accumulate iron using these achiral complexes, as well as from simple monohydroxamate analogs, at rates comparable to those of RA (Müller et al., 1985). Later Mössbauer spectroscopic results in *R. minuta* additionally revealed high intracellular levels of rhodotorulic acid, suggesting a membrane-mediated iron exchange mechanism between extracellular and intracellular RA (Matzanke et al., 1990). These results were also taken as a proof of a storage function of RA, as no ferritinlike iron pools could be detected in *Rhodotorula* cells. Several other related basidiomycetous fungi, such as *Leucosporidium, Rhodosporidium, Sporidiobolus,* and *Sporobolomyces* have been shown to produce RA (Atkin et al., 1970). Although no transport data are so far available, it may be assumed that ferric RA is taken up in a similar way as described in *Rhodotorula* species.

D. *Neurospora crassa*

The first studies on iron transport in Ascomycetes were performed with *Neurospora crassa* (arg-5 ota aga), an ornithine-deficient mutant (Winkelmann and Zähner, 1973). As ornithine is a necessary constitutent of all fungal siderophores, the omission of ornithine in the cultivation medium resulted in a siderophore-free culture. This made it possible to study the transport of various fungal siderophores without interference of interligand exchange. The transport data clearly showed that iron is taken up via coprogen, the principal siderophore of *N. crassa,* and also via ferricrocin, a minor siderophore; whereas ferrioxamines were ineffective. Using double-labeled $^{55}Fe/^{14}C$-coprogen it was demonstrated that the entire chelate molecule entered the cells (Winkelmann and Zähner, 1973). The presence of higher levels of ^{55}Fe and lower levels of ^{14}C in the mycelia indicated iron removal and egress of the ligand. Transport studies with this mutant also revealed that citrate can function as a siderophore (Winkelmann and Zähner, 1973; Winkelmann, 1979a). Later results showed that several other ferrichrome type siderophores were taken up, although ferrichrome A and ferrirubin were not recognized by the transport system of *N. crassa* (Winkelmann, 1974). A photoaffinity derivative of coprogen B, *p*-azidobenzylcoprogen B, has been synthesized and found to be a reversible inhibitor in the dark and an irreversible inhibitor after photolysis (Bailey et al., 1986). Although these experiments did not allow the isolation of the membrane proteins involved in coprogen transport, they showed that uptake of ^{55}Fe in the illuminated sample was always about 50% of that found for the nonilluminated controls. Moreover, the photoaffinity derivative of coprogen B inhibited both coprogen and ferrichrysin uptake, confirming earlier data obtained from kinetic studies that showed competitive inhibition of transport when both siderophore classes were added simultaneously to the cells, suggesting a common transport system (Huschka et al., 1985). A detailed analysis of the molecular recognition of siderophores in *N. crassa* revealed that the siderophore transport system recognizes structure and configuration of the iron-surrounding *N*-acyl residues of ferrichromes and coprogens in a different manner (Huschka et al., 1986). In the coprogen series, uptake decreased in the order coprogen, neocoprogen I, neocoprogen II; the reverse effect was observed in the ferrichrome series, where uptake decreased in the order ferrichrysin, asperchrome D1, asperchrome B1, and ferrirubin. Thus recognition of coprogen and ferrichromes was adversely affected by the presence of anhydromevalonic acid residues, being well accepted in the coprogen receptor and inhibiting transport in the ferrichrome receptor. Interestingly, the two structurally different siderophores, coprogen and ferrichromes, showed competition when added simultaneously to young mycelia of *N. crassa* (Huschka et al., 1985). This suggested different siderophore receptors but a common siderophore transport system. This view was confirmed by the finding of a high stereospecificity of the ferrichrome receptor, which recognizes the natural Λ-*cis* ferrichrome containing L-ornithine residues while excluding the synthetic Δ-*cis* ferrichrome containing D-ornithine residues (Winkelmann, 1979b). Attempts to isolate membrane proteins involved in fungal siderophore transport systems were unsuccessful so far. A comparison of plasma membrane proteins from *N. crassa* grown under iron-sufficient and iron-deficient conditions revealed no significant differences in SDS gel electophoretic profiles (Huschka and Winkelmann, 1989). Small differences in the content of single bands were presumably due to proteolytic degradation, but in no case could real overproduction of membrane proteins be observed, as can be seen in outer membrane profiles of gram-negative bacteria, which have been reviewed in a recent book on microbial iron chelates (Braun and Hantke, 1991). Though no overproduction of membrane proteins was seen in *N. crassa,* at least fivefold differences of transport rates were measured, suggesting that there is indeed a significant response to iron deficiency. As most fungi are grown from conidiospores, the utilization of conidial iron stores during growth of germ tubes and young mycelia seems to be apparent. Thus the possibility of constitutive expression of membrane siderophore transport systems together with cellular transport regulation systems has been proposed (Huschka and Winkelmann, 1989).

E. *Botrytis*

The grey mold *Botrytis cinerea* is a common plant pathogenic fungus that causes infections of all kinds of plant surfaces, crops, fruits, and seedlings. The genus *Botrytis* is the imperfect state (anamorph) of the perfect or sexual state (teleomorph) of Botryotinia, and the perfect state of *Botrytis cinerea* is designated *Botryotinia fuckeliana.* Isolation and identification of the siderophore of *B. cinerea* yielded ferrirhodin as the principal siderophore (Konetschny-Rapp et al., 1988), which had been isolated earlier from *Penicillium* and *Aspergillus* strains (Zähner et al., 1963). HPLC separation of the total siderophores showed some additional peaks that have not been identified. A comparison of iron transport activities of ferrirhodin and other fungal siderophores showed the following order of uptake in *B. cinerea:* ferrichrysin (100%), ferrirubin (57%), ferrirhodin (45%), coprogen (6%). Transport kinetics of ferrirhodin uptake revealed $K_M = 2.5$ mM and $V_{max} = 80$ pmol $\cdot$ min^{-1} mg^{-1}, suggesting the presence of a specific transport system for ferrichromes.

F. *Penicillium* and *Aspergillus*

Strains of *Penicillium* and *Aspergillus* are known to produce various ferrichrome type siderophores (Zähner et al., 1963; Charlang et al., 1981). Three strains of *Aspergillus, A. quadricinctus* (E. Yuill), *A. fumigatus* (Fresenius), and *A. melleus* (Yukawa), have been used for transport studies with different ferrichrome type siderophores (Wiebe and Winkelmann, 1975). All ferrichrome type siderophores showed good iron uptake, including ferrirubin, which has been shown to function as an inhibitor in *N. crassa* (Winkelmann, 1974; Huschka et al., 1985). Concentration dependent transport with ferrichromes revealed the typical saturation kinetics, which were sensitive to the respiratory inhibitor sodium azide, indicating a mediated and active transport of ferrichrome type siderophores in all *Aspergillus* strains. As with the findings for *N. crassa,* ferrioxamines were ineffective as siderophores in *Aspergillus. Aspergillus ochraceus* was found to produce a variety of other previously unknown ferrichrome type compounds with two or three dissimilar N^{δ}-acyl residues, named asperchromes (Jalal et al., 1984). The asperchromes, however, ranged only in quantities of 0.2–5%, while the major products were ferrichrysin and ferrirubin. This suggests that the asperchromes resulted from unspecificities of siderophore-biosynthesizing enzymes similar to the observed microheterogeneity of microbial peptide antibiotics (Kleinkauf and Döhren, 1990). In addition, a linear peptide siderophore was isolated from *A. ochraceus,* named des-(diserylglycyl)ferrirhodin (DDF), which consisted only of a triornithine backbone, lacking the commonly found -diserylglycyl- sequence of ferrirhodin (Jalal et al., 1985). While ferric DDF was not transported in *N. crassa* (Huschka et al., 1986), it was well transported in *Agaricus bisporus* (Eng-Wilmot et al., 1992). The latter fungus has been found to produce desferriferrichrome, DDF, and fusarinine C.

G. *Gliocladium*

The siderophores of *Gliocladium virens* have been analyzed in detail, consisting of the monohydroxamates *cis-* and *trans-*fusarinine, and the dihydroxamate dimerum acid, as well as minor quantities of desferricoprogen,

desferricoprogen B, and desferriferricrocin (Jalal et al., 1986, 1987). Transport of iron mediated by the monohydroxamates *cis*- and *trans*-fusarinine, and by the dihydroxamates dimerum acid and the exogenous rhodotorulic acid was more efficient than that of the trihydroxamate coprogen, although these hydroxamate type siderophores were closely related and composed of identical building blocks. Moreover, ferric mono- and dihydroxamates inhibited coprogen uptake, as measured with ^{14}C-coprogen. Ferricrocin, however, belonging to the ferrichrome type siderophores, behaved differently and showed good uptake activity. The respiratory inhibitors significantly inhibited Fe(III) uptake mediated by the fusarinines, dimerum acid, and the trihydroxamates (coprogen and ferricrocin). The actual mechanism of iron uptake in this fungus is still unclear as also is the question whether different transport systems are operative, i.e., a monohydroxamate transport system, a dihydroxamate transport system, and a trihydroxamate system. The differences between the transport of ferric mono-, di-, or trihydroxamates may also originate from the iron coordination geometries, e.g., Λ and Δ, or from the size, shape, and charge of the molecular complexes or their parts that interact with the receptors or transport assembly in the membrane.

H. *Fusarium*

Hydroxamates isolated from *Fusarium* species (Diekmann and Zähner, 1967; Sayer and Emery, 1968) were named fusarinines and were later found in several other genera, such as *Giberella, Nectria, Myrothecium, Trichothecium, Cylindrocarpon, Penicillium,* and *Aspergillus* (Diekmann, 1970). While the monomer was named fusarinine, the linear di- and trimer were named fusarinines A and B. The cyclic trimer was named fusarinine C or fusigen, and its triacetylated forms were named triacetylfusarinine C or triacetylfusigen. Although the fusarinines have been well characterized, their transport properties in the producing strains are not. Recent transport studies revealed that the iron transport activities of mono- and dihydroxamates in *Fusarium dimerum* were similar to those in *Gliocladium* (Jalal et al., 1987). However, while iron transport activities with fusarinines and dimerum acid were optimal, those with coprogen and ferricrocin were comparatively low.

I. *Stemphylium*

Stemphylium botyrosum is known as a producer of coprogen B, dimerum acid, and a monohydroxamate (Manulis et al., 1987). The strain *S. botyrosum* Walr. f. sp. *lycopersici* is the causal agent of leaf spot and foliage blight disease of the tomato. Iron transport studies showed that the siderophore activity of coprogen B and dimerum acid was nearly identical and was completely inhibited by 1 mM sodium azide. The exogenous dihydroxamate rhodotorulic acid was also taken up, and its transport activity was identical with dimerum acid, indicating similarities with *Gliocladium* and *Fusarium*. Moreover, coprogen was taken up in high amounts as an intact chelate molecule, as demonstrated by tritiated coprogen B (Manulis et al., 1987).

J. *Curvularia, Epicoccum, Alternaria*

Although the occurrence of coprogen and its transport activity is well known from *N. crassa* (Huschka and Winkelmann, 1987), strains of *Epicoccum* and *Curvularia* have been shown to produce additional coprogen type siderophores named triornicins and neocoprogens (Frederick et al., 1981, 1982; Hossain et al., 1987). Structure-activity relationships of these coprogen type compounds, however, have only been investigated in *N. crassa,* where iron uptake rate is in the order coprogen > neocoprogen I > neocoprogen II, indicating that the gradual replacement of *trans*-anhydromevalonoyl groups by acetyl groups reduced the uptake activity. Further derivatives of coprogen, such as *N*-dimethylcoprogens (Jalal et al., 1988) and hydroxycoprogens (Jalal et al., 1989) have been reported from *Alternaria longipes,* although transport data are still lacking.

K. *Trichoderma*

Strains of *Trichoderma* have been shown to produce coprogen, coprogen B, and ferricrocin as the major trihydroxamate siderophores (Anke et al., 1991). Two strains, *T. longibrachiatum* and *T. pseudokoningii,* additionally produce siderophores of the fusigen type. Moreover, a novel intracellular coprogen derivative was found that contained a palmitoyl instead of an acetyl group (Anke et al., 1991). Strains of *T. harzianum* and *T. hamatum* are well known as biocontrol agents used to control damping off of bean, tomato, and eggplant seedlings or black-root rot of strawberries (Chet, 1987). The role of siderophores in biocontrol are still unresolved, as the antagonistic properties of *Trichoderma* strains are not correlated to the type and the amount of siderophores (Anke et al., 1991). However, as *Pythium* is a non-hydroxamate producer (Thieken and Winkelmann, 1993), and as *Trichoderma* species are hydroxamate producers, the excretion of hydroxamate siderophores possessing high stability constants may well be a factor of ecological predominance.

L. *Mycelia sterilia*

While the cyclic triester of fusarinine, named fusarinine C or fusigen, has been shown to be easily hydrolyzed, the triacetylated form *N,N′,N″*-triacetylfusarinine C (TAFC) was more resistant to ester hydrolysis (Moore and Emery, 1976). TAFC has been isolated from strains of *Aspergillus* and *Penicillium* (Diekmann and Krezdorn, 1975; Anke, 1977; and Huschka and Winkelmann, 1987) and *Mycelia sterilia* (Adjimani and Emery, 1987). TAFC has been shown to be the predominant siderophore and a very efficient iron carrier in *Mycelia sterilia* EP-76 (Adjimani and Emery, 1988). Transport studies revealed that the iron transport system of this organism recognizes only the Λ coordination isomer as determined by the kinetically stable Cr(III)-triacetylfusarine C, even though the Δ configuration predominates in solution. It was therefore suggested that slow metal center isomerization is involved. After entrance of TAFC into the cells, iron is released by an esterolytic mechanism resembling that of enterobactin in enterobacteria. Although many other exogenous siderophores were able to donate iron to cells of *M. sterilia,* these siderophores seem to deliver iron to the cells by an indirect mechanism involving iron exchange into triacetylfusarinine C. Even for the stronger chelate, ferrichrome, almost quantitative exchange was observed *in vitro* after 12 h. However, although the *in vitro* rate of exchange of iron from ferrichrome into triacetylfusarinine C was significant, ferrichrome did not engage in an exchange mechanism *in vivo;* rather it entered the mycelia as the intact chelate, as shown by ^{14}C-labeling, suggesting the existence of a separate ferrichrome-specific transport system. Thus *M. sterilia* possesses at least three distinctive pathways of iron uptake: (i) a true siderophore system utilizing the ferric TAFC complex and involving cellular hydrolysis of the chelate for iron release; (ii) an exchange mechanism of iron from other complexes into TAFC, followed by uptake of ferric TAFC; and (iii) a third transport system for ferrichromes. Another interesting feature was observed with TAFC-mediated metal transport in *M. sterilia.* After uptake of ^{51}Cr(III)-labeled TAFC, the chromium was excreted again, complexed to the monomer, which provides the first evidence for metal excretion. According to the view of the authors, such a mechanism might be valuable in ridding the cells of metals such as aluminum and chromium, whose siderophore complexes are structurally similar to those of iron but which have no metabolic role and might even be toxic to the cells if accumulated.

M. *Microsporum* and *Trichophyton*

The availability of iron is a critical factor in pathogenicity of microorganisms invading living hosts. Dermatophytes are physiologically adapted for growth on keratin, and their infections are usually limited to the epidermis, causing superficial infections in humans and animals. The inability to invade deeper tissues has been attributed to the inhibitory effect of serum, which due to its transferrin content functions as a growth inhibitor of dermatophytes. *Microsporum gypseum* has been shown earlier to produce ferricrocin (Bentley et al., 1986). A more recent systematic study on hydroxamate siderophores in different strains of *Microsporum* and *Trichophyton* revealed that *M. gypseum, M. canis, M. audouini,* and *T. rubrum* produce ferricrocin and ferrichrome C, whereas *T. mentagrophytes* and *T. tonsurans* produce only ferrichrome (Mor et al., 1992). Thus the dermatophytic fungi resemble other imperfect fungi, such as *Aspergillus, Penicillium,* and *Neurospora,* in their siderophore production.

N. Zygomycetes

Until recently no siderophores could be isolated from fungi of the Zygomycetes group. However, using ion exchange columns, a critic acid containing compound was isolated from the fungus *Rhizopus microsporus* var. *rhizopodiformis,* which is known as an agent of Mucormycosis (Drechsel et al., 1991). The production of rhizoferrin has been observed in a variety of families (Mucoraceae, Thamnidiaceae, Choanephoraceae, Mortierellaceae, Basidiobolaceae) of the two orders Mucorales and Entomophthorales (Thieken and Winkelmann, 1992), confirming the view that rhizoferrin is in fact the principal siderophore of the class of Zygomycetes. From a taxonomical point of view only the higher taxa of fungi including the Basidio-, Asco, and Zygomycetes have been studied for siderophore production (Table 1). Thus within the lower fungi of the Mastigomycotina and Myxomycetes, no siderophores have been isolated so far. The transport properties of rhizoferrin have been investigated in *R. microsporus* var. *rhizopodiformis* in a time and concentration dependent manner (Drechsel et al., 1991, Thieken and Winkelmann, 1992), yielding K_M= 8 mM and V_{max} = $1.2 \cdot 10^9$ mol mg^{-1} min^{-1}. Interestingly, iron from ferrioxamines (B and E) was also taken up at similar rates (Drechsel et

Table 1 Occurrence of Siderophores in Fungal Classes

Classes	Siderophores
Myxomycetes	unknown
Mastigomycetes	unknown
Zygomycetes	polycarboxylates
Ascomycetes	hydroxamates
Basidiomycetes	hydroxamates
Deuteromycetes	hydroxamates

al., 1991) which might be the reason for the enhanced risk of Mucormycosis after treatment with Desferal® which is the methanesulfonate salt of ferrioxamine B (Boelaert et al., 1991).

O. *Lipomycetaceae*

Although an earlier report that *Cryptococcus melibiosum* produced an alanine containing ferrichrome type siderophore, named ferrichrome C (Atkin et al., 1970), suggested the occurrence of ferrichromes in the genus *Cryptococcus,* ferrichrome C production was absent in strains of *Cryptococcus neoformans* (Jacobson and Petro, 1987). *Cryptococcus melibiosum* is now reclassified as *Myxozyma melibiosi* Shifrine & Phaff belonging to the family of *Lipomycetaceae* (Van der Walt et al., 1981). Within the *Lipomycetaceae,* however, ferrichrome C and ferrichrome production was restricted to the genera *Dipodascopsis, Zygozyma,* and *Myxozyma* and was completely absent in the genera *Lipomyces* and *Waltomyces* (Van der Walt et al., 1990). Thus the soil-borne genera *Lipomyces* and *Waltomyces* can be chemotaxonomically distinguished from the entomophorous taxa *Dipodascopsis* and *Zygozyma* and from the genus *Myxozyma* (Van der Walt et al., 1981; Spaaij et al., 1990).

P. *Candida, Saccharomyces,* and Other Yeasts

There are several reports in the literature that *Candida albicans* secretes both hydroxamate and catecholate type siderophores when stressed for iron (Ismail et al., 1985; Ismail and Lupan, 1986). However, so far no proof of the chemical structure has been given, and it therefore remains doubtful if the spectral changes after the addition of iron salts to the culture filtrates are really indicative for siderophores. Especially after prolonged cultivation, a variety of cellular material from lysed cells may react positively in tests based on color formation. Recently, hydroxamates and a green pigment were reported to be secreted under iron stress in twelve strains of *C. albicans* (Sweet and Douglas, 1991). Also, in this report, no chemical structure was presented, and the test for hydroxamate siderophores was based only on the chrome azurol S assay (Schwyn and Neilands, 1987) and color formation. As the chrome azurole S assay reacts also positively with a variety of organic acids, the production of hydroxamates in *Candida* remains doubtful. However, *Candida albicans* is able to recognize a variety of hydroxamate siderophores, which has led to the development of a bioassay for the detection of natural and synthetic hydroxamate siderophores (Minnick et al., 1991).

Although *Saccharomyces cerevisiae* has been studied intensively for the production of siderophores (Neilands et al., 1987), no siderophores could be detected so far. The same applies to *Geotrichum candidum,* which was described as a nonsiderophore producer (Mor and Barash, 1990). Although *Cryptococcus neoformans* has been shown to respond to exogeneously supplied ferrioxamine B, the secretion of siderophoes under iron limitation has not been observed (Jacobson and Petro, 1987; Jacobson and Vartivarian, 1993). Various strains of *Cryptococcus* have beeen shown earlier to be unable to secrete hydroxamate siderophores (Atkin et al., 1970), with the exception of *C. melibiosum,* which is now transferred to the genus *Myxozyma melibiosi* (Van der Walt et al., 1981). Another yeastlike systemic opportunistic pathogen, *Histoplasma capsulatum,* however, has been shown to produce coprogen B as the princpal siderophore under iron limitation (Burt, 1982), which could be confirmed in an examination of further systemic fungal pathogens (Holzberg and Artis, 1983).

Q. Mycorrhizal Fungi

Earlier reports have focussed on the possibility of detecting hydroxamate siderophores in soil using a biotest assay with *Auerobacterium flavescens* JG9 as a hydroxamate-specific, siderophore auxotrophic organism (Szaniszlo et al., 1981). The bioassays allowed the study of the production of hydroxamate siderophores in axenic cultures of a variety of ectomycorrhizal fungi, although the produced siderophores had never been isolated. The first isolation and chemical characterization of hydroxamate siderophores from mycorrhizal fungi has been reported recently, showing that within the ericoid mycorrhizal fungi the ascomycete *Hymenoscyphus ericae* and the hyphomycete *Oidiodendron griseum* secreted ferricrocin, and the endophyte of *Rhodothamnus chamaecistus* secreted fusigen as their principal siderophores (Haselwandter et al., 1992).

VIII. CONCLUSIONS

Summarizing the results on the production of siderophores in fungi and their function as iron transporting agents, we realize that the manifold of structures and their cognate transport systems allow fungi to grow in diverse ecological habitats. While iron is plentiful in nature, its acquisition for cellular metabolism has been a constant challenge to all living organisms. Moreover, iron metabolism has to be carefully regulated in order to avoid toxic effects of free OH radicals generated by the Haber-Weiss-Fenton reaction. Therefore, in no case do Fe(III) or Fe(II) occur as free ionic species in the cellular metabolism, being always wrapped by complexing agents such as siderophores or iron-binding proteins. While all higher fungal taxa, like Ascomycetes and Basidiomycetes, produce hydroxamate siderophores, these seem to be absent in the Zygomycetes. Instead,

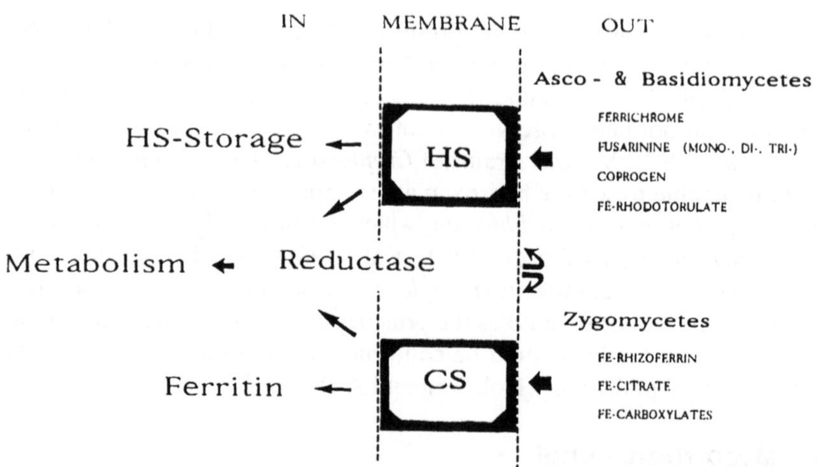

Figure 7 Scheme of events in iron delivery from siderophores to Ascomycetes, Basidiomycetes, and Zygomycetes via siderophore transport systems specific for ferric hydroxamates (HS) or ferric carboxylates (CS) and the subsequent distribution of iron into different compartments.

the Zygomycetes produce the polycarboxylate siderophore rhizoferrin (Drechsel et al., 1991, 1992). The current knowledge of fungal siderophores, the interacting transport systems, and the intracellular fate of iron is illustrated in Fig. 7. Interestingly, the Zygomycetes are the only fungal class where ferritins have been found so far. Thus the absence of hydroxamates with high formation constants and the presence of rhizoferrin possessing significantly lower formation constants may have led to the development of ferritin molecules for iron storage purposes (Winkelmann, 1992).

ACKNOWLEDGMENTS

The work in the laboratory of GW was supported by grants of the Deutsche Forschungsgemeinschaft (DFG) and that in the laboratory of DvdH by grants from the National Institutes of Health (GM 21822).

REFERENCES

Abu-Dari, K., Ekstrand, J. D., Freyberg, D. P., and Raymond, K. N. (1979). Coordination chemistry of microbial iron transport compounds, 14, Isolation and structural characterization of *trans*-tris(benzohydroxamato) chromium (III), *Inorg. Chem., 18:* 108.

Adjimani, J. P., and Emery, T. (1987). Iron uptake in *Mycelia sterilia* EP-76, *J. Bacteriol., 169:* 3664.

Adjimani, J. P., and Emery, T. (1988). Stereochemical aspects of iron transport in *Mycelia sterilia EP-76, J. Bacteriol., 170:* 1377.

Anke, H., Kinn, J., Bergquist, K.-E., and Sterner, O. (1991). Production of siderophores by strains of the genus *Trichoderma, BioMetals, 4:* 176–180.

Atkin, C. L., and Neilands, J. B. (1968). Rhodotorulic acid, a diketopiperazine dihydroxamic acid with growth-factor activity: Isolation and characterization, *Biochemistry, 7:* 3734.

Atkin, C. L., Neilands, J. B., and Phaff, H. (1970). Rhodotorulic acid from species of *Leucosporidium, Rhodosporidium, Rhodotorula, Sporidiobolus* and *Sporobolomyces* and a new alanine-containing ferrichrome from *Cryptococcus melibiosum, J. Bacteriol., 103:* 722.

Bailey, C. T., and Kime-Hunt, E. M., Carrano C. J., Huschka, H. G., and Winkelmann, G. (1986). A photoaffinity label for the siderophore-mediated iron transport system in *Neurospora crassa, Biochim. Biophys. Acta, 883:* 299.

Barnes, C. L., Hossain, M. B., Jalal, M. A. F., Eng-Wilmot, D. L., Grayson, S. L., Benson B. A., Agarwal, S. K., Mocherla, R., and van der Helm, D. (1985). Ferrichrome conformations: Ferrirubin, two crystal forms, *Acta Cryst., C41:* 341.

Bentley, M. D., Anderegg, R. J., Szaniszlo, P. J., and Davenport, R. F. (1986). Isolation and identification of the principal siderophore of the dermatophyte *Microsporum gypseum. Biochemistry, 25:* 1455.

Benz, G., Schröder, T., Kurz, J., Wünsche, C., Karl, W., Steffens, G., Pfitzner, J., and Schmidt, D. (1982). Konstitution der desferri-form der albomycin δ_1, δ_2 und ϵ. *Angew. Chem., 94:* 552.

Boelaert, J. R., Fenves, A.Z., and Coburn, J. W. (1991). Deferoxamine therapy and mucormycosis in dialysis patients; Report on an international registry, *Am. J. Kidney Diseases, 18:* 660.

Bossier, P., Hoefte, M., and Verstraete, W. (1988). Ecological significance of siderophores in soil, *Adv. Microbiol. Ecology, 10:* 385.

Braun, V., and Hantke, K. (1991). Genetics of iron transport, *Handbook of Microbial Iron Chelates* (G. Winkelmann, ed.), CRC Press, Boca Raton, FL, p. 107.

Budde D., and Leong, S. A. (1989). Characterization of siderophores from *Ustilago maydis, Mycopathologia, 108:* 125.

Burt, W. R. (1982). Identification of coprogen B and its breakdown products from *Histoplasma capsulatum, Infect. Immun., 35:* 990.

Carrano, C. J., and Raymond, K. N. (1978). Coordination chemistry of microbial iron transplant compounds: Rhodotorulic acid and iron uptake in *Rhodotorula pilimanae, J. Bacteriol., 136:* 69.

Charlang, G., Ng, B., Horowitz, N. H., and Horowitz, R. M. (1981). Cellular and extracellular siderophores of *Aspergillus nidulans* and *Penicillium chrysogenum, Mol. Cell. Biol., 1:* 94.

Chet, I. (1987). *Trichoderma*—Application, mode of action and potential as a biocontrol agent of soil-borne plant pathogenic fungi, *Innovative Approaches of Plant Disease Control* (I. Chet, ed.), John Wiley, New York, p. 137.

Crumbliss, A. L. (1991). Aqueous solution equilbrium and kinetic studies of iron siderophore and model siderophore complexes, *Handbook of Microbial Iron Chelates* (G. Winkelmann, ed.), CRC Press, Boca Raton, FL, p. 177.

Deml, G. (1985). Studies on heterobasidiomycetes, Part 34, A survey on siderophore formation in low-iron cultured smuts from the floral parts of *Polygonaceae, Systm. Appl. Microbiol., 6:* 23.

Deml, G., Voges, K., Jung, C., and Winkelmann, G. (1984). Tetraglycylferrichrome—The first heptapeptide ferrichrome, *FEBS Lett., 173:* 53.

Diekmann, H. (1967). Stoffwechselprodukte von Mikroorganismen, 56. Mitteilung, Fusigen—Ein neues Sideramin aus Pilzen, *Arch. Mikrobiol., 58:* 1.

Diekmann, H. (1968). Stoffwechselprodukte von Mikroorganismen, 68. Mitteilung, Die Isolierung und Darstellung von *trans*-5-Hydroxy-3-methylpenten-(2)-säure, *Arch. Microbiol., 62:* 322.

Diekmann, H., and Krezdorn, E. (1975). Stoffwechselproduckte von Mikroorganismen, 150. Mitteilung, Ferricrocin, Triacetylfusigen und anderes Sideramine aus Pilzen der Gattung *Aspergillus,* gruppe *Fumigatus, Arch. Microbiol., 106:* 191.

Dietrich, A., Powell, D. R., Eng-Wilmot, D. L., Hossain, M. B., and van der Helm, D. (1990). Structures of two isomeric hydroxamic acids: *N*-methyl-*p*-toluohydroxamic acid and *N*-(4-methyl-phenyl)acetohydroxamic acid, *Acta Cryst., C46:* 816.

Dietrich, A., Fidelis, K. A., Powell, D. R., van der Helm, D., and Eng-Wilmot, D. L. (1991). Crystal structrures of tris[*N*-(4-methyl-phenyl)acetohydroxamato]-iron(III) and tris (*N*-methyl-4-methyl-benzohydroxamato) iron (III) and gallium (III). Structure-stability relationships for the hydroxamato complexes of Fe^{3+} and Ga^{3+}, *J. Chem. Soc. Dalton Trans.*, 231.

Drechsel, H., Metzger, J., Freund, S., Jung, G., Boelaert, J. R., and Winkelmann, G. (1991). Rhizoferrin—A novel siderophore from the fungus *Rhizopus microsporus* var. *rhizopodiformis, BioMetals, 4:* 238.

Drechsel, H., Jung, G., and Winkelmann, G. (1992). Stereochemical characterization of rhizoferrin and identification of its dehydration products, *BioMetals, 5:* 141.

Drechsel, H., Thieken, A., Reissbrodt, R., Jung, G., and Winkelmann, G. (1993). a-Keto acids are novel siderophores in the genera *Proteus, Providencia, Morganella* and are produced by amino acid deaminases, *J. Bacteriol. 175:* 2727.

Ecker, D. J., and Emery, T. (1983). Iron uptake from ferrichrome A and iron citrate in *Ustilago sphaerogena, J. Bacteriol., 155:* 616.

Ecker, D. J., Lancaster, J. R., Jr., and Emery, T. (1982). Siderophore iron transport followed by electron paramagnetic resonance spectroscopy, *J. Biol. Chem., 257:* 8623.

Ecker, D. J., Passavant, C. W., and Emery, T. (1982). Role of two siderophores in *Ustilago sphaerogena.* Regulation of biosynthesis and uptake mechanisms, *Biochim. Biophys. Acta, 720:* 242.

Eide, D., and Guarente, L. (1992). Increased dosage of a transcriptional activator gene enhances iron-limited growth of Saccharomyces cerevisiae, *J. Gen. Microbiol., 138:* 347.

Emery, T. (1971a). Hydroxamic acids of natural origin, *Adv. Enzym., 35:* 135.

Emery, T. (1971b). Role of ferrichrome as a ferric ionophore in *Ustilago sphaerogena, Biochemistry, 10:* 1483.

Emery, T. (1980). Malinochrome, a new iron chelate from *Fusarium roseum, Biochim. Biophys. Acta, 629:* 382.

Emery, T. (1986). Exchange of iron by gallium in siderophores, *Biochem., 25:* 4629.

Emery, T. (1987). Reductive mechanisms of iron assimilation, *Iron Transport in Microbes, Plants and Animals* (G. Winkelmann, D. van der Helm, and J. B. Neilands, eds.), VCH, Weinheim, p. 235.

Emery, T. F., and Neilands, J. B. (1961). Structure of ferrichrome compounds, *J. Am. Chem. Soc., 83:* 1626.

Eng-Wilmot, D. L., Raman, A., Mendenhall, J. V., Grayson, S. L., and van der Helm, D. (1984). Molecular structure of ferric neurosporin, a minor siderophore-like compound containing N^{δ}-hydroxy-orinthine, *J. Am. Chem. Soc., 106:* 1285.

Eng-Wilmot, D., Adjimani, J. P., and van der Helm, D. (1992). Siderophore-mediated iron(III) transport in the mycelia of the cultivated fungus, *Agaricus bisporus, J. Inorg. Biochem., 48:* 183.

Ernst, J. F., and Winkelmann, G. (1977) Enzymatic release of iron from sideramines in fungi. NADH:sideramine oxidoreductase in *Neurospora crassa, Biochim. Biophys. Acta, 500:* 27.

Fidelis, K. A. (1990). Molecular dynamics and electron density studies of siderophores and peptides, Thesis, University of Oklahoma, p. 18.

Fidelis, K. A., Hossain, M. B., Jalal, M. A. F., and van der Helm, D. (1990). Structure and molecular mechanics of ferrirubin, *Acta Cryst., C46:* 1612.

Frederick, C. D., Bentley, M. D., and Shive, W. (1981). Structure of triornicin, a new siderophore, *Biochemistry, 20:* 2436.

Frederick, C. D., Bentley, M. D. and Shive, W. (1982). The structure of the fungal siderophore, isotriornicin. *Biochem. Biophys. Res. Commun., 105:* 133.

Hancock, R. D., and Martell, A. E. (1989). Ligand design for selective complexation of metal ions in aqueous solution, *Chem. Rev.,* 89: 1875.

Haselwandter, K., Dobernigg, B., Beck, W., Jung, G., Caniser, A., and Winkelmann, G. (1992). Isolation and identification of hydroxamate siderophores of ericoid mycorrhizal fungi, *BioMetals, 5:* 51.

Hesseltine, C. W., Pidacks, C., Whitehill, A. R., Bohonos, N., Hutchings, B. L., and Williams, J. H. (1952). Coprogen, a new growth factor for coprophilic fungi, *J. Chem. Soc., 74:* 1362.

Hider, R. C. (1984). Siderophore mediated absorption of iron, *Struct. Bonding, 58:* 25.

Hider, R. C., and Hall, A. D. (1991). Clinically useful chelators of tripositive elements. *Progress in Medicinal Chemistry* (G. P. Ellis and G. B. West, eds.), Vol. 28, Elsevier, New York. p. 43.

Holzberg, M., and Artis, W. M. (1983). Hydroxamate siderophore production by opportunistic and systemic fungal pathogens, *Infect. Immun., 40:* 1134.

Hossain, M. B., Eng-Wilmot, D. L., Loghry, R. A., and van der Helm, D. (1980). Ciruclar dichroism, crystal structure, and absolute configuration of the siderophore ferric *N,N′,N″*-triacetylfusarinine, *J. Am. Chem. Soc., 102:* 5766.

Hossain, M. B., Jalal, M. A. F., Benson, B. A., Barnes, C. L., and van der Helm, D. (1987). Structure and conformation of two coprogen-type siderophores: Neocoprogen I and neocoprogen II, *J. Am. Chem. Soc., 109:* 4948.

Huschka, H., and Winkelmann, G. (1987). Molecular recognition of siderophores in fungi, *Iron Transport in Microbes, Plants and Animals* (G. Winkelman, D. van der Helm, and J. B. Neilands, eds.), VCH, Weinheim, p. 317.

Huschka, H., and Winkelmann, G. (1989). Iron limitation and its effect on membrane proteins and siderophore transport in *Neurospora crassa, BioMetals, 2:* 108.

Huschka, H., Naegeli, H. U., Leuenberger-Ryf, H., Keller-Schierlein, W., and Winkelmann, G. (1985). Evidence for a common siderophore transport system but different siderophore receptors in *Neurospora crassa, J. Bacteriol., 162:* 715.

Huschka, H., Jalal, M. A. F., van der Helm, D., and Winkelmann, G. (1986). Molecular recognition of siderophores in fungi: Role of iron-surrounding *N*-acyl residues and the peptide backbone during membrane transport in *Neurospora crassa, J. Bacteriol., 167:* 1020.

Ismail, A., and Lupan, D. M. (1986). Utilization of siderophores by *Candida albicans, Mycopathologia, 96:* 109.

Ismail, A., Bedell, G. W., and Lupan, D. M. (1985). Effect of temperature on siderophore production by *Candida albicans, Biochem. Biophys. Res. Commun., 132:* 1160.

Jacobson, E., and Petro, M. J. (1987). Extracellular chelation in *Cryptococcus neoformans, J. Med. Vet. Mycol., 25:* 415.

Jacobson, E., and Vartivarian, S. E. (1993). Iron assimilation in *Cryptococcus neoformans, J. Med. Vet. Mycol., 30:* 443.

Jalal, M. A. F., and van der Helm, D. (1989). Siderophores of highly phytopathogenic *Alternaria longipes, BioMetals, 2:* 11.

Jalal, M. A. F., and van der Helm, D. (1991). Isolation and spectroscopic identification of fungal siderophores, *Handbook of Microbial Iron Chelates* (G. Winkelmann, ed.), CRC Press, Boca Raton, FL, p. 235.

Jalal, M. A. F., Mocharla, R., Barnes, C. L., Hossain, M. B., Powell, D. R., Eng-Wilmot, D. L., Grayson, S. L., Benson, B. A., and van der Helm, D. (1984). Extracellular siderophores from *Aspergillus ochraceus, J. Bacteriol., 158:* 683.

Jalal, M. A. F., Galles, J. L., and van der Helm, D. (1985). Structure of des(diserylglycyl)ferrirhodin, DDF, a novel siderophore from *Aspergillus ochraceus, J. Org. Chem., 50:* 5642.

Jalal, M. A. F., Love, S. K., and van der Helm, D. (1986). Siderophore mediated iron(III) uptake in *Gliocladium virens,* 1, Properties of *cis*-fusarinine, *trans*-fusarinine, dimerum acid, and their ferric complexes, *J. Inorg. Biochem., 28:* 417.

Jalal, M. A. F., Love, S. K., and van der Helm (1987). Siderophore mediated iron(III) uptake in *Gliocladium virens,* 2, Role of ferric mono- and dihydroxamates as iron transport agents, *J. Inorg. Biochem., 29:* 259.

Jalal, M. A. F., Love, S. K., and van der Helm, D. (1988). *N*-dimethylcoprogens, Three novel trihydroxamate siderophores from pathogenic fungi, *BioMetals, 1:* 4.

Keller-Schierlein, W. (1963). Stoffwechselprodukte von Mikroorganismen, 45. Mitteilung, Über die Konstitution von ferrirubin, ferrirhodin und ferrichrom, A. *Helv. Chim. Acta, 46:* 1920.

Kleinkauf, H., and von Döhren, H. (1990) Nonribosomal biosynthesis of peptide antibiotics, *Eur. J. Biochem., 192:* 1.

Konetschny-Rapp, S., Huschka, H. G., Winkelmann, G., and Jüng, G. (1988). High-performance liquid chromatography of siderophores from fungi, *BioMetals, 1,* 9.

Lankford, C. E. (1973). Bacterial assimilation of iron, *Crit. Rev. Microbiol., 2:* 273.

Leong, J., and Raymond, K. N. (1974). Coordination isomers of biological iron transport compounds, I, Models for the siderochromes, The geometrical and optical isomers of tris(*N*-methyl-l-methoxy-acethydroxamato) chromium(III), *J. Am. Chem. Soc., 96:* 1757.

Leong, J., and Raymond, K. N. (1974). Coordination isomers of biological iron transport compounds, II, The optical isomers of chromic desferriferrichrome and desferriferrichrysin, *J. Am. Chem. Soc., 96:* 6628.

Llinás, M., and Neilands, J. B. (1976). The structure of two alanine-containing ferrichromes: Sequence determination by proton magnetic resonance, *Biophys. Struct. Mech., 2:* 105.

Llinás, M., Klein, M. P., and Neilands, J. B. (1970). Solution conformation of ferrichrome, a microbial iron transport cyclohexapeptide, as deduced by high resolution proton magnetic resonance, *J. Mol. Biol., 52:* 399.

Manulis, S., Kashman, Y., and Barash, I. (1987). Identification of siderophores and siderophore-mediated uptake of iron in *Stemphylium botyrosum. Phytochemistry 26:* 1317.

Matzanke, B. (1987). Mössbauer spectroscopy of microbial iron uptake and metabolism, *Iron Transport in Microbes, Plants and Animals* (G. Winkelmann, D. van der Helm, and J. B. Neilands, eds.), VCH, Weinheim, p. 251.

Matzanke, B. F., Müller, G., and Raymond, K. N. (1989). Siderophore-mediated iron transport, *Iron Carriers and Iron Proteins* (T. M. Loehr, ed.), VCH, Weinheim, p. 1.

Matzanke, B. F., Bill, E., Trautwein, A. X., and Winkelmann, G. (1990). Siderophores as iron storage compounds in the yeasts *Rhodotorula minuta* and *Ustilago sphaerogena* detected by in vivo Mössbauer spectroscopy, *Hyperfine Interactions,* 58: 2359.

Minnick, A. A., Eizember, L. E., MyKee, J. A., Dolence, E. K., and Miller, M. (1991), Bioassay for siderophore utilization by *Candida albicans, Anal. Biochem., 194:* 223.

Monzyk, B., and Crumbliss, A. L. (1979). Mechanism of ligand substitution on high-spin iron (III) by hydroxamic acid chelators, Thermodynamic and kinetic studies on the formation and dissociation of a series of monohydroxamato iron(III) complexes, *J. Am. Chem. Soc., 101:* 6203.

Moore, R. E., and Emery, T. (1976). N^a-acetylfusarinines: Isolation, characterization and properties, *Biochemistry, 15:* 2719.

Mor, H., and Barash, I. (1990). Characterization of a siderophore-mediated iron transport in *Geotrichum candidum,* a non-siderophore producer, *BioMetals, 2:* 209.

Mor, H., Kashman, Y., Winkelmann, G., and Barash, I. (1992). Characterization of siderophores produced by different species of the dermatophytic fungi *Microsporum* and *Trichophyton, BioMetals, 5:* 213.

Müller, G., Barclay, S. J., and Raymond, K. N. (1985). The mechanism and specificity of iron transport in *Rhodotorula pilimanae* probed by synthetic analogs of rhodotorulic acid, *J. Biol. Chem., 260:* 13916.

Neilands, J. B. (1981). Microbial iron compounds, *Ann. Rev. Biochem., 50:* 715.

Neilands, J. B., Konopka, K., Schwyn, B., Coy, M., Francis, R. T., Paw, B. H., and Bagg, A. (1987). Comparative biochemistry of microbial iron assimilation, *Iron Transport in Microbes, Plants and Animals* (G. Winkelmann, D. van der Helm, and J. B. Neilands, eds.), VCH, Weinheim, p. 3.

Oosterhuis, W. T. (1974). The electronic state of iron in some natural compounds: Determination by Mössbauer and ESR spectroscopy, *Struct. Bonding, 20:* 59.

Powell, P. E., Cline, G. R., Reid, C. P. P., and Szaniszlo, P. J. (1980). Occurrence of hydroxamate iron chelators in soil, *Nature, 287:* 833.

Raymond, K. N., Müller, G., and Matzanke, B. F. (1984). Complexation of iron by siderophores, A review of their solution and structural chemistry and biological function, *Top. Curr. Chem., 123:* 49.

Sayer, J. M., and Emery, T. (1968). Structure of naturally occurring hydroxamic acids, fusarinines A and B, *Biochem., 7:* 184.

Scarrow, R. C., White, D. L., and Raymond, K. N. (1985). Ferric ion sequestering agents, 14, 1-hydroxy-2(1H)-pyridinone complexes: Properties and structure of a novel Fe-Fe dimer, *J. Am. Chem. Soc., 107:* 6540.

Schwyn, B., and Neilands, J. B. (1987). Universal chemical assay for the detection and determination of siderophores. *Anal. Biochem. 160:* 47.

Sheridan, R. P., Levy, R. M., and Englander, S. W. (1983). Normal mode paths for hydrogen exchange in the peptide ferrichrome, *Proc. Natl. Acad. Sci. USA, 80:* 5569.

Sivarama-Sastry, K., Adiga, P. R., Venkatasubramanyam, V., and Sarma, P. S. (1962). Interrelationships in trace element metabolism in metal toxicities in *Neurospora crassa, Biochem. J.,* 85: 486.

Spaaij, F., Weber, G., Van der Walt, J. P., and Oberwinkler (1990). *Myxozyma udenii* sp. nov. (*Candidaceae*), a new yeast isolated from the rhizosphere of *Magnifere indica, System. Appl. Microbiol., 13:* 182.

Szaniszlo, P. J., Powell, P. E., Reid, C. P. P., and Cline, G. R. (1981). Production of hydroxamate siderophore iron chelators by ectomycorrhizal fungi, *Mycologia, 73:* 1158.

Thieken, A., and Winkelmann, G. (1992) Rhizoferrin: A complexone type siderophore of the Mucorales and Entomophthorales (Zygomycetes), *FEMS Microbiol. Lett., 94:* 37.

Thieken, A., and Winkelmann, G. (1993). A novel bioassay for the detection of siderophores containing keto-hydroxy bidentate ligands, *FEMS Microbiol., 777:* 287.

van der Helm, D., Baker, J. R., Eng-Wilmot, D. L., Hossain, M. B., and Loghry, R. A. (1980). Crystal structure of ferrichrome and a comparison with the structure of ferrichrome A, *J. Am. Chem. Soc., 102:* 4224.

van der Helm, D., Baker, J.R., Loghry, R. A. and Ekstrand, J. D. (1981). Structures of alumichrome A and ferrichrome A at low temperature, *Acta Cryst., B37:* 323.

van der Helm, D., Jalal, M. A. F., and Hossain, M. B. (1987). The crystal structures, conformations and configurations of siderophores, *Iron Transport in Microbes, Plants and Animals* (G. Winkelmann, D. van der Helm, and J. B. Neilands, eds.), VCH, Weinheim, p. 135.

Van der Walt, J. P., Wejman, A. C. M., and Von Arx, J. A. (1981). The anamorphic yeast genus *Myxozyma* gen. nov. *Sydowia, Ann. Mycol.* Ser. II, *34:* 191.

Van der Walt, J. P., von Arx, J. A., Ferreira, N. P., and Richards, P. D. G. (1987). *Zygozyma* gen. nov., a new genus of the *Lipomycetaceae, System. Appl. Microbiol., 9:* 115.

Van der Walt, J. P., Botha, A., and Eicker, A. (1990). Ferrichrome production by Lipomycetaceae. *Syst. Appl. Microbiol. 13:* 131.

Wiebe, C., and Winkelman, G. (1975). Kinetic studies on the specificity of chelate-iron uptake in *Apsergillus, J. Bacteriol., 123:* 873.

Winkelmann, G. (1974). Metabolic products of microorganisms, 132, Uptake of iron by *Neurospora carassa,* III, Iron transport studies with ferrichrome-type compounds, *Arch. Microbiol., 98:* 39.

Winkelmann, G. (1979a). Surface iron polymers and hydroxy acids, A model of iron supply in sideramine-free fungi, *Arch. Microbiol., 121:* 43.

Winkelmann, G. (1978b). Evidence for stereospecific uptake of iron chelates in fungi, *FEBS Lett., 97:* 43.

Winkelmann, G. (1991). Specificity of iron transport in bacteria and fungi, *Handbook of Microbial Iron Chelates* (G. Winkelmann, ed.), CRC Press, Boca Raton, FL, p. 65.

Winkelmann, G, (1992). Structures and functions of fungal siderophores containing hydroxamate and complexone type iron binding ligands, *Mycol. Res., 96:* 529.

Winkelmann, G. (1993). Kinetics, energetics and mechanisms of siderophore iron transport in fungi, *Iron Chelation in Plants and Soil Microorganisms* (L. L. Barton, ed.), Academic Press, New York, p. 219.

Winkelmann, G., and Braun, V. (1981). Stereoselective recognition of ferrichrome by fungi and bacteria, *FEMS Microbiol. Lett., 11:* 237.

Winkelmann, G., and Zähner, H. (1973). Metabolic products of microorganisms, 115, Uptake of iron by *Neurospora crassa,* I, Specificity of iron transport, *Arch. Microbiol., 88:* 49.

Winkelmann, G., Barnekow, A., Ilgner, D., and Zähner, H. (1973). Metabolic products of Microorganisms, 120, Uptake of iron by *Neurospora crassa,* II, Regulation of the biosynthesis of sideramines and inhibition of iron transport by metal analogues of coprogen, *Arch. Microbiol., 92:* 285.

Wong, G. B., Kappel, M. J., Raymond, K. N., Matzanke, B., and Winkelmann, G. (1983). Coordination chemistry of iron transport compounds, 24, Characterization of coprogen and ferricrocin, two ferric hydroxamate siderophores, *J. Am. Chem. Soc., 105:* 810.

Xiao, G., van der Helm, D., Hider, R. C., and Dobbins, P. S. (1992). Structure-stability relationships of 3-hydroxypyridin-4-one complexes, *J. Chem. Soc. Dalton Trans.*, 1992: 3265.

Zähner, H., Keller-Schierlein, W., Hütter, R., Hess-Leisinger, K., and Dee'r, A. (1963). Stoffwechselprodukte von Mikroorganismen, 40. Mitteilung, Sideramine aus Aspergillaceen, *Arch. Microbiol. 45:* 119.

Zalkin, A., Forrester, J. D., and Templeton, D. H. (1966). Ferrichrome A tetrahydrate, Determination of crystal and molecular structure, *J. Am. Chem. Soc., 88:* 1810.

3

Enzymology of Siderophore Biosynthesis in Fungi

Hans Jürgen Plattner and Hans Diekmann
Institute of Microbiology, University of Hannover, Hannover, Germany

I. INTRODUCTION

In low iron environments, most fungi secrete iron chelating compounds known as siderophores. The biosynthesis and secretion of desferrisiderophores as well as the uptake of siderophore iron (III) complex is heavily increased during growth under iron deprivation.

Most known fungal siderophores are of the hydroxamate type and derivatives of L-ornithine. Their structures do not vary as widely as those from bacteria (Winkelmann, 1986, 1991). At first glance, the enzymology of fungal siderophore biosynthesis might seem to be quite simple. Emery (1974) has reviewed siderophore biosynthesis, but at that time little was known on microbial *N*-hydroxylation and nonribosomal peptide synthesis. Progress was enhanced by the fact that plasmid-coded aerobactin was shown to be a virulence factor of some bacteria (Williams, 1979; Waters and Crosa, 1991). Also, the interest in siderophores and their biosynthesis was stimulated by reports on many other biological activities, like induction of spore germination (Horowitz et al., 1976; Matzanke et al., 1987), enhanced iron uptake of mycorrhizal fungi in roots of vascular plants (Haselwandter et al., 1992), virulence of bacterial plant pathogens (Persmark et al., 1992), and infections caused by the opportunistic pathogen *Candida albicans* (Sweet and Douglas, 1991).

Despite the progress that has been made during the last 15 years, we are far from a complete understanding of all aspects of fungal siderophore biosynthesis. This review will not only report on results but likewise focus on questions to be answered. Progress will only be made if our knowledge of the molecular genetics of siderophore biosynthesis in fungi is comparable to what is known about bacteria. The study of *fur* regulation of gene expression in bacteria has revealed a great number of target sites (Braun and Hantke, 1991). Since we know that in fungi the concentration of iron ions regulates the biosynthesis of many enzymes, it would be worthwhile to study iron-regulated gene expression in fungi as well. Because of the great importance of iron ions to fungal metabolism, the study of the regulation of biosynthesis might also uncover many interesting aspects of fungal ecology, including symbiotic associations with man, animals, and plants.

Future progress made in the understanding of iron transport and function will contribute to our knowledge of the importance and function of trace elements in fungi (as before with bacteria: Silver and Walderhaug, 1992). It should be kept in mind that iron uptake is unique, since special low molecular weight chelating compounds are synthesized and their biosynthesis is regulated in parallel to the iron-regulated outer membrane proteins of bacteria. Siderophores provide a stroke of luck in trace element research because most of them are colored when loaded with Fe(III) and are redundantly overproduced under iron limitation even in wild strains. Specific chelators that function in sequestering and transport of other trace elements similar to iron have not yet been found.

II. ORIGIN AND UTILIZATION OF PRECURSORS

Precursors of siderophore biosynthesis that come from common primary metabolic pathways are the amino acids L-ornithine, L-glycine, L-serine, and L-alanine and the coenzyme A derivatives of acetic, malonic, anhydromevalonic, β-methylglutaric, 3-hydroxybutyric, and palmitic acid.

A. Ornithine Biosynthesis

In fungi, L-ornithine biosynthesis has been extensively studied, mainly in *Neurospora crassa* and *Saccharomyces cerevisiae* (Davis, 1986), as part of arginine metabolism. The anabolic pathway of ornithine synthesis starts with L-glutamate:

L-glutamate → *N*-acetylglutamate → *N*-acetyl-γ-glutamylphosphate → *N*-acetyl-γ-glutamate-semialdehyde → N^2-acetyl-ornithine → L-ornithine and takes place in the mitochondria. The first intermediate, *N*-acetyl-glutamate, is formed either by acetylation from acetyl coenzyme A or by acetyl-transfer from N^2-acetyl-ornithine. L-Ornithine, which is not a constituent of proteins, is either used for arginine biosynthesis or stored in the vacuole. (In *N. crassa,* 98% of the ornithine pool is to be found in the vacuole.) L-Ornithine can also be formed during catabolic degradation of arginine in the cytosol. The enzymes of anabolism are mainly regulated by arginine accumulation or starvation, those of catabolism by a general nitrogen catabolite or by an arginine-specific regulation (Davis, 1986).

The regulation of arginine/ornithine metabolism and the efflux of the amino acids from the vacuole that is likely to occur under iron limitation has not yet been investigated.

B. Biosynthesis of Anhydromevalonic Acids

Two isomeric 5-hydroxy-3-methyl-2-pentenoic acids (Δ^2-anhydromevalonic acids) were found in the siderophores ferrirhodin and ferrirubin (Keller-Schierlein, 1963), coprogen (Keller-Schierlein et al., 1964), and fusigen (Diekmann, 1967; Diekmann and Zähner, 1967) as well as in other metabolites from fungi (Fetz et al., 1965). When searching for these compounds in 23 strains of fungi (Diekmann, 1968) we found that only three of them contained both kinds of anhydromevalonic acid. In a study with *Fusarium cubense* it was shown that radioactivity from 2-^{14}C-mevalonic acid and U-^{14}C-leucine, but not from 1-^{14}C-leucine, was incorporated into *cis*-anhydromevalonic acid. Specifically labeled 2-^{3}H-mevalonic acids were stereospecifically incorporated to *cis*- or *trans*-anhydromevalonic acids in either fusigen-producing *F. cubense* or coprogen B–producing *Fusarium sp.*, respectively (Anke and Diekmann, 1971), while *Penicillium chrysogenum* apparently contained both stereospecific cis-eliminating dehydratases (Anke and Diekmann, 1974a). The substrate for the *cis*-Δ^2-anhydromevalonic acid hydratase was D-mevalonic acid (prepared by fermentation of *Endomycopsis fibuliger*), and the enzyme was studied in the cell-free extract (Anke and Diekmann, 1974a).

C. Precursor Studies and General Pathway of Biosynthesis

Before cell-free enzymatic studies were undertaken, the pathways of fungal siderophore biosynthesis were elucidated by incorporation studies. Labeled precursers used were N^5-hydroxy-L-ornithine and N^5-acetyl-N^5-hydroxy-L-ornithine (Emery, 1966; Anke and Diekmann, 1974b), L-ornithine (Diekmann, 1970), deuterated amino acids (Akers et al., 1972), glycine and serine

(Müller and Diekmann, 1977), mevalonic acid (Anke and Diekmann, 1971), and oxygen (Akers and Neilands, 1978).

From these studies a general biosynthetic pathway of fungal siderophores was derived that involved the following steps: (1) oxidation of ornithine to its N^5-hydroxy derivative, (2) acylation of N^5-hydroxy-L-ornithine (HO) leading to the monohydroxamic acid, and (3a) condensation of N^5-acyl-N^5-hydroxy-L-ornithine (AcylHO) to cyclic dipeptides and cyclic triesters, respectively, or (3b) condensation of AcylHO with amino acids forming cyclic peptides. Finally, (4) some siderophores are acetylated at the N^2-amino group of AcylHO. The general sequence is as shown in the table.

L-ornithine + ½ O_2	→	N^5-hydroxy-L-ornithine	(1)
N^5-hydroxy-L-ornithine + acyl-CoA[a]	→	N^5-acyl-N^5-hydroxy-L-ornithine	(2)
2 or 3 N^5-acyl-N^5-hydroxy-L-ornithine	→	siderophore[c]	(3a)
3 N^5-acyl-N^5-hydroxy-L-ornithine + 3(4) amino acids[b]	→	siderophore[d]	(3b)
Siderophore[e] + acetyl-CoA	→	N^2-acetylated siderophore[f]	(4)

[a]Acetyl, malonyl, anhydromevalonyl, β-methylglutaryl, 3-hydroxybutyryl, palmitoyl.
[b]Amino acids: glycine, serine, alanine.
[c]Rhodotorulic acid, dimerum acid, fusigen, coprogen B.
[d]Ferrichrome, ferricrocin, ferrichrysin, ferrirubin, ferrirhodin, ferrichrome A, ferrichrome C, malinochrome, tetraglycylferrichrome, asperchromes.
[e]Coprogen B, fusigen.
[f]Coprogen, triacetylfusigen, palmitoylcoprogen (Anke et al., 1991), triornicin, isotriornicin, neocoprogen I and II, neurosporin.

On the other hand, pulcherriminic acid, a hydroxamic acid—but not a siderophore—from fungi, is synthesized in *Metchnikowia pulcherrima* via a diketopiperazine cyclo-L-leucyl-L-leucyl with subsequent *N*-hydroxylation (MacDonald, 1965; Plattner and Diekmann, 1973).

III. ENZYMES OF FUNGAL SIDEROPHORE BIOSYNTHESIS

As the activity of enzymes for siderophore biosynthesis in cell-free extracts is low in wild strains and the enzymes are only expressed under iron limitation, microorganisms must be grown under controlled, reproducible conditions in order to maintain a high siderophore production rate. With filamentous fungi, high enzyme activities were achieved by inoculating growth media with conidia (10^8 conidia per L medium).

High amounts of biosynthetic enzymes suitable for isolation are only present for a few hours during the logarithmic growth phase. The fungal enzymes are soluble, cytoplasmic proteins as their activities are found in the 105,000 × g supernatant of cell-free extracts. Cells or mycelia can be broken by means of ultraturrax, sonication, or X-press disruption, depending on the enzyme being isolated.

This chapter is divided into three sections: (A) *N*-hydroxylases, (B) *N*-acyltransferases, and (C) siderophore synthetases.

The activation of precursor acids by acylCoA-synthetases (for instance *cis*-Δ^2-anhydromevalonic acid:CoA-ligase, Anke and Diekmann, 1974a) is not especially described. There is no indication, except for the activation of L-leucine in leucyl-leucyl-diketopiperazine biosynthesis, that the amino acids glycine, serine, alanine are activated by special kinases or by aminoacyl-tRNA synthetases.

A. L-Ornithine-N^5-Hydroxylase

The metabolic *N*-oxidation (hydroxylation) of nitrogenous compounds, e.g., xenobiotics, is well known in many biological systems. In general, two types of hydroxylases exist, cytochrome P-450 and the flavin containing monooxygenase, which should be taken into consideration for *N*-oxidations.

N-Hydroxylation of primary aromatic amines, especially 2-naphthylamine, was described by Miller and Miller (1966), and the enzyme from liver microsomes was characterized as a cytochrome P-450 dependent monooxygenase. On the other hand, a microsomal amine oxidase catalyzes the NADPH and O_2 dependent *N*-oxidation of secondary and tertiary amines, for example *N*-methyl aniline and *N,N*-dimethyl aniline, to the corresponding hydroxylamines and amine oxides. Primary amines like 1-naphthylamine or 2-naphthylamine are oxidized at lower rates (Massey and Hemmerich, 1975). Flavin dependent monooxygenases oxidize tertiary amines to *N*-oxides, whereas secondary and primary amines are oxidized through intermediate hydroxylamines to nitrones and oximes, respectively (Ziegler, 1988). Usually, these enzymes are involved in catabolic reactions, and they show low substrate specificities.

In contrast, *N*-hydroxylation of L-lysine (bacteria) and L-ornithine (fungi) in siderophore biosynthesis is an anabolic step, and the hydroxylating enzyme is expected to be substrate-specific.

The first evidence for the mechanism of microbial *N*-hydroxylation came from the work of Viswanatha and coworkers, who found membrane-associated activity of bacterial lysine-N^6-hydroxylase in cell-free extracts from *Aerobacter aerogenes* (*Klebsiella pneumoniae*) (Murray et al., 1977; Parniak et al., 1979; Jackson et al., 1984). The breakthrough came from the work of Braun (Gross et al., 1984, 1985) and Neilands (DeLorenzo et al., 1986; DeLorenzo and Neilands, 1986), who cloned the enzymes of aerobactin synthesis from plasmids pColV and constructed a gene map. The *N*-hydroxylase gene product was shown to be a 53 kDa protein (Engelbrecht and Braun, 1986), and the base sequence of the aerA (or iucD) gene was determined (Herrero et al., 1988). The mutant *Escherichia coli* EN222 expressed the *N*-hydroxylase efficiently, and using this mutant Plattner et al. (1989) were able to purify and characterize the lysine-N^6-hydroxylase as an FAD-dependent external monooxygenase. FAD is not tightly bound and can—in contrast to other flavoproteins—be removed by dialysis. Neither L- nor D-ornithine is hydroxylated by the bacterial enzyme.

Using an analogous activity test (Plattner et al., 1989), L-ornithine N^5-hydroxylating activity was found in the supernatant after 105,000×g centrifugation of dialyzed cell-free extracts from *Aspergillus quadricinctus* (ferrichrome-producing) and *Rhodotorula glutinis* (rhodotorulic acid-producing) (Diekmann and Plattner, 1991). As shown in the omission test, the reaction was strictly dependent on L-ornithine, FAD, and NADPH. NADH could not function as the external reductant. Reconstitution of N^5-hydroxylating activity in dialyzed crude extract in dependence on FAD concentration is shown in Fig. 1. The enzyme, an L-ornithine, NADPH:oxygen oxidoreductase (N^5-hydroxylating), was enriched by fractional ammonium sulphate precipitation (0.3–0.45 saturation) and FPLC on Superose 12.

The reaction sequence is

$$^{-}OOC\underset{|}{\overset{NH_3^+}{C}}H(CH_2)_3NH_2 + NADPH + O_2 \xrightarrow{FAD}$$

$$^{-}OOC\underset{|}{\overset{NH_3^+}{C}}H(CH_2)_3NHOH + NADP^+ + H_2O$$

B. *N*-Acyltransferase

N-Acetyltransferases are widely spread enzymes involved in the biosynthesis and detoxification of endogenous compounds. They transfer the acetyl group from acetyl coenzyme A to the amino group of acceptor compounds, e.g., primary aliphatic amino-, α-amino-, hydrazino-, arylamino-, and aryl sulfonamido groups.

In fungal siderophore biosynthesis, the primary hydroxylamine derivative

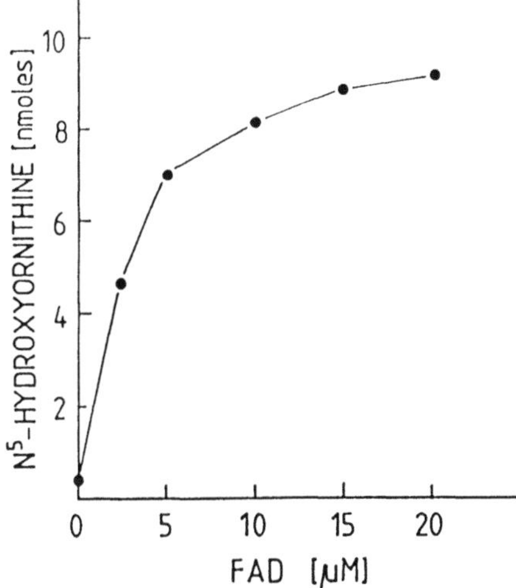

Figure 1 Reconstitution of ornithine N^5-hydroxylating activity by FAD in crude dialyzed extracts of *Rhodotorula glutinis*. 2.4 mg protein were incubated with 1.5 µmol L-ornithine and 1 µmol NADPH in 0.1 M potassium phosphate pH 7.0 at 37°C for 60 min.

of L-ornithine is transformed to hydroxamic acids by acyl transfer from acylCoA derivatives. As mentioned before, in fungi the acceptor is always N^5-hydroxy-L-ornithine. Acyl donors are acetyl CoA, succinyl CoA, malonyl CoA, and the CoA-activated Δ^2-anhydromevalonic acids. The reaction is catalyzed by acyl-CoA:N^5-hydroxy-L-ornithine *N*-acyl-transferases (EC 2.3.1.).

N-Acetyltransferase activity was first found in extracts of *Ustilago sphaerogena* (Ong and Emery, 1972) and *F. cubense* (Anke and Diekmann, 1974a). *N*-Acetyltransferases were isolated from *Aspergillus quadricinctus* (Kusters, 1984), *Rhodotorula glutinis* (Plattner, 1986), and *Rhodotorula pilimanae* (Neilands et al., 1987). The enzymes were purified by fractional ammonium sulphate precipitation, chromatography on DEAE-cellulose, Sephadex G200 size exclusion chromatography and hydroxylapatite. Affinity chromatography on reactive Blue 2-Sepharose was used in purification of *N*-acetyltransferase from *R. pilimanae*. Some properties of these enzymes are shown in Table 1. Data for the *N*-acetyltransferase from *E. coli* are included for comparison.

The *N*-acetyltransferase catalyzes the reaction

$$\text{AcetylCoA} + {}^-\text{OOC}\underset{}{\overset{\overset{\displaystyle NH_3^+}{|}}{\text{C}}}\text{H(CH}_2)_3\text{NHOH} \rightarrow {}^-\text{OOC}\overset{\overset{\displaystyle NH_3^+}{|}}{\text{C}}\text{H(CH}_2)_3\overset{\overset{\displaystyle HO}{|}}{\text{N}}-\overset{\overset{\displaystyle O}{\|}}{\text{C}}\text{CH}_3 + \text{CoA}$$

Table 1 Properties of Fungal *N*-Acetyltransferases Compared to Enzymes from *Escherichia coli* Strains

Source	Molecular weight (kDa)		Michaelis constant (M)		Reference
	SEC	SDS-PAGE	AcetylCoA	HO	
U. sphaerogena			2.2×10^{-4}	3.2×10^{-4}	a
A. quadricinctus	336	52	1.5×10^{-4}	1.1×10^{-4}	b
R. glutinis	190	48	6.5×10^{-5}	6.8×10^{-4}	c
R. pilimanae		40			d
E. coli pABN11	150–200	33	4.3×10^{-5}	3.2×10^{-4}*	e
E. coli GR128	70–140	35	n.d.	n.d.	f

**N*[6]-hydroxylysine
[a]Ong and Emery (1972)
[b]Kusters (1984)
[c]Plattner (1986)
[d]Neilands et al. (1987)
[e]Coy et al. (1986)
[f]Plattner (1987)

As acetyl CoA protects the enzyme from *R. glutinis* against rapid thermal loss of activity (10 min at 56°C results in 40% loss of activity), binding of the coenzyme prior to acyl transfer is assumed. CoA is only released from acetyl CoA when the acceptor is present (Fig. 2). The enzyme activity can be tested using ^{14}C-acetyl CoA (Kusters and Diekmann, 1984) or substrate dependent CoA release (dithionitrobenzoate method) (Coy et al., 1986). In regard to the acceptor compound, the *N*-acetyltransferases from *A. quadricinctus* and *R. glutinis* are not very specific *in vitro,* as they also act on N^6-hydroxylysine and *N*-methylhydroxylamine. The enzyme from *R. pilimanae* was reported to acetylate even inorganic hydroxylamine. Low substrate specificity was also found with the bacterial enzymes.

C. Siderophore Synthetases

When the product of the synthetase reaction is a dimer or trimer of the hydroxamic acid, the carboxyl group of the monomer has to be activated, and peptide bonds (as in rhodotorulic acid), ester bonds (as in fusigen), or both (as in coprogen) are formed. Activation and peptide and/or ester bond synthesis is catalyzed by the same multifunctional enzyme.

The task for ferrichrome synthetase is more elaborate. The hydroxamic acid monomer as well as amino acids are activated, oligopeptides are

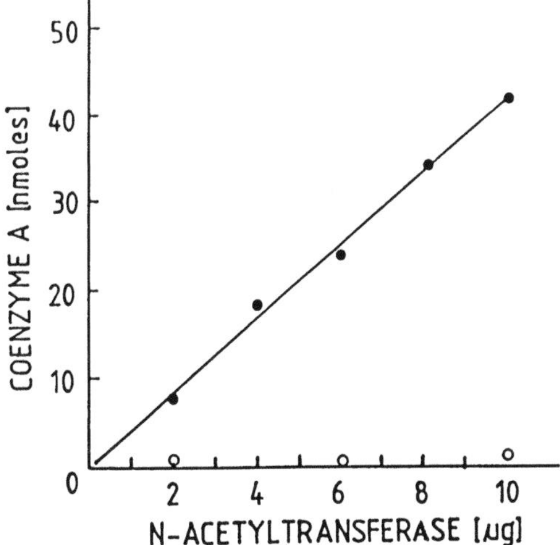

Figure 2 Release of coenzyme A from acetyl CoA in the presence (●) and absence (○) of *N*⁵-hydroxy-L-ornithine by *N*-acetyltransferase from *Rhodotorula glutinis*.

formed, and the cyclohexapeptide is built in a cyclization reaction. Reactions of this kind are accomplished by multienzyme complexes as reported for fatty acid synthetases and nonribosomal peptide synthetases (Kleinkauf and von Döhren, 1987).

1. *Rhodotorulic Acid Synthetase*

Rhodotorulic acid is the simplest siderophore in fungi. It is composed of two monomers, N^5-acetyl-N^5-hydroxy-L-ornithine (AHO), condensed by two peptide bonds to a diketopiperazine (Atkin and Neilands, 1968). The enzyme activity can be determined in a test using labeled HO or AHO or by ATP-^{32}P-pyrophosphate exchange in the presence of magnesium ions and mercaptoethanol. The pH optimum is near 8.5, and the optimal ATP concentration is 0.02 M. It is synthesized in the early logarithmic phase during growth of *Rhodotorula glutinis* (Anke and Diekmann, 1972).

2. *Fusigen Synthetase*

Fusigen synthetase was detected in cell-free extracts of *Fusarium cubense*. Peak activity was found in the middle of the logarithmic growth phase (Fig. 3). The activation of *cis*-fusarinine (*N*-*cis*-anhydromevalonyl-*N*-hydroxy-L-ornithine; prepared from fusigen by hydrolysis of the ester bonds) can be tested by ATP-^{32}P-pyrophosphate exchange (Anke et al., 1973). The enzyme was purified 240-fold by fractional ammonium sulphate precipitation

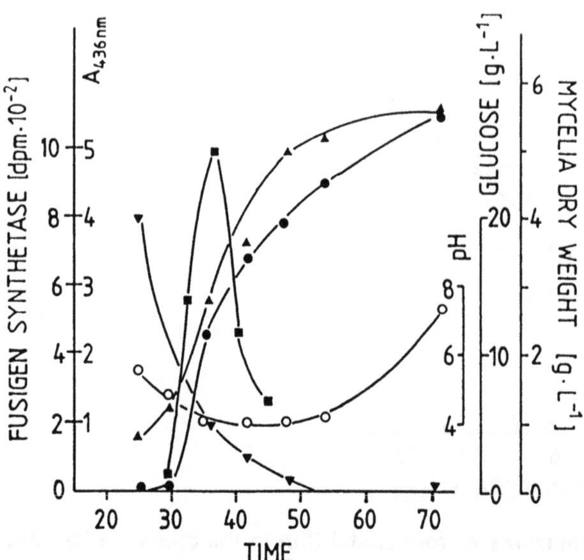

Figure 3 **Fusigen synthetase activity (incorporation of *cis*-fusarinine) of *Fusarium cubense* during growth under iron limitation. ■ – ■, fusigen synthetase; ● – ●, siderophore; ▲ – ▲, mycelial dry weight; ▼ – ▼, glucose; ○ – ○, pH. (*Source:* After Müller, 1976.)**

and chromatography on Bio-Gel A1.5m and DEAE-Sephadex A-25; its molecular weight is between 600,000 and 800,000 as determined by gel filtration. In the exchange reaction K_m for *cis*-fusarinine is 4.8×10^{-4} M at a Mg^{2+}/ATP ratio of 2.5, and the pH optimum is 7.0 (Müller, 1976). The exchange reaction was blocked by 0.1 mM pCMB.

Linear dimers or trimers of *cis*-fusarinine were not found. Therefore, it was concluded that the cyclic triester fusigen is synthesized from the monohydroxamic acid in a mechanism similar to the formation of enterochelin (Bryce et al., 1971), where the monomers appear to be bound to the enzyme as adenylates or thioesters (Bryce and Brot, 1973).

3. *Ferrichrome Synthetase*

When the siderophore production of fungi is investigated, usually a mixture of closely related products is found. In a study on siderophores from fungi of the genus *Aspergillus,* group *Fumigatus,* we found that *A. quadricinctus* CBS135.52 formed high amounts of almost pure ferrichrome (Diekmann and Krezdorn, 1975). After many unsuccessful attempts with other fungal strains, we succeeded in preparing cell-free extracts from this fungus capable of ferrichrome biosynthesis (Müller and Diekmann, 1977). The enzyme was enriched 20-fold and characterized as a large multienzyme complex of about 10^6 Da (Hummel and Diekmann, 1981). Further purification by DEAE cellulose, hydroxyapatite, and Bio-Gel A-5m chromatography was hampered by the fact that the enzyme was extremely labile and disintegrated into smaller complexes that were able to activate glycine and AHO but were no longer able to synthesize ferrichrome. It was the introduction of FPLC that finally allowed us to purify the enzyme and to obtain reliable molecular weight data (Siegmund et al., 1990, 1991).

Simultaneously, specimens of pure ferrichrome synthetase, prepared under conditions that avoided any disintegration, were investigated by electron microscopy. Projections of negatively stained ferrichrome synthetase revealed that the multienzyme complex consists of two different ring-shaped particles with a diameter of about 13 nm (type I) and 15 nm (type II) in a ratio of 1:1. While type II showed no detailed fine structure in its center, particle type I contained trapped negative stain with a diameter of about 3 nm (Siegmund, K. D., Johannsen, W., and Diekmann, H., unpublished). A similar structure was reported for GS2, the large subunit of gramicidin S-synthetase (Vater, 1990). It is supposed that after activation of the amino acids with ATP pyrophosphorylase, homotripeptides are formed, which are combined in the cyclization reaction to the hexapeptide.

IV. REGULATION AND GENETICS OF SIDEROPHORE BIOSYNTHETIC ENZYMES

The observation that a high concentration of iron suppresses the biosynthesis of siderophores was first reported by Garibaldi and Neilands (1956), and the fact that low iron concentrations favor the biosynthesis of siderophores has been used for the production of siderophores in many microorganisms. Likewise, the yield of enzyme depends on the iron concentration of the growth medium and the time of cell harvest. Enzymes are usually expressed during the early or middle exponential growth phase and are degraded very rapidly thereafter.

When the activity of biosynthetic enzymes could be measured in cell-free experiments, it was shown that the synthetases and hydroxylase were effectively repressed by iron *in vivo:* rhodotorulic acid synthetase (Anke and Diekmann, 1972), fusigen synthetase (Anke et al., 1973), ferrichrome synthetase (Hummel and Diekmann, 1981), and bacterial lysine-N^6-hydroxylase (Heydel et al., 1987). The iron concentrations used for repression were in the range of 10 μM to 70 μM. The dependence of rhodotorulic acid synthetase and *N*-acetyltransferase activities from *R. glutinis* on iron concentration of the medium is shown in Fig. 4. Both enzyme activities decreased rapidly with increasing iron concentration up to 15

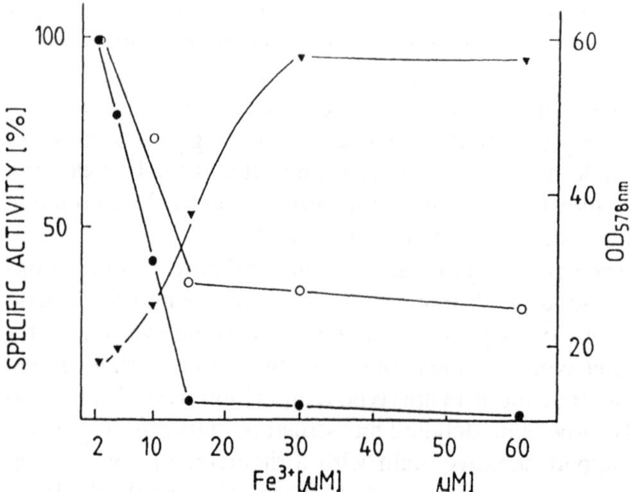

Figure 4 Activities of *N*-acetyltransferase and rhodotorulic acid synthetase of *R. glutinis* dependent on iron concentration in crude extract of cells grown for 20 h on glucose-asparagine medium. ○ – ○, *N*-acetyltransferase; ● – ●, rhodotorulic acid synthetase; ▼ – ▼, optical density (activities at 2 μM Fe^{3+} = 100 %).

μM. Although the growth was still limited by iron in the range of 15 to 30 μM, the synthetase was almost totally repressed, while the *N*-acetyltransferase was not repressed even with 60 μM. This was also observed in other cases, where the *N*-acetyltransferase activity was only reduced by about 50% of the maximal value under high iron concentrations (Kusters, 1984; Heydel et al., 1987). It is not known (and has not yet been studied in detail) whether there are two enzymes, one of which is iron-regulated and the other not, or if the enzyme synthesis is only partly repressed. Contrary to our results, *N*-acetyltransferase from *R. pilimanae* is repressed by just 10 μM iron in the medium (Neilands et al., 1987). Despite its effect *in vivo,* iron was not found to be inhibitory *in vitro.*

In contrast to the extensive molecular genetic analysis of bacterial siderophore biosynthesis (Crosa, 1989; Waters and Crosa, 1991), there are only a few papers on fungal genetics related to siderophores.

Early observations have shown that fungi can be siderophore auxotrophic (Hesseltine et al., 1953). First attempts to isolate mutants with a defect in siderophore biosynthesis (being therefore siderophore-auxotrophic) failed, probably because there are other pathways for iron transport. The first mutant to be used in siderophore transport studies (Winkelmann and

Zähner, 1973) was the arg-5, ota, aga mutant of *Neurospora crassa* (Davis et al., 1970), which is defective in ornithine biosynthesis and therefore cannot synthesize coprogen. Later on, Liu and Neilands (1984) reported on a mutant of *Rhodotorula pilimanae* with a defect in rhodotorulic acid biosynthesis and another mutant producing a new hydroxamate derived from N^5-hydroxyornithine. It was not before 1989 that Wang et al. (1989) isolated 11 mutants (from 35,000 colonies) of *Ustilago maydis* by a bioassay using the siderophore-auxotrophic *Salmonella typhimurium* LT-2 and the fact that siderophores can form complexes with aluminum (III).

The molecular biology and genetics of ferrichrome biosynthesis in *U. maydis* will be described elsewhere in this volume.

It is highly desirable to learn something about the presence and function of a gene domain that may function in a similar way to the *fur* box in bacteria (Bagg and Neilands, 1987; Crosa, 1989; Braun and Hantke, 1991). It can be expected that in fungi, as shown in bacteria, iron exerts its regulatory effect on siderophore biosynthesis by binding to a *fur*like protein that inhibits enzyme biosynthesis by blocking the operator region.

Siderophore biosynthesis may be controlled by different strategies like the binding of an effector to (and inactivation of) a transcriptional activator or by acting at a posttranscriptional level by altering the rate of mRNA turnover. To study these regulatory mechanisms, attention should be directed to intramolecular effectors, regulatory molecules, and transcriptional events.

Recently it was shown by Eide and Guarente (1992) that a transcriptional activator gene *FUP1* (ferric utilization proficient)—encoding a hydrophilic 43 kDa protein—enables the iron-limited growth of *Saccharomyces cerevisiae* by regulation of a membrane bound iron reductase. The gene is identical with a transcriptional activator gene (*MSN1*) involved in the regulation of carbon source utilization.

Further work should be directed toward the purification of fungal siderophore biosynthetic enzymes, in order to obtain sequence data or antibodies that in turn may help to recover the target mRNA. The aim should be to clone the genes using cDNA and expression systems.

On the other hand, the elucidation of the structure of the multienzyme siderophore synthetases may contribute to a better understanding of the mechanism of nonribosomal peptide synthesis.

ACKNOWLEDGMENT

Part of this work was supported by grants of the Deutsche Forschungsgemeinschaft.

REFERENCES

Akers, H.A., and Neilands, J.B. (1978). Biosynthesis of rhodotorulic acid and other hydroxamate type siderophores, *Biological Oxidation of Nitrogen* (J.W. Gorrod, ed.), Elsevier, Amsterdam, p. 429.

Akers, H.A., Llinás, M., and Neilands, J.B. (1972). Protonated amino acid precursor studies on rhodotorulic acid biosynthesis in deuterium oxide media, *Biochemistry, 11:* 2283.

Anke, H., and Diekmann, H. (1971). Metabolic products of microorganisms, 93, Biosynthesis of sideramines in fungi, Mevalonate as a precursor of *cis-* and *trans-*5-hydroxy-3-methyl-2-pentenoic acids, *FEBS Lett., 17:* 115.

Anke, T., and Diekmann, H. (1972). Metabolic products of microorganisms, 112, Biosynthesis of sideramines in fungi, Rhodotorulic acid synthetase from extracts of *Rhodotorula glutinis, FEBS Lett., 27:* 259.

Anke, H., and Diekmann, H. (1974a). Stoffwechselprodukte von Mikroorganismen, 125, Biosynthese von Sideraminen in Pilzen. *cis*-Δ^2-Anhydromevalonsäurehydratase und *cis*-Δ^2-Anhydromevalonsäure:CoA-ligase, *Arch. Mikrobiol., 95:* 213.

Anke, T., and Diekmann, H. (1974b). Stoffwechselprodukte von Mikroorganismen, 126, Einbau von δ-*N*-Hydroxy-L-ornithin und δ-*N*-Acyl-δ-*N*-hydroxy-L-ornithinen in Sideramine von Pilzen, *Arch. Mikrobiol., 95:* 227.

Anke, H., Anke, T., and Diekmann, H. (1973). Metabolic products of microorganisms, 127, Biosynthesis of sideramines in fungi, Fusigen synthetase from extracts of *Fusarium cubense, FEBS Lett., 36:* 323.

Anke, H., Kinn, J., Bergquist, K.-E., and Sterner, O. (1991). Production of siderophores by strains of the genus *Trichoderma,* Isolation and characterization of the new lipophilic coprogen derivative, palmitoylcoprogen, *Biol. Metals, 4:* 176.

Atkin, C. L., and Neilands, J. B. (1968). Rhodotorulic acid, a diketopiperazine dihydroxamic acid with growth-factor activity, 1, Isolation and characterization, *Biochemistry 7:* 3734.

Bagg, A., and Neilands, J. B. (1987). Molecular mechanism of regulation of siderophore-mediated iron assimilation, *Microbiol. Rev., 51:* 509.

Braun, V., and Hantke, K. (1991). Genetics of bacterial iron transport, *Handbook of Microbial Iron Chelates* (G. Winkelmann, ed.), CRC Press, Boca Raton, FL, p. 107.

Bryce, G. F., and Brot, N. (1973). Studies on the enzymatic synthesis of the cyclic trimer of 2,3-dihydroxy-*N*-benzoyl-L-serine in *Escherichia coli, Biochemistry, 11:* 1708.

Bryce, G. F., Weller, R., and Brot, N. (1971). Studies on the enzymatic synthesis of 2,3-dihydroxy-*N*-benzoyl-L-serine in *Escherichia coli, Biochim. Biophys. Res. Comm., 42:* 871.

Coy, M., Paw, B. H., Bindereif, A., and Neilands, J. B. (1986). Isolation and properties of N^6-hydroxylysine:acetyl coenzyme A N^6-transacetylase from *E. coli* pABN11, *Biochemistry, 25:* 2485.

Crosa, J. H. (1989). Genetics and molecular biology of siderophore-mediated iron transport in bacteria, *Microbiol. Rev., 53:* 517.

Davis, R. H. (1986). Compartmental and regulatory mechanisms in the arginine pathways of *Neurospora crassa* and *Saccharomyces cerevisiae, Microbiol. Rev., 50:* 280.

Davis, R. H., Lowless, M. B., and Port, L. A. (1970). Arginineless *Neurospora:* Genetics, physiology and polyamine synthesis, *J. Bacteriol., 102:* 299.

DeLorenzo, V., and Neilands, J. B. (1986). Characterization of iucA and iucC genes of the aerobactin system of plasmid ColV-K30 in *Escherichia coli, J. Bacteriol., 167:* 350.

DeLorenzo, V., Bindereif, A., Paw, B. H., and Neilands, J. B. (1986). Aerobactin biosynthesis and transport genes of plasmid ColV-K30 in *Escherichia coli* K-12, *J. Bacteriol. 165:* 570.

Diekmann, H. (1967). Stoffwechselprodukte von Mikroorganismen, 56, Fusigen—Ein neues Sideramin aus Pilzen, *Arch. Mikrobiol., 58:* 1.

Diekmann, H. (1968). Stoffwechselprodukte von Mikroorganismen, 68, Die Isolierung und Darstellung von *trans*-5-Hydroxy-3-methylpenten-(2)-säure, *Arch. Mikrobiol. 62:* 322.

Diekmann, H. (1970). Stoffwechselprodukte von Mikroorganismen, 83, Zur Biosynthese von Sideraminen in Pilzen, Einbau von Ornithin und Fusigen in Ferrirhodin, *Arch. Mikrobiol., 74:* 301.

Diekmann, H., and Krezdorn, E. (1975). Stoffwechselprodukte von Mikroorganismen, 150, Ferricrocin, Triacetylfusigen und andere Sideramine aus Pilzen der Gattung *Aspergillus,* Gruppe *Fumigatus, Arch. Microbiol., 106:* 191.

Diekmann, H., and Plattner, H. J. (1991). Microbial hydroxylases, *N-Oxidation of Drugs* (P. Hlavica and L. A. Damani, eds.), Chapman and Hall, London, p. 133.

Diekmann, H., and Zähner, H. (1967). Konstitution von Fusigen und dessen Abbau zu Δ^2-Anhydromevalonsäurelacton, *Eur. J. Biochem., 3:* 213.

Eide, D., and Guarente, L. (1992). Increased dosage of a transcriptional activator gene enhances iron-limited growth of *Saccharomyces cerevisiae, J. Gen. Microbiol., 138:* 347.

Emery, T. (1966). Initial steps in the biosynthesis of ferrichrome, Incorporation of δ-*N*-hydroxyornithine and δ-*N*-acetyl-δ-*N*-hydroxy-ornithine, *Biochemistry, 5:* 3694.

Emery, T. (1974). Biosynthesis and mechanism of action of hydroxamate-type siderochromes, *Microbial Iron Metabolism* (J. B. Neilands, ed.), Academic Press, New York, p. 107.

Engelbrecht, F., and Braun, V. (1986). Inhibition of microbial growth by interferences with siderophore biosynthesis. Oxidation of primary amino groups in aerobactin synthesis by *Escherichia coli, FEMS Microbiol. Lett., 33:* 223.

Fetz, E., Böhner, B., and Tamm, C. (1965). Die Konstitution von Verrucarin J, *Helv. Chim. Acta, 48:* 1669.

Garibaldi, J. A., and Neilands, J. B. (1956). Formation of iron binding compounds by microorganisms, *Nature* (London), *177:* 526.

Gross, R., Engelbrecht, F., and Braun, V. (1984). Genetic and biochemical characterization of the aerobactin synthesis operon on pColV, *Mol. Gen. Genet., 196:* 74.

Gross, R., Engelbrecht, F., and Braun, V. (1985). Identification of the genes and their polypeptide products responsible for aerobactin synthesis by pColV plasmids, *Mol. Gen. Genet., 201:* 204.

Haselwandter, K., Dobernigg, B., Beck, W., Jung, G., Cansier, A., and Winkelmann, G. (1992). Isolation and identification of hydroxamate siderophores of ericoid mycorrhizal fungi, *BioMetals, 5:* 51.

Herrero, M., DeLorenzo V., and Neilands, J. B. (1988). Nucleotide sequence of the iucD gene of the pColV-K30 aerobactin operon and topology of its product studied with phoA and lacZ gene fusions, *J. Bacteriol., 170:* 56.

Hesseltine, C. W., Whitehill, A. R., Pidacks, C., Hagen, M. T., Bohonos, N., Hutchings, B. L., and Williams, J. H. (1953). Coprogen, a new growth factor present in dung, required by *Pilobolus* species, *Mycologia, 45:* 7.

Heydel, P., Plattner, H., and Diekmann, H. (1987). Lysine-*N*-hydroxylase and *N*-acetyltransferase of the aerobactin system of pColV plasmids in *Escherichia coli, FEMS Microbiol. Lett., 40:* 305.

Horowitz, N. H., Charlang, G., Horn, G., and Williams, N. P. (1976). Isolation and identification of the conidial growth factor of *Neurospora crassa, J. Bacteriol., 127:* 135.

Hummel, W., and Diekmann, H. (1981). Preliminary characterization of ferrichrome synthetase from *Aspergillus quadricinctus, Biochim. Biophys. Acta, 657:* 313.

Jackson, G. E. D., Parniak, M. A., Murray, G. J., and Viswanatha, T. (1984). Stimulation by glutamine on the formation of N^6-hydroxylysine in a cell free extract from *Aerobacter aerogenes* 62-1, *J. Cell. Biochem., 24:* 395.

Keller-Schierlein, W. (1963). Stoffwechselprodukte von Mikroorganismen, 45, Über die Konstitution von Ferrirubin, Ferrirhodin und Ferrichrom A, *Helv. Chim. Acta, 46:* 1920.

Keller-Schierlein, W., Prelog, V., and Zähner, H. (1964). Siderochrome (Natürliche Eisen(III)-trihydroxamat-Komplexe), *Fortschr. Chem. Org. Naturst., 22:* 279.

Kleinkauf, H., and von Döhren, H. (1987). Biosynthesis of peptide antibiotics, *Ann. Rev. Microbiol., 41:* 259.

Kusters, J. (1984). Untersuchung zur Biosynthese von Siderophoren in *Aspergillus quadricinctus* und *Klebsiella pneumoniae.* Vergleich der Acetyl-Coenzym A:Alkylhydroxylamin *N*-acetyltransferase, Dissertation, Universität Hannover.

Kusters, J., and Diekmann, H. (1984). Assay for acetyl coenzyme A:alkylhydroxylamine *N*-acetyltransferase in *Klebsiella pneumoniae, FEMS Microbiol. Lett., 23:* 309.

Liu, A., and Neilands, J. B. (1984). Mutational analysis of rhodotorulic acid synthesis in *Rhodotorula pilimanae, Structure and Bonding, 58:* 97.

MacDonald, J. C. (1965). Biosynthesis of pulcherriminic acid, *Biochem, J., 96:* 533.

Massey, V., and Hemmerich, P. (1975). Flavin and pteridine monooxygenases, *The Enzymes* (P. D. Boyer, ed.), Academic Press, New York, *12:* 191.

Matzanke, B. F., Bill, E., Trautwein, A. X., and Winkelmann, G. (1987). Role of siderophore in iron storage in spores of *Neurospora crassa* and *Aspergillus ochraceus*, *J. Bacteriol.*, *169:* 5873.

Miller, E. C., and Miller, J. A. (1966). Mechanisms of chemical carcinogenesis: Nature of proximate carcinogens and interactions with macromolecules, *Pharmacol. Rev.*, *18:* 805.

Müller, H.-G. (1976). Untersuchung der Biosynthese des cyclischen Esters Fusigen mit Protoplasten und zellfreien Extrakten von *Fusarium cubense* und Versuche zur Synthese von Hexapeptidsideraminen in vivo und in vitro, Dissertation, Universität Tübingen.

Müller, H.-G., and Diekmann, H. (1977). Einbau von Glycin und Serin in Sideramine vom Ferrichrom-Typ bei Pilzen der Gattung *Aspergillus* in vivo und in vitro, *Arch. Mikrobiol.*, *113:* 243.

Murray, G. J., Clark, G. E. D., Parniak, M. A., and Viswanatha, T. (1977). Effect of metabolites on ε-*N*-hydroxylysine formation in cellfree extracts of *Aerobacter aerogenes* 62-1, *Can. J. Biochem.*, *55:* 625.

Neilands, J. B., Konopka, K., Schwyn, B., Coy, M., Francis, R. T., Paw, B. H., and Bagg, H. (1987). Comparative biochemistry of microbial iron assimilation, *Iron Transport in Microbes, Plants and Animals* (G. Winkelmann, D. van der Helm, and J. B. Neilands, eds.), VCH, Weinheim, p. 3.

Ong, D. E., and Emery, T. F. (1972). Ferrichrome biosynthesis: Enzyme catalyzed formation of the hydroxamic acid group, *Arch. Biochem. Biophys.*, *148:* 77.

Parniak, M. A., Jackson, G. E. D., Murray, G. J., and Viswanatha, T. (1979). Studies on the formation of N^6-hydroxylysine in cell-free extracts of *Aerobacter aerogenes* 62-1, *Biochim. Biophys. Acta*, *569:* 99.

Persmark, M., Expert, D., and Neilands, J. B. (1992). Ferric iron uptake in *Erwinia chrysanthemi* mediated by chrysobactin and catechol-type compounds, *J. Bacteriol.*, *174:* 4783.

Plattner, H. J. (1986). Isolation and properties of acetyl coenzyme A: N^5-hydroxyornithine-*N*-acetyltransferase from *Rhodotorula glutinis*, Abstracts of the VAAM-meeting in Münster (Germany), March 1986.

Plattner, H. J. (1987). Isolation of acetyl coenzyme A: alkylhydroxylamine *N*-acetyltransferase from *Escherichia coli* GR128, Abstracts of the VAAM-meeting in Konstanz (Germany), March 1987.

Plattner, H., and Diekmann, H. (1973). Stoffwechselprodukte von Mikroorganismen, 124, Einbau von L-Leucin in cyclo-L-Leucyl-L-leucyl, dem Intermediärprodukt der Biosynthese von Pulcherriminsäure, in zellfreien Extrakten von *Candida pulcherrima*, *Arch. Mikrobiol.*, *93:* 363.

Plattner, H. J., Pfefferle, P., Romaguera, A., Waschütza, S., and Diekmann, H. (1989). Isolation and some properties of lysine N^5-hydroxylase from *Escherichia coli* strain EN222, *Biol. Metals*, *2:* 1.

Siegmund, D., Plattner, H. J., and Diekmann, H. (1990). Determination of the molecular weight of the ferrichrome synthetase and its subunits from *Aspergillus quadricinctus* by fast protein liquid chromatography (FPLC), *Zbl. Bakt.* *272*(2): 358.

Siegmund, K.-D., Plattner, H.J., and Diekmann, H. (1991). Purification of ferrichrome synthetase from *Aspergillus quadricinctus* and characterization as a phosphopantetheine containing multienzyme complex, *Biochim. Biophys. Acta. 1076:* 123.

Silver, S., and Walderhaug, M. (1992). Gene regulation of plasmid- and chromosomal-determined inorganic ion transport in bacteria, *Microbiol. Rev., 56:* 195.

Sweet, S. P., and Douglas, L. J. (1991). Effect of iron deprivation on surface composition and virulence determinants of *Candida albicans, J. Gen. Microbiol., 137:* 859.

Vater, J. (1990). Gramicidin S synthetase, *Biochemistry of Peptide Antibiotics* (H. Kleinkauf and H. von Döhren, eds.), Walter de Gruyter, Berlin, p. 33.

Wang, J., Budde, A. D., and Leong, S. A. (1989). Analysis of ferrichrome biosynthesis in the phytopathogenic fungus *Ustilago maydis:* Cloning of an ornithine-N^5-oxygenase gene, *J. Bacteriol., 171:* 2811.

Waters, V. L., and Crosa, J. H. (1991). Colicin V virulence plasmids, *Microbiol. Rev., 55:* 437.

Williams, P. H. (1979). Novel iron uptake system specified by ColV plasmids: An important component in the virulence of invasive strains of *Escherichia coli, Infect. Immun., 26:* 925.

Winkelmann, G. (1986). Iron complex products (Siderophores), *Biotechnology* (H.-J. Rehm and G. Reed, eds.), VCH, Weinheim, *4,* p. 215.

Winkelmann, G. (1991). *Handbook of Microbial Iron Chelates,* CRC Press, Boca Raton, FL.

Winkelmann, G., and Zähner, H. (1973). Stoffwechselprodukte von Mikroorganismen, 115, Eisenaufnahme bei *Neurospora crassa,* I, Zur Spezifität des Eisentransportes, *Arch. Mikrobiol., 88:* 49.

Ziegler, D. M. (1988). Flavin-containing monooxygenases. Catalytic mechanism and substrate specificities, *Drug Metab. Rev., 19:* 1.

4

Molecular Biology of Iron Transport in Fungi

Baigen Mei
University of Wisconsin, Madison, Wisconsin

Sally Ann Leong
United States Department of Agriculture and University of Wisconsin, Madison, Wisconsin

I. INTRODUCTION

Iron, a trace element required for cellular functions such as DNA synthesis and respiration, is essential for virtually all organisms with few exceptions (Neilands et al., 1987; Archibald, 1983). Although abundant in nature, it occurs in the extremely insoluble form of ferric hydroxide complexes in the presence of oxygen and water (Williams, 1990), making iron availability a biological problem. Therefore, under such iron-limiting conditions microorganisms have developed different solutions to the problem of assimilation of ferric iron from the environment. In most aerobic and facultative anaerobic microbial species, the siderophore-mediated high affinity iron uptake systems seem to be of great significance for survival. The major role of siderophores is to bind iron(III) very strongly and participate in the transport of iron into the microbial cells under conditions of cellular iron deprivation (Neilands, 1981a). In the yeast *Saccharomyces cerevisiae*, iron uptake under aerobic condition depends on a transmembrane electron transport system that uses an externally directed ferric reductase to reduce ferric iron external to the cell (Lesuisse et al., 1987; Dancis et al., 1992) (see Chap. 5). In addition, *Saccharomyces cerevisiae* can use siderophores secreted by other microbes as iron sources (Lesuisse and Labbe, 1989), but this yeast does not produce a siderophore (Neilands et al., 1987).

The first siderophore discovered was ferrichrome, which is produced by the basidiomycete *Ustilago sphaerogena* (Neilands, 1952). Since then, a growing number of siderophores with diverse chemical structures have been isolated from numerous prokaryotes and eukaryotes (Winkelmann, 1991). Fungal siderophores have been detected from different fungal classes including *Basidiomycetes, Ascomycetes,* and *Deuteromycetes.* A complete list is provided by Winkelmann (1991, 1992).

An understanding of the mechanism for siderophore-mediated iron transport and its regulation at the molecular level would be of obvious importance to research on iron metabolism in the microbial world. Most of the progress to date has been made with prokaryotes. The first cloning experiment of siderophore biosynthetic genes was accomplished for the siderophore enterobactin from *E. coli* (Laird et al., 1980a, 1980b). Results from cloning of different bacterial siderophore genes indicated that some of the siderophore genes are chromosomally encoded, such as the enterobactin gene cluster in *Salmonella typhimurium* and *E. coli* (Neilands, 1981b) as well as the aerobactin iron transport system of *E. coli* K1 (Valvano and Crosa, 1988), while others are plasmid-borne, e.g., the aerobactin operon of pColV-K30 in clinical isolates of *E. coli* (Williams, 1979) and the anguibactin genes located on pJMI in the fish pathogen *Vibrio anguillarum* (Tolmasky et al., 1988). Although fungal siderophores have been the subject of many investigations in past years, detailed studies of their biosynthesis and regulation at the molecular level have just begun. This situation can be attributed to the difficulty in isolation of relevant fungal mutants and the lack of suitable cloning systems for these organisms. *Ustilago maydis,* one of the best understood plant pathogenic fungi at the molecular level (Saville and Leong, 1992), offers an attractive system in which to study the molecular genetics of siderophore biosynthesis and transport. The organism grows as a haploid yeast on defined laboratory media, mutants are readily generated by UV or chemical mutagenesis, stable diploids can be constructed for mitotic recombination and complementation analysis, and the fungus is amenable to Mendelian genetic analysis (Holliday, 1974). *U. maydis* produces ferrichrome and ferrichrome A (Budde and Leong, 1989), the same two hydroxamate siderophores isolated from *Ustilago sphaerogera* (Neilands, 1952; Garibaldi and Neilands, 1955). Moreover, except for rhizoferrin, a complexone-type siderophore of the Mucorales (Drechsel et al., 1991), most fungal siderophores are hydroxamate-type compounds (Winkelmann, 1992). Therefore, the information from studies on siderophore genes in *U. maydis* will be of particular interest, since it may serve as a model for the corresponding processes in other fungi.

In this chapter, recent progress on the molecular biology of siderophore biosynthesis and its regulation in *U. maydis* is presented. Relevant information from the study of other filamentous fungi is also included.

II. BIOCHEMICAL AND GENETIC ANALYSIS OF SIDEROPHORE BIOSYNTHESIS

Hydroxamate siderophores produced by most fungal genera can be divided into three families: (1) the ferrichromes; (2) the coprogens; and (3) the fusarinines (Jalal et al., 1991). Ferrichrome-type siderophores have a cyclohexapeptide ring containing three residues of δ-*N*-acyl-δ-*N*-hydroxyornithine and a tripeptide of neutral amino acids (Fig. 1). The structures of fungal siderophores have been compiled recently (Jalal et al., 1991; Winkelmann and Huschka, 1987). Despite the structural diversity of hydroxamate siderophores, their biosynthesis almost certainly follows a similar pattern, since δ-*N*-hydroxyornithine is a basic unit of these compounds (Hider, 1984).

A. Biochemical Studies

Significant contributions to elucidate the mechanism for siderophore biosynthesis have been made with the use of appropriately labeled precursors. The synthesis of ferrichrome and ferrichrome A was examined in the culture of *U. sphaerogena* using incorporation of ^{14}C-labeled precursors (Emery, 1966).

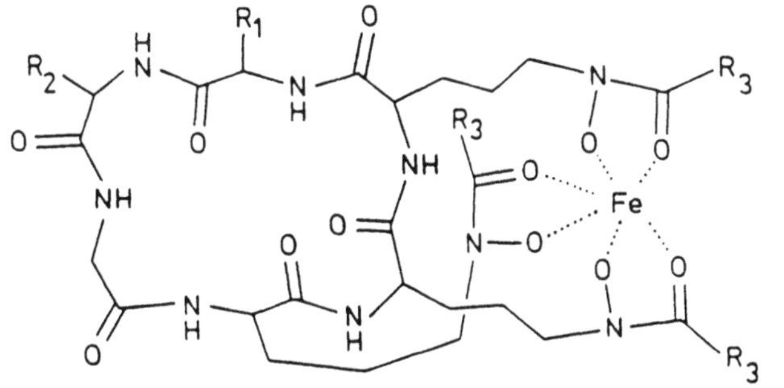

Figure 1 **Ferrichrome and ferrichrome A. Both siderophores have a cyclohexapeptide ring containing three residues of δ-*N*-acyl-*N*-hydroxy-ornithine and a tripeptide of neutral amino acids consisting of three glycine residues in ferrichrome ($R_1 + R_2 = H$) or one glycine and two serine residues in ferrichrome A ($R_1 + R_2 = CH_2\text{-}OH$). $R_3 = CH_3$ in ferrichrome and R_3 = trans-β-methylglutaconic acid in ferrichrome A.**

Based on these results, a biosynthetic pathway for ferrichrome and ferrichrome A was proposed as shown in Fig. 2 (Emery, 1971a). The biosynthesis of ferrichrome involves (1) hydroxylation of ornithine; (2) acetylation of δ-*N*-hydroxyornithine to form δ-*N*-acetyl-δ-*N*-hydroxyornithine; and (3) condensation of the latter compounds with glycine to form ferrichrome. The first step in ferrichrome and ferrichrome A biosynthesis is likely shared, since both siderophores contain δ-*N*-hydroxyornithine. The biosynthesis of rhodotorulic acid (Fig. 3), a fungal siderophore produced by *Rhodotorula pilimanae* (Liu and Neilands, 1984), follows the same pathway, indicating that hydroxylation of ornithine may be the first committed step for biosynthesis of hydroxamate-type siderophores in fungi. Experiments with ^{18}oxygen showed that the hydroxylamino oxygen arises from molecular oxy-

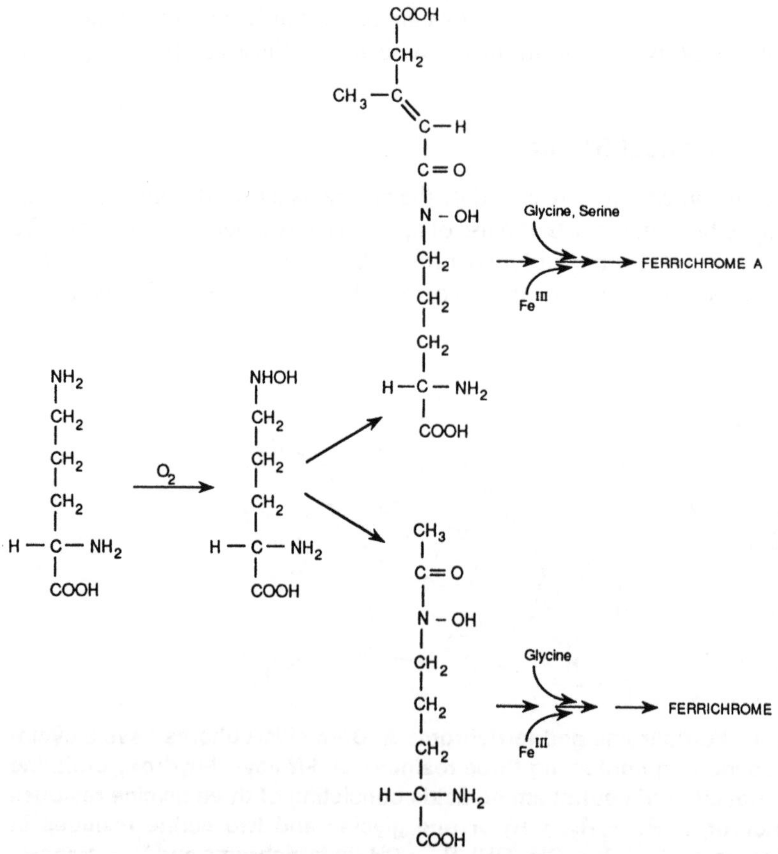

Figure 2 Biosynthetic pathways for ferrichrome and ferrichrome A.

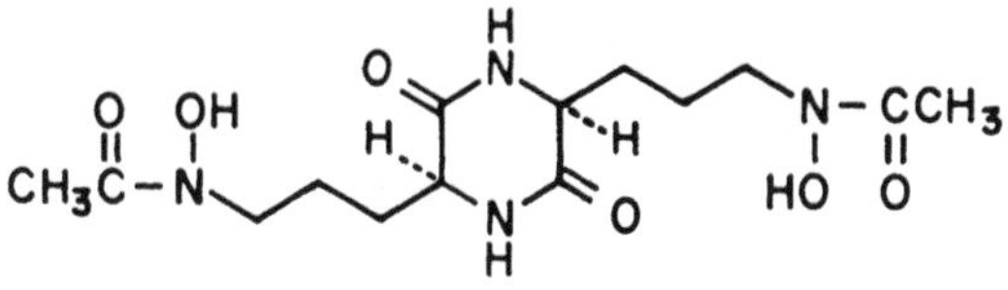

Figure 3 Rhodotorulic acid.

gen (Akers and Neilands, 1978). The next step, acylation of δ-*N*-hydroxyornithine, is variable with the δ-*N*-acyl groups being derived from various carboxylic acids. Ferrichrome has an acetyl group, while ferrichrome A contains trans-β-methylglutaconic acid as the δ-*N*-acyl group (Hider, 1984). The final step in the synthesis of siderophores might involve amino acid activation, peptide bond formation, and cyclization (Hummel and Diekmann, 1981). The enzymes catalyzing these reactions are L-ornithine-N^5-oxygenase, δ-*N*-hydroxyornithine:acetylCoA δ-*N*-transacetylase, and ferrichrome synthetase, respectively. L-Ornithine-N^5-oxygenase activity was detected in crude extracts from *U. maydis* cells grown in low iron medium (Leong et al., 1990). The *sid*1 gene encoding this enzyme has been cloned and characterized from this fungus (Wang et al., 1989; Mei et al., 1993) (see below). An enzyme that catalyzes the acetylation of δ-*N*-hydroxyornithine has been partially purified from *U. sphaerogena* (Ong and Emery, 1972) and from *R. pilimanae* (Neilands et al., 1987). Ferrichrome synthetase isolated from *Aspergillus quadricinctus* has been characterized as a phosphopantetheine-containing multienzyme complex (Siegmund et al., 1991). The enzymology of siderophore biosynthesis in fungi has been described in detail in Chap. 3. Here we will only present information related to L-ornithine-N^5-oxygenase of *U. maydis* studied in this laboratory.

To facilitate the study of *U. maydis* L-ornithine-N^5-oxygenase, we have established an enzyme assay for this enzyme based on the method developed for the *E. coli* lysine-N^6-hydroxylase (Plattner et al., 1989), which catalyzes the hydroxylation of lysine, the first step in biosynthesis of aerobactin (Gross et al., 1985; Lorenzo et al., 1986). The reaction mixtures were prepared with 50 mM sodium phosphate buffer (pH 7.0), L-ornithine as a substrate, cofactors (NADPH and FAD), and an appropriate amount of the crude extract from *U. maydis*. The products of the reaction were determined using a modified Csaky test (Tomlinson et al., 1971). The activity of L-ornithine-N^5-oxygenase was present in the crude extract of *U. maydis* cells grown under iron-limiting medium, but not in crude extract of cells grown in medium amended with iron. Enzyme activity is dependent on both NADPH and FAD. Lysine could not be substituted for ornithine as substrate. Boiling the crude extract for 1 min completely destroyed its activity. Preliminary efforts to purify this enzyme have resulted in a 200-fold increase in specific activity after two chromatographic steps (Table 1). A band of 65 kd was enriched in an SDS-PAGE analysis of this preparation. The molecular weight of this protein is consistent with the predicted size of L-ornithine-N^5-oxygenase determined by cDNA sequence analysis of the *sid*1 gene, the structural gene for this enzyme (see below). Many attempts to purify further the enzyme have been hampered by low activity and instability of the enzyme. However, the availability of the *sid*1 gene now provides an opportunity to overexpress and purify the enzyme to homogeneity.

B. Genetic Analysis

A classical genetic approach to study siderophore biosynthesis and transport has involved the isolation of mutants by exposing cells to NTG (*N*-methyl-*N'*-nitro-*N*-nitroso-guanidine) or ultraviolet light. The characterization of metabolic products of mutants in the pathway should establish the number and order of genes involved in biosynthesis of siderophores.

In initial studies, Davis *et al.* (1970) isolated a triple mutant of *Neurospora crassa* (*arg*-5, *ota, aga*), which is blocked in the biosynthesis of ornithine. This mutant does not produce siderophores when grown under iron deprivation, but the addition of ornithine to low iron medium restored the ability of the mutant to produce siderophore, indicating that ornithine or a product thereof is a constituent of the fungal siderophore (Winkelmann and Huschka, 1987). A mutational analysis of rhodotorulic acid biosynthesis in *R. pilimanae* has been reported (Liu and Neilands, 1984). In this investigation, 12 mutants were found from a total of 5000 colonies screened, and one mutant defective in biosynthesis of rhodotorulic acid was identified that produced an aberrant hydroxamate. This compound was characterized by paper electrophoresis and proton NMR. The δ-*N*-hydroxy-ornithine residue was detected in this hydroxamate, suggesting that the mutant is probably blocked in a late step of the biosynthetic pathway, possibly at the cyclization step (Liu and Neilands, 1984).

Table 1 Partial Purification of L-Ornithine-N^5-Oxygenase from *U. maydis*

Fraction	Protein (mg)	Activity		Yield	Purification (fold)
		Specific	Total		
Crude extract	390	0.11	41.7	100	—
DEAE-Sepharose	6.3	4.55	28.6	69	42
Sepharose 6B	0.7	20.9	14.6	35	195

To begin an analysis of siderophore biosynthesis in *U. maydis,* several kinds of mutants were isolated by NTG mutagenesis (Wang et al., 1989). Class I mutants no longer produce ferrichrome and retain the ability to produce ferrichrome A. Class II mutants are blocked in the production of both ferrichrome and ferrichrome A. Class III mutants, a kind of deregulated mutant, do not respond normally to the iron concentration of growth medium and produce siderophores constitutively. For class I mutants, addition of δ-*N*-hydroxyornithine to growth medium does not restore the biosynthesis of ferrichrome. However, all class II mutants are capable of converting δ-*N*-hydroxyornithine, but not ornithine, to both siderophores when grown under iron deprivation conditions. Genetic analysis using crosses between these mutants and a sexually compatible wild-type strain has suggested that in each mutant the lesion is caused by a recessive defect at a single locus. Together with the results from biochemical complementation experiments, we conclude that class II mutants are defective in the first step of ferrichrome and ferrichrome A biosynthesis, the hydroxylation of L-ornithine, while class I mutants are blocked in later step(s) of the biosynthesis of ferrichrome, after the hydroxylation of L-ornithine (Fig. 2). Finally, class III mutants carry a genetic lesion at a locus that regulates the biosynthesis of siderophores (Wang et al., 1989).

III. CLONING AND CHARACTERIZATION OF GENES INVOLVED IN SIDEROPHORE BIOSYNTHESIS AND REGULATION

A. Application of Molecular Genetic Techniques

The development of molecular genetic tools has provided the groundwork for rapid progress in our understanding of iron transport in microorganisms, especially in prokaryotic systems. Although this methodology has been equally applied to selected eukaryotic organisms such as the yeast *Saccharomyces cerevisiae,* it has been necessary to develop specific methods that enable us to clone and transfer genes involved in iron transport in other fungi.

Stable, integrative transformation systems (Wang et al., 1988; Keon et al., 1991) have been developed for *U. maydis.* In this laboratory, a selectable marker for transformation of *U. maydis* was constructed by transcriptional fusion of the coding region of the gene for *E. coli* hygromycin B phosphotransferase (Gritz and Davies, 1983) with a *U. maydis hsp* 70 gene promoter (Wang et al., 1988). The transforming DNA is integrated into chromosomal DNA and recombines homologously at the *hsp* 70 locus in about 30% of *U. maydis* transformants. Linearization of the vector increases the transformation frequency yielding about 1000 transformants per μg linearized vector. The *U. maydis* transformation vector pHL1 can also be stably transformed into *Ustilago hordei* and *Ustilago nigra* (Holden et al., 1988). Tsukuda et al. (1988) have inserted a *U. maydis* ARS (autonomously replicating sequence) into pHL1 to generate the transformation vector pCM54, which self-replicates in fungal cells. A self-replicating cosmid vector pJW42 has been constructed by inserting this ARS into a cosmid derivative of pHL1 (Leong et al., 1990). A genomic DNA library for *U. maydis,* constructed in pJW42, has been successfully employed to isolate the *a* mating-type locus (Froeliger and Leong, 1991). An expression vector for *U. maydis* has been constructed by insertion of a GAPDH (glycer*a*ldehyde-3-*p*hosphate *deh*ydrogenase) promoter from *U. maydis* (Smith and Leong, 1990) into an autonomously replicating vector (Kinal et al., 1991). Finally, high-frequency cotransformation of two DNA molecules in *U. maydis* has been successfully employed to locate the *sid*1 gene and *a* mating-type genes on cosmid clones (Froeliger and Leong, 1991; Wang et al., 1989).

Homologous integration of cloned DNAs provides the opportunity for developing powerful genetic techniques such as gene replacement and gene disruption. In this manner, a null mutation can be created and used to confirm phenotypically the cloning of a gene by allelism tests, to test the essentiality of a gene for cell viability, or to eliminate a gene function. Protocols for single step gene replacement and gene disruption in *U. maydis* have been developed (Fotheringham and Holloman, 1989; Kronstad et al., 1989). With the use of these techniques, roles of several *U. maydis* genes including *pyr*6 (Kronstad et al., 1989), *LEU1* gene (Fotheringham and Holloman, 1989), an *hsp*70 gene (Holden et al., 1989), two *b* mating-type alleles (Gillissen et al., 1992; Kronstad and Leong, 1989), and two *a* mating-type alleles (Bolker et al., 1992; Froeliger and Leong, 1991) have been characterized.

The availability of mutants and the development of a gene-transfer system in *U. maydis* has made it possible to clone genes involved in siderophore biosynthesis and regulation. Two genes, *sid*1, a gene coding for L-ornithine-N^5-oxygenase, and *urbs*1, a gene involved in iron-regulation of siderophore biosynthesis, have been cloned in this manner.

B. *sid*1, a Gene Coding for L-Ornithine-N^5-Oxygenase

1. Cloning and Characterization

To clone *sid*1, a genomic DNA library was constructed from wild-type *U. maydis* and used in complementation analysis with *sid*1 mutants (class II mutants) (Wang et al., 1989). Two positive cosmid clones, pSid1 and pSid2, were independently isolated from transformants by a sib-selection procedure (Vollmer and Yanofsky, 1986) or by *in vitro* packaging (Yelton et al., 1985) of cosmid DNA from complemented *U. maydis* transformants (Wang et al., 1989).

For a clone isolated by complementation of a mutation, it is possible to clone suppressors of various sorts instead of the gene being sought. Demonstration that the correct gene has been cloned must be made. The confirmation of the identity of *sid*1 came from several experiments as follows. First, a 6.5 kb *Ssp*I-*Hind*III fragment (Fig. 4) from pSid1 can complement *sid*1 mutant S023 when subcloned into pCM54, an autonomously replicating vector. Thus complementation is assessed *in trans*. Although the 8.1 kb *Hind*III and 3.2 kb *Ssp*I-*Eco*RI fragments (Fig. 4) were able to restore siderophore production in the *sid*1 mutant S023 when integrated into the genome, both fragments gave negative results when tested for complementation *in trans*. These results indicated that the entire *sid*1 gene is

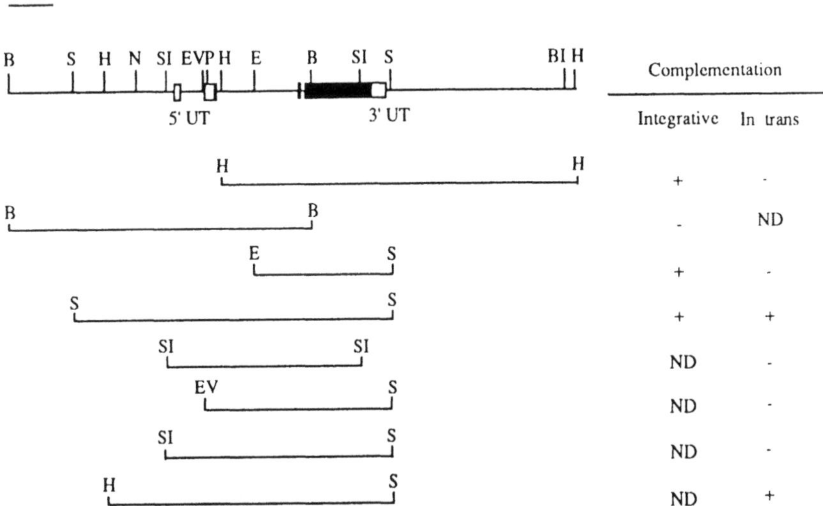

Figure 4 **Structure and restriction map of *sid*1. Only sites used for subcloning are indicated (B, *Bgl*II; BI, *Bgl*I; E, *Eco*RI; EV, *Eco*RV; H, *Hind*III; N, *Nco*I; P, *Pvu*II; S, *Ssp*I; SI, *Sst*I). Exons are boxed. Coding regions are indicated as black boxes, whereas 5′ and 3′ untranslated regions (UT) found in the cDNA clone are indicated by open boxes. The fragments shown were subcloned into pHL1 and pCM54, and tested for complementation by transformation of *U. maydis* S023.**

located on the 6.5 kb *Ssp*I-*Hind*III fragment and not on these other fragments (Fig. 4). Second, gene disruption was used to generate a *sid*1 mutant by inserting a hygromycin B resistance gene (Wang et al., 1988) into a unique *Bgl*II site of 6.5 kb *Ssp*I-*Hind*III fragment (Fig. 4). Such a mutation confers the expected phenotype, i.e., abolishment of siderophore production and L-ornithine-N^5-oxygenase activity (Mei et al., 1993). Third, an allelism test demonstrated that the allele mutated by NTG is the same as that generated by gene disruption (Mei et al., 1993). Fourth, a 2.3 kb full-length cDNA of *sid*1 (Mei et al., 1993) was constructed from a cDNA library and rapid amplification of the 5′ end of the mRNA using the RACE technique (Frohman et al., 1988). The predicted *sid*1 protein of 570 amino acids shows significant similarity to L-lysine-N^6-hydroxylase (Fig. 5). L-ornithine-N^5-oxygenase and lysine-N^6-hydroxylase are functional analogs; both convert an aliphatic primary amino acid to a δ-*N*-hydroxyamino compound, the first step of hydroxamate siderophore biosynthesis (Herrero et al., 1988). Such structural similarities most likely provide the basis for the functional equivalence of the proteins and may represent an important functional domain for a putative FAD-binding sequence.

2. *Structure of* sid1

A comparison of the cDNA and genomic DNA sequences showed that *sid*1 is interrupted by three introns (Fig. 4). The sizes of introns are 654, 2183,

```
                               30        40        50        60
sid1      VESPLAASTSSLRAMNMVSSHTTVAKDEIYDLLGIGFGPAHLALSISLRESSEANETNFK
                                        |..|.| ||..|... .|..  |    ...
iucD                                 MKKSVDFIGVGTGPFNLSIA-ALSHQIE----ELD
                                           10        20             30

           70        80        90       100       110       120
sid1      AHFLEKRGHFAWHPALLLPGSQLQVSPLKDLVTLRDPASTYSFYNYLHSHGRLARYINKE
            |.... ||.|||..|.|....|.  |||||.  .|...||| ||| .| .. |....
iucD      CLFFDEHPHFSWHPGMLVPDCHMQTVFLKDLVSAVAPTNPYSFVNYLVKHKKFYRFLTSR
                  40        50        60        70        80        90

          130       140       150       160       170
sid1      QGVPSRREWTSYLAWAARRMNQAVSYGQDVISIEPLALASASP
            . || |...|| |||  ||  . ....| .|.
iucD      LRTVSREEFSDYLRWAAEDMNN-LYFSHTVENIDFDKKRRLFL
                 100        110      120       130
```

Figure 5 Sequence comparison of the predicted *sid*1 protein and *E. coli* lysine-N^6-hydroxylase encoded by *iuc*D. Lines between aligned amino acids designate identities, and dots denote conservative replacements. The putative FAD-binding sequences are underlined.

and 95 bp, a size much larger than that of most filamentous fungal introns identified to date, which are in the range of 48 to 398 bp (Ballance, 1991). The first intron separates the single 5′-noncoding exon from the coding regions. The exon-intron splice junctions are in good agreement with the consensus splice site sequences for introns of filamentous fungi (Table 2). However, a "lariat formation" consensus sequence is found only in intron 2. Interestingly, a second, less abundant cDNA species was found that is located in intron 2 and overlaps exon 2 of *sid*1 (Mei and Leong, unpublished observation). A genbank data search analysis showed that the sequence of the second cDNA contains a putative open reading frame (ORF) with some amino acid sequence similarity to an ORF of the transposable element Tgm5 of soybean (Rhodes and Vodkin, 1988). This finding could account for the unusually large 2.2 kb intron in *sid*1 and may indicate that the gene has evolved to its present state through insertion of a foreign DNA element.

The nucleotide sequence of the 5′-flanking DNA of *sid*1 is shown in Fig. 6. The two transcription initiation sites of the gene were defined by primer extension analysis (Mei et al., 1993). The promoter region does not contain a canonical TATA box, but a CAAT box is present 350 bp from the start sites of transcription. Typically, a CAAT consensus is found approximately 70–90 bp upstream of the transcription initiation site of eukaryotic genes (Ballance, 1991). Two pyrimidine-rich regions, one between positions 8 and 19 and another positions 134 and 145 upstream of the transcription initiation site, are found. Polypyrimidine sequences have also been found in other filamentous fungal promoters and play a role in promoting initiation of transcription (Ballance, 1991). The sequences TATC and TGATAC found at the *sid*1 promoter region could be important in determining the gene's regulation by iron (see below). These sequences are characteristic of genes regulated by transcription factors of GATA family (Evans et al., 1988; Martin et al., 1989). Furthermore, two inverted repeat sequences are

Table 2 Conserved Intronic Sequences of *sid*1

	Intron	5′ site	Internal consensus	3′ site
*sid*1	1	GTATGT	—	CAG
	2	GTAAGT	aACTAAC	TAG
	3	GTACGT	—	CAG
Filamentous fungi Consensus[a]		GTAAGT	tg/aCTAAC	c/tAG

[a]From Ballance (1991).

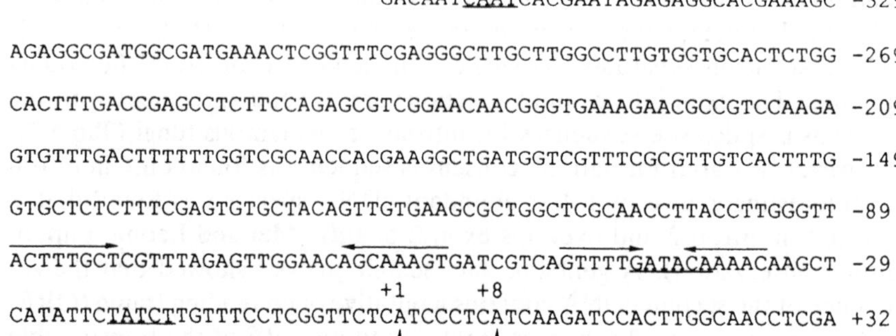

Figure 6 Nucleotide sequence of the 5′-flanking region of *sid1*. Nucleotide residues are numbered from +1 at the first transcription start site, and nucleotides in the 5′ untranscribed region are indicated by negative numbers. The transcription initiation sites are indicated as vertical arrows. The putative CAAT box and GATA sequences are underlined. Horizontal arrows are positioned above the inverted repeat sequences.

found in the region immediately upstream of the transcriptional start sites, and one contains a GATA sequence in its loop region.

3. *Regulation of* sid1 *by Iron*

Siderophore biosynthesis is tightly regulated by the amount of iron in the growth medium (Neilands, 1981a). In *U. maydis,* extracellular ferrichrome and ferrichrome A are not produced in mineral medium containing 10 μM $FeSO_4$ (Budde and Leong, 1989). Activity of the L-ornithine-N^5-oxygenase, like siderophore production, was absent in the growth medium containing similar levels of iron. The expression of the cloned *sid*1 gene was also found to be negatively regulated by iron (Mei et al., 1993).

In most prokaryotic systems, the expression of biosynthetic genes of siderophores is regulated by iron at the transcriptional level (Bindereif and Neilands, 1985; O'Sullivan and O'Gara, 1991; Expert et al., 1992). To address the question whether this model could be applied to the regulation of siderophore biosynthetic genes in *U. maydis,* the levels of *sid*1 mRNA were measured in total RNA isolated from the wild type and a constitutive mutant grown under iron-rich and iron-limiting conditions. Northern hybridization analysis showed a high level of a 2.3 kb transcript in the wild-type cells grown in low iron medium, whereas a less abundant 2.7 kb transcript was found in cells grown in high iron medium (Fig. 7). By contrast, the same high levels of the 2.3 kb mRNA were observed in the total RNA isolated from the *urbs*1-disrupted mutant strain irrespective of the iron concentration of the medium (Mei et al., 1993). These results suggest

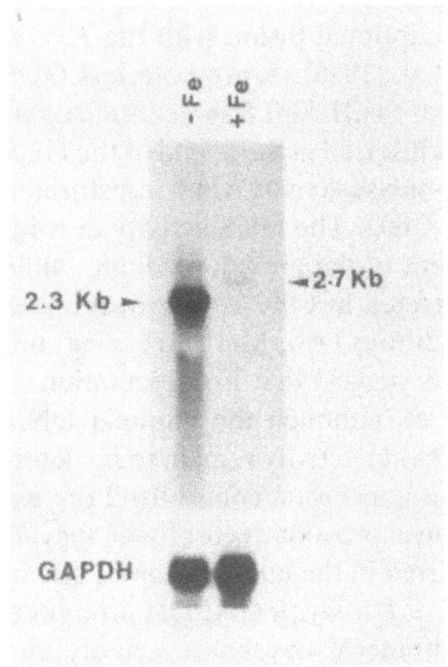

Figure 7 Northern blot analysis of *sid1* mRNA. Total RNA was isolated from a wild-type strain grown in low iron medium without 10 μM $FeSO_4$ (−) and in LI medium with 10 μM $FeSO_4$ (+). Ten μg of total RNA from each sample were electrophoresed, blotted and probed with ^{32}P-labeled 7.2 kb *SspI* genomic DNA (*upper*). The same blots were stripped and reprobed with GAPDH DNA (Smith and Leong, 1990) as an internal standard (*lower*).

that expression of *sid*1 may be both transcriptionally and posttranscriptionally regulated by iron. The regulatory effect of iron could be mediated via a system involving a regulator protein (such as Urbs1) that may repress initiation of transcription under iron-replete conditions. The *urbs*1 encodes a putative regulatory protein that is related to transcription factors of the GATA family (see below). Furthermore, the promoter region of the *sid*1 gene has two GATA boxes, the recognition sequence for this class of transcription regulatory proteins, suggesting that *sid*1 may be one of the Urbs1 target genes in the biosynthesis of siderophores in *U. maydis*. The presence of the longer transcript in wild-type cells grown in iron-rich medium may indicate that iron-mediated regulation of *sid*1 is also at the level of mRNA splicing. Another possibility is that two iron-regulated promoters differentially control expression of *sid*1. Alternatively, *sid*1 mRNA stability may be influenced by iron, as in the case of htR (human transferrin receptor) mRNA, but no sequence motifs similar to the IRE (iron regulatory element) sequence of htR (Mullner et al., 1989) were found in the *sid*1 cDNA.

The 5′ flanking region of *sid*1 was analyzed for its promoter activity and regulation by the construction of a transcriptional fusion with the *E. coli* GUS (β-glucuronidase) gene (Jefferson et al., 1986). A promoterless GUS gene was cloned into pCM54, and the 2.8 kb *Pvu*II-*Ssp*I fragment that overlaps exon 1 and intron 1 of *sid*1 (Fig. 4) was inserted at the 5′ end of the GUS gene. This construct gave rise to β-glucuronidase activity when transformed into wild-type cells and the *urbs*1 mutant C002. The GUS activity in wild-type cells was regulated by the iron content of the growth medium, while similar levels of enzyme activity were detected in C002 cells grown under either iron-deficient or iron-sufficient conditions (An, Mei and Leong, unpublished data). These preliminary results suggest that iron-regulation of *sid*1 may occur at the transcriptional level, although the minimal DNA sequences required for iron-regulated promoter activity remain to be determined. Further evidence that supports this conclusion comes from the expression of *sid*1 cDNA in plasmid pUXV, an expression vector for *U. maydis* (Kinal et al., 1991). The *sid*1 cDNA, inserted in the unique *BamH*I site of pUXV, is under the transcriptional control of *U. maydis* GAPDH promoter. The resulting plasmid can restore L-ornithine-N^5-oxygenase activity and siderophore production when transformed into the *sid*1 mutant S023 (Mei and Leong, unpublished data). The enzyme activity was found in crude extracts of the transformant strain irrespective of the iron concentration in the medium, while ferrichrome and ferrichrome A were only detected in the cells grown in low iron medium. Two tentative conclusions can be drawn from these results. First, the promoter region of *sid*1 contains specific sequence that responds to iron regulation. Second, as in the case of *E. coli* (Bagg and Neilands, 1987), other genes besides *sid*1, involved in biosynthesis of siderophores in *U. maydis,* may be regulated by iron.

4. *Siderophores and Pathogenicity*

In the case of animal and human pathogenic microbes, siderophores are virulence factors for some pathogens (Bullen and Griffiths, 1987). However, the role of siderophores in the infection of plants by microbial pathogens is poorly understood (Neilands and Leong, 1986). Results from studies of a number of bacterial phytopathogens including *Agrobacterium tumefaciens* (Leong and Neilands, 1981), *Pseudomonas syringae* (Cody and Gross, 1987), and *Erwinia carotovora* (Ishimaru and Loper, 1992) have shown that siderophore-mediated iron uptake systems do not have a determinative role in phytopathogenicity. By contrast, Enard et al. (1988) have provided direct evidence to correlate plant infection with the production of the siderophore chrysobactin by *Erwinia chrysanthemi,* a pectinolytic bacterial pathogen causing soft rot on a wide range of plants. *U. maydis* is the causal agent of corn smut disease, and infection is generally initiated by the fusion of haploid cells of opposite mating type to form a pathogenic dicaryon (Saville and Leong, 1992). Experiments were conducted to determine the role of siderophore production in pathogenicity. A mixture of haploid, NTG-induced *sid*1^{-} mutant progenies of compatible mating-type were used to inoculate maize seedlings, and this resulted in disease development without a significant difference in disease ratings from those obtained from infection with wild-type strains (Leong et al., 1988). Results from the use of *sid*1 null mutants constructed by gene disruption also revealed no correlation between phytopathogenicity and the biosynthetic system for ferrichrome siderophores in *U. maydis* (Mei et al., 1993). It should be noted that the conditions used in the laboratory to assess phytopathogenicity do not mimic those found in the field. Furthermore, we have not ruled out the possibility that another iron uptake system is utilized by *U. maydis* in host plants. Finally, a number of plant species can utilize various microbial siderophores (Wallace, 1982); maize cells can acquire Fe from ferrioxamine B (Bar-Ness et al., 1992). At this time, it is difficult to conclude what roles siderophores play in the relationship between *U. maydis* and its host plant. However, siderophores exhibit iron-storage functions in mycelia and spores of certain fungi (Matzanke et al., 1987), in addition to the role they play in the solubilization and transport of iron into the organism under conditions of cellular iron deprivation. For example, siderophores have been shown to be essential for spore germination of *N. crassa* (Horowitz et al., 1976). *U. maydis* sporidia and teliospores contain ferrichrome (Budde and Leong, unpublished findings). Thus siderophores may be important to spore germination and teliospore survival in *U. maydis,* a subject that we are currently investigating.

C. *urbs*1, a Gene Involved in Iron Regulation of Siderophore Biosynthesis in *U. maydis*

1. *Identification of Regulatory Mutants*

As in most microorganisms, the production of siderophores by *U. maydis* is regulated by iron availability in the growth medium. As a first step to understanding the mechanism that regulates the biosynthesis of siderophores in *U. maydis,* a novel bioassay approach was developed to obtain fungal mutants defective in this regulatory process (Leong et al., 1987). This method utilizes the growth of *Salmonella typhimurium* mutant strain TA2701 (*fhu*A^{-}, *ent*$^{-}$), which is defective in the ability to transport ferrichrome as well as in the biosynthesis of enterobactin (Luckey et al., 1972), as an indicator of constitutive siderophore production by *U. maydis* in iron-replete medium. The TA2701 mutant will only grow when enterobactin or free iron is present in the medium. Fungal mutants, deregulated for siderophore production and grown under an iron-rich medium, produce and secrete large amounts of siderophores, which in turn chelate free iron in the medium and thus inhibit the growth of the mutant bacterial indicator. By contrast, wild-type *U. maydis,* which does not produce detectable amounts of siderophores on an iron-rich medium, does not inhibit the growth of *S. typhimurium* TA2701 (Fig. 8). Besides ferrichrome and ferrichrome A, several other fungal and bacterial siderophores are also capable of preventing the growth of *S. typhimurium* TA2701 (Wang, 1988). Thus it is possible to extend this bioassay to isolate mutants deregulated for siderophore biosynthesis in other organisms.

Following this method, three mutants constitutive for siderophore production were identified among a population of 200,000 NTG-mutagenized colonies of *U. maydis* (Wang, 1988). These mutants were confirmed to

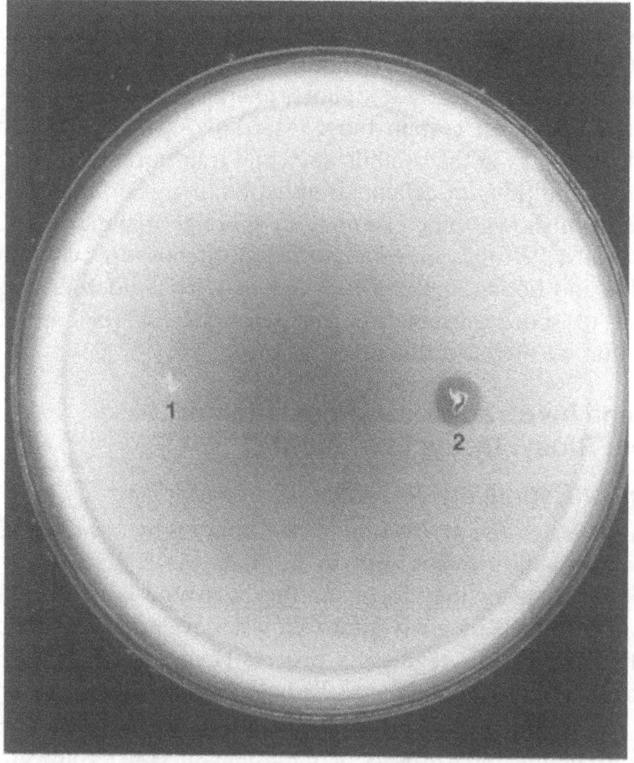

Figure 8 Inhibition of growth of *S. typhimurium* mutant TA2701 by *U. maydis* mutant that produces siderophores constitutively. (1) The wild-type strain; (2) the constitutive mutant. (From J. Wang, 1988, Ph.D. diss.)

produce siderophores in liquid medium containing 10 μM $FeSO_4$. Genetic analysis of one mutant, UMC002, indicated that the regulatory defect is controlled by a single genetic locus and that the mutation is recessive.

2. *Isolation of* urbs1

A cosmid, pSidCB2, capable of complementing the three deregulated mutants, was isolated from a cosmid library prepared from DNA of an *U. maydis* wild-type strain in pCU3 (Voisard et al., in press). From this cosmid an 11.9kb *Bam*HI fragment was identified that was able to restore normal regulation of siderophore biosynthesis in UMC002. Further subcloning experiments have shown that a 4.5 kb *Xba*I-*Bgl*II fragment from this region can fully complement *in trans* UMC002 as well as two other constitutive mutants when subcloned into the autonomously replicating vector pCM54. This clone was presumed to harbor a gene designated as *urbs*1 for *Ustilago* *r*egulator of *b*iosynthesis of *s*iderophores.

To confirm that the cloned sequence contains *urbs*1, a gene disruption experiment was performed by inserting a hygromycin B resistance gene at the unique *Xho*I site of the 4.5 kb *Xba*I-*Bgl*II fragment (Voisard et al., in press). The disruption mutant produced siderophores constitutively (Voisard et al., in preparation). A cross between the disruption mutant UMC013 and a wild-type strain gave a 1:1 segregation of $urbs1^-$ and $urbs1^+$ haploid progeny as expected for alternate alleles at a locus. Furthermore, diploids constructed with the NTG-induced mutants and the gene-disrupted strain are constitutive for siderophore production, indicating that the NTG-induced mutations are allelic to the mutation created by gene disruption (Voisard et al., in press). Gap repair experiments have revealed that the mutation of the three constitutive mutants caused by NTG are within a 2 kb region of the 4.5 kb *Xba*I-*Bgl*II fragment (McEvoy and Leong, unpublished data). These results, together with DNA sequence analysis (see below), indicate that the cloned DNA fragment contains the structural gene for *urbs*1.

3. *Analysis of* urbs1

Structure of urbs*1*. Unlike *sid*1, there is no indication of an intron within the 4.5 kb *Xba*I-*Bgl*II genomic DNA fragment encompassing *urbs*1. This has been investigated by comparison of a partial cDNA sequence with that of genomic DNA, and comparison of products amplified from the mRNA and the corresponding genomic DNA by PCR (polymerase chain reaction) (Voisard et al., in press). Primer extension analysis defined the two transcription initiation sites for *urbs*1 (Voisard et al., in press). The upstream region of the mRNA contains a CAAT sequence, while a TATA box is absent. Similar low levels of a 4.2 kb transcript were detected in

RNA isolated from wild-type cells grown in low iron and high iron media by Northern hybridization (Voisard et al., in press).

Characteristics of the Predicted Urbs1 *Protein.* Nucleotide sequence analysis of 3987 bp of *urbs*1 genomic DNA revealed a single open reading frame encoding a protein of 950 amino acids (Voisard et al., in press). Comparison of the predicated amino acid sequence of Urbs1 protein with sequences in Genbank showed that two regions of the Urbs1 protein have 60–70% identity with the vertebrate erythroid transcriptional factors GATA-1 of human (Tsai et al., 1989), rat (Whitelaw et al., 1990), and chicken (Evans et al., 1988) as well as the regulatory proteins Gln3 of *Saccharomyces cerevisae* (Minehart and Magasanik, 1991), Nit-2 of *Neurospora crassa* (Fu and Marzluf, 1990a), and AreA of *Aspergillus nidulans* (Kudla et al., 1990). The vertebrate proteins contain two closely related putative zinc finger structures consisting of Cys-x-x-Cys-x_{17}-Cys-x-x-Cys, while the ascomycete factors have only one such motif in each protein (Fig. 9). The Urbs1 protein contains two motifs separated by 100 amino acids, while the finger motifs of the vertebrate proteins follow each other directly. All are predicted to play a role in DNA binding and transcriptional regulation; in a number of cases direct evidence for DNA binding has been demonstrated (Fu and Marzluf, 1990b; Martin et al., 1989; Tsai et al., 1989; Evans et al., 1988). Recently, Omichinski et al. (1993) reported that a synthesized 66-amino acid residue single 'Finger' peptide derived from the

```
          codon
cGATA1    108    RECVNCGATATPLWRRDGTGHILCNACGLYHRLNGQNRPLYRPKKRLLVSKR
cGATA1    162    TVCSNCQTSTTTLWRRSPMGDPVCNACGLYYKLHQVNRPLTMRKDGIQTRNR
hGATA1    202    RECVNCGATATPLWRRDRTGHYLCNACGLYHKMNGQNRPLIRPKKRLIVSKR
hGATA1    256    TQCTNCQTTTTTLWRRNASGDPVCNACGLYYKLHQVNRPLTMRKDGIQTRNR
mGATA1    202    RECVNCGATATPLWRRDRTGHYLCNACGLYHKMNGQNRPLIRPKKRMIVSKR
mGATA1    256    TQCTNCQTTTTTLWRRNASGDPVCNACGLYFKLHQVNRPLTMRKDGIQTRNR
AreA      514    TTCTNCFTQTTPLWRRNPEGQPLCNACGLFLKLHGVVRPLSLKTDVIKKRNR
Gln3      304    IQCFNCKTFKTPLWRRSPEGNTLCNACGLFQKLHGTMRPLSKSDVIKKRISK
Nit-2     740    TTCTNCFTGTTPLWRRNPDGGPLCNACGLFLKLHGVVRPLSLKTDVIKKRNR

Urbs1     337    MRCSNCGVTSTPLWRRAPDGSTICNACGLYIKSHSTHRSASNRLSGSDASPP
Urbs1     481    LRCTNCQTTTTPLWRRDEDGNNICNACGLYHKLHGTHRPIGMKKSVIKRRKR

Consensus        ..C.NC..T.T.LWRR...G...CNACGLY.KLHG..RPL..........R
```

Figure 9 Alignment of the putative zinc finger motif of various transcription factors. Bold letters indicate conserved amino acids. cGATA1, chicken erythroid specific transcription factor; hGATA1, human erythroid specific transcription factor; mGATA1, murine erythroid specific transcription factor; Nit-2, *Neurospora crassa* nitrogen regulatory protein; AreA, *Aspergillus nidulans* regulatory protein; Gln3, *Saccharomyces cerevisiae* regulatory protein; Urbs1, *Ustilago maydis* regulator of biosynthesis of siderophores.

regulatory factor GATA-1 binds tightly and specifically to the GATA target sequence and this interaction is dependent on the metals Zn^{2+} or Fe^{2+}.

In addition to the putative zinc finger motif, the predicted Urbs1 protein contains a larger number of S(T)PXX motifs, a sequence motif frequently found in regulatory proteins (Suzuki, 1989). Another intriguing feature of the primary amino acid sequence is the existence of a polyhistidine stretch of 21 residues at the C terminus (Voisard et al., in press). The Fur (ferric uptake regulation) protein of *E. coli* also contains 12 histidines toward the carboxyl end, and these have been implicated as playing a role in sensing the iron status of the cell (Saito et al., 1991).

4. *Is Urbs1 a Negative Regulator for Siderophore Biosynthesis?*

Iron uptake systems of microorganisms are regulated by iron-dependent mechanisms. In bacteria, regulation of a wide variety of genes including at least five separate iron transport systems and a series of other iron-regulated genes are governed by the Fur gene that encodes an iron-dependent repressor (Bagg and Neilands, 1987). The Fur protein acts as a global iron-responsive transcriptional repressor that turns off these genes in the presence of iron. Negative transcriptional control of iron transport mediated by an iron-responsive two-component system has been characterized in the phytopathogen *Erwinia chrysanthemi,* which produces a catechol-type siderophore (Expert et al., 1992). In addition, there are also genes that are either unrelated or only indirectly related to iron transport, which are regulated by iron. For example, expression of the *tox* gene in *Corynebacterium diphtheriae,* the causative agent of diphtheria, is regulated by iron via an iron-dependent diphtheria toxin repressor (Schmitt et al., 1992). However, information on the regulation of iron transport in eukaryotic microorganisms is lacking. The cloning and characterization of *urbs*1 in *U. maydis* provides an opportunity to study the mechanism of iron-dependent regulation of iron uptake in fungi.

Results from studies on *sid*1 and *urbs*1 as described above indicate that biosynthesis of siderophores in *U. maydis* is regulated by an iron-dependent mechanism and that the product of *urbs*1 may act as a negative regulator of siderophore biosynthetic genes. First, the *urbs*1 mutants, either induced by NTG or by gene disruption, are constitutive for production of siderophores. Moreover, the *urbs*1 mutant phenotype correlates with the constitutive expression of *sid*1, which is normally inducible in low-iron growth conditions (Mei et al., 1993).

The nature of this regulatory role is suggested by the analysis of the deduced amino acid sequence of Urbs1, which has revealed the presence of putative zinc finger and S(T)PXX motifs, sequences found in transcriptional regulatory proteins. For the GATA family of regulatory proteins, the

DNA binding site is characterized by the core sequence (A/T)GATA(A/G) (Hannon et al., 1991). The recognition elements of Nit-2 contain versions of a sequence with a consensus of TATCTA (Fu and Marzluf, 1990b), a sequence similar to the core recognition sequence of GATA-1. Furthermore, the bacterial Fur protein target consensus motif "iron box," 5′-GATAATGATAATCATTATC-3′ (Bagg and Neilands, 1987), contains repeated GATA sequences. From these comparisons, it is reasonable to expect that the DNA binding site of the *urbs*1 protein contains this sequence. The two GATA boxes in the promoter region of *sid*1 (Mei et al., 1993) may represent Urbs1 binding sites. A test of the effect of iron on the expression of GUS gene under control of the *sid*1 promoter in the wild-type and constitutive mutant C002 suggests that the *sid*1 promoter region is a site of regulation by iron. Additional evidence comes from the observation that expression of the *sid*1 cDNA driven by the GAPDH promoter is constitutive irrespective of the iron concentration of the growth medium. Although more work needs to be done, these results support the hypothesis that *urbs*1 acts directly or indirectly to repress the expression of siderophore genes in *U. maydis*.

Finally, since *urbs*1 is expressed constitutively (Voisard et al., in press), it is possible that Urbs1 activity is allosterically regulated by iron, a putative corepressor. The stretch of 21 histidine residues of the protein could be involved in binding iron. Alternatively, the finger motifs could interact with ferrous ion as has been shown for GATA-1 (Omichinski et al., 1993). It is intriguing to note that GATA-1 regulates globin gene expression in erythroid cells and that globin constitutes 75% of the total iron in the human body (Emery, 1991). Thus *urbs*1 and GATA-1 both play important roles in regulating iron metabolism and iron homeostasis in eukaryotic cells.

IV. SIDEROPHORE-MEDIATED IRON TRANSPORT IN FUNGI

A siderophore-mediated high affinity iron uptake system consists of siderophores and their cognate membrane-bound transport system. Progress on the molecular biology of siderophore biosynthesis and regulation in fungi has been presented as above. Here, we would like to give a brief discussion of iron(III)-siderophore transport into fungi. Details of siderophore-mediated iron transport in fungi have been described in Chap. 2. For a more comprehensive description of iron transport in microorganisms, the reader is referred to a recent review (Winkelmann, 1991).

The accumulated knowledge on iron transport in fungi has mainly come from biochemical studies using ^{55}Fe- or ^{59}Fe-labeled iron complexes and ^{3}H-

or ^{14}C-labeled ligands to evaluate transport kinetics of siderophores (Winkelmann, 1991). Iron uptake studies have been carried out with *Ustilago sphaerogena* (Emery, 1971b; Leong et al., 1974; Ecker et al., 1982), *Rhodotorula pilimanae* (Carrano and Raymond, 1978), *Neurospora crassa* (Bailey et al., 1986), *Fusarium roserum* (Emery, 1965), and various *Aspergillus* strains (Wiebe and Winkelmann, 1975). The transport process has been shown to be energy consuming and dependent on the overall structure, configuration, and stereospecificity of the chelate molecules (Winkelmann and Huschka, 1987). From these studies, different transport mechanisms (e.g., "shuttle mechanism," "taxi mechanism," and "reductive taxi mechanism") have been proposed for different fungal species (Winkelmann and Huschka, 1987). Despite considerably different transport mechanisms, the ability of fungi to take up iron from the different siderophores is dependent on the presence of specific membrane-associated receptor proteins (Winkelmann and Huschka, 1987). The characterization of such receptors and siderophore permeation of membranes at the molecular level are not well advanced in fungi at present. However, progress in the study of bacterial transport systems (Braun and Hantke, 1991) has provided clues to study the transport process in fungi.

In *E. coli,* a number of outer membrane receptor proteins have been identified that are responsible for iron transport mediated by different siderophores (i.e., Iut for aerobactin, FepA for enterobactin, FhuA for ferrichrome, FhuE for coprogen, FecA for ferric dicitrate) (Braun and Winkelmann, 1987; Hantke, 1983). Furthermore, it is known that genes for siderophore biosynthesis and the corresponding outer membrane receptor are often located on the same gene cluster and are expressed concomitantly in low iron medium (Braun and Hantke, 1991). Comparison of the amino acid sequences of different receptor proteins has shown that there are two similar regions that could be involved in the binding of the ferric siderophores (Braun and Hantke, 1991). In addition to the receptor, the transport of ferric hydroxamate compounds requires the FhuD protein in the periplasmic space and the FhuB and FhuC proteins in the cytoplasmic membrane. The tonB gene, which encodes a cytoplasmic membrane–located protein of 36 kD, is essential for all the ferric siderophore uptake systems (Plastow and Holland, 1979; Braun and Hantke, 1991). Interestingly, bacteria use not only the siderophores aerobactin and enterobactin they produce, but also the siderophores ferrichrome, coprogen, and rhodotorulic acid produced by fungi (Braun and Winkelmann, 1987). It is possible that the receptor for hydroxamate siderophores in fungi may be similar to that for the fungal siderophores in *E. coli,* although experimental evidence is lacking to support this hypothesis. On the other hand, since fungi possess a single plasma membrane usually surrounded by a hyphal wall, unlike gram-negative bacteria, which are surrounded by two membranes (the outer membrane and the cytoplasmic membrane), it is speculated that the transport mechanism in fungi should be different from that in bacteria. For instance, it is well known that bacteria can overproduce iron-regulated membrane-bound receptors under iron limitation (Braun and Hantke, 1991). However, the SDS gel electrophoretic analysis of membrane proteins isolated from *N. crassa* grown under iron-deficient and iron-sufficient media revealed no differences in the plasma membrane protein pattern (Huschka and Winkelmann, 1989), suggesting that the corresponding transport systems in this fungus may be constitutive rather than inducible by iron deficiency. It should be noted that membrane-bound proteins involved in siderophore transport in fungi have not yet been identified. However, mutants defective in uptake of exogenous ferricrocin and coprogen have been isolated from *N. crassa* (Perkins et al., 1982). With the availability of a gene transfer system for this fungus (Vollmer and Yanofsky, 1986), the cloning of genes for these transport functions should be straightforward.

V. FUTURE DIRECTIONS

Application of molecular genetic analysis to study iron transport and its regulation in prokaryotic systems has expanded rapidly in the past decade. However, our understanding of siderophore-mediated iron transport in fungi is still in a preliminary state, even though considerable progress on the elucidation of fungal siderophore structures and their transport mechanisms has been made through biochemical and physiological studies (Winkelmann, 1991; Winkelmann and Huschka, 1987). Thus much still remains to be learned about these processes.

The first challenging problem is to determine molecular details for the remaining steps leading to siderophore production in fungi. We have cloned *sid*1, a gene initiating the biosynthesis of ferrichrome and ferrichrome A in *U. maydis*. As most fungal siderophores contain a δ-*N*-hydroxyornithine unit (Winkelmann, 1992), this gene must be widespread in the fungi. To define other genes required for siderophore biosynthesis, efforts have been directed toward collecting mutants blocked at various steps in the pathway. Mutants have been isolated that are specifically defective in ferrichrome biosynthesis and are blocked after the hydroxylation of ornithine (Wang et al., 1989). These may be affected in the acetylation of δ-*N*-hydroxyornithine or in peptide synthesis or cyclization. Once these mutants are further characterized, the genes involved in ferrichrome biosynthesis can be cloned using the same strategy that was employed for *sid*1. It may also be possible to clone the genes for ferrichrome A biosynthesis by probing a genomic DNA library with ferrichrome biosynthetic genes. A similar approach has been used successfully to clone genes involved in polyketide biosynthesis from different *Streptomycetes* (Malpartida et al., 1987). Another approach could involve the purification of fungal enzymes that carry out various steps of siderophore biosynthesis. The purification of ferrichrome synthetase from *A. quadricinctus* (Siegmund et al., 1991) would facilitate molecular cloning of the gene. Since this enzyme is a multienzyme complex, the different subunits of the enzyme could be encoded by different genes. With pure protein in hand, it will be possible to determine a partial amino acid sequence of each subunit and then synthesize corresponding oligonucleotides for use as probes to a genomic or cDNA library of this fungus. Alternatively, antibodies against the pure protein could be used to screen a cDNA expression library and then cDNA clones could be isolated from immunopositive phage. *HTS*I, a gene encoding a multifunctional cyclic peptide synthetase involved in biosynthesis of *Cochliobolus carbonum* HC-toxins, has been cloned by this approach (Panaccione et al., 1992). Finally, degenerated oligonucleotides could be synthesized based on the conserved amino acid sequences of a protein, and the corresponding gene could be amplified by PCR with the oligonucleotides as primers and genomic DNA as template. Jonathan Walton and coworkers (personal communication) have used this approach to pull out cyclic peptide synthetase genes from *Cochliobolus* and other filamentous fungi.

Since hydroxylation of ornithine to δ-*N*-hydroxyornithine is a unique enzymatic reaction that does not occur in plant or animal cells, L-ornithine-N^5-oxygenase is of considerable interest, as this enzyme could provide a target for development of antifungal agents to cripple fungal pathogens that produce hydroxamate-type siderophores as virulence factors. The availability of *sid*1 will permit us to overproduce and purify this enzyme and to study its biochemical properties. Expression of *sid*1 cDNA under control of the GAPDH promoter has resulted in a fivefold increase in specific activity of the enzyme (Mei and Leong, unpublished data). For overexpression of *sid*1 cDNA, a heterologous protein expression system, for instance, the T7 RNA polymerase system in *E. coli* (Tabor and Richardson, 1985) or other fungal expression systems (Davies, 1991), might also be considered.

Research on a number of prokaryotic systems has provided evidence that regulation of iron uptake is controlled by a variety of mechanisms, ranging from the very simple operon-like organization of the aerobactin gene system, where the Fur protein acts as global iron-responsive transcriptional repressor (Bagg and Neilands, 1987), to the more complex regulatory circuit found in the *Vibrio anguillarum* iron uptake system, in which positive and negative regulatory factors control the biosynthesis of siderophore and membrane transport proteins (Crosa, 1989). The cloning of *U. maydis urbs*1 has opened avenues to unravel the regulatory mechanism of iron transport in fungi. Several general approaches could be used. For example, a reporter gene (e.g., GUS gene) has been used to locate the minimal sequences of *sid*1 promoter that is required for responding to iron regulation. Once the Urbs1 protein is available, gel retardation assays using the upstream region of *sid*1 will be performed to directly demonstrate whether Urbs1 acts as a transcriptional repressor and whether iron acts as corepressor. If Urbs1 does bind to *sid*1 upstream regions, the sites of binding will be defined by DNase I protection experiments. It will be interesting to alter the consensus sequences of *urbs*1 fingers by site-directed mutagenesis or deletion and to study the physiological consequences. However, fungal regulatory pathways are often complex and involve numerous activators and repressors that control a set of genes (Davis and Hynes, 1991). Additional mutants altered in siderophore regulation would be helpful to understand the regulatory mechanism governing regulation of iron uptake in fungi.

Another important aspect of iron uptake in fungi is the identification of membrane-bound transport components. The first efforts will be directed toward the isolation of mutants defective in transport steps. Once this is achieved, it will be possible to clone genes involved in the transport systems. Iron transport mutants of fungi could be detected by colony autoradigraphy (Raetz, 1975), which has been used to isolate inositol transport mutants of *S. cerevisiae* (Nikawa et al., 1983), by suicide mutagenesis (Finklstein et al., 1977) using radioactive siderophore iron complexes, or by selecting for resistance to siderophore analogs. For example, ferrichrome uptake mutants of *E. coli* were isolated using the antibiotic albomycin that shares structural features with ferrichrome (Fecker and Braun, 1983). If fungal and bacterial receptor proteins for the same siderophore have some structural similarity, one could isolate the fungal transport gene by Southern hybridization with the corresponding cloned bacterial gene.

The role of fungal siderophores in the infection of plants and animals is still an exciting area. Knowledge of the molecular mechanisms of microbial pathogenicity mediated by iron will provide the framework for developing novel strategies of antimicrobial chemotherapy (Dionis et al., 1991).

It is hoped that we will achieve the same level of understanding of iron uptake and regulation in fungi as has been achieved in bacterial systems. Future breakthroughs, like past ones, will be dependent on a multidisciplinary endeavor, and indeed, molecular biology will play a key role.

ACKNOWLEDGMENTS

We would like to thank A. Budde, J. McEvoy, B. Saville, and C. Voisard for critical review of the manuscript. The work from our laboratory described in this chapter was supported by the USDA and NIH grant ROI-GM33716.

REFERENCES

Akers, H. A., and Neilands, J. B. (1978). Biosynthesis of rhodotorulic acid and other hydroxamate type siderophores, in *Biological Oxidation of Nitrogen* (J. W. Gorrod, ed.), Elsevier/North-Holland Biomedical Press, Amsterdam, pp. 429–436.

Archibald, F. (1983). *Lactobacillus plantarum,* an organism not requiring iron, *FEMS Microbiol. Lett., 19:* 29–32.

Bagg, A., and Neilands, J. B. (1987). Molecular mechanism of regulation of siderophore-mediated iron assimilation, *Microbiol. Rev., 51:* 509–518.

Bailey, T. C., Kime-Hunt, E. M., Carrano, C. J., Huschka, H., and Winkelmann, G. (1986). A photoaffinity label for the siderophore-mediated iron transport system in *Neurospora crassa, Biochim. Biophys. Acta, 883:* 299–305.

Ballance, D. J. (1991). Transformation systems for filamentous fungi and an overview of fungal gene structure, in *Molecular Industrial Mycology* (S. A. Leong and R. M. Berka, eds.), Marcel Dekker, New York, pp. 1–30.

Bar-Ness, E., Hadar, Y., Chen, Y., Romheld, V., and Marschner, H. (1992). Short-term effects of rhizosphere microorganisms on Fe uptake from microbial siderophores by maize and oat, *Plant Physiol., 100:* 451–456.

Bindereif, A., and Neilands, J. B. (1985). Promoter mapping and transcriptional regulation of the iron assimilation system of plasmid ColV-K30 in *Escherichia coli* K-12, *J. Bacteriol., 162:* 1039–1046.

Bolker, M., Urban, M., and Kahmann, R. (1992). The *a* mating type locus of *U. maydis* specifies cell signaling components, *Cell, 68:* 441–450.

Braun, V., and Hantke, K. (1991). Genetics of bacterial iron transport, in *Handbook of Microbial Iron Chelates* (G. Winkelmann, ed.), CRC Press, Boca Raton, FL, pp. 107–138.

Braun, V., and Winkelmann, G. (1987). Microbial iron transport. Structure and function of siderophores, *Prog. Clin. Biochem. Med., 5:* 67–99.

Budde, A. D., and Leong, S. A. (1989). Characterization of siderophores from *Ustilago maydis, Mycopathol., 108:* 125–133.

Bullen, J. J., and Griffiths, E. (1987). *Iron and Infection,* John Wiley, New York.

Carrano, C. J., and Raymond, K. N. (1978). Coordination chemistry of microbial iron transport compounds: Rhodotorulic acid and iron uptake in *Rhodotorula pilimanae, J. Bacteriol., 136:* 69–74.

Cody, Y. S., and Gross, D. C. (1987). Outer membrane protein mediating iron uptake via pyoverdin$_{pss}$, the fluorescent siderophore produced by *Pseudomonas syringae* pv. *syringae, J. Bacteriol., 169:* 2207–2214.

Crosa, J. (1989). Genetics and molecular biology of siderophore-mediated iron transport in bacteria, *Microbial Rev., 53:* 517–530.

Dancis, A., Roman, D. G., Anderson, G. J., Hinnebusch, A. G., and Klausner, R. D. (1992). Ferric reductase of *Saccharomyces cerevisiae:* Molecular characterization, role in iron uptake, and transcriptional control by iron, *Proc. Natl. Acad. Sci. USA, 89:* 3869–3873.

Davies, R. W. (1991). Molecular biology of a high-level recombinant protein production system in *Aspergillus,* in *Molecular Industrial Mycology* (S. A. Leong and R. M. Berka, eds.), Marcel Dekker, New York, pp. 45–81.

Davis, M. A., and Hynes, M. J. (1991). Regulatory circuits in *Aspergillus nidulans,* in *More Gene Manipulations in Fungi* (J. W. Bennett and L. L. Lasure, eds.), Academic Press, San Diego, pp. 151–189.

Davis, R. H., Lawless, M. B., and Port, A. (1970). Arginaseless *Neurospora*: Genetics, physiology, and polyamine synthesis, *J. Bacteriol., 102:* 299–305.

Dionis, J. B., Jenny, H.-B., and Peter, H. H. (1991). Therapeutically useful iron chelators, in *Handbook of Microbial Iron Chelates* (G. Winkelmann, ed.), CRC Press, Boca Raton, FL, pp. 339–356.

Drechsel, H., Metzger, J., Freund, S., Jung, G., Boelaert, J., and Winkelmann, G. (1991). Rhizoferrin—A novel siderophore from the fungus *Rhizopus microsporus* var. *rhizopodiformis, Biol. Metals, 4:* 238–243.

Ecker, D. J., Passavant, C. W., and Emery, T. F. (1982). Role of two siderophores in *Ustilago sphaerogena.* Regulation of biosynthesis and uptake mechanisms, *Biochim. Biophys. Acta, 720:* 242–249.

Emery, T. F. (1965). Isolation, characterization, and properties of fusarinine, a δ-hydroxamic acid derivative of ornithine, *Biochem., 4:* 1410–1417.

Emery, T. F. (1966). Initial steps in the biosynthesis of ferrichrome. Incorporation of δ-*N*-hydroxyornithine and δ-*N*-acetyl-δ-*N*-hydroxyornithine, *Biochem., 5:* 3694–3701.

Emery, T. F. (1971a). Hydroxamic acids of natural origin, *Adv. Enzymol., 35:* 135–185.

Emery, T. F. (1971b). Role of ferrichrome as a ferric ionophore in *Ustilago sphaerogena, Biochem., 10:* 1483–1488.

Emery, T. F. (1991). *Iron and Your Health: Facts and Fallacies,* CRC Press, Boca Raton, FL.

Enard, C., Diolez, A., and Expert, D. (1988). Systemic virulence of *Erwinia chrysanthemi* 3937 requires a functional iron assimilation system, *J. Bacteriol., 170:* 2419–2426.

Evans, T., Reitman, M., and Felsenfeld, G. (1988). An erythrocyte-specific DNA-binding factor recognizes a regulatory sequence common to all chicken globin genes, *Proc. Natl. Acad. Sci. USA, 85:* 5976–5980.

Expert, D., Sauvage, C., and Neilands, J. B. (1992). Negative transcriptional control of iron transport in *Erwinia chrysanthemi* involves an iron-repressive two-factor system, *Mol. Microbiol., 6:* 2009–2017.

Fecker, L., and Braun, V. (1983). Cloning and expression of the *fhu* genes involved in iron(III)-hydroxamate uptake by *Escherichia coli, J. Bacteriol., 156:* 1301–1314.

Finkelstein, M. C., Slayman, C. W., and Adelberg, E. A. (1977). Tritium suicide selection of mammalian cell mutants defective in the transport of neutral amino acids, *Proc. Natl. Acad. Sci. USA, 74:* 4549–4551.

Fotheringham, S., and Holloman, W. K. (1989). Cloning and disruption of *Ustilago maydis* genes, *Mol. Cell. Biol., 9:* 4052–4055.

Froeliger, E. H., and Leong, S. A. (1991). The *a* mating-type alleles of *Ustilago maydis* are idiomorphs, *Gene, 100:* 113–122.

Frohman, M. A., Dush, M. K., and Martin, G. R. (1988). Rapid production of full-length cDNAs from rare transcripts: Amplification using a single gene-specific oligonucleotide primer, *Proc. Natl. Acad. Sci. USA, 85:* 8998–9002.

Fu, Y.-H., and Marzluf, G. A. (1990a). *nit*-2, the major nitrogen regulatory gene of *Neurospora crassa,* encodes a protein with a putative zinc finger DNA-binding domain, *Mol. Cell. Biol., 10:* 1056–1065.

Fu, Y.-H., and Marzluf, G. A. (1990b). *nit*-2, the major positive-acting nitrogen regulatory gene of *Neurospora crassa,* encodes a sequence-specific DNA-binding protein, *Proc. Natl. Acad. Sci. USA, 87:* 5331–5335.

Garibaldi, J. A., and Neilands, J. B. (1955). Isolation and properties of ferrichrome A, *J. Am. Chem. Soc., 77:* 2429–2430.

Gillissen, B., Bergemann, J., Sandmann, C., Schroeer, B., Bolker, M., and Kahmann, R. (1992). A two-component regulatory system for self/non-self recognition in *Ustilago maydis,* Cell, 68: 647–657.

Gritz, L., and Davies, J. (1983). Plasmid-encoded hygromycin B resistance: The sequence of hygromycin B phosphotransferase gene and its expression in *Escherichia coli* and *Saccharomyces cerevisiae, Gene, 25:* 179–188.

Gross, R., Engelbrecht, F., and Braun, V. (1985). Identification of the gene and their polypeptide products responsible for aerobactin synthesis by pColV plasmids, *Mol. Gen. Genet., 201:* 204–212.

Hannon, R., Evans, T., Felsenfeld, G., and Gould, H. (1991). Structure and promoter activity of the gene for the erythroid transcription factor GATA-1, *Proc. Natl. Acad. Sci. USA, 88:* 3004–3008.

Hantke, K. (1983). Identification of an iron uptake system specific for coprogen and rhodotorulic acid in *Escherichia coli, Mol. Gen. Genet., 191:* 301–306.

Herrero, M., Lorenzo, V. de, and Neilands, J. B. (1988). Nucleotide sequence of the *iuc*D gene of the pColV-K30 aerobactin operon and topology of its product studied with phoA and *LacZ* gene fusions, *J. Bacteriol., 170:* 56–64.

Hider, R. C. (1984). Siderophore mediated absorption of iron, *Struct. Bonding, 58:* 26–87.

Holden, D., Wang, J., and Leong, S. A. (1988). DNA-mediated transformation of *Ustilago hordei* and *Ustilago nigra, Physiol. Mol. Pl. Pathol., 33:* 235–239.

Holden, D. W., Kronstad, J., and Leong, S. A. (1989). Mutation in a heat-regulated *hsp*70 gene of *Ustilago maydis, EMBO J., 8:* 1927–1934.

Holliday, R. (1974). *Ustilago maydis,* in *Handbook of Genetics* (R. C. King, ed.) Plenum Press, New York, vol. 1, pp. 575–595.

Horowitz, N. H., Charlang, G., Horn, G., and Williams, N. P. (1976). Isolation and identification of the conidial germination factor of *Neurospora crassa, J. Bacteriol., 127:* 135–140.

Hummel, W., and Diekmann, H. (1981). Preliminary characterization of ferrichrome synthetase from *Aspergillus quadricinctus, Biochim. Biophys. Acta, 657:* 313–320.

Huschka, H.-G., and Winkelmann, G. (1989). Iron limitation and its effect on membrane proteins and siderophore transport in *Neurospora crassa, Biol. Metals, 2:* 108–113.

Ishimaru, C. A., and Loper, J. E. (1992). High-affinity iron uptake systems present in *Erwinia carotovora* subsp. *carotovora* include the hydroxamate siderophore aerobactin, *J. Bacteriol., 174:* 2993–3003.

Jalal, M. A. F., and van der Helm, D. (1991). Isolation and spectroscopic identification of fungal siderophores, in *Handbook of Microbial Iron Chelates* (G. Winkelmann, ed.), CRC Press, Boca Raton, FL, pp. 235–269.

Jefferson, R. A., Burgess, S. M., and Hirsh, D. (1986). β-Glucuronidase from *Escherischia coli* as a gene-fusion marker, *Proc. Natl. Acad. Sci. USA,* 8447–8451.

Keon, J. P. R., White, G. A., and Hargreaves, J. A. (1991). Isolation, characterization, and sequence of a gene conferring resistance to the systemic fungicide carboxin from the maize smut pathogen, *Ustilago maydis, Curr. Genet., 19:* 475–481.

Kinal, H., Tao, J., and Bruenn, J. A. (1991). An expression vector for the phytopathogenic fungus, *Ustilago maydis, Gene, 98:* 129–134.

Kronstad, J., and Leong, S. A. (1989). Isolation of two alleles of the *b* locus of *Ustilago maydis, Proc. Natl. Acad. Sci. USA, 86:* 978–982.

Kronstad, J. W., Wang, J., Covert, S. F., Holden, D. W., McKnight, G. L., and Leong, S. A. (1989). Isolation of metabolic genes and demonstration of gene disruption in the phytopathogenic fungus *Ustilago maydis, Gene, 79:* 97–106.

Kudla, B., Caddick, M. X., Langdon, T., Martini-Rossi, N. M., Bennett, C. F., Sibley, S., Davies, R. W., and Arst, H. N., Jr. (1990). The regulatory gene *are*A mediating nitrogen metabolite repression in *Aspergillus nidulans.* Mutations affecting specificity of gene activation alter a loop residue of a putative zinc finger, *EMBO J., 9:* 1355–1364.

Laird, A. J., Ribbons, D. W., Woodrow, G. C., and Young, I. G. (1980a). Bacteriophage Mu-mediated gene transposition and in vitro cloning of the enterochelin gene cluster of *Escherichia coli, Gene, 11:* 347–357.

Laird, A. J., and Young, I. G. (1980b). Tn5 mutagenesis of the enterochelin gene cluster of *Escherichia coli, Gene, 11:* 359–366.

Leong, J., Neilands, J. B., and Raymond, K. N. (1974). Coordination isomers of biological iron transport compounds. III. Transport of Λ-*cis* chromic desferriferrichrome by *Ustilago sphaerogena, Biochim. Biophys. Res. Commun., 60:* 1066–1071.

Leong, S. A., and Neilands, J. B. (1981). Relationship of siderophore-mediated iron assimilation to virulence in crown gall disease, *J. Bacteriol., 147:* 482–491.

Leong, S. A., Wang, J., Budde, A., Holden, D., Kinscherf, T., and Smith, T. (1987). Molecular strategies for the analysis of the interaction of *Ustilago maydis* and maize, in *Molecular Strategies for Crop Protection* (C. Artzen and C. Ryan, eds.), Alan R. Liss, New York, pp. 95–106.

Leong, S. A., Kronstad, J., Wang, J., Budde, A., Russin, W., and Holden, D.

(1988). Identification and molecular characterization of genes which control pathogenic growth of *Ustilago maydis* in maize, in *Molecular Genetics of Plant-Microbe Interactions* (D. P. Verma and R. Palacios, eds.), APS Press, St. Paul, Minn., pp. 241–246.

Leong, S. A., Wang, J., Kronstad, J., Holden, D., Budde, A., Froeliger, E., Kinscherf, T., Xu, P., Russin, B., Samac, D., Smith, T., Covert, S., Mei, B., and Voisard, C. (1990). Molecular analysis of pathogenesis in *Ustilago maydis,* in *Molecular Strategies of Pathogens and Host Plants* (S. S. Patil, D. Mills, and C. Vance, eds.), Springer-Verlag, New York, pp. 107–118.

Lesuisse, E., and Labbe, P. (1989). Reductive and non-reductive mechanisms of iron assimilation by the yeast *Saccharomyces cerevisiae, J. Gen. Microbiol., 135:* 257–263.

Lesuisse, E., Raguzzi, F., and Crichton, R. R. (1987). Iron uptake by the yeast *Saccharomyces cerevisiae:* Involvement of a reduction step, *J. Gen. Microbiol., 133:* 3229–3236.

Liu, A., and Neilands, J. B. (1984). The mutational analysis of rhodotorulic acid synthesis in *Rhodotorula pilimanae, Struct. Bonding, 58:* 97–106.

Lorenzo, V. de, Bindereif, A., and Neilands, J. B. (1986). Aerobactin biosynthesis and transport genes of plasmid ColV-K30 in *Escherichia coli* K-12, *J. Bacteriol., 165:* 570–578.

Luckey, M., Pollack, J. R., Wayne, R., Ames, B. N., and Neilands, J. B. (1972). Iron uptake in *Salmonella typhimurium:* Utilization of exogenous siderophores as iron carriers, *J. Bacteriol., 111:* 731–738.

Malpartida, F., Hallam, S. E., Kieser, H. M., Motamedi, H., Hutchinson, C. R., Butler, M. J., Sugden, D. A., Warren, M., McKillop, C., Bailey, C. R., Humphreys, G. O., and Hopwood, D. A. (1987). Homology between Streptomyces genes coding for synthesis of different polyketides used to clone antibiotic biosynthetic genes, *Nature* (London), *325:* 818–821.

Martin, D. I. K., Tsai, S. F., and Orkin, S. H. (1989). Increased γ-globin expression in a nondeletion HPFH mediated by an erythroid-specific DNA-binding factor, *Nature* (London), *338:* 435–438.

Matzanke, B. F., Bill, E., Trautwein, A. X., and Winkelmann, G. (1987). Role of siderophores in iron storage in spores of *Neurospora crassa* and *Aspergillus ochraceus, J. Bacteriol., 169:* 5873–5876.

Mei, B., Budde, A. D., and Leong, S. A. (1993). *sid*1, a gene initiating siderophore biosynthesis in *Ustilago maydis:* Molecular characterization, regulation by iron, and role in phytopathogenicity, *Proc. Natl. Acad. Sci. USA, 90:* 903–907.

Minehart, P. L., and Magasanik, B. (1991). Sequence and expression of GLN3, a positive nitrogen regulatory gene of *Saccharomyces cerevisiae* encoding a protein with a putative zinc finger DNA-binding domain, *Mol. Cell. Biol., 11:* 6216–6228.

Mullner, E., Neupert, B., and Kuhn, L. C. (1989). A specific mRNA factor regulates the iron-dependent stability of cytoplasmic transferrin receptor mRNA, *Cell, 58:* 373–382.

Neilands, J. B. (1952). A crystalline organo-iron pigment from a rust fungus (*Ustilago sphaerogena*), *J. Am. Chem. Soc., 74:* 4846–4847.

Neilands, J. B. (1981a). Iron absorption and transport in microorganisms, *Ann. Rev. Nutr., 1:* 27–46.

Neilands, J. B. (1981b). Microbial iron compounds, *Ann. Rev. Biochem., 50:* 715–731.

Neilands, J. B., and Leong, S. A. (1986). Siderophores in relation to plant growth and disease, *Ann. Rev. Plant Physiol., 37:* 187–208.

Neilands, J. B., Konopka, K., Schwyn, B., Coy, M., Francis, R. T., Paw, B. H., and Bagg, A. (1987). Comparative biochemistry of microbial iron assimilation, in *Iron Transport in Microbes, Plants and Animals* (G. Winkelmann, D. van der Helm, and J. B. Neilands, eds.), VCH, Weinheim, pp. 1–33.

Nikawa, J.-I., Nagumo, T., and Yamashita, S. (1982). *myo*-Inositol transport in *Saccharomyces cerevisiae, J. Bacteriol., 150:* 441–446.

Omichinski, J., Trainor, C., Evans, T., Gronenborn, A., Clore, M., and Felsenfeld, G. (1993). A small single-'finger' peptide derived from the regulatory factor GATA-1 binds specifically to its DNA site when complexed with either zinc or iron, *Proc. Natl. Acad. Sci. USA, 90:* 1676–1680.

Ong, D. E., and Emery, T. F. (1972). Ferrichrome biosynthesis: Enzyme catalyzed formation of the hydroxamic acid group, *Arch. Biochem. Biophys., 148:* 77–83.

O'Sullivan, D. J., and O'Gara, F. (1991). Regulation of iron assimilation: Nucleotide sequence analysis of an iron-regulated promoter from a fluorescent pseudomonod, *Mol. Gen. Genet., 228:* 1–8.

Panaccione, D., Scott-Craig, J. S., Pocard, J.-A., and Walton, J. D. (1992). A cyclic peptide synthetase gene required for pathogenicity of the fungus *Cochliobolus carbonum* on maize, *Proc. Natl. Acad. Sci. USA*, 89: 6590–6594.

Perkins, D., Radford, A., Newmeyer, D., and Bjorkman, M. (1982). Chromosomal loci of *Neurospora crassa, Microbiol. Rev., 46:* 426–570.

Plastow, G. S., and Holland, I. B. (1979). Identification of an inner membrane polypeptide specified by a λ-*ton*B transducing bacteriophage, *Biochem. Biophys. Res. Commun., 90:* 1007–1014.

Plattner, H. J., Pfefferle, P., Romaguera, A., Waschutza, S., and Diekmann, H. (1989). Isolation and some properties of lysine N^6-hydroxylase from *Escherichia coli* strain EN222, *Biol. Metals,* 2: 1–5.

Raetz, C. R. H. (1975). Isolation of *Escherichia coli* mutants defective in enzymes of membrane lipid synthesis, *Proc. Natl. Acad. Sci. USA, 72:* 2274–2278.

Rhodes, P. R., and Vodkin, L. O. (1988). Organization of *Tgm* family of transposable elements in soybean, *Genet., 120:* 597–604.

Saito, T., Duly, D., and Williams, J. P. (1991). The histidines of the iron-uptake regulation protein, Fur. *Eur. J. Biochem., 197:* 39–42.

Saville, B. J., and Leong, S. A. (1992). The molecular biology of pathogenesis in *Ustilago maydis,* in *Genetic Engineering* (J. K. Setlow, ed.), Plenum Press, New York, vol. 14, pp. 139–162.

Schmitt, M. P., Twiddy, E. M., and Holmes, R. K. (1992). Purification and characterization of the diphtheria toxin repressor, *Proc. Natl. Acad. Sci. USA, 89:* 7576–7580.

Siegmund, K.-D., Plattner, H. J., and Diekmann, H. (1991). Purification of ferrichrome synthetase from *Aspergillus quadricinctus* and characterization as a phosphopantetheine containing multienzyme complex, *Biochim. Biophys. Acta, 1076:* 123–129.

Smith, T. L., and Leong, S. A. (1990). Isolation and characterization of a *Ustilago maydis* glyceraldehyde-3-phosphate dehydrogenase-encoding gene, *Gene, 93:* 111–117.

Suzuki, M. (1989). SPXX, a frequent sequence motif in gene regulatory proteins, *J. Mol. Biol., 207:* 61–84.

Tabor, S., and Richardson, C. C. (1985). A bacteriophage T7 RNA polymerase/promoter system for controlled exclusive expression of specific genes, *Proc. Natl. Acad. Sci. USA, 82:* 1074–1078.

Tolmasky, M. E., Actis, L. A., and Crosa, J. H. (1988). Genetic analysis of the iron uptake region of *Vibrio anguillarum* plasmid pJMI: Molecular cloning of genetic determinants encoding a novel trans activator of siderophore biosynthesis, *J. Bacteriol., 170:* 1913–1919.

Tomlinson, G., Cruickshank, W. H., and Viswanatha, T. (1971). Sensitivity of substituted hydroxylamines to determination by iodine oxidation, *Anal. Biochem., 44:* 670–679.

Tsai, S. F., Martin, D. I., Zon, L. I., D'Andrea, A. D., Wong, G. G., and Orkin, S. H. (1989). Cloning of cDNA for the major DNA-binding protein of the erythroid lineage through expression in mammalian cells, *Nature* (London), *339:* 446–450.

Tsukuda, T., Carleton, S., Fotheringham, S., and Holloman, W. K. (1988). Isolation and characterization of an autonomously replicating sequence from *Ustilago maydis, Mol. Cell. Biol., 8:* 3703–3709.

Valvano, M., and Crosa, J. H. (1988). Molecular cloning, expression, and regulation in *Escherichia coli* K-12 of a chromosome-mediated aerobactin iron transport system from a human invasive isolate of *E. coli* K1, *J. Bacteriol., 170:* 5529–5538.

Voisard, C., Wang, J., McEvoy, J. L., Xu, P., and Leong, S. A. *Urbs1,* a gene regulating siderophore biosynthesis in *Ustilago maydis* encodes a protein with structural similarity to erythroid transcription factor GATA-1, *Mol. Cell. Biol.,* in press.

Vollmer, S. J., and Yanofsky, C. (1986). Efficient cloning of genes of *Neurospora crassa, Proc. Natl. Acad. Sci. USA, 83:* 4869–4873.

Wallace, A. (1982). Historical landmarks in progress relating to iron chlorosis in plants, *J. Plant Nutr., 5:* 277–288.

Wang, J., Holden, D. W., and Leong, S. A. (1988). Gene transfer system for the phytopathogenic fungus *Ustilago maydis, Proc. Natl Acad. Sci. USA, 85:* 865–869.

Wang, J., Budde, A. D., and Leong, S. A. (1989). Analysis of ferrichrome biosynthesis in the phytopathogenic fungus *Ustilago maydis:* Cloning of an ornithine-N^5-oxygenase gene, *J. Bacteriol., 171:* 2811–2818.

Whitelaw, E., Tsai, S.-F., Hogben, P., and Orkin, S. H. (1990). Regulated expression of globin chains and the erythroid transcription factor (GF-1/NF-EI/Eryf1) during erythropoesis in the developing mouse, *Mol. Cell. Biol., 10:* 6595–6606.

Wiebe, C., and Winkelmann, G. (1975). Kinetic studies on the specificity of chelate iron uptake in *Aspergillus, J. Bacteriol., 123:* 837–842.

Williams, P. H. (1979). Novel iron uptake system specified by ColV plasmid: An important component in the virulence of invasive strains of *Escherichia coli, Infect. Immun., 26:* 925–932.

Williams, R. J. P. (1990). An introduction to the nature of iron transport and storage, in *Iron Transport and Storage* (P. Ponka, H. M. Schulman, and R. C. Woodworth, eds.), CRC Press, Boca Raton, FL, pp. 2–15.

Winkelmann, G. (1991). Specificity of iron transport in bacteria and fungi, in *Handbook of Microbial Iron Chelates* (G. Winkelmann, ed.), CRC Press, Boca Raton, FL, pp. 65–105.

Winkelmann, G. (1992). Structure and function of fungal siderophores containing hydroxamate and complexone type iron binding ligands, *Mycol. Res., 96:* 529–534.

Winkelmann, G., and Huschka, H.-G. (1987). Molecular recognition and transport of siderophores in fungi, in *Iron Transport in Microbes, Plants and Animals* (G. Winkelmann, D. van der Helm, and J. B. Neilands, eds.), VCH, Weinheim, pp. 317–336.

Yelton, M. M., Timberlake, W. E., and van den Hondel, C. A. M. J. J. (1985). A cosmid for selecting genes by complementation in *Aspergillus nidulans:* Selection of the developmentally regulated yA locus, *Proc. Natl. Acad. Sci. USA, 82:* 834–838.

5

Reductive Iron Assimilation in *Saccharomyces cerevisiae*

Emmanuel Lesuisse and Pierre Labbe
Institut Jacques Monod, University of Paris VII, Paris, France

I. INTRODUCTION

Iron is an essential element for virtually all forms of life. It is involved in such vital reactions as the transport, storage, and activation of oxygen, electron transport, detoxification, and synthesis of deoxyribonucleotides. Iron's ability to pass readily between the divalent and trivalent states is its most important biological property. However, this same property poses a problem for life processes: iron tends to oxidize, hydrolyze, and polymerize into insoluble hydroxides and oxyhydroxides under aerobic conditions so that the bioavailability of this abundant metal is paradoxically extremely low. Soluble Fe(II) and Fe(III) may, in the presence of oxygen, be toxic for cells by catalyzing the production of free radicals via the Fenton and Haber-Weiss reactions.

These properties have made it necessary for organisms to develop specific strategies for taking up iron and storing it intracellularly. Many microorganisms—bacteria and fungi—respond to iron limitation by excreting siderophores, which are low molecular weight compounds, mainly hydroxamates and catecholates, that have a high affinity for ferric iron.

These iron-chelating molecules remove iron from its insoluble forms and make it available for transport (for a review, see Bagg and Neilands, 1987). Another strategy common in dicotyledonous plants involves reduction of ferric iron either at the cell surface via a trans–plasma membrane redox system or in the extracellular medium via excreted reducing compounds. The iron is then taken up by the cells as soluble Fe^{2+} ions (for a review, see Bienfait, 1985).

The first biochemical study of iron metabolism in *Saccharomyces cerevisiae* was made by Elvehjem in 1931. However, when we started to work on iron metabolism in *S. cerevisiae* in 1986, nothing was known about the mechanism of iron uptake by this yeast. This was surprising, since *S. cerevisiae* is a valuable model system that can be easily manipulated both metabolically and genetically, and since the uptake and intracellular distribution of many other metal ions had been already investigated. These included Mg^{2+} (Conway and Beary, 1958); Zn^{2+}, Co^{2+}, Ni^{2+} (Fuhrmann and Rothstein, 1968); Cd^{2+} and Co^{2+} (Norris and Kelly, 1977); Mn^{2+} (Okorokov et al., 1977); Ca^{2+} and Sr^{2+} (Roomans et al., 1979); Zn^{2+} (White and Gadd, 1987); Ca^{2+} (Kovac, 1985); Al^{3+}, Cr^{3+} (Kunst and Roomans, 1985) (for reviews, see Jones and Greenfield, 1984; Jones and Gadd, 1990). Several laboratories are now working on iron metabolism in *S. cerevisiae,* and this is probably linked to the fact that the plasma membrane reductase systems of animal cells are now receiving considerable attention (Nunez et al., 1983; Sun et al., 1987; Thorstensen and Romslo, 1988; for reviews, see Crane et al., 1990; Low et al., 1990). The results obtained with yeast as a cellular model will undoubtedly be of great value for understanding the different aspects of iron metabolism in higher eukaryotes, essentially mammals, as claimed recently by Dancis et al. (1990): "It is tempting to speculate that the iron uptake process in the simple eucaryote *S. cerevisiae* . . . provides a model for the missing link in mammalian cellular iron metabolism. A reductase-ferrous transport system may well explain iron uptake from the gut, from transferrin in the acidic endosome, and from non-transferrin carriers."

The yeast *S. cerevisiae* is now accepted as an ideal eukaryotic microorganism for biological studies. It has many characteristics that have made it perfect for studies of molecular genetics and molecular and cellular biology, including a rapid growth and a high growth yield, a budding pattern resulting in dispersed cells, easy replica plating and mutant isolation, a well-defined genetic system, and, most important, a highly versatile DNA transformation system (Sherman, 1991). In addition, *S. cerevisiae* is a facultative aerobe, allowing the identification and study of all types of respiratory-deficient mutants, such as hemeless mutants (Labbe-Bois and Labbe, 1990).

II. IRON UPTAKE BY THE CELLS

A. Reductive and Nonreductive Mechanisms of Iron Uptake

The yeast *S. cerevisiae* grows easily in synthetic (Yeast Nitrogen Base, Difco) or complex media (YPG: *y*east extract, bacto*p*eptone, *g*lucose), which contain 1 μM and 10–15 μM iron, respectively. These iron concentrations are within the range of the optimal concentrations for the growth of this yeast (Jones and Gadd, 1990); under these growth conditions, the total amount of intracellular iron ranges from about 0.4 to 2 nmol/mg dry weight (Jones and Gadd, 1990), depending on the strain, the growth medium, the carbon source, the growth phase, and the growth conditions; in these conditions, total heme iron accounts for 0.05 to 0.3 nmol/mg dry weight (Labbe-Bois and Labbe, 1990).

S. cerevisiae is able to use a great variety of iron compounds present in extracellular media to fulfill its nutritional requirements. The rate of iron uptake by resting cells depends on several factors, including the way the cells were cultured (growth phase, growth conditions, i.e., composition of the growth medium, carbon source, iron content, etc.) and the conditions of the iron assay itself (energy source, nature of the iron chelate used). Under basal conditions (growth on complete medium), the maximal uptake activity occurs in cells grown on ethanol or pyruvate (fully aerobic metabolism) harvested in log phase and supplied, in the assay, with a fermentable carbon source such as glucose (Lesuisse et al., 1987). Eide and Guarente (1992) showed that a gene (*MSN1*) encoding a transcriptional activator implicated in carbon source regulation could also regulate synthesis of gene products involved in iron uptake. These authors suggest that carbon source utilization and iron uptake are, to some extent, coregulated processes, since the specific requirements for respiration (cytochromes, iron-sulfur proteins) involve an increased need for iron by cells grown on a nonfermentable carbon source. As for the uptake of other ions (see Borst-Pauwels, 1981), several nonspecific factors—transmembrane potential, surface potential, or the presence of nonspecific binding sites at the cell surface—probably affect the rate and the amount of iron uptake by the cells. However, of all the factors affecting the iron uptake capacity of resting cells, two are of outstanding importance. One is the iron content of the medium in which the cells were grown, and the other is the oxidation state of iron presented to the cells during the assay. Cells grown under basal conditions take up iron far faster when the iron is presented as Fe^{2+} than when it is presented as Fe^{3+} (Lesuisse et al., 1987; Eide et al., 1992). The spectacular effects of these two factors are due to *S. cerevisiae* having a reductive mechanism of iron transport, the components of which are induced by iron-limited growth. The following section outlines the evidence for such a reductive mechanism of iron assimilation—which means that iron has to be reduced prior to uptake.

Many microorganisms secrete siderophores in response to iron limitation, but there is no report of *S. cerevisiae* secreting a siderophore. For instance, Atkin et al. (1970), during "an examination of 142 strains with 19 genera of yeasts and yeastlike organisms for formation of hydroxamic acids in low-iron culture" concluded that *S. cerevisiae* does not excrete hydroxamates in response to iron limitation. More recently, using the sensitive Chrom Azurol S-test (Schwyn and Neilands, 1987), Neilands et al. (1987) concluded that this yeast does not form a siderophore (hydroxamate or catecholate) under the conditions of growth used. However, *S. cerevisiae* does excrete large amounts of Krebs cycle intermediates, in particular citric, malic, and succinic acids, whose concentrations in extracellular medium can be greater than 1 mM under laboratory or industrial growth conditions (Lupianez et al., 1974; Whiting, 1976). These compounds, especially citric acid, are potential ligands of iron. All *Saccharomyces* species can also grow anaerobically, and tend strongly to acidify their growth media. These features may be of great advantage to these organisms for iron uptake in their ecological niche. It should be pointed out that in *S. cerevisiae,* the cell wall itself can bind relatively large amounts of ferric iron (Lesuisse et al., 1987). As suggested by Winkelmann (1979) for *Neurospora crassa,* such an "extracellular iron pool" on the cell surface could be gradually solubilized by excreted citrate and malate, to provide iron for the cells *in vivo.* We tested the capacity of *S. cerevisiae* to take up iron from siderophores secreted by other microorganisms and found that ferrioxamine B, ferricrocin, and rhodotorulic acid were all suitable iron sources for this yeast (Lesuisse et al., 1987; Lesuisse and Labbe, 1989). Wright and Honek (1989) also showed that addition of desferrioxamine B to the culture medium of *S. cerevisiae* did not affect its growth or the cytochrome P450 content of the cells. Desferrioxamine B is a hydroxamate siderophore excreted by bacteria and actinomycetes, in particular *Streptomyces pilosus* (Bickel et al., 1960). Hydroxamate siderophores produced by actinomycetes are abundant in nature: their concentrations in the soils can reach 0.1 μM (Powell et al., 1980). It is therefore probable that naturally occurring ferrioxamine B can act as an iron source for *S. cerevisiae* in its ecological niche(s).

The assimilation of iron from a ferric chelate requires intra- or extracellular dissociation of iron from its ligand. The great differences in the affinities of siderophores for ferrous and ferric iron offers a mechanism for removing the Fe^{3+} from siderophores (Neilands, 1957). Siderophore reductase activities have been described in cell-free extract of many microorganisms (Ernst and Winkelmann,[1] 1977; McReady and Ratledge, 1979; Straka and Emery, 1979; Arceneaux and Byers, 1980; Cox, 1980; Gaines et al., 1981; Moody and Dailey, 1985; Huyer and Page, 1989; Fischer et al., 1990; Le Faou and Morse, 1991; Andrews et al., 1992; Hallé and Meyer, 1992). Initial experiments with *S. cerevisiae* showed that whole cells of this yeast can reduce extracellular ferrioxamine B in the presence of an iron(II) chelator at a rate more than sufficient to account for iron uptake. Reciprocally, strong iron(II) chelators like bathophenantroline sulfonate or ferrozine strongly inhibit iron uptake from ferrioxamine B by resting cells. We also used ^{59}Fe-labeled and ^{14}C-labeled ferrioxamine B to show that iron is taken up by the cells, while the desferri compound remains essentially extracellular (Lesuisse et al., 1987). These data indicate that an extracellular reductive dissociation is involved in iron uptake from ferrioxamine B by *S. cerevisiae.* A comparable situation has been reported by Carrano and Raymond (1978) for iron uptake from ferrirhodotorulic acid by *Rhodotorula pilimanae,* and by Emery (1987) for iron uptake from ferrichrome A by *Ustilago sphaerogena.* However, more recent results have shown that the situation is more complex. While the iron(II)-trapping reagent ferrozine strongly inhibits (70–80%) iron uptake at high ferrioxamine B concentrations (360 μM), there is much less inhibition (0–20%) when the concentration of ferrioxamine B is decreased to 7 μM. Moreover, the rate of iron uptake by whole cells of heme-deficient mutants, which have no significant ferrireductase activity (see below) is unaffected when the concentration of ferrioxamine B is low (7 μM). Spectrophotometric analysis showed that these cells accumulate undissociated ferrioxamine B intracellularly (Lesuisse and Labbe, 1989). These observations demonstrate that *S. cerevisiae* uses more than a single mechanism to extract iron from ferrioxamine B. Experiments with the gallium analog of ferrioxamine B (Ga^{3+}-desferrioxamine B) showed that a nonreductive mechanism can contribute to the uptake of the siderophore by yeast cells. As shown by Emery (Emery and Hoffer, 1980; Emery, 1987), gallium is a valuable probe for studying microbial iron transport. Ga^{3+} forms complexes with siderophore ligands that are isomorphous with the Fe^{3+} complexes but are not reducible. Ga^{3+} can also displace iron from siderophores under reducing conditions (Emery, 1986). We observed that the iron uptake from ferrioxamine B by a wild-type strain is biphasic. When the ferrioxamine B is present at low concentration (<7 μM), it is taken up via a high-affinity mechanism ($K_M = 3$ μM) and Ga^{3+}-desferrioxamine B acts as a competitive inhibitor ($K_i = 4$

[1]An iron concentration of 1.6 μM in the growth medium was shown to be sufficient to account for iron deficiency in *N. crassa*. In our hands, such a situation is attained for *S. cerevisiae* at an iron concentration < 0.1 μM.

μM). As the ferrioxamine B concentration is raised, deviations from Michaelis-Menten are observed, indicating that saturation was not reached as rapidly as would be expected if a single high-affinity mechanism were involved. Increasing the ferrioxamine B concentration also caused the effect of Ga^{3+} ions to change from inhibition to slight stimulation. In contrast, heme-deficient cells show simple uptake kinetics, indicating a single high-affinity transport system that is competitively inhibited by Ga^{3+}-desferrioxamine B but is insensitive to iron(II) chelators and Ga^{3+} ions (Lesuisse and Labbe, 1989). These data demonstrate that *S. cerevisiae* can take up iron from ferrioxamine B by both a specific high-affinity transport system, by which the siderophore is internalized as such, and by a low-affinity transport system, by which iron is dissociated from the ligand by reduction prior to uptake. As the ferrioxamine B concentration is raised, the high-affinity transport system becomes saturated and the contribution of the reductive mechanism increases. Consequently, uptake becomes more sensitive to iron(II) chelators. The extracellular oxygen concentration could also influence the mode of transport of ferrioxamine B: iron uptake from ferrioxamine B is less sensitive to an iron(II) chelator when the cell suspension is saturated with oxygen than under standard conditions (Lesuisse and Labbe, 1989). Emery (1987) reported a similar finding for the uptake of ferrichrome A by *Ustilago sphaerogena.* Ferrichrome A, which can be taken up by this fungus via a reductive mechanism, enters the cell without prior dissociation in the presence of excess oxygen. Both the siderophore-mediated transport system and the reductive mechanism are induced in wild-type cells grown under iron-limitation and repressed in cells grown in iron-rich medium. A plasma membrane protein that binds ferrioxamine B with high affinity can be visualized as a single band by autoradiography of a native gel loaded with total plasma membrane proteins (solubilized with 1% Triton X100) and incubated, after electrophoresis, with ^{55}Fe-ferrioxamine B (Lesuisse and Labbe, unpublished data). Plasma membrane proteins with ferrireductase activity can also be shown by native-PAGE (see Chap. 3). We surveyed a variety of other iron sources (ferric citrate, ferricrocin, ferrichrome, rhodotorulic acid, etc.), but we found no other specific ferric transport system in *S. cerevisiae.* All these ferric chelates are reduced and efficiently used as iron sources by wild-type strains but not by heme-deficient mutants. Therefore, since this yeast does not itself excrete desferrioxamine B, it appears that the main iron uptake system of *S. cerevisiae* involves, at least under laboratory growth conditions, extracellular nonspecific reduction of ferric chelates, followed by internalization of the free ferrous ions. Figure 1 shows a model of iron acquisition by *S. cerevisiae.*

In plants, the role played by excreted reducing compounds and/or a membrane-bound redox system in reductive iron uptake has been ques-

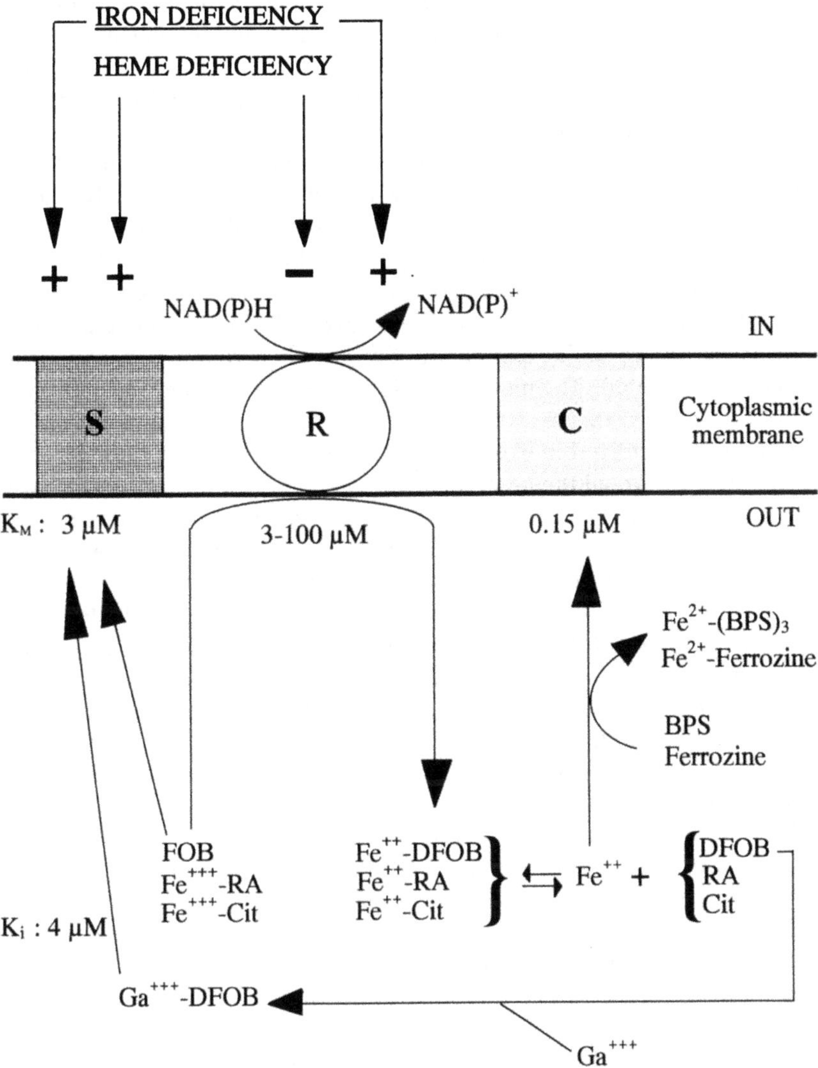

Figure 1 Schematic representation of the mechanisms involved in iron uptake by *Saccharomyces cerevisiae*. S, inducible siderophore-mediated transport system; R, inducible ferrireductase system; C, Fe^{2+} transport system; BPS, bathophenantroline sulfonate; Cit, citrate; (D)FOB, (desferri)ferrioxamine B; RA, rhodotorulic acid.

tioned by several authors (Marschner et al., 1986; Bienfait, 1987; Romheld, 1987). Several plants and microorganisms release fluorescent molecules, very often small phenolic compounds, when grown under iron-limited conditions. Those molecules are not siderophores, but they could facilitate the transport of iron into the cell in several ways. They could keep iron in the Fe^{2+} reduced state after reduction of ferric iron at the plasma membrane and/or they could solubilize iron in the growth medium or even directly reduce certain ferric species (Olsen et al., 1981; Romheld, 1987). *S. cerevisiae* also excretes such reducing/chelating compounds (Lesuisse et al., 1987). The main molecules are anthranilate and 3-hydroxyanthranilate (Lesuisse et al., 1992). However, the excretion of these compounds does not increase when the cells are grown under iron-limitation, and their role in reductive iron uptake seems limited. The main component of yeast cell reducing activity is bound to the cells; this is a trans–plasma membrane redox system whose activity can be detected in isolated plasma membranes (see Chap. 3).

As stated above, all the heme-deficient mutants tested to date have very little ferrireductase activity, so that they are unable to take up iron efficiently from several ferric chelates (except ferrioxamine B). Although these mutant strains turned out to be good tools for iron uptake studies, they are not "true" ferrireductase mutants, i.e., mutants affected in a structural gene of the ferrireductase system. Such a mutant was first described by Dancis et al. (1990). After mutagenesis and screening for the lack of ferrireductase activity, these authors isolated a clone of *S. cerevisiae* that was deficient in ferrireductase activity and had corresponding defect in iron uptake capacity from ferric citrate. Both deficiencies resulted from a single mutation (*fre1-1*). The wild-type allele of the mutant gene was called *FRE1*. The ferrous uptake system of the *fre1-1* mutant is not affected by the mutation: Dancis et al. (1990) determined that a high-affinity (K_M = 5 μM) ferrous uptake system was present in both wild-type and mutant cells. In an earlier study (Lesuisse et al., 1987), we described a ferrous uptake system that was sensitive to competitive inhibition by Co^{2+} and had a low affinity for Fe^{2+} (apparent K_M of about 0.3 mM). More recent studies by Dancis et al. (1991) and Eide et al. (1992) have shown that Fe^{2+} can enter the cell by both a low-affinity and a high-affinity transport system, the latter probably being the only one acting *in vivo*. The high-affinity system described by Eide et al. (1992) is highly specific for Fe^{2+} (for which they measured a K_M of 0.15 μM): these authors observed no inhibition of Fe^{2+} uptake by other divalent metals such as Ni^{2+}, Co^{2+}, Mn^{2+}, or Zn^{2+}. They showed that this ferrous uptake system is regulated in response to iron concentration in the medium independently of the ferrireductase activity of the cells. While regulation of the Fe(II) transporter activity is closely linked to the intracellular iron content, ferrireductase activity may be controlled by factors other than iron status (see Chap. 3). Thus iron uptake by the cells can be limited by the activity of either the ferrireductase or the Fe(II) transporter according to the growth conditions. As pointed out by Eide et al. (1992), differential regulation of ferrireductase and transporter may allow cells to control precisely iron accumulation in response to a variety of environmental conditions.

B. Intracellular Fate of Iron

The iron taken up by the cells is detected very early in the vacuolar compartment. The rate of iron accumulation in vacuoles is maximal right from the moment the cells are incubated with iron (Raguzzi et al., 1988). A great number of divalent cations accumulate, mainly in vacuoles of yeast where they could be bound to polyphosphates (Lichko et al., 1982; Kunst and Roomans, 1985). This compartmentation may be important for maintaining the cytoplasmic homeostasis of physiologically important cations such as Ca^{2+} (Eilam et al., 1985), Mg^{2+} (Lichko et al., 1982), and Zn^{2+} (White and Gadd, 1987) and for the storage of potentially toxic cations that can be accumulated by the yeast cell (Sr^{2+}, Pb^{2+}, Cr^{3+}, etc). The vacuoles are also the major storage compartment for iron. This is the only compartment whose iron concentration increases when yeast cells are subjected to an increased extracellular iron concentration. In contrast, the iron content of the cytosolic fraction is relatively stable; in particular, the iron content of ferritin is not representative of the total cellular iron content (Raguzzi et al., 1988). It seems, however, that the vacuole is not simply a passive storage site for excess iron in the cell. Eide et al. (1992) showed that excess accumulated iron can be used by the cells when they are transferred to an iron-limiting environment. Preliminary observations also suggest that iron can move from the vacuole to the mitochondrial compartment, where it is required for heme synthesis (Raguzzi et al., 1988). Preliminary unpublished experiments showed that iron can be mobilized *in vitro* from intact isolated vacuoles by the combined action of an iron(II) chelating agent and a reductant (ascorbate). However, we do not know whether a reductive mechanism plays a role in vacuolar iron accumulation/mobilization *in vivo*. The movement of iron between different compartments could be facilitated by high intracellular concentrations (millimolar) of citric and malic acids in *S. cerevisiae* (Lesuisse and Labbe, 1992). *Neurospora crassa* also contains high concentrations of citric and malic acids (Ernst and Winkelmann, 1977), whose Fe(II)-chelates are good iron donors for mitochondria (Winkelmann, 1979).

Ferrochelatase, the last enzyme of the heme biosynthesis pathway, is located at the M side of the inner mitochondrial membrane and is responsible for inserting Fe^{2+} into protoporphyrin to form heme (see Chap. 15). Since Fe^{2+} is the metal substrate of ferrochelatase and not Fe^{3+}, a membrane-bound iron-reducing process may also be expected to operate in the mitochondria. Intact functional yeast mitochondria show energy-dependent iron-reducing activity (Lesuisse et al., 1990). The reduction process is maximally stimulated by external NADH, while NADPH and succinate are inefficient electron donors. Several ferric chelates can be reduced, but not ferrioxamine B, suggesting that this siderophore cannot be used directly as an iron source by mitochondria when it is accumulated by the cell without prior dissociation. The hypothetical role of an "external" mitochondrial NADH dehydrogenase in yeast intracellular iron assimilation will require further investigation. Nothing is known about the transport and utilization of iron by yeast mitochondria *in vivo.* The mitochondrial iron transport process, if it does exist, could be unspecific to some extent since Zn^{2+} can also be incorporated into protoporphyrin IX *in vivo.* A "free" iron pool (nonheme, non-Fe-S) has been quantified by Tangeras et al. (1980) in intact rat liver mitochondria; the size of that pool frequently reaches 1 nmol/mg protein and is equally divided between the matrix and the inner membrane; half of the iron is in the reduced form. Tangeras (1985) later showed that the whole mitochondrial pool of "free" iron has access to ferrochelatase and is readily chelatable. Using isolated unpurified yeast mitochondrial membranes, Camadro and Labbe (1982) and Labbe (Lesuisse and Labbe, unpublished data) observed that a substantial amount of "free" iron is associated with these membranes. That "free" iron ranges from 0.5 to 2 nmol/mg protein, and Fe^{2+} accounts for 40 to 70% of this iron (depending on the strain and growth conditions). The "free" reduced iron is the substrate of ferrochelatase, is not readily oxidized by oxygen, but is easily chelated by EDTA. This suggests that Fe^{2+} ions are loosely bound to the surfaces of the membranes, possibly by ionic interactions with the heads of the phospholipids. These apparently small amounts of free ferrous iron are in fact not negligible, since these may be millimolar concentrations in the microenvironment of the mitochondrial membranes, i.e., saturating concentrations for ferrochelatase activity measured *in vitro.* Such a situation cannot prevail in intact mitochondria, or of course *in vivo* where the enzyme never functions at its V_{max}. Therefore the question arises whether this pool of "free" reduced iron in yeast mitochondrial membranes is present *in vivo* or whether it is an artifact arising from the disruption of the cells and the subsequent redistribution of iron from other places, such as the vacuoles (Raguzzi et al., 1988; see Chap. 15).

Soluble cytosolic ferrireductase activities may participate in iron assimilation in many microorganisms and may account for the intracellular dissociation of siderophores (Ernst and Winkelmann, 1977; McReady and Ratlege, 1979; Straka and Emery, 1979; Arceneaux and Byers, 1980; Cox, 1980; Gaines et al., 1981; Moody and Dailey, 1985; Huyer and Page, 1989; Fischer et al., 1990; Le Faou and Morse, 1991; Andrews et al., 1992; Hallé and Meyer, 1992). Because *S. cerevisiae* is able to use iron from ferrioxamine B after internalization of the siderophore, the question also arises how iron is dissociated from its ligand at the intracellular level in this yeast. Heme-deficient cells become red-brown when grown in the presence of ferrioxamine B. Apparently the siderophore taken up by these mutant cells does not dissociate (Lesuisse and Labbe, 1989). In addition, the siderophore-mediated transport system of the mutants remains induced in all growth conditions, while it is repressed in wild-type cells grown in iron-rich (as ferrioxamine B or ferric citrate) medium. We believe that the deficiency of plasma membrane ferrireductase activity of heme-deficient cells is accompanied by a deficiency in intracellular ferrireductase activity that may be required for the removal—and subsequent utilization—of iron from the internalized siderophore. The need for cytosolic ferrireductase activities might however exceed the simple function of removing iron from ferrioxamine B. Several nonspecific diaphorases with iron-reducing properties could play a role in iron metabolism by allowing the direct delivery of iron in a soluble ferrous form to iron-requiring metabolic systems. Such a nonspecific oxidoreductase has been purified from cytosolic extracts of *S. cerevisiae* (Lesuisse et al., 1990). The purified protein, a 40 KDa flavoprotein (FAD), uses NADPH as electron donor and not NADH to reduce several ferric species but not ferrioxamine B. A NADPH-dependent ferrioxamine B reductase activity has also been detected in the cytosol, but the activity is very unstable. The fact that ferrireductase activities (especially the NADPH-dependent activity) detected in cytosolic extracts are increased when the cells are grown in iron-limited conditions (Lesuisse et al., 1990) suggests that they could play a physiological role in iron metabolism. However, confirmation of this hypothesis will require more investigation.

III. THE TRANS–PLASMA MEMBRANE REDOX SYSTEM: PHYSIOLOGICAL, BIOCHEMICAL, AND GENETICAL ASPECTS

Whole cells of *S. cerevisiae* were first shown to be able to reduce external nonpermeant ferricyanide by Crane et al. (1982); that reducing ability was further investigated by Yamashoji and Kajimoto (1986a,b) and Yamashoji et al. (1991). This property is not specific to *S. cerevisiae;* whole cells of *Ustilago sphaerogena* also reduce ferricyanide at a very high rate (Emery, 1983). Crane et al. (1982) suggested that oxygen was the natural electron acceptor for this plasma membrane-bound redox system, which is maximal in log phase of growth. Previously, Ainsworth et al. (1980,a,b) purified an NAD(P)H oxidoreductase from a nonmitochondrial membrane fraction of *S. cerevisiae:* this FMN-dependent protein (55 kDa) containing heme (as P420) was shown to use both NADH and NADPH as electron donor and oxygen as low affinity electron acceptor; oxygen uptake was cyanide insensitive. However, we do not know whether this protein originates in the endoplasmic reticulum or the plasma membrane. Further investigations (Lesuisse et al., 1987; Lesuisse and Labbe, 1989; Dancis et al., 1990) demonstrated that the plasma membrane–bound redox system is involved in reductive iron assimilation by *S. cerevisiae.*

A. Reductase Activity of Whole Cells and Isolated Plasma Membranes: One or Several Reductase System(s)?

Glucose strongly stimulates iron reduction by resting cells from aerobic cultures (3- to 5-fold). The effect is biphasic with a slight increase in the rate of iron reduction immediately after glucose is added, and the maximum reduction rate[2] is reached after a lag of 2 to 3 min. Replacing glucose by ethanol as energy source produces only a small, rapid stimulatory effect (Lesuisse and Labbe, 1992). Anaerobically grown cells are practically devoid of activity, but the bulk of iron in anaerobic cultures is present as Fe(II). The ferrireductase activity of whole aerobic cells is strongly inhibited (80–100% inhibition) by 50–100 μM uncouplers (CCCP), 25–50 μM H^+-ATPase inhibitors (vanadate, DES, SW26), permeabilizing agents (toluene-ethanol, digitonine) (Lesuisse and Labbe, 1992), micromolar concentrations of Pt(II) salts (Eide et al., 1992),[3] low concentrations (20–50 μM) of ionic detergents like CTAB, SDS, or Z16,[4] and 2 mM concentration of the sulfhydryl agent mersalyl. Other sulfhydryl agents such as iodoacetic acid, *p*-chloromercuribenzoate, $HgCl_2$, or DTNB inhibit the ferrireductase to a lesser extent (10–40% inhibition at millimolar concentrations). NaN_3 (5 mM) inhibits ferrireductase activity, while KCN (2–10 mM) has no effect. The effect of H^+-ATPase inhibitors on cell ferri-

[2]Around 1 nmol min^{-1} mg^{-1} (dry weight) for log phase cells grown in YNB-glucose media.
[3]Ferrireductase activity itself was not measured; Pt(II) ions inhibited iron accumulation by yeast cells from Fe(III) and not from Fe(II).
[4]During a systematic search for a possible latent ferrireductase activity requiring detergent for its optimal activity as observed for plant reductases (Buckhout and Luster, 1991), we found that ionic detergents having very different critical micellar concentrations (CMC) like CTAB, Z16, or SDS strongly inhibit the ferrireductase activity at very low concentration (20–50 μM), while nonionic detergents like Tween 80 or Triton X100 were ineffective even at millimolar concentration (Lesuisse et al., unpublished data).

reductase activity should be interpreted carefully; preliminary results suggest that vanadate can be reduced by the cells, so that its inhibitory effect could at least be partly due to a competitive effect. In this respect, it is worth noting that the uptake of vanadium by *S. cerevisiae* seems to involve extracellular reduction of vanadate to vanadyl (Bode et al., 1990). DES could also act nonspecifically to some extent, by making the plasma membrane labile and permeable (Lesuisse and Labbe, unpublished data). The strong inhibitory effect of permeabilizing agents, especially when cells have an induced ferrireductase activity, suggests that the redox process requires that the integrity of the plasmalemma be preserved. Therefore the data obtained with isolated plasma membranes should also be interpreted carefully. Both NADH- and NADPH-dependent ferrireductase activities are detected in purified plasma membranes. Using ferrioxamine B as electron acceptor, the K_M for NADH is 30 μM, and the K_M for NADPH is 2 μM; only the NADPH-dependent activity in plasma membranes from iron-starved cells is 2–3-fold higher than in those from iron-replete cells. However, this increased ferrireductase activity in purified plasma membranes is considerably less than that measured with intact cells (Lesuisse and Labbe, 1992). Assuming that the plasma membrane proteins represent 5% of total cell proteins, the specific ferrireductase activity in isolated plasma membranes should be much higher than that actually measured (20- to 100-fold higher for NADH-dependent activity and 3- to 10-fold for NADPH-dependent activity). The inability of isolated plasma membranes to reflect whole cells' ferrireductase activity is also indicated by the observation that while the *frel-1, heml,* or *rasl* mutant intact cells have much less ferrireductase activity than the wild-type strains (see Dancis et al., 1990; Lesuisse and Labbe, 1989; Lesuisse et al., 1991), the reductase activities of plasma membranes isolated from these strains are similar (Anderson et al., 1992).

One striking characteristic of the plasma membrane redox system is its lack of specificity. Emery (1987) suggested that the reductive systems of iron assimilation could have purposely evolved to be nonspecific, so that they could recognize iron octahedrally coordinated to oxygen ligands rather than to a specific siderophore. Actually, even nonferric compounds are efficiently reduced by both intact cells and isolated plasma membranes, and compounds as different as ferric citrate, ferric-EDTA, ferricyanide, nitroprusside, ferrioxamine B, Cu^{2+}, vanadate, cytochrome *c*, and resazurin can accept electrons from the plasma membrane redox system. We have observed that yeast cells, which are known to reduce permeant methylene blue dye, also reduce nonpermeant (blue) resazurin dye to its highly fluorescent red derivative resorufin. Toluidine blue, which is a nonpermeant molecule distantly related to methylene blue and resazurin, is not reduced by yeast cells (Ito, 1977). That finding prompted us to devise a very sensitive assay of the reducing properties of yeast cells in suspension by recording the increase in resorufin fluorescence (λ_{exc} = 560 nm, λ_{em} = 585 nm). Resazurin is both a pH and a redox indicator ($E'_0 = -50$ mV) that can be used as a redox dye by buffering the medium at pH = 6.5–7.5 (Jacob, 1970). A resazurin assay was first introduced by Guilbault and Kramer (1964) for measuring dehydrogenase activities and then used to quantify NADP(P)H by Cartier (1968).[5] Our assay is based on the fact that the ferrireductase of whole cells has a very low affinity for resazurin (K_M = 285 μM) and that the initial resazurin concentration in the fluorimetric assays should be very low (10 μM) to avoid quenching the highly fluorescent resorufin. It is then relatively simple to measure the rate of reduction of resazurin itself *and* that of another electron acceptor, such as ferricyanide added during the assay that competes with resazurin (see Fig. 2D). In this way we were able to measure indirectly the K_M of ferricyanide and nitroprusside (Lesuisse and Labbe, unpublished data). Typical time-course assays are shown in Figs. 2D and 2E. The reduction of ferric chelates such as Fe(III)-citrate, Fe(III)-EDTA, or ferrioxamine B is easily measured spectrophotometrically by following the formation of the ferrous bathophenantroline sulfonate complex (λ_{max} = 540 nm), while reduction of ferricyanide to ferrocyanide is followed directly by recording the decrease in absorbance at 400 nm. The reduction of Cu^{2+} can be measured by following the formation of the cuprous bathocuproine sulfonate complex (λ_{max} = 478 nm). All our tests showed a constant proportionality between the reduction rates of the various substrates used (mainly resazurin, ferricyanide, ferric citrate, ferric EDTA, and Cu^{2+}) for all the strains tested under all the growth conditions used. For example, no cells showed a low ferric citrate reductase activity and a high ferricyanide or resazurin reductase activity, or vice versa. In particular, the mutant strains bearing the mutations *fre1.1, heml,* or *ras1* are deficient for *all* the reductase activities. Reciprocally, *all* these activities are elevated to the same degree when wild-type cells are grown in iron-deficient conditions (Lesuisse and Labbe, unpublished data). *All* activities are inhibited by low concentrations of DES and ionic detergents, or by 5 mM NaN_3. These observations indicate that a single enzyme system is responsible for all the reductase activities measured. However, some experimental data do not favor this simple hypothesis: (i) reduction of resazurin is only partly inhibited by high concentration (10 mM) of mersalyl; (ii) kinetic studies suggest that reduction of ferric citrate and ferricyanide should be accounted for—at least partly—by

[5]Cox used this assay to measure the membrane-bound diaphorases in his work on the iron reductases of *Pseudomonas aeruginosa* (1980).

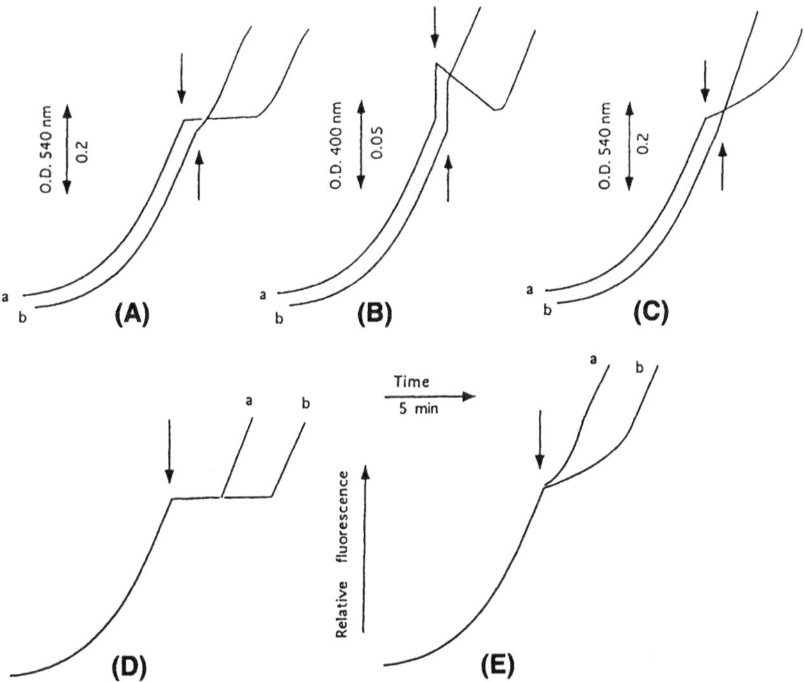

Figure 2 Spectrophotometric (A–C) and fluorometric (D, E) measurements of the reductase activity of a cell suspension. For spectrophotometric measurements, the cells (10 mg/mL in 50 mM Na-citrate buffer, pH 6.5) were incubated at 30°C with constant magnetic stirring in medium containing 5% glucose, 1 mM bathophenantroline sulfonate (BPS), and either 360 μM Fe(III)-EDTA (a) or 360 μM Fe(III)-citrate (b); final volume, 1 mL. The reaction was initiated by adding glucose and iron simultaneously. The absorbance changes of the Fe(II)-(BPS)3 complex was recorded at either 540 nm (A, C) or 400 nm (B) with a Kontron Uvikon 860 spectrophotometer. The arrows indicate the addition of either 50 μM ferricyanide (A, B) or 50 μM Cu^{2+} (C). Parts (A) and (B) thus show the same experiment followed at two different wavelengths: the absorbance of ferricyanide (λ_{max} = 400 nm) at 540 nm is negligible, while ferrous BPS has a significant absorbance at both 540 nm (λ_{max}) and 400 nm.

For fluorimetric measurements, the cells (2 mg/mL in 50 mM Na-citrate buffer, pH 6.5) were incubated at 30°C with constant magnetic stirring in medium containing 5% glucose and 10 μM resazurin. Fluorescence emission at 585 nm (excitation at 560 nm) was recorded (Jobin-Yvon JY3D spectrofluorimeter). In (D), the arrow indicates the addition of 0.5 μM (a) or 1 μM (b) ferricyanide; in (E), the arrow indicates the addition of 0.5 μM Fe(III)-citrate (a) or 0.5 μM Fe(III)-EDTA (b).

two separate mechanisms, as deduced from the recordings shown in Fig. 2. Figures 2A–C show the effect of added ferricyanide or Cu^{2+} on the rate of reduction of ferric EDTA (a) and ferric citrate (b) by a cell suspension. Ferricyanide results in an immediate inhibition of ferric EDTA reduction, which lasts until practically all the ferricyanide has been reduced to ferrocyanide (as confirmed by direct measurement of ferricyanide reduction by the cells). This strong competitive effect is not observed when ferric EDTA is replaced by ferric citrate or ferrioxamine B at the same concentrations (Figs. 2A, 2B). Copper(II) has a less pronounced competitive effect on ferric EDTA reduction (and has no effect on ferric citrate or ferrioxamine B reduction) than has ferricyanide (Fig. 2C). These data indicate that any single reductive system should have a greater affinity for ferricyanide than for ferric EDTA, and a greater affinity for ferric citrate than for ferricyanide and copper(II) (Fe(III)-citrate > ferricyanide > Fe(III)-EDTA, Cu^{2+}). When resazurin is used as the reference electron acceptor however, the apparent order of affinity becomes ferricyanide > Fe(III)-EDTA > Fe(III)-citrate (Figs. 2D, 2E). Clearly, interpretation of these data is incompatible with a single enzyme having different affinities for various electron acceptors. Direct measurements of the reduction of ferric EDTA, ferric citrate, and ferricyanide by whole cells showed comparable K_M values for Fe(III)-EDTA and Fe(III)-citrate (50–100 μM), while ferricyanide reduction was biphasic (*two* apparent K_M's, one about 1–3 μM and the other 50 μM). Similar *double* K_M's were obtained when the rate of ferricyanide reduction was calculated from resazurin competition experiments, as shown in Fig. 2D, but a single K_M (50 μM) was found for nitroprusside. This complex situation also occurs with isolated plasma membranes: ferric EDTA, ferric citrate, and ferrioxamine B are reduced with NADPH as preferential electron donor (Lesuisse and Labbe, 1992), while NADH is the best electron source for the reduction of resazurin. Both NADH and NADPH efficiently reduce ferricyanide (Lesuisse and Labbe, unpublished data). Thus the results obtained with intact cells and isolated plasma membranes indicate that the cell surface bears either several reductase systems or a complex multicomponent reductase system. Some authors have postulated that plant cells have several oxidoreductase enzymes in the plasma membrane (for a review, see Crane et al. 1991). One of them is a constitutive "standard" system that can act as an NADH-oxygen oxidoreductase and use ferricyanide as artificial electron acceptor; another is a "turbo" system functioning with NADPH as electron donor and ferric chelates as natural electron acceptors; a third is an "external" oxidase able to transfer electrons from extracellular added NAD(P)H to oxygen. The situation could be similar in yeast: the standard system may reduce electron acceptors of high redox potential such as ferricyanide, or resazurin, with NADH as electron donor. The "turbo" system, induced in iron-deficient conditions, could reduce ferric chelates in addition to artificial electron acceptors, with NADPH as electron donor. The plasma membrane may also bear an "external" oxidase, since we have observed that *S. cerevisiae* intact cells can transfer electrons from externally added NAD(P)H to oxygen in a cyanide-insensitive way (Lesuisse and Labbe, unpublished data). However, such a simple model involving independent redox systems does not account for the apparent undissociability of the various reductase activities observed. In particular, it is difficult to understand how a single mutation (*fre1-1*) affecting a structural component of the redox system (Dancis et al., 1990, 1992) could result in a deficiency in several reductase systems. We believe that the plasma membrane redox system should be regarded as a complex chain of electron carriers that can be bypassed by some electron acceptors/donors under certain conditions.

B. Genetic and Biochemical Characterization of Ferrireductase

Dancis et al. (1990) have shown that a single mutation in the *FRE1* gene results in a deficiency of the ferrireductase activity of whole cells (5–10% of the parental strain activity). They cloned the *FRE1* gene by complementation of the ferrireductase deficiency of a *fre1-1* mutant strain. The predicted FRE1 protein has a molecular weight of 78.8 kDa and contains 686 amino acids. The protein has hydrophobic regions compatible with transmembrane domains. Interestingly, the predicted FRE1 protein is significantly similar to the large subunit of human cytochrome b_{558}, a component of the respiratory burst oxidase of neutrophils: the proteins are 17.9% identitical and 62.2% similar over the carboxy terminal 402 amino acids of the FRE1 protein (Dancis et al., 1992). At the biochemical level, an NADPH-dependent ferrireductase has been partially purified from isolated solubilized (Tween 80) plasma membranes (Lesuisse et al., 1990). It catalyzes the transfer of electrons from NADPH (K_M = 1.5 μM) to various ferric chelates or to artificial electron acceptors (ferricyanide, dichlorophenolindophenol) but not to oxygen. NADH is not used as an electron donor. The enzyme requires FMN for optimal activity, and the most purified fraction does not contain detectable heme. The partially purified enzyme can be resolved into several bands by SDS-PAGE, but attempts at further purification result in complete loss of activity, which suggests that several components are required for reductase activity. However, native PAGE of the purified enzyme preparation followed by detection of the NADPH-dependent reductase activity (incubation with Fe(III)-EDTA, NADPH, and bathophenantroline sulfonate) identified three bands with reductase activity, at 175 kDa, 80 kDa, and 50 kDa (Lesuisse and Labbe, unpublished data). Total plasma membrane proteins (solubilized in Triton X-100) from either iron-deficient or iron-rich cells were analyzed by native PAGE. The two bands at 175 kDa and 50 kDa were present with the same intensity in both types of membranes; the band of NADPH-dependent reductase activity at 80 kDa was much more pronounced in plasma membranes from iron-deficient cells; and the plasma membranes from iron-deficient cells, but not from iron-rich cells, showed an additional band with NADPH-dependent activity at 68 kDa and a band with NADH-dependent activity at 90 kDa (Lesuisse and Labbe, unpublished data). It is interesting to note that one of the bands whose activity was increased in plasma membranes from iron-deficient cells had a molecular weight (80 kDa) close to that of the predicted *FRE1* gene product. Recently, Yamashoji et al. (1991) described a yeast plasma membrane NADH-menadione oxidoreductase that had properties reminiscent of those of ferrireductase, e.g., an 80 kDa flavoprotein, more expressed in aerobic conditions, more abundant in log phase cells, and inhibited by 20 μM DES (whole cells). However, the physiological role of this oxidoreductase was not discussed. All these data, taken together with the findings of Ainsworth et al. (1980a,b, see above), clearly indicate that a complete biochemical characterization of the yeast plasma membrane redox system(s) is needed. This, together with the genetic approach, would provide a clearer picture of the situation in higher eukaryotes, and especially in dicotyledonous plants.

C. Regulation of Ferrireductase Activity

Whole cell ferrireductase activity is dramatically increased in response to iron limitation during growth. This increase is probably, at least partly, related to an increase in the catalytic capacity of the transmembrane electron transport system. The NADPH-dependent ferrireductase activity of plasma membranes isolated from iron-deficient cells is 2–3-fold higher than that from iron-sufficient cells (Lesuisse and Labbe, 1992), and the level of the *FRE1* gene transcript is regulated by iron in the same way as the reductase activity (Dancis et al., 1990). In addition, *FRE1* nontranscribed sequences can transfer this iron-dependent regulation to the *E. coli lac Z* gene. An 85-base-pair segment of the *FRE1* 5′ noncoding sequence contains a binding site for the transcriptional activator *RAP1* and a repeat sequence (TTTTTGCTCAYC) conferring iron-repressible transcriptional activity on heterologous downstream promoter elements (Dancis et al., 1992).

The increased reducing capacity of iron-deficient cells could also partly result from an increased synthesis of intracellular reduced pyridine nucleotides, as suggested for plants (Sijmons et al., 1984). If NADPH is the electron donor, the large increase in ferrireductase activity of iron-deficient cells could be partly due to an increase in the glucose flux through the NADPH-generating hexose monophosphate pathway, and/or from an increased flux of carbon through citrate. This would lead to increased NADPH production via mitochondrial aconitase and cytosolic $NADP^+$-linked isocitrate dehydrogenase. Our results do not confirm this. The intracellular levels of NADPH, NADH, and organic acids of the tricarboxylic acids cycle in cells with increased ferrireductase activity are unchanged, and the activity of NADPH-generating enzymes is not increased (Lesuisse and Labbe, 1992). Definitive conclusions are, however, difficult to draw from such determinations of intracellular metabolite concentrations. These are steady-state levels, and only direct measurements of the actual rates of synthesis and consumption can provide more accurate conclusions.

The ferrireductase activity of whole cells in batch culture is not dependent solely on the initial iron concentration in the growth medium: even in iron-rich media there is a peak of ferrireductase activity—though less pronounced than in iron-deficient conditions—in the late exponential growth phase (Lesuisse et al., 1990). Eide et al. (1992) have also observed that the cell ferrireductase activity is subject to complex regulation. Cell ferrireductase activity may be partially controlled via the RAS/cAMP pathway (Lesuisse et al., 1991): strains with disrupted *ras1* or *ras2* genes have significantly lower ferrireductase activity than their wild-type counterpart when grown under iron limitation, and the ferrireductase activities of *cdc25* and *cdc35* mutants drop when the cells are shifted to nonpermissive temperature. The ferrireductase activity of a *ras2* mutant grown under iron limitation is restored by the *bcy1* mutation, affecting the regulatory subunit of the cAMP-dependent protein kinase. However, the ferrireductase activity of strains with either *bcy1* or $RAS2^{val19}$ mutations, which both constitutively activate the cAMP pathway, remains repressed under iron-rich conditions. Therefore it seems that a cAMP-mediated activation of protein kinase A may be necessary fully to derepress the ferrireductase activity of the cells, while a strong nutritional cAMP-independent control is probably exerted under repression conditions (high iron concentration). In this respect, it is worth noting that there is both a transient increase in intracellular cAMP (Mbonyi et al., 1990) and a strong increase in ferrireductase activity (Lesuisse and Labbe, 1992) a few minutes after glucose is added to the resting cells. Heme may also be involved in the regulation of ferrireductase activity, since heme-deficient mutants lack ferrireductase activity whatever the growth conditions (Lesuisse and Labbe, 1989). However, it is always possible that the ferrireductase enzymatic system contains a hemoprotein, despite the fact that heme is not detected in isolated plasma membranes.

The loss of a component essential for activity during the preparation of purified plasma membranes could account for the inability of plasma membranes to reflect intact cell reductase. The presence of hemoproteins in yeast plasma membranes is controversial; difference spectra analysis by Schneider et al. (1979) showed *b*-type cytochrome(s), while Ramirez et al. (1984) found no cytochrome in a purified preparation having an NADH-dependent ferricyanide reductase activity.

Experimental conditions can be produced in which the cell ferrireductase activity is induced very early in iron-deficient growth medium (maximum activity after 3 to 6 h of growth) and remains very low (for at least 9 h) in iron-rich media (Lesuisse and Labbe, 1992). Under such conditions, the level of cell ferrireductase activity drops 10-fold when the iron concentration in the growth medium is increased from 0.1 to 50 μM. The activity is induced under iron-deficient conditions before any defect can be detected in cell growth or in cell oxidative/fermentative activities. In addition, the cell ferrireductase activity decreases much more slowly as a function of the intracellular iron content than as a function of extracellular iron concentration. Cells with an intracellular initial iron content of either 4 nmol (mg dry weight)$^{-1}$ (preculture in the presence of 500 μM ferric citrate) or 0.05 nmol (mg dry weight)$^{-1}$ (preculture in the presence of 0.1 μM ferric citrate) both increase their ferrireductase activity up to 7 nmol min^{-1} mg^{-1} (70% of maximal induction) and 10 nmol min^{-1} mg^{-1} (100% induction), respectively, when transferred to an iron-deficient medium (Lesuisse and Labbe, 1992).

Agents other than Fe^{3+} can affect cell ferrireductase activity. The effect of copper is spectacular: a growth medium concentration of 10 nM results in a 50% decrease in cell ferrireductase activity; this is a 100-fold greater repressive effect than that of iron. In contrast, Zn^{2+} is stimulatory, with a maximum around 1 μM. The presence of 50 μM menadione in the growth medium inhibits cell ferrireductase activity by 80%, and continuous air flushing during culture inhibits the activity by 65% (Lesuisse and Labbe, 1992). Thus the cell ferrireductase activity is not governed simply by the iron status of the cells. It can be controlled by a number of factors, several of which could act by modifying the cell redox status. Adding Fe^{3+} to a suspension of iron-deprived cells has the same effect as an oxidative stress: progressive depletion of the intracellular reduced glutathione pool, rapid oxidation of intracellular pyridine nucleotides, oxidation of exofacial sulfhydryl groups (Lesuisse and Labbe, 1992). In this respect, the repressive effect of Fe^{3+} (or Cu^{2+}, or excess oxygen, or menadione) on cell ferrireductase activity can be regarded as an additional response to oxidative stress leading to increased protection of the cells against the loss of intracellular reducing equivalents. Conversely, cells protected against oxidative conditions (Fe^{3+}) during growth not only show a greater ferrireductase activity but also contain more reduced gluthatione and have a much higher capacity to expose exofacial sulfhydryl groups (Lesuisse and Labbe, 1992). The net result is that the cells have an increased capacity to transfer reducing equivalents from the intracellular to the extracellular compartment.

It thus seems possible that the plasma membrane ferrireductase system could play a role in adapting the redox potential of the cells to the growth conditions and be regulated accordingly. Such a hypothesis is not incompatible with iron (or copper) having a specific role in regulating the ferrireductase synthesis by modulating the transcription of the corresponding gene(s) via an activator/repressor whose activity would depend on the ratios Fe^{3+}/Fe^{2+} and/or Cu^{2+}/Cu^{+} and/or cellular redox state, as postulated for the FNR protein in *E. coli* (for a review, see Guest, 1992).

The cell ferrireductase activity could also be modulated to some extent via changes in the plasma membrane potential. The addition of glucose to resting cells is followed within a few minutes by a hyperpolarization of the cells and a large increase in their reducing capacity. Both the glucose-induced polarization of the cells and their ferrireductase activity are significantly increased by adding Zn^{2+} to the growth medium and decreased by adding Cu^{2+} or Co^{2+} (Lesuisse and Labbe, 1992). More experiments are required to determine whether the changes in cell plasma membrane polarizability are actually responsible for the changes in ferrireductase activity.

D. Cell Ferrireductase Activity and Proton Release

Trans–plasma membrane electron transfer in plant cells is accompanied by an increased efflux of protons. Several models have been proposed for the coupling between reductase activity and the accompanying pH changes (for reviews, see Prins and Elzenga, 1991; Rubinstein and Stern, 1991).

Some authors suggest that the reductase could act as an alternative electrogenic H^{+} pump, but it is generally accepted that the primary action of the redox system is the transport of electrons across the plasmalemma, while protons and other cations follow secondarily.

In yeast, there is also a link between the reducing activity of the cells and their ability to acidify the extracellular medium; such a link can readily be shown by plating cells of different strains (ferrireductase deficient mutants and their wild-type counterparts) on a solid medium containing a pH indicator like bromophenol blue. These experiments show that wild-type strains acidify the surrounding medium (yellow halo around the clones) more than do mutant strains bearing the *fre1-1, hem1,* or *ras1* mutation (ferrireductase deficient mutants) (Lesuisse and Labbe, unpublished data).

The addition of Fe^{3+} to a cell suspension results in a rapid initial depolarization of the plasma membrane, followed, in iron-rich cells only, by a phase of slow depolarization. In iron-deficient cells, the rapid depolarization is followed by slow repolarization (Lesuisse and Labbe, 1992). There must therefore be some mechanism that prevents extensive plasma membrane depolarization in iron-deprived cells during trans–plasma membrane electron transfer to extracellular Fe^{3+}. This compensatory mechanism could depend, as proposed for plants (Trockner and Marré, 1988), on the activation of the ATP-driven plasmalemma proton pump, which could be activated by both the initial depolarization and the decrease in intracellular pH that should occur during the transfer of electrons across the plasmalemma.

We have confirmed the report by Crane et al. (1982) that adding ferricyanide to a resting cell suspension results in a measurable decrease in the extracellular pH; the $\Delta H^+/\Delta e^-$ ratio is close to 1 with iron-deprived cells. The thermosensitive ATPase-deficient mutant strain RS373 (Cid and Serrano, 1988) also shows similar rates of decrease in proton excretion and ferricyanide reduction by cells transferred from 26 to 37°C (Lesuisse and Labbe, unpublished data).

The evidence for a contribution of the redox system to acidification is, however, not unequivocal. The addition of ferric EDTA instead of ferricyanide does not cause a titratable H^+ efflux. A similar observation was reported for plants by Lass et al. (1986), who suggested that this was caused by the binding of H^+ to the EDTA. As pointed out by Rubinstein and Stern (1991), reduction of ferricyanide, an anion with a charge of −3, to ferrocyanide, an anion with a charge of −4, results in an additional negative charge that must be balanced by a species with a net charge of +1. This species may be an H^+ derived from some external source. H^+ ions that are detected by this reaction are called "scalar" by Rubinstein and Stern (1991) and are contrasted with "vectorial" H^+, which contribute to the pH of the extracellular medium by moving protons across the plasmalemma from the cytosol.

It is not clear whether the extracellular acidification that occurs during ferricyanide reduction by *S. cerevisiae* is of scalar or of vectorial origin. However, preliminary results (not shown) suggest that the cell surface redox activity could contribute to proton excretion even in the absence of an exogenous artificial electron acceptor, but the mechanism remains to be determined.

IV. CONCLUSIONS AND PERSPECTIVES

The yeast *S. cerevisiae* possesses a plasma membrane electron transport system that is far from simple. It probably consists of a multicomponent system that is very probably regulated by a number of chemical and physical factors. Many prokaryotes and unicellular eukaryotes can adapt their need for iron according to their environment, as they do for other nutrients. Therefore the diversity of iron acquisition mechanisms found in these organisms is accounted for by the physicochemical properties of iron itself, together with ecological factors (ecological niche and competition for iron between species). A unicellular organism, unlike tissue-organized eukaryotes, should be able to fulfill its absolute requirement for iron under environmental conditions that are subjected to rapid and substantial changes. It is therefore likely that these organisms have to concentrate several of the mechanisms of iron acquisition/deposition that occur in different tissues of higher eukaryotes into a single type of cell. Such a diversity is illustrated by the facultative aerobe *S. cerevisiae;* this organism can take up iron not only via an inducible ferrireductase system and a high-affinity ferrous iron transporter but also via a siderophore-mediated transport system, even though *S. cerevisiae* itself does not excrete any siderophore. This type of situation has never been described in other microorganisms, probably because it has never been looked for.

We believe that the plasma membrane electron transport system of *S. cerevisiae* could be involved in physiological processes other than iron assimilation *sensu stricto:* it may play a role in copper and vanadium assimilation and even participate in the regulation of the cell's redox equilibrium and/or the transport of metabolites. Plasmalemma electron transport systems have been implicated in several physiological processes in plants, including hormonal control, regulation of cell division, peroxidative defense mechanisms, and ion transport. Conway's initial (1953) thermodynamic concept of a membrane-bound redox pump was further substantiated using *S. cerevisiae* yeast cells as a model (Conway and Kernan, 1955). However, there is still no evidence for the direct involvement of plasma membrane oxidoreductases in ion transport either in yeast or in plants. More recently, it has been suggested that plasmalemma redox systems may influence the regulation of ion transport by altering the redox state of sulfhydryl groups located at or near the catalytic or regulatory sites of ion transport proteins (for a review, see Kochian and Lucas, 1991).

Electron transport in the plasmalemma of eukaryotic cells has received little attention in the past, but this relatively new area of research is rapidly expanding (Crane et al., 1990; Robertson, 1991). The many advantages offered by *S. cerevisiae* and the increasing power of techniques of molecular biology will undoubtedly mean that the yeast plasma membrane redox system will be further dissected and analyzed. These new findings will contribute greatly to a better understanding of the plasmalemma redox systems' physiological role, not only in yeast but also in plants and animals.

ACKNOWLEDGMENTS

This work was supported by grants from Paris VII Université and the Centre National de la Recherche Scientifique. We thank Monique Simon-Casteras for her expert assistance with some experiments and Dr. O. Parkes for checking the manuscript.

ABBREVIATIONS

CCCP: carbonyl cyanide 3-chlorophenylhydrazone
CTAB: cetyl-trimethyl ammonium bromide
DES: diethylstilbestrol
DTNB: 5-5′-dithio-bis(2-nitrobenzoic acid)
Mersalyl: (*o*-[(3-hydroxymercuri-2-methoxypropyl)-carbamoyl]-phenoxyacetic acid
SDS: sodium dodecyl sulfate
SW26: 2,2,2-trichloroethyl 3,4-dichlorocarbanilate
Resazurin: 7-hydroxy-3H-phenoxazin-3-one-10-oxide
Triton X100: polyethylene glycol-*p*-isooctylphenyl ether
Tween 80: polyoxyethylene sorbitane monooleate
Z16:*N*-hexadecyl-*N*,*N*-dimethyl-3-ammonio-propanesulfonate

BIBLIOGRAPHY

Ainsworth, P. J., Ball, A. J. S., and Tustanoff, E. R. (1980a). Cyanide resistant respiration in yeast, I. Isolation of a cyanide insensitive NAD(P)H oxidoreductase, *Arch. Biochem. Biophys., 202:* 172.

Ainsworth, P. J., Ball, A. J. S., and Tustanoff, E. R. (1980b). Cyanide resistant respiration in yeast, II. Characterization of a cyanide insensitive NAD(P)H oxidoreductase, *Arch. Biochem. Biophys., 202:* 187.

Anderson, G. J., Lesuisse, E., Dancis, A., Roman, D. G., Labbe, P., and Klausner, R.D. (1992). Ferric iron reduction and iron assimilation in *Saccharomyces cerevisiae, J. Inorg. Biochem., 47:* 249.

Andrews, S. C., Shipley, D., Keen, J. N., Findlay, J. B. C., Harrison, P. M., and Guest, J. R. (1992). The haemoglobin-like protein (HMP) of *Escherichia coli* has ferrisiderophore reductase activity and its C-terminal domain shares homology with ferredoxin NADP$^+$ reductases, *FEBS Lett., 302:* 247.

Arceneaux, J. E. L., and Byers, B. R. (1980). Ferrisiderophore reductase activity in *Bacillus megaterium, J. Bacteriol., 141:* 715.

Atkin, C. L., Neilands, J. B., and Phaff, H. J. (1970). Rhodotorulic acid from species of *Leucosporidium, Rhodosporidium, Rhodotorula, Sporidiobolus,* and *Sporobolomyces,* and a new alanine-containing ferrichrome from *Cryptococcus melibiosum, J. Bacteriol., 103:* 722.

Bagg, A., and Neilands, J. B. (1987). Molecular mechanism of regulation of siderophore-mediated iron assimilation, *Microbiol. Reviews, 51:* 509.

Bickel, H., Hall, G. E., Keller-Schierlein, W., Prelog, V., Vischer, E., and Wettstein, A. (1960). Stoffwechselprodukte von Actinomycetes, *Helv. Chim. Acta, 43:* 2129.

Bienfait, F. (1985). Regulated redox processes at the plasmalemma of plant root cells and their function in iron uptake, *J. Bioenerg. Biomembranes, 17:* 73.

Bienfait, F. (1987). Biochemical basis of iron efficiency reactions in plants, *Iron Transport in Microbes, Plants, and Animals* (G. Winkelmann, D. Van der Helm, and J. B. Neilands, eds.), VCH, Weinheim, p. 339.

Bode, H. P., Friebel, C., and Fuhrmann, G. F. (1990). Vanadium uptake by yeast cells, *Biochim. Biophys. Acta, 1022:* 613.

Borst-Pauwels, G. W. F. H. (1981). Ion transport in yeast, *Biochim. Biophys. Acta, 650:* 88.

Buckhout, T. J., and Luster, D. G. (1991). Pyridine nucleotide-dependent reductases of the plant plasma membrane, *Oxidoreduction at the Plasma Membrane: Relation to Growth and Transport, Vol. II, Plants* (F. L. Crane, D. J. Morré, and H. E. Low, eds.), CRC Press, Boca Raton, FL, p. 61.

Camadro, J. M., and Labbe, P. (1982). Kinetic studies of ferrochelatase in yeast. Zinc and iron as competing substrates, *Biochim. Biophys. Acta, 707:* 280.

Carrano, C. J., and Raymond, K. N. (1978). Coordination chemistry of microbial transport compounds: Rhodotorulic acid and iron uptake in *Rhodotorula pilimanae, J. Bacteriol., 136:* 69.

Cartier, P. H. (1968). Dosage des pyridine nucléotides oxydés et réduits dans le sang et les tissus animaux, *Eur. J. Biochem., 4:* 247.

Cid, A., and Serrano, R. (1988). Mutations of the yeast plasma membrane H^+-ATPase which cause thermosensitivity and altered regulation of the enzyme, *J. Biol. Chem., 263:* 14134.

Conway, E. J. (1953). A redox pump for the biological performance of osmotic work, and its relation to the kinetics of free ion diffusion across membranes, *Intern. Rev. Cytol., 2:* 419.

Conway, E. J., and Beary, M. E. (1958). Active transport of magnesium across the yeast cell membrane, *Biochem. J., 69:* 275.

Conway, E. J., and Kernan, R. P. (1955). The effect of redox dyes on the active transport of hydrogen. Potassium and sodium ions across the yeast cell membrane, *Biochem. J., 61:* 32.

Cox, C. D. (1980). Iron reductases from *Pseudomonas aeruginosa, J. Bacteriol., 141:* 199.

Crane, F. L., Roberts, H., Linnane, A. W., and Low, H. (1982). Transmembrane ferricyanide reductase by cells of the yeast *Saccharomyces cerevisiae, J. Bioenerg. Biomembranes, 14:* 191.

Crane, F. L., Low, H. E., and Morré, J. D. (1990). Historical perspectives, *Oxidoreduction at the Plasma Membrane: Relation to Growth and Transport, Vol. I, Animals* (F. L. Crane, D. J. Morré, and H. E. Low, eds.), CRC Press, Boca Raton, FL, p. 1.

Crane, F. L., Morré, D. J., Low, H. E., and Bottger, M. (1991). The oxidoreductase enzymes in plant plasma membranes, *Oxidoreduction at the Plasma Membrane: Relation to Growth and Transport, Vol. II, Plants* (F. L. Crane, D. J. Morré, and H. E. Low, eds.), CRC Press, Boca Raton, FL, p. 21.

Dancis, A., Klausner, R. D., Hinnebusch, A. G., and Barriocanal, J. G. (1990). Genetic evidence that ferric reductase is required for iron uptake in *Saccharomyces cerevisiae, Mol. Cell Biol., 10:* 2294.

Dancis, A., Roman, D. G., Anderson, G., Hinnebusch, A. G., and Hlausner, R. D. (1992). Ferric reductase of *Saccharomyces cerevisiae:* Molecular characterization, role in iron uptake, and transcriptional control by iron, *Proc. Natl. Acad. Sci. USA, 89:* 3

Eide, D., and Guarente, L. (1992). Increased dosage of a transcriptional activator gene enhances iron-limited growth of *Saccharomyces cerevisae, J. Gen. Microbiol., 138:* 347.

Eide, D., Davis-Kaplan, S., Jordan, I., Sipe, D., and Kaplan, J. (1992). Regulation of iron uptake in *Saccharomyces cerevisiae.* The ferrireductase and Fe(II) transporter are regulated independently, *J. Biol. Chem., 267:* 20774.

Eilam, Y., Lavi, H., and Grossowicz, N. (1985). Cytoplasmic Ca^{2+} homeostasis maintained by a vacuolar Ca^{2+} transport system in the yeast *Saccharomyces cerevisiae, J. Gen. Microbiol., 131:* 623.

Elvehjem, C. A. (1931). The role of iron and copper in the growth and metabolism of yeast, *J. Biol. Chem., 90:* 111.

Emery, T. (1983). Reductive mechanism for fungal iron transport, *Microbiology 1983* (D. Schlessinger, ed.), American Society for Microbiology, Washington, D.C., p. 293.

Emery, T. (1986). Exchange of iron by gallium in siderophores, *Biochemistry, 25:* 4629.

Emery, T. (1987). Reductive mechanisms of iron assimilation, *Iron Transport in Microbes, Plants and Animals* (G. Winkelmann, D. Van der Helm, and J. B. Neilands, eds.), VCH, Weinheim, p. 235.

Emery, T., and Hoffer, P. B. (1980). Siderophore-mediated mechanism of gallium uptake demonstrated in the microorganisms *Ustilago sphaerogena, J. Nucl. Med., 21:* 935.

Ernst, J., and Winkelmann, G. (1977). Enzymatic release of iron from sideramines in fungi. NADH:sideramine oxidoreductase in *Neurospora crassa, Biochim. Biophys. Acta, 500:* 27.

Fischer, E., Strehlow, B., Hartz, D., and Braun, V. (1990). Soluble and membrane-bound ferrisiderophore reductases of *Escherichia coli*-K12, *Arch. Microbiol., 153:* 329.

Furhmann, G. F., and Rothstein, A. (1968). The transport of Zn^{2+}, Co^{2+} and Ni^{2+} into yeast cells, *Biochim. Biophys. Acta, 163:* 325.

Gaines, C. G., Lodge, J. S., Arceneaux, J. E. L., and Byers, B. R. (1981). Ferrisiderophore reductase activity associated with an aromatic biosynthetic enzyme complex in *Bacillus subtilis, J. Bacteriol., 148:* 527.

Guest, J. R. (1992). Oxygen-regulated expression in *Escherichia coli, J. Gen. Microbiol., 138:* 2253.

Guilbaut, G. G., and Kramer, D. N. (1964). New direct fluorometric method for measuring dehydrogenase activity, *Anal. Chem., 36:* 2497.

Hallé, F., and Meyer, J. M. (1992). Ferrisiderophore reductases of *Pseudomonas.* Purification, properties and cellular location of the *Pseudomonas aeruginosa* ferripyoverdine reductase, *Eur. J. Biochem., 209:* 613.

Huyer, M., and Page, W. J. (1989). Ferric reductase activity in *Azotobacter vinelandii* and its inhibition by Zn^{2+}, *J. Bacteriol., 171:* 4031.

Ito, T. (1977). Toluidine blue: The mode of photodynamic action in yeast cells, *Photochem. Photobiol., 25:* 47.

Jacob, H. E. (1970). Redox potential, *Methods in Microbiology* (J. R. Norris and D. W. Ribbons, eds.), Vol. 2, Academic Press, New York, p. 91.

Jones, R. P., and Gadd, G. M. (1990). Ionic nutrition of yeast, Physiological mechanisms involved and implication for biotechnology, *Enzyme Microb. Technol., 12:* 402.

Jones, R. P., and Greenfield, P. F. (1984). A review of yeast ionic nutrition, Part 1: Growth and fermentation requirements, *Process Biochemistry, 19:* 48.

Kochian, V. L., and Lucas, W. J. (1991). Do plasmalemma reductases play a role in plant mineral ion transport? *Oxidoreduction at the Plasma Membrane: Relation to Growth and Transport, Vol. II, Plants* (F. L. Crane, D. J. Morré, and H. E. Low, eds.), CRC Press, Boca Raton, FL, p. 149.

Kovac, L. (1985). Calcium and *Saccharomyces cerevisiae, Biochim. Biophys. Acta, 840:* 317.

Kunst, L., and Roomans, G. M. (1985). Intracellular localisation of heavy metals in yeast by X-ray microanalysis, *Scanning Electron Microscopy, 1:* 191.

Labbe-Bois, R., and Labbe, P. (1990). Tetrapyrrole and heme biosynthesis in the yeast *Saccharomyces cerevisiae, Biosynthesis of Heme and Chlorophylls* (H. A. Dailey, ed.), McGraw-Hill, New York, p. 235.

Lass, B., Thiel, G., and Ullrich-Eberius, C. I. (1986). Electron transport across the plasmalemma of *Lemna Gibba* G-1, *Planta, 169:* 251.

Le Faou, A. E., and Morse, S. A. (1991). Characterization of a soluble ferric reductase from *Neisseria gonorrhoeae, Biol. Metals, 4:* 126.

Lesuisse, E., and Labbe, P. (1989). Reductive and non-reductive mechanisms of iron assimilation by the yeast *Saccharomyces cervisiae, J. Gen. Microbiol., 135:* 257.

Lesuisse, E., Crichton, R. R., and Labbe, P. (1990). Iron-reductases in the yeast *Saccharomyces cerevisiae, Biochim. Biophys. Acta, 1038:* 253.

Lesuisse, E., and Labbe, P. (1992). Iron reduction and trans-plasma membrane electron transfer in the yeast *Saccharomyces cerevisiae, Plant Physiol., 100:* 769.

Lesuisse, E., Raguzzi, F., and Crichton, R. R. (1987). Iron uptake by the yeast *Saccharomyces cerevisiae:* Involvement of a reduction step, *J. Gen. Microbiol., 133:* 3229.

Lesuisse, E., Horion, B., Labbe, P., and Hilger, F. (1991). The plasma membrane ferrireductase activity of *Saccharomyces cerevisiae* is partially controlled by cyclic AMP, *Biochem. J., 280:* 545.

Lesuisse, E., Simon, M., Klein, R., and Labbe, P. (1992). Excretion of anthranilate and 3-hydroxyanthranilate by *Saccharomyces cerevisiae:* Relationship to iron metabolism, *J. Gen. Microbiol., 138:* 85.

Lichko, L. P., Okorokov, L. A., and Kulaev, I. S. (1982). Participation of vacuoles in regulation of levels of potassium ion and ortho phosphate ion in cytoplasm of the yeast *Saccharomyces cerevisiae, Arch. Microbiol., 132:* 289.

Low, H. E., Crane, F. L., Morré, J. D., and Sun, I. L. (1990). Oxidoreductase enzymes in the plasma membrane, *Oxidoreduction at the Plasma Membrane: Relation to Growth and Transport, Vol. I, Animals* (F. L. Crane, D. J. Morré, and H. E. Low, eds.), CRC Press, Boca Raton, FL, p. 29.

Lupianez, J. A., Machado, A., Nunez de Castro, I., and Mayor, F. (1974). Succinic acid production by yeasts grown under different hypoxic conditions, *Mol. Cell Biochem., 3:* 113.

McReady, K. A., and Ratledge, C. (1979). Ferrimycobactin reductase activity from *Mycobacterium smegmatis, J. Gen. Microbiol., 113:* 67.

Marschner, H., Romheld, V., and Kissel, M. (1986). Different strategies in higher plants in mobilization and uptake of iron, *J. Plant Nutrition, 9:* 695.

Mbonyi, K., Van Aelst, L., Arguelles, J. C., Jans, A. W. H., and Thevelein, J. M. (1990). Glucose-induced hyperaccumulation of cyclic AMP and defective glucose repression in yeast strains with reduced activity of cyclic AMP-dependent protein kinase, *Mol. Cell Biol., 10:* 4518.

Moody, M. D., and Dailey, H. A. (1985). Ferric iron reductase of *Rhodopseudomonas spheroides, J. Bacteriol., 163:* 1120.

Neilands, J. B. (1957). Some aspects of microbial iron metabolism, *Bacteriol. Rev., 21:* 101.

Neilands, J. B., Konopka, K., Schwyn, B., Coy, M., Francis, R. T., Paw, B. H., and Bagg, A. (1987). Comparative biochemistry of microbial iron assimilation, *Iron Transport in Microbes, Plants and Animals* (G. Winkelmann, D. Van der Helm, and J. B. Neilands, eds.), VCH, Weinheim, p. 3.

Norris, P. R., and Kelly, D. P. (1977). Accumulation of cadmium and cobalt by *Saccharomyces cerevisiae, J. Gen. Microbiol., 99:* 317.

Nunez, M. T. R., Cole, E. S., and Glass J. (1983). The reticulocyte plasma membrane pathway of iron uptake as determined by the mechanism of α, α'-dipyridyl inhibition, *J. Biol. Chem., 258:* 1146.

Okorokov, L. A., Lichko, L. P., Kadomtseva, V. M., Khodolenko, V. P., Titovsky, V. T., and Kulaev, I. S. (1977). Energy-dependent transport of manganese into yeast cells and distribution of accumulated ions, *Eur. J. Biochem., 75:* 373.

Olsen, R. A., Clark, R. B., and Bennett, J. H. (1981). The enhancement of soil fertility by plant roots, *Am. Sci., 69:* 378.

Powell, P. E., Cline, G. R., Reid, C. P. P., and Szaniszlo, J. (1980). Occurrence of hydroxamate siderophore iron chelates in soils, *Nature, 287:* 833.

Prins, H. B. A., and Elzenga, J. T. M. (1991). Electrical effects of the plasma membrane bound reductase on cell membrane potential, *Oxidoreduction at the Plasma Membrane: Relation to Growth and Transport, Vol. II, Plants* (F. L. Crane, D. J. Morré, and H. E. Low, eds.), CRC Press, Boca Raton, FL, p. 149.

Raguzzi, F., Lesuisse, E., and Crichton, R. R. (1988). Iron storage in *Saccharomyces cerevisiae, FEBS Lett., 231:* 253.

Ramirez, J. M., Gallego, G. G., and Serrano, R. (1984). Electron transfer constituents in plasma membrane fractions of *Avena sativa* and *Saccharomyces cerevisiae, Plant Sci. Lett., 34:* 103.

Robertson, R. N. (1991). Introduction: Early days of oxidoreduction and transport hypotheses in plants, *Oxidoreduction at the Plasma Membrane: Relation to Growth and Transport, Vol. II, Plants* (F. L. Crane, D. J. Morré, and H. E. Low, eds.), CRC Press, Boca Raton, FL, p. 1.

Romheld, V. (1987). Existence of two different strategies for the acquisition of iron in higher plants, *Iron Transport in Microbes, Plants and Animals* (G. Winkelmann, D. Van der Helm, and J. B. Neilands, eds.), Weinheim, p. 353.

Roomans, G. M., Theuvenet, A. P. R., Van den Berg, T. P. R., and Borst-Pauwels, G. W. F. H. (1979). Kinetics of Ca^{2+} and Sr^{2+} uptake by yeast. Effects of pH, cations and phosphate, *Biochim. Biophys. Acta, 551:* 187.

Rubinstein, B., and Stern, A. I. (1991). Proton release and plasmalemma redox in plants, *Oxidoreduction at the Plasma Membrane: Relation to Growth and Transport, Vol. II, Plants* (F. L. Crane, D. J. Morré, and H. E. Low, eds.), CRC Press, Boca Raton, FL, p. 167.

Schneider, H., Furhmann, G. F., and Fiechter, A. (1979). Plasma-membrane from *Candida tropicalis* grown on glucose or hexadecane, II. Biochemical properties and substrate induced alterations, *Biochim. Biophys. Acta, 554:* 309.

Schwyn, B., and Neilands, J. B. (1987). Universal chemical assay for the detection and determination of siderophores, *Anal. Biochem., 160:* 47.

Sherman, F. (1991). Getting started with yeast, *Guide to Yeast Genetics and Molecular Biology, Methods in Enzymology, Vol. 194* (C. Guthrie and G. R. Fink, eds.), Academic Press, San Diego, p. 3.

Sijmons, P. C., Van den Briel, W., and Bienfait, H. F. (1984). Cytosolic NADH is the electron donor for extracellular iron III reduction in iron deficient bean *Phaseolus-vulgaris* cultivar prelude roots, *Plant Physiol., 75:* 219.

Straka, J. G., and Emery, T. (1979). The role of ferrichrome reductase in iron metabolism of *Ustilago sphaerogena, Biochim. Biophys. Acta, 569:* 277.

Sun, I. L., Navas, P., Crane, F. L., Morré, J. D., and Low, H. (1987). NADH diferric transferrin reductase in liver plasma membrane, *J. Biol. Chem., 262:* 15915.

Tangeras, A. (1985). Mitochondrial iron not bound in heme and iron-sulfur centers and its availability for heme synthesis in vitro, *Biochim. Biophys. Acta, 843:* 199.

Tangeras, A., Flatmark, T., Backstrom, D., and Ehrenberg, A. (1980). Mitochondrial iron not bound in heme and iron-sulfur centers. Estimation, compartmentation and redox state, *Biochim. Biophys. Acta, 589:* 162.

Thorstensen, K., and Romslo, I. (1988). Uptake of iron from transferrin by isolated hepatocytes. A redox-mediated plasma membrane process? *J. Biol. Chem., 263:* 8844.

Trockner, V., and Marré, E. (1988). Plasmalemma redox chain and H^+ extrusion, II. Respiratory and metabolic changes associated with fusicoccin induced and ferricyanide induced H^+ extrusion, *Plant Physiol., 87:* 30.

Van Dijken, J. P., and Scheffers, W. A. (1986). Redox balances in the metabolism of sugars by yeasts, *FEMS Microbiol. Lett., 32:* 199.

White, C., and Gadd, G. M. (1987). The uptake and cellular distribution of zinc in *Saccharomyces cerevisiae, J. Gen. Microbiol., 133:* 727.

Whiting, G. C. (1976). Organic acid metabolism of yeasts during fermentation of alcoholic beverages—A review, *J. Ind. Brew., 82:* 84.

Williams, R. J. P. (1982). Free manganese(II) and iron(II) cations can act as intracellular cell controls, *FEBS Lett., 140:* 3.

Winkelmann, G. (1979). Surface iron polymers and hydroxy acids. A model of iron supply in sideramine-free fungi, *Arch. Microbiol., 121:* 43.

Wright, G. D., and Honek, J. F. (1989). Effects of iron binding agents on *Saccharomyces cerevisiae* growth and cytochrome P450 content, *Can. J. Microbiol., 35:* 945.

Yamashoji, S., and Kajimoto, G. (1986a). Catalytic action of vitamin K3 on ferricyanide reduction by yeast cells, *Biochim. Biophys. Acta, 849:* 223.

Yamashoji, S., and Kajimoto, G. (1986b). Decrease of NADH in yeast cells by external ferricyanide reduction, *Biochim. Biophys. Acta, 852:* 25.

Yamashoji, S., Ikeda, T., and Yamashoji, K. (1991). Extracellular generation of active oxygen species catalyzed by exogenous menadione, *Biochem. Biophys. Acta, 1059:* 99.

6

Iron Storage in Fungi

Berthold F. Matzanke
University of Tübingen, Tübingen, Germany

I. INTRODUCTION

For oxygen-dependent organisms, but also for most anaerobes, iron is an indispensable element. This transition metal is involved in a great variety of enzymatic processes including electron transfer in the respiratory chain, redox reactions with inorganic substrates (oxygenases, hydrogenases, nitrogenases), acid-base reactions (aconitase), DNA synthesis (ribonucleotide reductase), and DNA cleavage (endonuclease III). Why iron has gained such an essential role in the course of biological evolution remains open to speculation (Wächtershäuser, 1988).

In order to prevent growth-limiting effects, a sufficient supply of iron must be warranted. This can be achieved either by specific iron uptake systems (siderophores, see Chap. 2) or by intracellular iron pools compiled under iron-sufficient growth conditions. Therefore it is not surprising that nature has evolved systems for intracellular iron storage. One class of intracellular iron storage compounds is represented by ferritin in eukaryotes and bacterioferritin in prokaryotes (Secs. III, IV.B, IV.C) Harrison, 1989; Crichton, 1989). The second class are siderophores (Sec. IV.D), which have been identified recently as intracellular iron storage compounds in various fungal systems (Matzanke, 1988).

Although iron is an essential element, it can pose a serious hazard to any organism. Randomly distributed and free accessible iron catalyzes the formation of toxic oxygen free radicals via the Haber-Weiss-Fenton reaction (Sec. II) (Dunford, 1987). In order to prevent such a process, intracellular excess iron must be secured. Thus iron storage compounds (ferritins and siderophores) are also iron detoxifiers.

A very convenient method to analyze iron metabolism, and in particular iron storage, is Mössbauer spectroscopy. Not only isolated molecules can be examined but also whole cells (Secs. III.C and IV). Employing *in vivo* Mössbauer spectroscopy under suitable conditions, number, quantity, and time-dependent changes (kinetics) of the main components of iron metabolism can be analyzed without any isolation procedure (Matzanke, 1987).

II. THE NOXIOUS POTENTIAL OF INTRACELLULAR IRON

In aerobically growing cells, H_2O_2 is a normal metabolite. Its steady-state concentration is in the range 10^{-8}–10^{-9} M (Filho and Meneghini, 1984). Instead, the hydroxyl radical ($OH\cdot$) reacts with almost every type of molecule found in living cells: sugars, amino acids, phospholipids, DNA bases and organic acids (Halliwell and Gutteridge, 1984; Minotti and Aust, 1987). The hydroxyl radical attacks molecules with a rate constant usually of the order of 10^9 M^{-1} s^{-1} (Dorfman and Adams, 1973). Cellular iron is potentially harmful when released from its natural environment because it catalyzes a reaction cycle that generates the $OH\cdot$ radical. A simple mechanism for the generation of oxygen free radicals involves the reduction of H_2O_2 by divalent metal ions, particularly Fe^{2+}, with the formation of hydroxyl radicals (Fenton reaction):

$$H_2O_2 + Fe^{2+} \rightarrow Fe^{3+} + OH^- + OH\cdot \qquad (1)$$

When the Fe^{2+} concentration is limiting, Fe^{3+} reduction is required for the continued formation of hydroxyl radicals. This can be accomplished by the superoxide anion, which functions in addition as a chain carrier of the radical reaction:

$$0_2^-\cdot + Fe^{3+} \rightarrow Fe^{2+} + O_2 \qquad (2)$$

The net reaction of Eqs. (1) and (2) has been termed the iron-catalyzed Haber-Weiss reaction. Due to this reaction, iron may participate in the oxidative damage of cells (Fridovich, 1978; Halliwell and Gutteridge, 1984). The resultant free oxygen radicals have been hypothesized to be causative factors in aging (Harman, 1981), carcinogenesis (Harman, 1981; Ames, 1983), and radiation injury and to be a contributory factor in tumor promotion. In fact, it was demonstrated that oxygen free radicals generated by Fe^{2+} in aqueous solution are mutagenic (Loeb et al., 1988). Therefore virtually all organisms, growing either aerobically or anaerobically, must have evolved well-designed iron-containing enzymes, as well as low or high molecular weight iron transport and storage compounds sufficiently inert with respect to iron-catalyzed Haber-Weiss-Fenton reactions. In fact, ferritins and bacterioferritins are repositories in cells for excess iron, and sequestration of iron within the ferritin molecule serves to protect the cell from oxidative damage caused by free iron. In order to achieve a similar protection, the strategy of high-affinity ferric iron chelators (intracellular storage siderophores) has evolved in various fungal systems. Moreover, stable intracellular ferrous iron carriers must be postulated enabling sheltered iron transfer. Such carriers still remain to be discovered (Matzanke et al., 1989, 1991, 1992). Nevertheless, iron complexes do not preclude reactions with activated oxygen species. Some chelates, e.g., the Fe-EDTA complex, are reactive in the iron-catalyzed Haber-Weiss reaction, whereas others, e.g., ferrioxamine, are not (Halliwell and Gutteridge, 1984; Dreyer and Dervan, 1985; Inoue and Kawanishi, 1987; Basile et al., 1987). Ferrous NADH and NADPH complexes also appear to act as Fenton reagents (Rowley and Halliwell, 1982), as do AMP, ADP, and ATP complexes (Rush and Koppenol, 1986; Rowley and Halliwell, 1982; Floyd and Lewis, 1983).

III. IRON STORAGE VIA FERRITINS

As early as 1894 the isolation of an iron-rich protein from pig liver was reported. In 1937 for the first time an iron-rich protein was isolated and crystalized from horse spleen by Laufberger, who coined the term ferritin for this compound. Now, half a century later, animal ferritin has been observed ubiquitously in many cells, from lower invertebrates (e.g., worms and flies), amphibians, and fish to higher mammalian systems including the human body (Harrison et al., 1987). In plants, phytoferritin is found in plastids (Seckback, 1982; Theil and Hase, 1993). Bacterioferritins (BF) have been detected in *Mycoplasma* (Bauminger et al., 1980), in various gram-negative bacteria (e.g., Dickson and Rottem, 1979; Bauminger et al., 1980) and filament-forming gram-positive Streptomycetes (Winkler et al., 1993). Surprisingly, little information is available on mycoferritins. The species from which mycoferritins have been isolated, *Phycomyces blakesleeanus* (Peat and Banbury, 1968; David and Easterbrook, 1971) and *Mortierella alpina* (Bozarth and Goenaga, 1972) belong to the Zygomycetes.

Within the last decade we have witnessed an explosive progress in our knowledge about ferritins. A vast body of information was obtained on ferritin structures, mechanisms of iron deposition and mobilization, and regulation of its biosynthesis (e.g., Theil and Hase, 1993). These data were mainly derived from mammalian ferritin employing genetic methods (e.g., site-directed mutagenesis) and modern spectroscopic techniques (x-ray diffraction, EXAFS, Mössbauer spectroscopy, EPR).

A. Structures of Ferritins

Ferritin is composed of a protein shell and of iron-containing microcrystals within the protein cavity. The soluble molecule exhibits a molecular mass between 400 kDa and 550 kDa depending on the organism chosen and the iron content. The relatively spherical protein shell (432 point symmetry) with an outer diameter of approximately 120 Å in hydrated crystals is composed of 24 subunits (Fig. 1) (Ford et al., 1984). The horse spleen apoferritin subunit is composed of 174 amino acids yielding a mass of 19.824 kDa (Heusterspreute and Crichton, 1981); the bacterioferritin subunit comprises 158 amino acids giving a mass of 18,495 kDa (Andrews

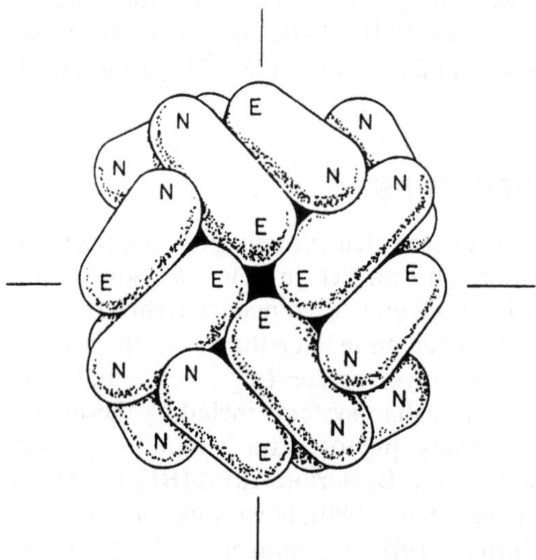

Figure 1 Schematic drawing of the ferritin molecule, showing the symmetrical arrangement of 24 equivalent protein subunits. Each subunit is represented by a "sausage." The sausage is composed of a bundle of four long helices lying along its length and a short helix E lying across the end of the bundle. N depicts the approximate position of the N terminus and E that of the E helix (refer to Fig. 2). (Reprinted with permission of VCH Publishers from *Iron Carriers and Proteins*, T. M. Loehr, ed., 1989, p. 185.)

et al., 1989). The apparent molecular masses found in SDS-PAGE gels range from 15 kDa in *E. coli* (Yariv et al., 1981) to 28 kDa in pea (Laulhere et al., 1988). Within the fungal ferritins, the molecular masses are close to each other: 19.3 kDa in *Mortierella alpina* (Bozarth and Coenaga, 1972) and 18.5 kDa in *Phycomyces blakesleeanus* (LaBombardi et al., 1982).

Variations of the primary structure are found not only between species but also within species (Harrison and Lilley, 1989; Crichton and Charloteaux-Wauters, 1987), e.g., mammalian ferritin (MF) exhibits two isoforms, a 19-kDa subunit (L-form predominant in liver and spleen) and a 21-kDa subunit (H-form, predominant in heart). Each of the 24 subunits consists of a four-helix barrel (A–D) and a fifth shorter helix (E) which forms a cylinder (Fig. 2).

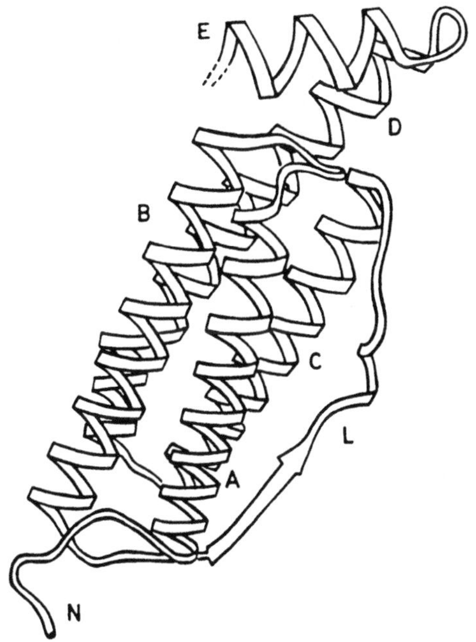

Figure 2 Ribbon diagram of apoferritin subunit backbone. The main body of the subunit is a bundle of four long helices, A, B, C, and D, with a fifth shorter helix, E, lying at an acute angle to the bundle. The N terminus, N, lies at the end of the bundle opposite the C terminus of the polypeptide chain. The last two residues at the C terminus are ill-defined in the electron density maps. A long loop L lying on the outside surface of the molecule joins the C terminal end of helix B to the N terminal end of helix C. (Reprinted with permission of VCH Publishers from *Iron Carriers and Proteins*, T. M. Loehr, ed., 1989, p. 184.)

The holoprotein is pierced by eight channels lined by hydrophilic residues along 3-fold symmetry axes and by six channels with hydrophobic properties along 4-fold axes (Harrison et al., 1986). These channels provide access to the protein interior (presumably for electrons, protons, small ions, and molecules) connecting the cavity to the external surface (Ford et al., 1984; Harrison et al., 1986). The hydrophilic channels were recently identified as the likely sites of iron entry into apo-mammalian ferritin (Stefanini et al., 1989).

The protein cavity is 70–80 Å in diameter harboring the iron core, which is variable in size and can attain a maximum of 4000 and 4500 ferric ions (Fig. 3). The iron core of mammalian ferritin corresponds to the approximate formula $(FeOOH)_8 \cdot (FeOH_2PO_4)$. The structure of the mineral core is similar to the protocrystalline mineral ferrihydrite, in which Fe(III) ions have sixfold oxygen coordination and oxygens are hexagonally close packed (Ford et al., 1984; Taft et al., 1993).

In *E. coli* (Bauminger et al., 1980), *Azotobacter vinelandii* (Stiefel and Watt, 1979), *Pseudomonas aeruginosa* (Moore et al., 1986), *Azotobacter chroococcum* (Chen and Crichton, 1982), and *Nitrobacter winogradskii* (Kurokawa et al., 1989), bacterioferritins (BF) have been detected that are

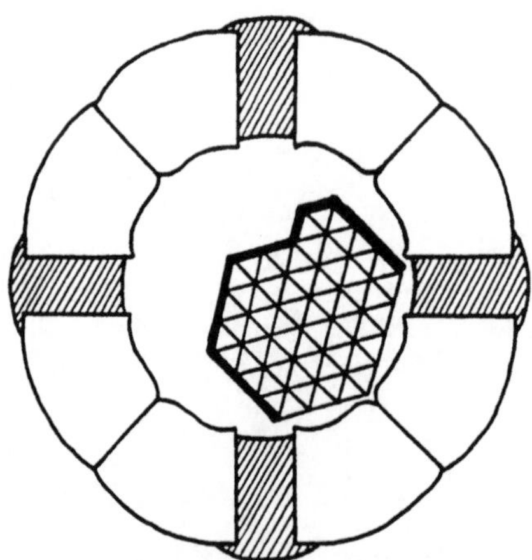

Figure 3 Schematic view of ferritin molecule. The diagram shows a section through the molecule with a symmetrical arrangement of protein subunits forming a hollow shell. Inside the shell there is a microcrystalline iron core of ferrihydrite phosphate. Molecules contain space for 4500 Fe(III), but typical ferritin preparations contain iron-free apoferritin as well as molecules that are only partially occupied by mineral. (Reprinted with permission of VCH Publishers from *Iron Carriers and Proteins*, T. M. Loehr, ed., 1989, p. 126.)

clearly different from the ferritins discussed above, since they contain protoporphyrin IX heme groups (cytochrome $b_{557.5}$ of *A. vinelandii, E coli, A. chroococcum,* cytochrome b_{557} of *P. aeruginosa,* cytochrome b_{559} of *N. winogradskyi*). In *E. coli* and *A. winelandii,* 1 heme per pair of ferritin subunits was found, while there is 1 heme per 5 subunits in *P. aeruginosa* (Moore et al., 1986; Watt et al., 1986; Yariv, 1981). In the case of *E. coli* the identity of bacterioferritin and cyctochrome b_1 has been shown (Yariv 1983; Smith et al., 1988). In *P. aeruginosa* ferritin the axial ligands of heme were identified as thioether side chains of two methionine residues employing EPR and near IR-CD spectroscopy (Cheesman et al., 1990). It has been suggested, assuming the same structure of BF as that of horse spleen L ferritin, that the porphyrin IX is placed in the hydrophilic groove of the inner surface of BF. Two Met-59 ligands related by a twofold axis bind protoporhyrin IX iron at axial positions (Cheesman et al., 1990).

There exists a second feature in which bacterioferritins differ from mammalian ferritins. The iron core of BF contains much more phosphate than MF, suggesting the presence of a different mineral phase (Mann et al., 1987; Watt et al., 1986). This phenomenon will be discussed in more detail in Sec. III.B. Despite the differences of BF and MF, a striking similarity between these proteins has been found. Derivation of a global alignment showed that virtually all key residues specifying the unique structural motifs of eukaryotic ferritin are at least conservatively substituted if not conserved in the sequence of *A. vinelandii* and *E. coli* BF (Grossman et al.. 1992).

B. The Variation of Ferritin Iron Cores Analyzed by Mössbauer Spectroscopy

Mössbauer spectroscopy is an excellent tool for analyzing the electronic state of iron. The commonly observed multipeak Mössbauer spectra have their origin in the interaction between the shell electrons and the ^{57}Fe nucleus. Information about the redox state of the metal center, its spin configuration (high or low spin), symmetry of the ligand field, type of ligand (e.g., O, N, S), and about magnetic properties and particle sizes of magnetic material can be in general derived from the Mössbauer spectra. Therefore, ^{57}Fe Mössbauer spectroscopy is a suitable technique for investigating the iron cores of ferritins. Mössbauer spectra of ferritins display a typical temperature dependence. In Fig. 4, temperature-dependent spectra are depicted of bacterioferritin in *Streptomyces olivaceus* (Winkler et al., in press). At high temperature a quadrupole doublet is observed that is typical of ferric iron in the $^6S_{5/2}$ spin state. Below 15 K, the Mössbauer spectrum broadens, yielding essentially a six-line pattern when the temperature is

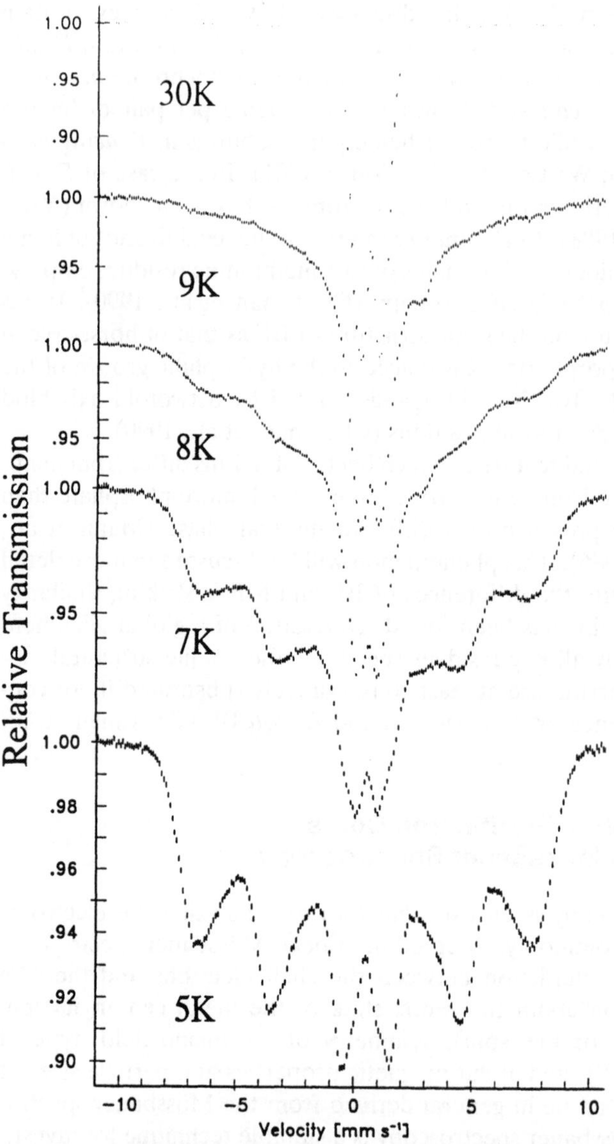

Figure 4 **Mössbauer spectra of gram-positive filament forming *Streptomyces olivaceus* cells grown for 3 days in a mineral salt medium supplemented with 200 μM ^{57}Fe citrate (1/20). The spectra were recorded at various temperatures in a field of 20 mT, applied perpendicularly to the γ-beam.**

further lowered. These spectral changes can be understood within the frame of Neél's theory of superparamagnetism of ultrafine particles (St. Pierre et al., 1986; Bell et al., 1984; Rimbert et al., 1985; Kaufman et al., 1980; Mørup et al., 1980).

The internal field at the iron nucleus depends on the number of unpaired electrons surrounding the nucleus. Whether the internal magnetic hyperfine field splitting, i.e., the six-line pattern, is observable by Mössbauer spectroscopy depends on the relaxation time of the unpaired electron spins. If this relaxation time is much longer than the Larmor precession time of nuclear spin, a magnetically split six-line spectrum is observed. If the relaxation time is shorter than this characteristic time, the average effective field observed at the nucleus is zero, and no magnetic structure is observed in the Mössbauer spectrum. In addition, due to magnetic exchange interaction among the metal sites, the particles exhibit macroscopic magnetization, which, however, may fluctuate, and which at higher temperatures (e.g., 77K in the case of mammalian ferritin) averages to zero within the lifetime of ^{57}Fe or the precession time of its nuclear spin. The superparamagnetic relaxation time τ is the time for the magnetization direction of each particle to flip between the various magnetic easy axes. τ is given by an expression of the form

$$\tau = \tau_0 \exp\left(\frac{KV}{k_B T}\right), \tag{3}$$

K is the magnetic anisotropy constant per unit volume, V is the volume of the particle, k_B is the Boltzmann constant, and τ_0 is a temperature-independent constant that can be evaluated in a number of ways (e.g., Mørup et al., 1980). The value of τ is a function of the volume and magnetic anisotropy constant of the particles and of the temperature. At high temperatures where $KV << k_B T$, the relaxation time τ will be much smaller than the Larmor precession time. The resulting Mössbauer spectrum will display a quadrupole-split two-line pattern. Below the blocking temperature ($KV > k_B T$), a Zeeman split six-line spectrum is expected. However, in practice there is always a distribution of particle volumes due to the microheterogeneity, and this leads to a distribution of blocking temperatures and the characteristic coexistence of doublet and sextet components in the spectra (Bell et al., 1984). The blocking temperature T_B is defined as the temperature at which the two types of components exist with equal probability. In the case of *Streptomyces olivaceus*, T_B is 7K. The separation of such a spectrum into its magnetic and nonmagnetic components is not trivial due to the distribution of hyperfine fields resulting from the particle size distribution.

Table 1 summarizes both the magnetic fields and the blocking temperatures of various ferritins and bacterioferritins. The mycoferritins of sporangia from Zygomycetes (*Phycomyces, Chaetostylum*) display blocking temperatures of approximately 17K, indicating half the core size of mammalian ferritin. The blocking temperatures of all of the bacterioferritins ranging from 20K to 3K are smaller than those of mammalian ferritins and display considerable variations of T_B. As mentioned above, the iron cores of bacterioferritins differ from ferritins in that they are phosphate rich. In fact, the P:Fe ratios in iron cores of bacterioferritins from *P.aeruginosa* and *A.vinelandii* are very high (1:1.71 and 1:1.91, respectively) compared with 1:11.0 and 1:21.0, respectively, in horse and human thalassaemia spleen ferritins (Mann et al., 1987; Treffry et al., 1987). Obviously, hydrated ferric phosphate is the mineral phase in bacterioferritins, whereas the mineral phase in ferritin is hydrated iron oxide (Mann et al., 1987; Rohrer et al., 1990; Watt et al., 1992). In general the higher the phosphate content, the greater the structural disorder in the mineral core.

With the exception of *A.vinelandii* BF (Stiefel and Watt, 1979; Mann et al., 1987) all bacterioferritins analyzed so far display considerably smaller internal fields (41–43T) than ferritins (48–50T). Moreover, for *E. coli*—and

Table 1 Mössbauer Parameters of Various Ferritins, Bacterioferritins, and Mycoferritins

Organism	T (K)	δ (mm/s)	ΔE_Q (mm/s)	T_B (K)	B (T)	Ref.
Native horse	96	0.46	0.70	38		Watt et al., 1986
	4.2				50.2	
Native *Azotobacter*	80	0.48	0.78	20		Mann et al., 1987
BF	4.2	0.45			49	
E.coli (isolated	4.2	0.52	0.65	3		Bauminger et al., 1980
BF and cells)		0.08			43	
Mycoplasma	4.2	0.51	0.63	2.5		Bauminger et al., 1980
capricolum (cells)	0.88				43	
P. aeruginosa BF	78	0.5	0.70	3		St. Pierre et al., 1986
	1.3				42	
S. olivaceus (cells)	77	0.51	0.86	7		Winkler et al., 1993
	1.5				43	
Phycomyces (spores)	77	0.30	0.76	17		Spartalian et al., 1975
	4.2				48.3	
Chaetostylum (spores)	4.2	0.49			49	Matzanke et al., 1993
Chaetostylum	77	0.53	0.64			
(mycelia)	4.2	0.52	0.58			Matzanke et al., 1993

similarly for *Proteus mirabilis* (Dickson and Rottem, 1979), *Mycoplasma capricolum* (Bauminger et al., 1980b), and *Pseudomonas aeruginosa* (St. Pierre et al., 1986)—magnetic ordering is observed below 4.2K only, though the diameter of the electron dense core seems to be fairly large. In the case of the isolated protein of *E. coli,* electron microscopy has revealed a diameter of 60 Å of the iron core (Bauminger et al., 1980a,b). In a subsequent paper the same group (Yariv et al., 1981) reported that there are features discernible in some electron micrographs of the core, in which the electron-dense material has granular appearance with noncontiguity of the substructure. This indicates that the smaller internal fields of many bacterioferritins can be explained in terms of either smaller sizes of the iron cores or of a distribution of smaller iron cores within a single holoferritin molecule.

C. Mechanisms of Iron Deposition and Mobilization in Ferritin

An important role in iron deposition in apoferritin has been attributed to phosphate (Cheng and Chasteen, 1991). However, phosphate is not involved in the initial steps of iron nucleation. Lately the mechanisms of oxidation-coupled iron deposition in ferritins were analyzed in detail by reconstitution studies *in vitro* from apoferritin, Fe(II), and O_2 or another suitable oxidant. These investigations were made possible in particular by Mössbauer spectroscopy, x-ray crystallography, and site-directed mutagenesis of H chain ferritin and overexpression in *E. coli.* By these means, iron nucleation was studied in various human H chain recombinant ferritins (Bauminger et al., 1989, 1991a). It was demonstrated that the first step in ferrihydrite deposition is Fe(II) oxidation catalyzed by apoferritin. Moreover, the oxidation occurs at H chain residues at the postulated ferroxidase center (Lawson et al., 1989, 1991) and not as previously assumed in the 3-fold intersubunit channels (Jacobs et al., 1989). The initially formed isolated Fe(III) and Fe(III) μ-oxo-bridged dimers are located on H chains at the ferroxidase centers (Bauminger et al., 1991a). This solvent-accessible site resembles that of ribonucleotide reductase in its ligands (Nordlund et al., 1990). The residues responsible for Fe binding are Glu-23, Glu-58, His-61, Glu-104, and Gln-137. UV-Vis experiments (Treffry et al., 1992) suggest that the dimer formation is coupled to fast oxidation by a two-electron transfer from two Fe(II) to dioxygen yielding H_2O_2. A sequence alignment and structural comparison of eukaryotic and prokaryotic ferritins suggests the residues of the ferroxidase center in the *A.vinelandii* (or *E.coli*) sequence to be Glu, Glu, His, Glu, and Leu(Met) (Grossman et al., 1992). Of the five key eukaryotic residues, four are conserved. From this it has been predicted that BF should possess the ferroxidase activity of the H subunit (Grossman et al., 1992).

The initial oxidation process is followed by Fe(III) dimer dissociation and migration into the cavity. The migration is both an intra molecular and an intermolecular process enabling ferric iron transfer between ferritin molecules (Bauminger et al., 1991b). However, the mechanisms of migration and of iron core formation are not clear yet. L chain apoferritin seems to be more efficient in forming iron cores than H chain apoferritin. Once stable clusters are formed, they provide an alternative surface for Fe(II)oxidation, enabling the oxidation of 4 Fe(II) per dioxygen yielding water (Ford et al., 1984).

In summary, intrasubunit ferroxidase centers are located between the four helices of H subunits at which (Fe(II) oxidation is catalyzed so that Fe(III) is provided rapidly for ferrihydrite nucleation (Levi et al., 1988; Lawson et al., 1989). L chains lack these centers. However, their cavity faces favor nucleation because they are relatively rich in carboxylic amino acids (Levi et al., 1992).

Mobilization of iron from ferritin can be achieved by complex formation with chelators that are normally present in cells (citrate, sugars, thiols, siderophores). However, the complex formation is rather slow (Funk et al., 1986). Moreover, citrate and sugars exhibit very low complex formation constants, thus making the ferric ion susceptible for reactions of the Haber-Weiss-Fenton cycle (see Sec. II). Direct transfer of ferric iron between ferritin molecules has been observed as well (Bauminger et al., 1991b). Iron is removed from the protein more rapidly by reductants and Fe(II) chelators (Funk et al., 1985; Crichton and Charloteaux-Wauters, 1987). Numerous reducing agents, including dithionite, thiols, ascorbate, dihydroflavins, and superoxide have been shown to enable reductive release. However, in the absence of a chelator all the reduced iron is retained in the protein shell, and for each reduced iron ion two protons are transferred to the core (Watt et al., 1985; Frankel et al., 1987). These findings suggest that iron mobilization from ferritin requires Fe(III) reduction and chelation, and that the Fe(II) chelate is the form in which the iron is exported through the channels. Surprisingly, large, low-potential reductants (flavoproteins and ferredoxins), which cannot enter the ferritin cavity, react anaerobically with both mammalian and bacterial ferritin to produce quantitatively Fe^{2+} in the ferritin cores (Watt et al., 1988). In addition, the oxidation of Fe(II) ferritin by large protein oxidants (e.g., cytochrome *c*) also occurs readily. The path by which electrons may reach or leave the core under these conditions remains enigmatic.

D. Regulation of Ferritin Biosynthesis

The induction of ferritin subunit synthesis by iron is translationally regulated in response to changes in iron availability (Aziz and Munro, 1987;

Hentze et al., 1987; Klausner and Harford, 1989). The human ferritin H chain messenger RNA contains a specific iron-responsive element (IRE) in its 5∩ untranslated region, which mediates regulation by iron of ferritin translation. The human H chain IRE can form a characteristic stem-loop structure. The stem contains one or more unpaired bases including an unpaired cytidine six-nucleotide 5′ region of the loop. The sequence of the loop of all known ferritin IREs is CAGUGX (Aziz and Munro, 1987; Hentze et al., 1987; Rouault et al., 1988). The IREs of these messenger RNAs function as recognition sites for binding of a cytosolic protein, named IRE binding protein (IRE-BP) (Rouault, 1988). In iron-depleted cells (e.g., after desferrioxamine B treatment), the IRE binding protein has a high affinity for IREs. When the cell is iron replete (e.g., after Fe(II) treatment), a decrease is observed in the amount of IRE-BP that binds IREs with high affinity (Constable et al., 1992). Under these conditions, translation of ferritin proceeds. Thus the cytosolic IRE-binding protein acts

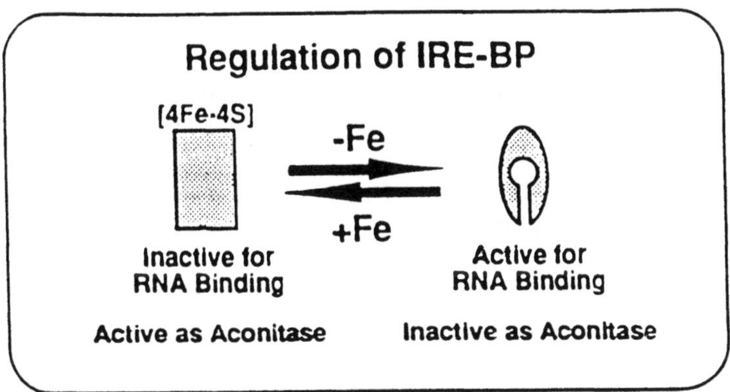

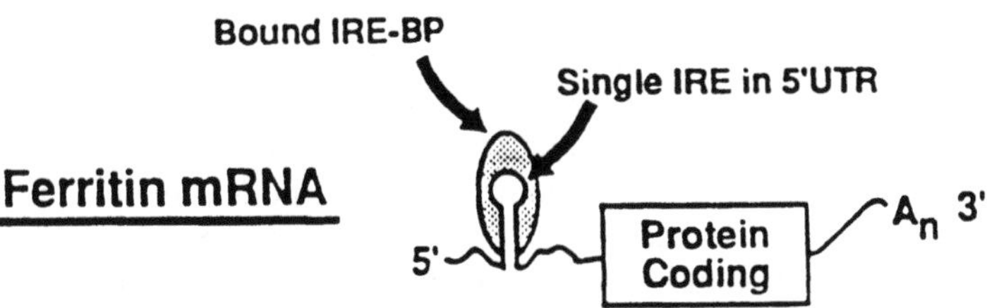

Figure 5 **Coordinate regulation of ferritin biosynthesis is achieved through alteration of the iron-sulfur cluster of the IRE-BP. IRE-BP that is active for RNA binding simultaneously represses translation of ferritin. (Reprinted with permission of Oxford Rapid Communications from *BioMetals, 5:* 135, 1992.)**

as an essential sensor and regulator in the cell. Very recently, it has been shown that the IRE-BP is a cytosolic iron-sulfur cluster protein that is highly related to mitochondrial aconitase (Hirling et al., 1992; Rouault et al., 1990, 1992; Philpott et al., 1991; Kennedy et al., 1992). Aconitase contains a labile Fe-S cluster that has been shown to be readily interconverted *in vitro* between cubane [4Fe-4S] and [3Fe-4S] forms (Surerus et al., 1989). However, in the case of cytosolic aconitase(IRE-BP), the total disassembly of the cubane iron-sulfur cluster is required for high-affinity RNA binding (Haile et al., 1992). The corresponding model is shown in Fig. 5.

IV. DIVERSITY OF IRON STORAGE IN FUNGI: MYCOFERRITINS, ZYGOFERRITINS, AND SIDEROPHORES

In this section, mainly new data will be presented on iron storage in various Zygomycetes, Ascomycetes, and Heterobasidiomycetes. *In vivo* Mössbauer spectroscopy was employed in these studies, enabling a time-dependent monitoring of the main components of iron metabolism. The results indicate a remarkable variability of iron storage strategies in fungi.

A. Analysis of Iron Storage in Fungi by *in Vivo* Mössbauer Spectroscopy

Information on the intracellular fate of accumulated iron obtained with radioactive labels or photometrically with compounds like ferrozine is limited. In particular, the determination of cellular ferrous iron is not a trivial task. Commonly used colorimetric reagents like bathophenanthroline or ferrozine exhibit redox potentials near 1 V. The ferric/ferrous redox couple is 0.77 V. Therefore the ferrous ion complexing agents may function as reducing agents, yielding an artificial formation of ferrous iron. Hence wrong positive results must always be taken into account. Fortunately, Mössbauer spectroscopy permits an unambiguous determination of ferrous iron, thus circumventing the flaws of colorimetric methods. Given a suitable resonance nuclide concentration, Mössbauer spectroscopy allows a nondestructive investigation of iron metabolism *in vivo*. In many studies we have demonstrated that Mössbauer spectroscopy of whole cells yields valuable information about main components of iron metabolism. *In vivo* Mössbauer spectroscopy enables a qualitative and quantitative assignment of these components in a parallel mode. Thus the time dependent changes of these species, i.e., the kinetics of metabolic changes, can be monitored without any isolation procedure.

In order to analyze the routes and dynamics of iron metabolism and storage in fungi by means of *in vivo* ^{57}Fe Mössbauer spectroscopy and EPR, we have exploited a specific iron accumulation route. Most microorganisms growing in their natural habitat are faced with conditions of low-iron stress. This is the driving force for the excretion of siderophores by microbes (see Chap. 2) and the expression of high-affinity siderophore transport systems, which provide a tightly controlled and specific cellular iron supply. In long term studies, ^{57}Fe-labeled siderophores are added with the inoculum to a low-iron culture. Kinetics of siderophore uptake and metabolic utilization of the metal are most easily analyzed in cultures grown for many generations under iron-deficient conditions. The kinetics of iron uptake and metabolic turnover of the metal can in principle be observed with a time resolution merely limited by the time required for sample preparation. Depending on the experimental setup, the lower limit can be as small as 1 min. Details of the methodology and of the experimental design are described in previous review articles (Matzanke, 1987, 1991).

B. Zygoferritin Iron Storage in Rhizoferrin Producing Zygomycetes

Iron uptake and iron storage of Zygomycetes differ considerably from other fungi analyzed so far: (i) zygomycetes are the one and only class of fungi of which synthesis of mycoferritins has been reported (David and Easterbrook, 1971; Spartalian et al., 1975); (ii) whereas many Ascomycetes and Basidiomycetes produce hydroxamate-type siderophores under iron-deficient growth conditions, no hydroxamates were found in culture supernatants of Zygomycetes. In contrast, a novel polycarboxylate-type siderophore has been detected in various families of the Mucorales; in *Rhizopus microsporus, Mucor mucedo, Phycomyces nitens, Chaetostylum fresenii, Cunninghamella elegans, Mortierella vinacea,* and *Basidiobulus microsporus* (Thieken and Winkelmann, 1992). This siderophore, named rhizoferrin, is composed of two units of citric acid linked by diaminobutane (Drechsel et al., 1992) (see Chap. 2).

In order to study the mode of specific iron uptake and metabolism in Mucorales we have analyzed ^{57}Fe rhizoferrin accumulation in *Chaetostylum fresenii* and *Rhizopus arrhizus* by *in vivo* Mössbauer spectroscopy and EPR (Matzanke et al., 1993). Figure 6 represents a Mössbauer spectrum of rhizoferrin. This spectrum is typical of high-spin ferric iron in the slow relaxation limit. The internal field of 51.7 T is smaller than those of hydroxamate and catecholate siderophores (53–56 T). Moreover, no distribution of hyperfine fields can be observed which is typical of ferritins. Therefore rhizoferrin can be discriminated easily from other siderophores, MF and BF. The spectrum in Fig. 6 shows spores of *C. fresenii* that were

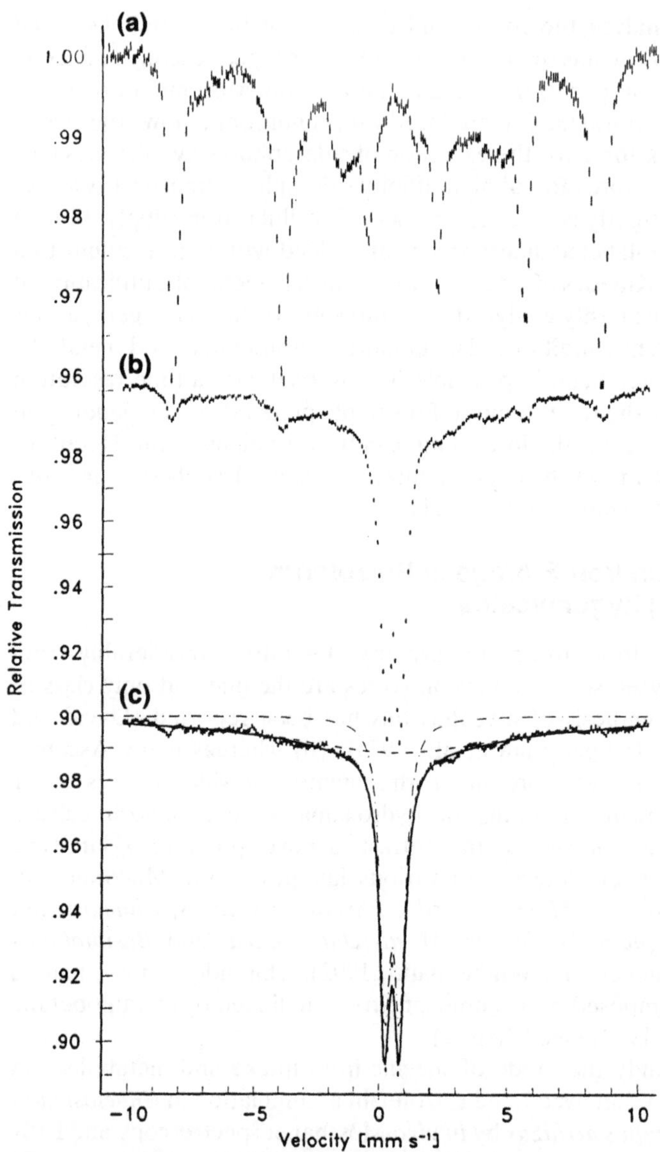

Figure 6 Mössbauer spectra of a frozen aqueous solution of (a) ^{57}Fe rhizoferrin and of (b, c) *C. fresenii* recorded at 4.2K in a field of 20 mT, applied perpendicularly to the c-beam. Spores were allowed to germinate and grow for 20 h; then mycelia were incubated for 25 min with 15μM ^{57}Fe rhizoferrin. Spectrum (c) corresponds to spectrum (b) after subtraction of rhizoferrin (18% of the absorption area of (b)).

allowed to germinate and grow for 24 h in a low-iron medium. Then the mycelia were incubated for 24 min with rhizoferrin. The spectrum contains a magnetically split subspectrum and a doublet species. The six-line pattern is identical to that of rhizoferrin. In fact, spectrum a could be subtracted from spectrum b yielding a rhizoferrin contribution to the total cellular iron pool of 18(±5)%. Spectrum (c) of Fig. 6 corresponds to spectrum (b) from which component a has been subtracted (18%). The spectrum represents a single, magnetically broadened doublet, the Mössbauer parameters of which are close to bacterioferritin found in *E. coli* (see Table 1). It is the sole and major iron metabolite after 25 min of incubation, accounting for 82(±5)% of the total absorption area. In contrast to other fungal and bacterial systems investigated so far, no ferrous iron could be detected. The bacterioferritinlike compound obviously is the immediate intracellular target of rhizoferrin-bound iron. In order to confirm an iron storage function of the BFlike species found in the cell spectrum, *in vivo* EPR spectroscopy was employed. In an early EPR study on ferritin (Boas and Troup, 1971; Weir et al., 1985) an extremely broad resonance centered around $g = 2$ has been observed in the temperature range between 77K and 295K. At 4.2K this very broad resonance could not be detected. This characteristic feature was explained by superparamagnetic properties of ferritin. If ferric iron particles are antiferromagnetically ordered (as is the iron core of ferritin), their ESR signals will broaden and decrease in intensity below the Neél (ordering) temperature. The line width ΔH is given by

$$\Delta H = AKV \exp\left(\frac{KV}{k_B T}\right) \tag{4}$$

where A is a constant dependent upon the core size. Therefore, the lower the temperature, the broader the EPR signal, and no resonance will be observed at very low temperatures.

Figure 7 displays EPR spectra of cells grown under the same conditions as those in Fig. 6. The spectra were recorded at X-band in the temperature range 20–200K. At 20K two signals can be discerned, one at $g = 4.3$ and one at $g = 2$. The signal at $g = 4.3$ is typical of high-spin ferric iron which we attribute to rhizoferrin. The signal at $g = 2$ is due to a free radical. This signal of unkown origin is found in all the EPR spectra of cells we have analyzed so far. At 100K and 200K an additional, very broad signal typical of superparamagnetic resonance can be observed. From these EPR spectra is becomes clear that the magnetically broadened doublet of the Mössbauer spectrum (Fig. 5c) doubtless is a bacterioferritin-type iron storage compound for which we shall use the term zygoferritin (ZF). A similar Mössbauer pattern was found in cells of *C. fresenii* at various growth and transport conditions.

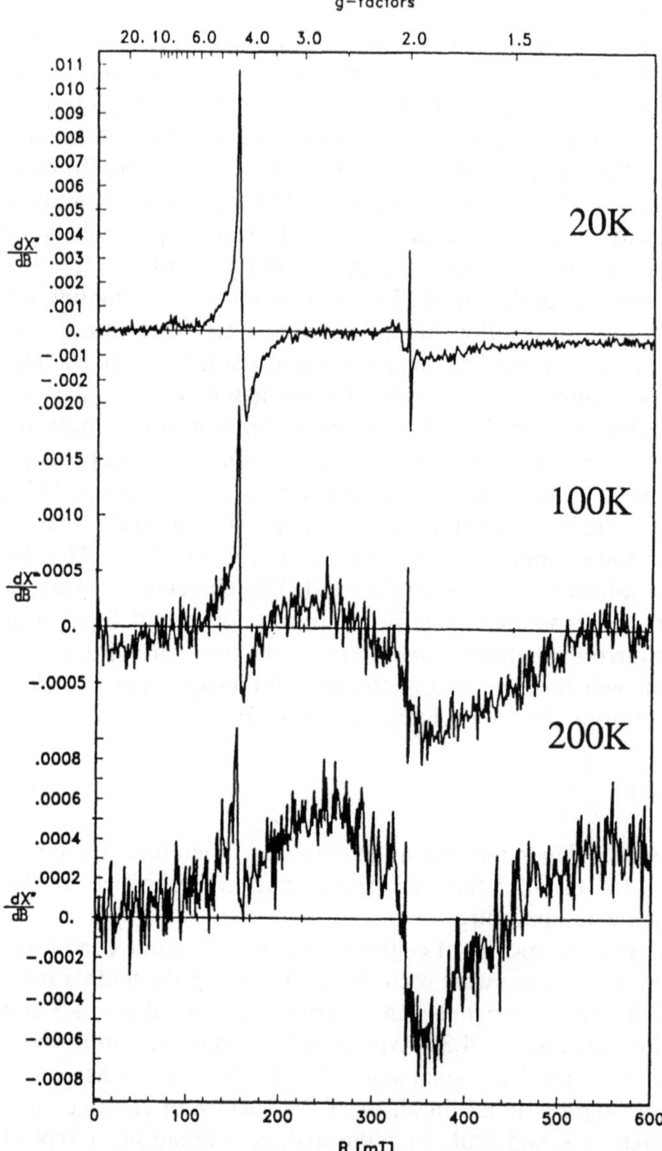

Figure 7 EPR spectra of frozen cells of *C. fresenii*. Preparation conditions were the same as described in Fig. 6. The spectra were measured at 20K, 100K, and 200K (20 μW/40 dB, 9.4 GHz, 10.0 G modulation amplitude).

From these findings we can conclude that (i) the siderophore rhizoferrin is transported as an intact complex through the cell membrane and accumulated to some extent inside the cell; (ii) in contrast to bacteria and other fungal systems, ferric iron is directly transferred to zygoferritin. ZF represents obviously the iron storage pool in mycelia of *C. fresenii*. Zygoferritin-type iron storage could be also detected in *Rhizopus*.

Several investigators have demonstrated a wide range of cross-reactivity between ferritins isolated from different species. For example, within the bacterioferritins, immuno-cross-reactivity of *Azotobacter*-BF and of *Pseudomonas*-BF with *E. coli*-BF antisera was observed, indicating the structural similarities and common surface epitopes of bacterioferritins (Andrews et al., 1991). No reactivity of *E. coli*-BF antisera to horse spleen ferritin or to pea seed was found. From mycelia of *Phycomyces blakesleeanus* grown in iron-rich ([Fe] = 270 μM) media a ferritin was isolated that was immunologically unrelated to horse spleen ferritin (LaBombardi et al., 1982). Therefore, mycelial ferritin from *Phycomyces* exhibits features that distinguish it from ferritins. Actually, tryptic digests of *Phycomyces* ferritin yielded 17 ninhydrinpositive spots as compared to 26 for horse spleen ferritin trypic digest. Thus the term zygoferritin for these fungal ferritins is justified.

C. Iron Storage in Spores of Zygomycetes: Mycoferritins

Figure 8 shows a Mössbauer spectrum of spores of the Zygomycete *Chaetostylum fresenii* (Matzanke et al., 1993). The spores were harvested from agar cultures supplemented with 100 μM ferric citrate (Fe/citrate = 1/20). Two features are visible in the spore spectrum: a quadrupole doublet (25(1)% of the total absorption area) and a dominant broad six-line pattern (75(1)% of the iron content of spores) exhibiting a magnetic splitting of 489 kG. The magnetic split species cannot be fitted with single Lorentzian lines indicating a distribution of magnetic sextets due to a distribution in hyperfine fields resulting from a size distribution of magnetic particles. Very similar spectra have been reported earlier for spores of another Zygomycete, *Phycomyces blakesleeanus,* harvested from ^{57}Fe-enriched agar cultures (Spartalian et al., 1975). Temperature-dependent Mössbauer spectra of *Phycomyces* spores (not shown) revealed, in accordance with the theory of superparamagnetism, an increasing amount of a magnetic subspectrum at temperatures at and below 20K. A blocking temperature of about 17K was determined, which is about half that of horse ferritin (38K) (Mann et al., 1987; see also Table 1). Based on Eq. (3), the average particle size has been estimated to be approximately 40 Å. Therefore the iron core of *Phycomyces* ferritin is about half the size of that for

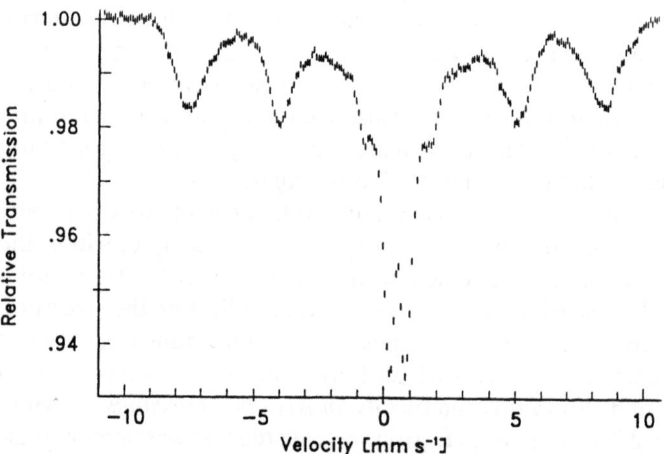

Figure 8 Mössbauer spectrum of frozen spores of *C. fresenii* at 4.2K in a magnetic field H_{app} = 20 mT perpendicular to the γ-rays. ^{57}Fe-enriched spores were harvested from YMG agar cultures with 50 μM ^{57}Fe added as the citrate complex.

horse ferritin. This is in agreement with determination of the particle size by sedimentation (David and Easterbrook, 1971). Quadrupole splitting and hyperfine field (483 ± 10 kG) are in the same range as mammalian ferritins (see Table 15.3). This mammalian ferritinlike iron storage compound is named mycoferritin.

Mössbauer spectra of spores from both *Chaetostylum* and *Phycomyces* show an inner doublet, which, contradicting Neél's theory, does not disappear when the temperature is lowered to 4.2K. The Mössbauer parameters of this species (δ = 0.53(2) mm/s, ΔE_Q = 0.64(2) mm/s) are the same as those for zygoferritin, which is the dominant species in mycelia of *C. fresenii* and close to *E. coli*-type BF (see Table 1). An analysis of the mycelia and various parts of the sporangiophores of *phycomyces* reveal that in the bottom sections of sporangiophores zygoferritin is the major species. There is progressively less zygoferritin in going from mycelia to the sporangiophores to the sporangia themselves. Therefore, we relate the doublet species to zygoferritin. On the way to the sporangia, iron bound to zygoferritin is obviously transformed into mycoferritin. It should be mentioned that the zygoferritin iron species was not recognized as such in the early *phycomyces* study (Spartalian et al., 1975).

In summary, spores of various Zygomycetes exhibit two different iron pools: large amounts of mycoferritin and minor amounts of zygoferritin. It then can be suggested that mycoferritins are true iron storage pools in spores required for germination, thus preventing iron from becoming a germination-limiting factor.

D. Siderophores Are Iron Storage Compounds in Hydroxamate Siderophore Producing Fungi

1. *Iron Storage via Coprogen and Ferricrocin in the Ascomycete* N. crassa

Under conditions of iron deficiency, the ascomycete *N. crassa* synthesizes and excretes several hydroxamate-type siderophores. The major component (70%) is coprogen (Wong et al., 1984). Besides other minor siderophore fractions, traces of the cyclic hexapeptide ferricrocin can be detected in the culture filtrate. Both coprogen and ferricrocin have been isolated from cell extracts of *N. crassa 74A* (Horowitz et al., 1976). These siderophores are transported across the cytoplasmic membrane by a process that involves stereospecific recognition by a siderophore receptor and energy-dependent uptake (see Chap. 2).

In order to analyze the intracellular role of coprogen and ferricrocin, uptake and metabolically induced metal transfer from and between these siderophores in wild-type *Neurospora crassa* 74A and in the siderophore-free triple mutant *N. crassa* (*arg-5 ota aga*) were followed by Mössbauer spectroscopy (Matzanke et al., 1987b; Matzanke et al., 1988). The corresponding contributions to the total absorption area of the Mössbauer spectra are summarized in Table 2. In Fig. 9, Mössbauer spectra of frozen *N. crassa arg-5 ota aga* cells are shown at various phases of coprogen-mediated iron accumulation (0.5, 4, 27, and 65 h). For comparison, Fig. 10 shows the same spectra after subtraction of the siderophore contributions. All Mössbauer spectra exhibit coprogen contributions indicating that this siderophore is accumulated inside the cell as an entity. From parallel ^{55}Fe measurements, it could be derived that coprogen is accumulated in the cytoplasm (Matzanke et al., 1987b). A continuous but slow decrease of coprogen concentration with time is found. Moreover, after 3 days of iron starvation the rate of metabolic turnover is much higher (45% in 27 h) than after 1 day of iron starvation (25% in 44 h). From these observations it can be concluded that metal release from coprogen is regulated by the iron requirement of the cell. The slow metal release from coprogen indicates an additional function of this siderophore, namely, the role as an intermediate-term iron storage compound. After 27 h of growth in mycelia supplied with ^{57}Fe-coprogen, the siderophore ferricrocin can be observed in the cell spectra. This is unexpected, since *N. crassa arg-5 ota aga* is unable to synthesize ornithine. Ferricrocin is obviously synthesized by the use of coprogen degradation products.

Table 2 **^{57}Fe coprogen and ^{57}Fe ferricrocin uptake in *N. crassa* 74A and in Mutant Strain *N. crassa arg-5 ota aga*, which is unable to synthesize siderophores.**

Time	Percentage of intracellular iron species			
	Coprogen	Ferricrocin	Fe^{2+}	Fe^{3+} (?)
A. ^{57}Fe-COPROGEN-UPTAKE IN *N. CRASSA* 74A				
0.5 h	65%	28%	—	7%
4 h	36%	47%	?	17%
27 h	0	73%	9%	18%
B. ^{57}Fe-COPROGEN UPTAKE IN *N. CRASSA* "OTA"				
0.5 h	85%		15%	
4 h	68%		16%	16%
27 h	55%	12%	18%	15%
65 h	23%	26%	16%	35%
C. ^{57}Fe-COPROGEN UPTAKE IN *N. CRASSA* "OTA" (1 DAY DEF.)				
4 h	80%		12%	8%
30 h	75%		12%	13%
D. ^{57}Fe-FERRICROCIN UPTAKE IN *N. CRASSA* "OTA"				
0.5 h	—	99	1	?
4 h	—	89	?	11
65 h		90	?	10
E. ^{57}Fe-FERRICROCIN UPTAKE IN *N.CRASSA* 74A				
0.5 h		95%	3%	2%(?)
4 h		93%	—	7%
27 h		90%	—	10%

Cells were grown for 24 h (C) and 72 h (A,B,D,E) in iron-depleted minimal medium and incubated for 1 h with ^{57}Fe-siderophore. The percentages of subspectra contributing to the total absorption area of cell spectra in Fig. 5 were obtained by stripping (coprogen, ferricrocin) and least squares fit procedures of the residual spectra (Fig. 6).

Wild-type *N. crassa* 74A was analyzed under the same conditions as the mutant. Neither ferritin nor bacterioferritin could be detected. Rather, the long-term iron storage compound was found to be ferricrocin (see Table 2). Within 27 h coprogen is completely metabolized. The mechanism of metal

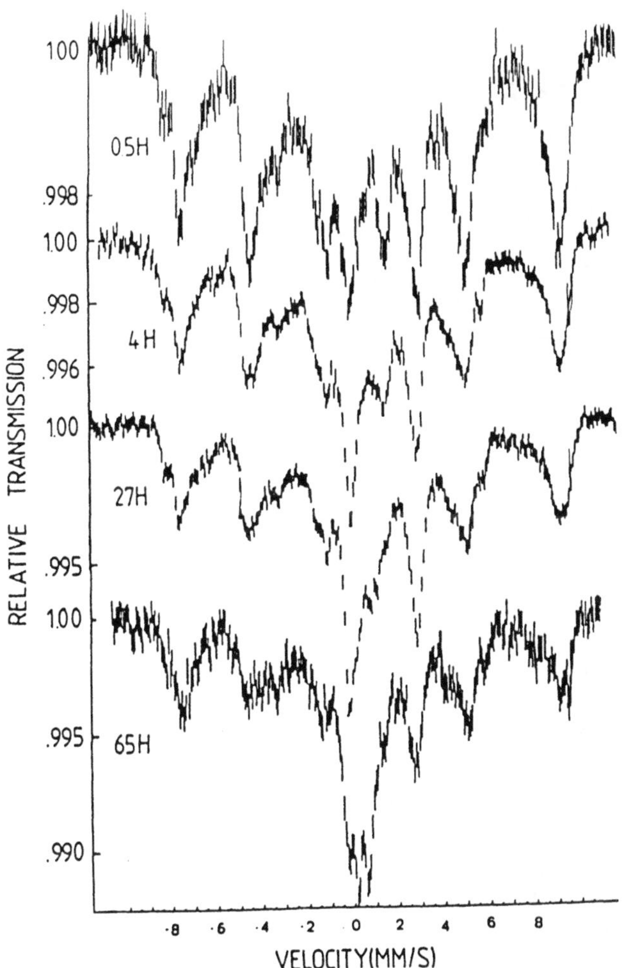

Figure 9 Mössbauer spectra of frozen *N. crassa arg-5 ota aga* cells at 4.2K in a magnetic field H_{app} = 20 mT perpendicular to the γ-rays. The cells have been cultured over a 72-h period in a minimum medium with no iron added. After 72 h the cell suspension was incubated for 1 h with ^{57}Fe coprogen, washed, and grown for additional time periods in fresh minimum medium. Metabolization was stopped by freezing samples after 0.5 h, 4 h, 27 h, and 65 h of additional growth.

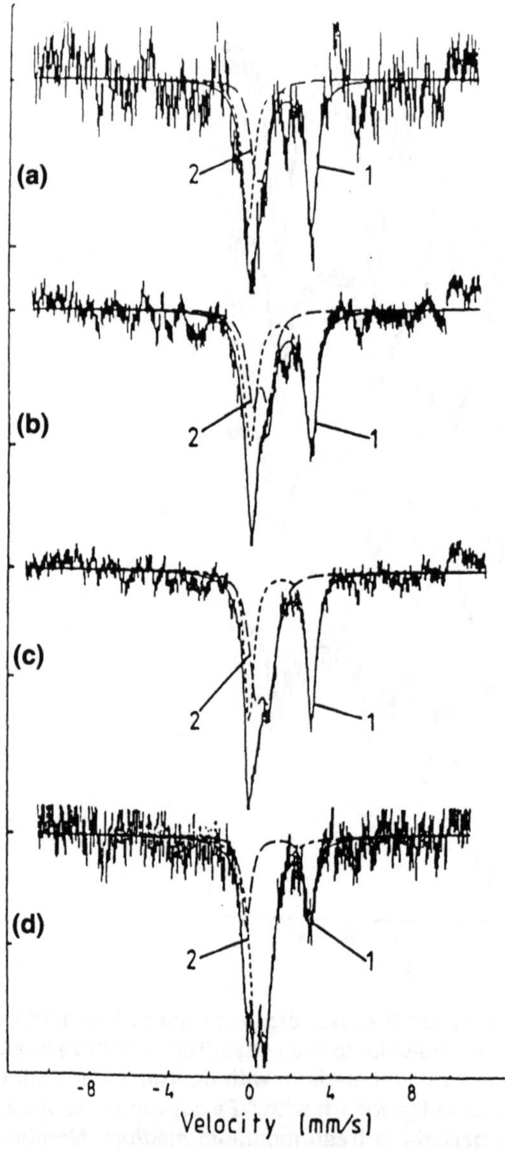

Figure 10 Mössbauer spectra of frozen *N. crassa arg-5 ota aga* cells after stripping. Experimental conditions as in Fig. 9: 0.5 h (a), 4 h (b), 27 h (c), and 65 h (d). Contributions of coprogen and, when necessary, of ferricrocin (spectra c and d) were subtracted from the cell spectra. The resulting metabolite spectra were fitted with two quadrupole doublets of Lorentzian line shape (see text). The corresponding percentages of the absorption areas are listed in Table 8.

transfer from coprogen to desferriferricrocin can be explained in terms of simple ligand exchange followed by reexport of desferricoprogen. In a third set of Mössbauer experiments, ferricrocin uptake was monitored in both *74A* and *ota* (Matzanke et al., 1988). Again growth and uptake conditions were kept similar to the experiments described above. Surprisingly, removal of ^{57}Fe from its host ferricrocin did not exceed 11% of the total absorption area (see Table 2), indicating that iron is scarcely shunted to other metabolic functions. All these observations point to a special intracellular function of ferricrocin. The metal transfer to ferricrocin and the scarcity of ligand exchange of ferricrocin-bound Fe^{3+} indicate a quasi inert iron storage function of this siderophore in mycelia of *N. crassa.*

In addition, we have explored iron pools in spores of *Neurospora* and of *Aspergillus ochraceus,* because little is known about iron storage in fungal spores. Again, *in vivo* Mössbauer spectroscopy was employed (Matzanke et al., 1987). Spores of *Neurospora crassa* are lacking in ferritinlike iron storage compounds. Ferricrocin constituted 47% of the total iron content in spores. In spores of *A. ochraceus,* 74% of the total iron content was bound by ferrichrome-type siderophores. When spores of *N. crassa* were allowed to germinate and to grow for a time period of 12 h, the ferricrocin iron pool disappeared, indicating that the metal was utilized. From this it has been concluded that the siderophore ferricrocin serves as an iron storage peptide in spores of *N. crassa.* In view of the large siderophore iron storage pools in spores, the intracellular accumulation of ferricrocin in mycelia seems to warrant an iron pool sufficiently large for sporulation. In fact, many nutritional factors contribute to the induction of fungal sporulation, and this process is associated with tremendous metabolic activity. Macromolecules must be synthesized to provide sporulation-specific structures. It is known that a number of fungal spores have prepackaged RNA and proteins that are required for the preliminary steps of spore germination (Brambl et al., 1978; Van Etten et al., 1981; Dahlberg and Van Etten, 1982). In addition, it has been suggested that mycelial vesicles of *N. crassa* which contain basic amino acids serve as a reservoir for precursors required during sporulation (Brody, 1981). Thus the rationale for high amounts of ferricrocin and its inertness in mycelia must be sought in terms of survival and propagation.

2. *Iron Storage Siderophores in the Yeasts* Rhodotorula minuta *and* Ustilago sphaerogena

The biosynthesis of the dihydroxamate rhodotorulic acid (RA), (17), is characteristic of heterobasidiomycetous yeasts and of *Microbotryum* species parasitic in the anthers of Caryophyllaceae while all other Ustilaginaceae produce siderophores of the ferrichrome type (Deml and Oberwinkler, 1982). Transport of ferric RA in *Rhodotorulla minuta* var. ex *texensis* (formerly termed *Rhodotorula pilimanae*) and of ferrichrome A in *Ustilago sphaerogena* has been studied by means of radioactive labeled complexes (Ecker et al., 1982; Müller, 1985a, 1985b). It is important to note that in *Rhodotorula* the $Fe_2(RA)_3$ complex is not transported across the cell membrane but that rather ligand exchange occurs.

Mössbauer measurements of frozen aqueous Fe_2RA_3 yielded a typical siderophore spectrum (Fig. 11a). Figure 11b depicts a Mössbauer spectrum of frozen *Rhodotorula minuta* cells that had been grown in a salt medium supplied with 15μM $^{57}Fe_2(RA)_3$ and harvested at $OD_{660} = 0.85$. The spectrum contains a magnetically split subspectrum that closely resembles $Fe_2(RA)_3$. From this spectrum the experimental ferric rhodotorulate spectrum (Fig. 11a) was stripped, yielding an $Fe_2(RA)_3$ contribution to the envelope spectrum of 55($\pm$5)%. The identity of RA was confirmed by isolation of $Fe_2(RA)_3$ from cells grown in the same way as in the Mössbauer experiment. The remaining metabolite spectrum after stripping is shown in Fig. 11c. Two metabolites can be discerned: an Fe^{2+} high-spin species (δ = 1.36(8) mm/s, ΔE_Q = 2.84(9) mm/s) and an additional component, representing Fe^{3+} in a high-spin state (δ = 0.43(5) mm/s, ΔE_Q = 1.07(6) mm/s). The contribution to the envelope spectrum is 18($\pm$3)% for Fe^{2+} and 27($\pm$3)% for Fe^{3+}. Prolonged growth to $OD_{660} = 1.5$ yields a similar pattern: $Fe_2(RA)_3$ 55($\pm$6)%, Fe^{2+} 34($\pm$2)%, and Fe^{3+} 11($\pm$4)%. If $^{57}Fe_2(RA)_3$ (15μM) is added to iron depleted cultures of *R. minuta* at $OD_{660} = 0.8$, 60% of the metal is accumulated within 30 min. Figure 11d represents the corresponding Mössbauer spectrum. The $Fe_2(RA)_3$ contribution is 68%. The residual contribution is solely due to an Fe^{2+} high-spin species.

No ferritinlike iron pools could be detected. Surprisingly, the main intracellular iron component in *Rhodotorula minuta* is the siderophore RA. Since energy dependent ligand exchange occurs at the membrane level, ligand exchange is not a simple exchange between $Fe_2(RA)_3$ and excess desferri-RA. Rather, the process of metal transfer from extracellular RA to intracellular iron storage RA requires an additional mediator. Because the major intracellular source of iron in this yeast is a siderophore, and ferritinlike structures are not at all present, we assign to RA an additional function of intracellular main iron storage compound.

In *Ustilago sphaerogena,* iron accumulation is different from that of *Rhodotorula* and *N. crassa.* Iron transport is achieved via ferrichrome A, whereas the iron storage siderophore is ferrichrome. Ferrichrome A is not accumulated. Ferrichrome accounts for 50% of the cellular iron pool. A second ferric iron component exhibits Mössbauer parameters (δ = 0.51 mm/s, ΔE_Q = 0.67 mm/s) similar to *E. coli* BF. However, EPR spectra of the cell suspension recorded at X-band in the temperature range 2.8–200K (not depicted) did not show a ferromagnetic resonance signal. Thus

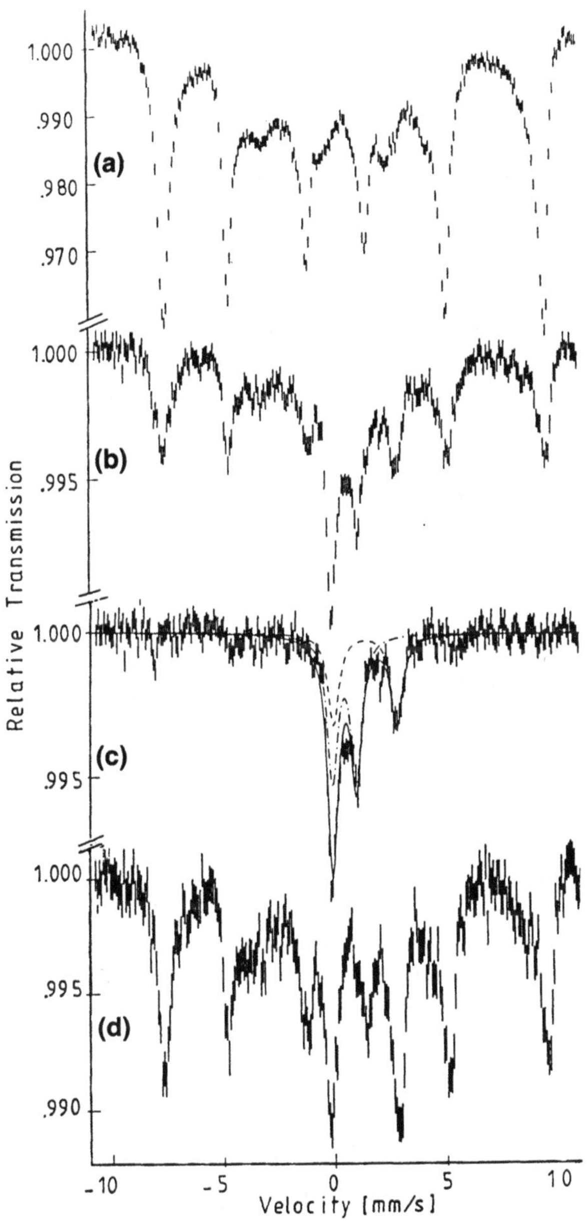

Figure 11 Mössbauer spectra of ferric rhodotorulate uptake in the yeast *Rhodotorula minuta*. Spectrum (a) corresponds to frozen aqueous $Fe_2(RA)_3$. Spectrum (b) corresponds to frozen *Rhodotorula minuta* cells, grown in a salt medium supplied with 15μM $^{57}Fe_2(RA)_3$ and harvested at $OD_{660} = 0.85$. The magnetically split subspectrum is due to $Fe_2(RA)_3$. 1(c) shows 1(b) after stripping 1(a). This metabolic pattern is similar to those found in gram-negative bacteria (see Fig. 8). If $^{57}Fe_2(RA)_3$ (15μM) is added to iron-depleted cultures of *R. minuta* at $OD_{660} = 0.8$, 60% of the metal is accumulated within 30 min. Fig. 1(d) represents the corresponding Mössbauer spectrum.

superparamagnetic particles and hence a bacterioferritinlike iron storage compound can be excluded for *Ustilago*. Again, a hydroxamate-type siderophore represents the exclusive iron storage compound.

V. SUMMARY AND OUTLOOK

Recently it has been postulated that ferritins are potentially common to all aerobic organisms (Grossman, et al., 1992). However, taking into account the emerging diversity of fungal iron storage, this postulate seems to be questionable.

For example, the Mössbauer studies of siderophore uptake in various Ascomycetes and Heterobasidiomycetes revealed that siderophores have not only the role of solubilization of iron and transfer of this metal to the cell but also important intracellular functions. In mycelia and spores of various ascomycetes, hydroxamate-siderophores represent the major iron storage pool. The same holds true for a variety of yeasts. No ferritins were detected. Based on these findings it is safe to conclude that intracellular siderophores replace ferritins in these organisms. This further extends the biological role of siderophores. Also in spores of hydroxamate producing ascomycetes siderophores serve as iron storage compounds. The observation of large inert siderophore pools in mycelia can be rationalized as iron reservoirs for sporulation. Whether genes for ferritin or bacterioferritin are present in these organisms remains an open question. However, ferritins definitely play no appreciable role in iron storage of these fungi.

In the yeast *S. cerevisiae* a ferritinlike molecule was found. However, the low iron content of this protein (less than 100 iron atoms per molecule), which is not related to the overall cellular iron content, excludes its role as a major iron storage compound (Raguzzi et al., 1988). Rather, in the same study it has been inferred, but not yet proven, that iron storage in *S. cerevisiae* might occur in polyphosphate-containing vacuoles.

The α-hydroxycarboxylate siderophore rhizoferrin and the mode of iron storage distinguishes iron metabolism of Zygomycetes from many that of Ascomycetes and Heterobasidiomycetes. In mycelia of Zygomycetes iron storage is achieved by zygoferritin, a protein that exhibits iron cores similar to bacterioferritins. Spores of Zygomycetes are packed with mycoferritin as a major component and zygoferritin as a minor constituent. Mycoferritin is a true iron storage compound, because iron is required for germination and mycoferritin is formed in the sporangiophores during the process of sporulation. At this point it is not yet clear why a mycelia-specific form of ferritin and a spore-specific form of ferritin is found in Zygomycetes. In addition, the question that remains to be elucidated is whether the difference between the two ferritins is due simply to variations of the iron core or also to diverse protein shells. Moreover, whether the biosyntheses of zygoferritin and mycoferritin are regulated in accordance with the scheme outlined in Section III.C is also questionable. The transfer of iron from zygoferritin to mycoferritin occurring in the sporangiophores of Zygomycetes suggests diverse pathways of iron storage metabolism requiring a regulatory apparatus that is not simply controlled by lack or presence of the metal. The rationale and mechanisms of these processes, and the characterization of mycoferritin and zygoferritin, remain topics of continuous research in our laboratory.

Our current knowledge about iron storage in fungi is still incomplete. Systematic investigations will be required to obtain a clear view about the diversity of iron storage in fungal systems. Mössbauer spectroscopy of ^{57}Fe-enriched cells and EPR will be of great aid for such investigations. The different strategies of iron storage by fungi may turn out to be useful traits in taxonomic classification, evolutionary analysis, and environmental monitoring.

REFERENCES

Ames, B. N. (1983). Dietary carcinogens and anticarcinogens, *Science, 221:* 1256–1264.

Andrews, S. C., Smith, J. M. A., Guest, J. R., and Harrison, P. M. (1989). Amino acid sequence of the bacterioferritin (cytochrome b_1) of *Escherichia coli*-K12, *Biochem. Biophys. Res. Commun., 158:* 489–496.

Andrews, S. C., Findlay, J. B. C., Guest, J. R., Harrison, P. M., Keen, J. N., and Smith, J. M. A. (1991). Physical, chemical and imunological properties of bacterioferritins of *Escherichia coli, Pseudomonas aeruginosa* and *Azotobacter vinelandii. Biochim. Biophys. Acta, 1078:* 111–116.

Aziz, N., and Munro, H. N. (1987). Iron regulates ferritin mRNA translation through a segment of its 5 ∩ untranslated region, *Proc. Natl. Acad. Sci. USA, 84:* 8478.

Basile, L. A., Raphael, A. L., and Barton, J. K. (1987). Metal activated hydrolytic cleavage of DNA, *J. Am. Chem. Soc, 109:* 7550–7551.

Bauminger, E. R., Cohen, S. G., Ofer, S., and Rachmilewitz, E. A. (1979). Quantitative studies of ferritinlike iron in erythrocytes of thalassemia, sickle-cell anemia, and hemoglobin Hammersmith with Mössbauer spectroscopy, *Proc. Natl. Acad. Sci. USA, 76:* 939–943.

Bauminger, E. R., Cohen, S. G., Ofer, S., and Bachrach, U. (1980a). Study of storage iron in cultured chick embryo fibroblasts and rat glioma cells, using Mössbauer spectroscopy, *Biochim. Biophys. Acta, 623:* 237–242.

Bauminger, E. R., Cohen, S. G., Dickson, D. P. E., Levy, A., Ofer, S., and Yariv, J. (1980b). Mössbauer spectroscopy of *Escherichia coli* and its iron-storage protein, *Biochim. Biophys. Acta, 623:* 237–242.

Bauminger, E. R., Cohen, S. G., Labenski de Kanter, F., Levy, A., Ofer, S.,

Kessel, M. and Rottem (1980c). Iron storage in *Mycoplasma capricolum,* S, *J. Bacteriol. 141:* 378–381.

Bauminger, E. R., Harrison, P. M., Nowik, I., and Treffry, A. (1989). Mössbauer spectroscopic study of the initial stages of iron-core formation in horse spleen apoferritin: Evidence for both isolated Fe(III) atoms and oxo-bridged Fe(III) dimers as early intermediates, *Biochemistry, 28:* 5486–5493.

Bauminger, E. R., Harrison, P. M., Hechel, D., Nowik, I., and Treffry, A. (1991a). Mössbauer spectroscopic investigation of structure-function relations in ferritin, *Biochim. Biophys. Acta, 1118:* 48–58.

Bauminger, E. R., Harrison, P. M., Hechel, D., Nowik, I., and Treffry, A. (1991b). Iron(II) can be transferred between ferritin molecules, *Proc. R. Soc. London Ser. B, 244:* 211–217.

Bell, S. H., Weir, M. P., Dickson, D. P. E., Gibson, J. F., Sharp, G. A., and Peters, T. J. (1984). Mössbauer spectroscopic studies of human haemosiderin and ferritin, *Biochim. Biophys. Acta, 787:* 227–236.

Boas, J. F., and Troup, G. J. (1971). Electron spin resonance and Mössbauer effect studies of ferritin, *Biochim. Biophys. Acta, 229:* 68–74.

Bozarth, R. F., and Goenaga, A. (1972). Purification and properties of mycoferritin from *Mortierella alpina, Can. J. Microbiol., 18:* 619–622.

Brambl, S., Dunkle, L. D., and Van Etten, J. L. (1978). Nucleic acid and protein synthesis during fungal spore germination, *The Filamentous Fungi* (J. E. Smith and D. R. Berry, eds.), Arnold, London, pp. 94–118.

Brody S. (1981). Genetic and biochemical studies on *Neurospora* conidia germination and formation, *The Fungal Spore: Morphogenetic Controls* (G. Turian and H.R. Hohl, eds.), Academic Press, New York, pp. 605–626.

Cheesman, M. R., Thomson, A. J., Greenwood, C., Moore, G. R., and Kadir, F. (1990). Bis-methionine axial ligation of haem in bacterioferritin from *Pseudomonas aeruginosa, Nature, 346:* 771–773.

Chen, M., and Crichton, R. R. (1982). *Biochim. Biophys. Acta, 707:* 1–6.

Cheng, Y. G., and Chasteen, N. D. (1991). Role of phosphate in initial iron deposition in apoferritin, *Biochemistry, 30:* 2947–2953.

Constable, A., Quick, S., Gray, N. K., and Hentze, M. W. (1992). Modulation of the RNA-binding activity of a regulatory protein by iron *in vitro:* Switching between enzymatic and genetic function? *Proc. Natl. Acad. Sci. USA, 89:* 4554.

Crichton, R. R. (1989). Proteins of iron storage and transport, *Advances in Protein Chemistry.*

Crichton, R. R., and Charloteaux-Wauters, M. (1987). Iron transport and storage, *Eur. J. Biochem., 164:* 485–506.

Dahlberg, K. R., and Van Etten, J. L. (1982). Physiology and biochemistry of fungal sporulation, *Ann. Rev. Phytopathol., 20:* 281–301.

David, N., and Easterbrook, K. (1971). Ferritin in the fungus phycomyces, *J. Cell. Biol., 48:* 15.

Deml, G., and Oberwinkler, F. (1982). Studies in heterobasidiomycetes, part 22. A survey on siderophore formation in low-iron cultured anther smuts of *caryophyllaceae, Zbl. Bakt. Hyg., I. Abt., Orig., C3:* 475–477.

Dickson, D. P. E., and Rottem, S. (1979). Mössbauer spectroscopic studies of iron in *Proteus mirabilis, Eur. J. Biochem, 101:* 291–295.

Dorfman, L. M., and Adams, G. E. (1973). Reactivity of the hydroxyl radical in aqueous solutions. National Standards Reference Data System. National Bureau of Standards, Bulletin No. 46.

Drechsel, H., Metzger, J., Freund, S., Jung, G., Boelaert, J. R., and Winkelmann, G. (1992). Rhizoferrin—A novel siderophore from the fungus *Rhizopus microsporus* var. *rhizopodiformis, Biol. Metals, 4:* 238–243.

Dreyer, G., and Dervan, P. B. (1985). Sequence-specific cleavage of single-stranded DNA: Oligonucleotide-EDTA ·FE(II), *Proc. Natl. Acad. Sci. USA, 82:* 968–972.

Dunford, H. B. (1987). Free radicals in iron-containing systems, *Free Radic. Biol. Med., 3:* 405–421.

Ecker, D. J., Lancester, J. R., Jr., and Emery, T. (1982). *J. Biol. Chem., 257:* 8623.

Filho, A. C. M., and Meneghini, R. (1984). In vivo formation of single-strand breaks in DNA by hydrogen peroxide is mediated by the Haber-Weiss reaction, *Biochim. Biophys. Acta, 781:* 56–63.

Floyd, R. A., and Lewis, C. A. (1983). Hydroxyl free radical formation from hydrogen peroxide by ferrous iron-nucleotide complexes, *Biochemistry, 22:* 2645–2649.

Ford, G. C., Harrison, P. M., Rice, J. M. A., Smith, A., Treffry, J. L., White, J. L., and Yariv, J. (1984). Ferritin: Design and formation of an iron-storage protein, *Philos. Trans. R. Soc. London Ser. B., 304:* 551–565.

Frankel, R. B. Papaefthymiou, G. C., and Watt, G. D. (1987). Binding of Fe^{2+} by mammalian ferritin, *Hyperf. Interact., 33:* 233–240.

Fridovich, I. (1978). The biology of oxygen radicals, *Science, 201:* 875–880.

Funk, F., Lenders, J. P., Crichton, R. R., and Schneider, W. (1985). Reductive mobilization of ferritin iron, *Eur. J. Biochem., 154:* 167–172.

Funk, F., Lecrenier, C., Lesuisse, E., Crichton, R. R., and Schneider, W. (1986). A comparative study on iron sources for mitochondrial haem synthesis including ferritin and models of transit pool species, *Eur. J. Biochem., 157:* 303–309.

Grossman, M. J., Hinton, S. M., Minak-Bernero, V., Slaughter, C., and Stiefel, E.I. (1992). Unification of the ferritin family of proteins, *Proc. Natl. Acad. Sci. USA, 89:* 2419.

Haile, D. J., Rouault, T. A., Harford, J. B., Kennedy, M. C., Blondin, G. A., Beinert, H., and Klausner, R. D. (1992). Cellular regulation of the iron-responsive element binding protein: Disassembly of the cubane iron-sulfur cluster results in high-affinity RNA binding, *Proc. Natl. Acad. Sci. USA, 89:* 11735.

Halliwell, B., and Gutteridge, J. M. C. (1984). Oxygen toxicity, oxygen radicals, transition metals and disease, *Biochem. J., 219:* 1–14.

Harman, D. (1981). The aging process, *Proc. Natl. Acad. Sci. USA, 78:* 7124–7128.

Harrison, P. M., and Lilley, T. H. (1989). Ferritin, *Iron Carriers and Proteins* (T. M. Loehr, ed.), VCH, New York, pp. 123–238.

Harrison, P. M., Treffry, A., and Lilley, T. H. (1986). Ferritin as an iron storage protein: Mechanism of iron uptake, *J. Inorg. Biochem. 27:* 287–293.

Harrison, P. M., Andrews, S. C., Ford, G. C., Smith, J. M. A., Treffry, A., and

White, J. L. (1987) Ferritin and bacterioferritin: Iron sequestering molecules from man to microbe, *Iron Transport in Microbes, Plants and Animals* (G. Winkelmann, D. van der Helm, and J. B. Neilands, eds.), Verlag Chemie, Weinheim.

Hentze, M. W., Caughman, S. W., and Rouault, T. A. (1987). Identification of the iron-responsive element for the translational regulation of human ferritin mRNA, *Science, 244:* 1570–1573.

Heusterspreute, M., and Crichton, R. R. (1981). Amino acid sequence of horse spleen ferritin, *FEBS Lett., 129:* 322–327.

Hirling, H., Emery-Goodman, A., Thompson, N., Neupert, B., Seiser, C., and Kuhn, L. C. (1992). Expression of active iron regulatory factor from a full-length human cDNA by *in vitro* transcription/translation, *Nucleic Acids Res., 20:* 33–39.

Horowitz, N. H., Charlang, G., Gorn, G., and Williams, N. P. (1976). Isolation and identification of the conidial germination factor of *Neurospora crassa, J. Bacteriol., 127:* 135–140.

Inoue, S., and Kawanishi, S. (1987). Hydroxyl radical production and human DNA damage induced by ferric nitrilotriacetate and hydrogen peroxide, *Cancer Res., 47:* 6522–6527.

Jacobs, D., Watt, G. D., Frankel, R. B., and Papaefthymiou, G. C. (1989). Fe^{2+} binding to apo and holo mammalian ferritin, *Biochemistry, 28:* 9216–9221.

Kaufman, K. S., Papaefthymiou, G. C., Frankel, R. B., and Rosenthal, A. (1980). Nature of iron deposits on the cardiac walls in β-thalessimia by Mössbauer spectroscopy, *Biochim. Biophys. Acta, 629:* 522–529.

Kennedy, M. C., Mende-Mueller, L., Blondin, G. A., and Beinert, H. (1992). Purification and characterization of cytosolic aconitase from beef liver and its relationship to the iron-responsive element binding protein, *Proc. Natl. Acad. Sci., 89:* 11730–11732.

Klausner, R. D., and Harford, J. B. (1989). *Cis-trans* models for post-transcriptional gene regulation, *Science, 246:* 870–872.

Kurokawa, K., Fukumori, Y., and Yamanaka, T. (1989). *Biochim. Biophys. Acta, 976:* 135–139.

LaBombardi, V. J., Pisano, M. A., and Klavins, J. V. (1982). Isolation and characterization of *Phycomyces blakesleeanus* ferritin, *J. Bacteriol., 150:* 671–675.

Laulhere, J.-P., Lescure, A.-M., and Briat, J.-F. (1988). Purification and characterization of ferritins from maize, pea, and soybean seeds, *J. Biol. Chem., 263:* 10289–10294.

Lawson, D. M., Treffry, A., Artymiuk, P. J., Harrison, P. M., Yewdall, S. J., Luzzago, A., Cesareni, G., Levi, S., and Arosio, P. (1989). Identification of the ferroxidase centre in ferritin, *FEBS Lett., 254:* 207–210.

Lawson, D. M., Artymiuk, P. J., Yewdall, S. J., Smith, J. M. A., Livingstone, J. C., Treffry, A., Luzzago, A., Levi, S., Arosio, P., Cesareni, G., Thomas, C. D., Shaw, W. V., and Harrison, P. M. (1991). Solving the structure of human H ferritin by genetically engineering intermolecular crystal contacts, *Nature, 349:* 541–544.

Levi, S., Luzzago, A., Cesareni, G., Cozzi, A., Franceschinelli, F., Albertini, A., and Arosio, P. (1988). Mechanism of ferritin iron uptake: Activity of the H-chain and deletion mapping of the ferro-oxidase site, *J. Biol. Chem., 263:* 18086–18092.

Levi, S., Yewdall, S. J., Harrison, P. M., Santambrogio, P., Cozzi, A., Rovida, E., Albertini, A., and Arosio, P. (1992). Evidence that H- and L-chains have co-operative roles in the iron-uptake mechanism of human ferritin, *Biochem. J., 288:* 591.

Loeb, L. A., James, E. A., Waltersdorph, A. M., and Klebanoff, S. J. (1988). Mutagenesis by the autoxidation of iron with isolated DNA, *Proc. Natl. Acad. Sci. USA, 85:* 3918–3922.

Mann, S., Williams, J. M., Treffry, A., and Harrison, P. (1987). Reconstituted and native iron-cores of bacterioferritin and ferritin. *J. Mol. Biol., 198:* 405–416.

Matzanke, B. F. (1987). Mössbauer spectroscopy of microbial iron uptake and metabolism, *Iron Transport in Microbes, Plants and Animals* (G. Winkelmann, D. van der Helm, and J. B. Neilands, eds.), Verlag Chemie, Weinheim.

Matzanke, B. F., Bill, E., Müller, G. I., Winkelmann, G., and Trautwein, A. X. (1987a). Metabolization of ^{57}Fe-coprogen in *N.crassa.* An in vivo Mössbauer study, *Eur. J. Biochem., 162:* 643–650.

Matzanke, B. F., Bill, E., Trautwein, A. X., and Winkelmann, G. (1987b). Role of siderophores in iron storage in spores of *N. crassa* and *A. ochraceus. J. Bacteriol., 169:* 5873–5876.

Matzanke, B. F., Bill, E., Trautwein, A. X., and Winkelmann, G. (1988). Ferricrocin functions as the main intracellular iron-storage compound in mycelia of *Neurospora crassa, Biol. Metals, 1:* 18–25.

Matzanke, B. F., Müller, G., Bill, E., and Trautwein, A. X. (1989). Iron metabolism of *E. coli* studied by Mössbauer spectroscopy and biochemical methods, *Eur. J. Biochem., 183:* 371–379.

Matzanke, B. F., Bill, E., Trautwein, A. X., and Winkelmann, G. (1990a): Siderophores as storage compounds in the yeasts *Rhodotorula minuta* and *Ustilago sphaerogena* detected by in vivo Mössbauer spectroscopy, *Hyperfine Interact., 58:* 2359–2364.

Matzanke, B. F., Bill, E., and Trautwein, A. X. (1990b), Fur participates in the regulation of iron redox states in *Escherichia coli, Forum Mikrobiologie, 13:* 20 (V73).

Matzanke, B. F. (1991). Structures, coordination chemistry and functions of microbial iron chelates, *Handbook of Microbial Iron Chelates (Siderophores)* (G. Winkelmann, ed.), CRC Press, Boca Raton, FL, pp. 15–60.

Matzanke, B. F., Bill, E., and Trautwein, A. X. (1992). Main components of iron metabolism in microbial systems—Analyzed by in vivo Mössbauer spectroscopy, *Hyperfine Interact., 71:* 1259–1262.

Matzanke, B. F., Bill, E., Thieken, A., Trautwein, A. X., and Winkelmann, G. (1993). Rhizoferrin transport and iron storage in Mucorales studies by Mössbauer spectroscopy and EPR, International Conference on the Application of the Mössbauer Effect, Vancouver (Canada), August 9–13.

Minotti, G., and Aust, S. D. (1987). The role of iron in the initiation of lipid peroxidation, *Chem. Phys. Lipid, 44:* 191–208.

Moore, G. R., Mann, S., and Bannister, J. V. (1986). Isolation and properties of the complex nonheme-iron containing cytochrome b_{557} (bacterioferritin) from *Pseudomonas aeruginosa, J. Inorg. Biochem., 28:* 329–336.

Mørup, S., Dumesic, J. A., and Topsøe, H. (1980). Magnetic microcrystals, *Applications of Mössbauer Spectroscopy, Vol. II* (Cohen, R.L., ed.), New York: Academic Press, pp. 1–53.

Müller, G., Barclay, S. J., and Raymond, K. N. (1985a). The mechanism and specificity of iron transport in *Rhodotorula pilimanae* probed by synthetic analogs of rhodotorulic acid, *J. Biol. Chem., 280:* 13916.

Müller, G., Isowa, Y., and Raymond, K. N. (1985b). Stereospecificity of siderophore-mediated iron uptake in *Rhodotorula pilimanae* as probed enantiorhodotorulic acid and isomers of chromic rhodotorulate, *J. Biol. Chem., 260:* 13921.

Nordlund, P., Sjöberg, B.-M., and Eklund, H. (1990). Three-dimensional structure of the free radical protein of ribonucleotide reductase, *Nature, 345:* 593–598.

Peat, A., and Banbury, G. H. (1968). Occurrence of ferritin-like particles in a fungus, *Planta* (Berlin), *79:* 268.

Philpott, C. C., Rouault, T. A., and Klausner R. D. (1991). Sequence and expression of the murine iron-responsive element binding protein, *Nucleic Acids Res., 19:* 6333–6338.

Raguzzi, F., Lesuisse, E., and Crichton, R. R. (1988). Iron storage in *Saccharomyces cerevisiae, FEBS Lett., 231:* 253–258.

Rimbert, J. N., Dumas, F., Kellershohn, C., Girot, R., and Brissot, P. (1985). Mössbauer spectroscopy of iron overloaded livers, *Biochimie, 67:* 663–668.

Rohrer, J. S., Islam, Q. T., Watt, G. D., Sayers, D. E., and Theil, E. C. (1990). Iron environment in ferritin with large amounts of phosphate, from *Azotobacter vinelandii* and horse spleen, analyzed using extended x-ray absorption fine structure (EXAFS), *Biochemistry, 29:* 259–264.

Rouault, T. A., Hentze, M. W., Caughman, S. W., Harford, J. B., and Klausner, R. D. (1988). Binding of a cytosolic protein to the iron-responsive element of human ferritin messenger RNA, *Science, 241:* 1207.

Rouault, T. A., Tang, C. K., and Kaptain, S. (1990). Cloning of the cDNA encoding and RNA regulatory protein—The human iron-responsive element-binding protein, *Proc. Natl. Acad. Sci. USA, 87:* 7958–7962.

Rouault, T. A., Haile, D. J., Downey, W. E., Philpott, C. C., Tang, C., Samaniego, F., Chin, J., Paul, I., Orloff, D., Harford, J. B., and Klausner, R. D. (1992). An iron-sulfur cluster plays a novel regulatory role in the iron-responsive element binding protein, *BioMetals, 5:* 131.

Rowley, D. A., and Halliwell, B. (1982). Superoxide dependent formation of hydroxyl radicals from NADH and NADPH in the presence of iron salts, *FEBS Lett., 142:* 39–41.

Rush, J. D., and Koppenol, W. H. (1986). Oxidizing intermediates in the reaction of ferrous EDTA with hydrogen peroxide. Reactions with organic molecules and ferrocytochrome *c, J. Biol. Chem., 261:* 6730–6733.

St. Pierre T., Bell, S. H., Dickson, D. P. E., Mann, S., Webb, J., Moore, G. R., and Williams, R. J. P. (1986). Mössbauer spectroscopic studies of the cores of human, limpet and bacterial ferritins, *Biochim. Biophys. Acta, 870:* 127–134.

Seckback, J. (1982). Ferreting out the secrets of plant ferritin—A review, *J. Plant Nutr., 5:* 369–394.

Smith, J. M. A., Quirk, A. V., Plank, R. W. H., Diffin, F. M., Ford, G. C., and Harrison, P. M. (1988). The identity of *Escherichia coli* bacterioferritin and cytochrome b_1, *Biochem. J., 255:* 737–740.

Spartalian, K., Oosterhuis, W. T., and Smarra N. (1975). Mössbauer effect studies in the fungus *Phycomyces, Biochim. Biophys. Acta, 399:* 203–212.

Stefanini, S., Desideri, A., Vecchini, P., Drakenberg, T., and Chiancone, E. (1989). Identification of the entry channels in apoferritin. Chemical modification and spectroscopic studies, *Biochemistry, 28:* 378–382.

Stiefel, E. I., and Watt, G. D. (1979). *Azotobacter vinelandii* cytochrome $b_{557.5}$ is a bacterioferritin, *Nature, 279:* 81–83.

Surerus, K. K., Kennedy, M. C., Beinert, H., and Münck, E. (1989). Mössbauer study of the inactive Fe_3S_4 and Fe_3Se_4 and the active Fe_4S_4 forms of beef heart aconitase, *Proc. Natl. Acad. Sci. USA, 86:* 9846–9850.

Taft, K. L., Papaefthymiou, G. C., and Lippard, S. J. (1993). A mixed-valent polyiron oxo complex that models the biomineralization of the ferritin core, *Science, 259:* 1302.

Theil, E. C., and Hase, T. (1993). Plant and microbial ferritins, *Iron Chelation in Plants and Soil Microorganisms* (L.L. Barton and B.C. Hemming, eds.), Academic Press, New York, pp. 133–157.

Thieken, A., and Winkelmann, G. (1992). Rhizoferrin: A complexone type siderophore of the Mucorales and Entomophtorales (Zygomycetes). *FEMS Microbiol. Lett.*, 94: 37–42.

Treffry, A., Harrison, P. M., Cleton, M. I., de Bruijn, W. C., and Mann, S. (1987). A note on the composition and properties of ferritin iron cores, *J. Inorg. Biochem., 31:* 1–6.

Treffry, A., Hirzmann, J., Yewdall, S. J., and Harrison, P. M. (1992). Mechanism of catalysis of Fe(II) oxidation by ferritin H chains, *FEBS Lett., 302:* 108–112.

Van Etten, J. L., Dahlberg, K. R., and Russo, G. M. (1981) Nucleic acids, *The Fungal Spore: Morphogenetic Controls* (G. Turian and H. R. Hohl, eds.), Academic press, New York, pp. 277–302.

Wächtershäuser, G. (1988). Before enzymes and templates: Theory of surface metabolism, *Microbiol. Rev., 52:* 452–484.

Watt, G. D., Frankel, R. B., and Papaefthymiou, G. C. (1985). Reduction of mammalian ferritin, *Proc. Natl. Acad. Sci. USA, 82:* 3640–3643.

Watt, G. D., Frankel, R. B., Papaefthymiou, G. C., Spartalian, K., and Stiefel, E. I. (1986). Redox properties and Mössbauer spectroscopy of *Azotobacter vinelandii* bacterioferritin, *Biochemistry, 25:* 4330–4336.

Watt, G. D., Jacobs, D., and Frankel, R. B. (1988). Redox reactivity of bacterial and mammalian ferritin: Is reductant entry into the ferritin interior a necessary step for iron release? *Proc. Natl. Acad. Sci. USA, 85:* 7457–7461.

Watt, G. D., Frankel, R. B., Jacobs, D., Huang, H., and Papaefthymiou, G. C. (1992). Fe^{2+} and phosphate interactions in bacterial ferritin from *Azotobacter vinelandii. Biochemistry, 31:* 5672–5679.

Weir, M. P., Peters, T. J., and Gibson, J. F. (1985). Electron spin resonance studies of splenic ferritin and haemosiderin, *Biochim. Biophys. Acta, 828:* 298–305.

Winkler, H., Matzanke, B. F., Bill, E., and Trautwein, A. X. (in press). Iron storage in *Streptomyces olivaceus. Hyperfine Interact.*

Wong, B., Kappel, M. J., Raymond, K. N., Matzanke, B., and Winkelmann, G. (1983). Coordination chemistry of microbial iron transport compounds, 24. Characterization of coprogen and ferricrocin, two ferric hydroxamate siderophores, *J. Am. Chem. Soc., 105:* 810–815.

Yang, C.-Y., Meagher, A., Huynh, B. H., Sayers, D. E., and Theil, E. C. (1987). Iron(III) bound to horse spleen apoferritin: An x-ray absorption and Mössbauer spectroscopy study that shows that iron nuclei can form on the protein, *Biochemistry, 26:* 497–503.

Yariv, J. (1983). The identity of bacterioferritin and cytochrome b_1, *Biochem. J. 211:* 527.

Yariv, J., Kalb, A. J., Sperling, R., Bauminger, E. R., Cohen, S. G., and Ofer, S. (1981). The composition and the structure of bacterioferritin of *Escherichia coli, Biochem. J., 197:* 171–175.

7

Manganese: Function and Transport in Fungi

Georg Auling
University of Hannover, Hannover, Germany

I. INTRODUCTION

Since the overwhelming majority of fungal species has never been cultivated nor grown in defined media, our knowledge of the requirements for trace elements within fungi is rather limited. It is also obvious that in contrast to other transition metals the role of manganese in fungal nutrition has been less studied (Garraway and Evans, 1984). No unique and universal function of manganese for fungi has been presented. The purpose of this chapter, then, is to stimulate research in this area, to encourage industrial microbiologists to find more rational designs of media for growth of and production with fungi, and to provide mycologists with the current knowledge on the role of manganese in both fungal metabolism and differentiation.

Another intention is to highlight the diversity of energy-dependent manganese transport systems in filamentous fungi by comparing the different routes used for entry of this cation. Obviously, transport of manganese (Auling, 1989) and of most other (except iron) transition metal cations has been neglected in fungi and higher eukaryotes. The role of Ca^{2+} channels in transmembrane signaling and the function of voltage-controlled Na^+ and K^+ channels in transmission of nerve impulses have attracted more attention in higher eukaryotic systems and fungi (Harold, 1986). It is self-evident that manganese homeostasis in fungi cannot be omitted, because as compartmentalized microorganisms they appear to control carefully their internal manganese content over a wide range of external concentrations. Finally, a minireview of the special methods supporting trace element (manganese) research, which are scattered over the biological, chemical, and physical literature, is provided.

II. SPECIAL METHODS OF TRACE ELEMENT (MANGANESE) RESEARCH

Manganese as an essential micronutrient is provided in defined media in quantities of a few mg per liter or less. However, there are many pitfalls associated with media preparation. As a result of precipitation due to alkaline pH or autoclaving, the dissolved concentration of this trace element in the final incubation medium may differ from that needed. The bioavailability of manganese may also be reduced in rich, nonmineral growth media by complexation with ligands (Hughes and Poole, 1991). In order to avoid these pitfalls, media based on buffers without complexing properties are required. Due to their negligible metal-complexing capacity, the hydrogen ion buffers of Good et al. (1966) are the buffers of choice for basic studies (kinetics, specificity, inducibility) on manganese transport systems in fungi. However, their application may be limited by the observation that the particular fungus under study does not maintain its biological activity during centrifugation and resuspension after harvest from a growth medium.

It is obvious that the manganese content of complex media fluctuates greatly depending on the source of raw materials or the technical production process. For large-scale industrial production with fungi it is attractive to use complex media based on cheap plant raw materials (molasses), which generally have elevated levels of Mn^{2+} derived from the manganese-containing photosystem II. For removal of manganese in molasses (cf. Sec. V), precipitation with potassium hexacyanoferrate is employed (Clark et al., 1966).

Calculation of the actual concentration of manganese is difficult even with defined media, i.e., the inherent manganese contamination of carbohydrates, which are often used in high amounts as substrates for production of fungal metabolites, is easily overlooked. Whereas a solution with less than 0.2 μM Mn^{2+} is usually considered manganese-free in chemistry, biologists should know that fungal metabolism is sometimes dramatically affected by manganese concentrations below this level (cf. Sec. V). "Manganese-free" defined mineral media with less than 0.2 μM Mn^{2+} can be obtained by passage of commercial-grade carbohydrate substrates through a cation exchange resin, containing, for example, iminodiacetic acid as a powerful chelating agent, and by treatment of the glassware with hot nitric acid (Hockertz et al., 1987a). A very sensitive bioassay for manganese using *Aspergillus niger* has been proposed in the early "nutritional" era of microbial growth and physiology (Sulochana and Lakshmanan, 1968).

Reliable determination of the residual contamination in growth media purported to be "manganese-free" is now available by inductively coupled plasma (ICP) mass-spectrometry (Brockaert, 1990). This method determines total manganese in a sample and is very favorable for this element because manganese is a 100% isotope. ICP analysis of complex and defined media has been helpful in studies on the specific manganese requirement for patulin production with *Penicillium urticae* (cf. Sec. III.C) and indicated ferric chloride as a trace metal component that contributes a significant additional amount of manganese (Scott et al., 1986a).

Determination of the total manganese content of fungal biomass is often necessary. Atomic absorption spectroscopy has been used for this purpose (Seehaus and Auling, submitted). However, the detection level of this method is less sensitive for manganese than for other transition metals. Differentiation of free and protein-bound manganese in biomass is sometimes required (cf. Sec. VI). This can be achieved with the characteristic six-line EPR (electron paramagnetic resonance) signal of Mn^{2+} (Ash and Schramm, 1982). We have readily measured the broad EPR spectrum centered at $g = 2.00$ in Mn-supplied bacteria and fungi as well (Follmann et al., 1986; Auling et al., unpublished results). A less sophisticated approach for determination of free manganese from biological samples would be ion trapping with advanced chromatographic materials (Bonn et al., 1990) followed by AAS determination.

In order to study localization and distribution of manganese and other transition elements in filamentous fungi (cf. Sec. VI.E), x-ray microanalysis (Galpin et al., 1978) and the Oxford scanning proton microprobe (Gadd et al., 1988), originally introduced for the measurement of calcium gradients along hyphae, may also be used. However, in analogy to their extensive use in calcium-binding studies, it would be highly desirable to synthesize fluorescent stains specifically designed for intracellular detection of manganese.

III. MANGANESE REQUIREMENTS

A. Mn Requirement for Growth

A long list of predominantly bacterial manganese enzymes has been compiled (Archibald, 1986). However, too little is known about manganese metabolism in fungi to permit a comprehensive coverage. Consequently, it is intended in this and the two following sections to place the diversity of cellular and metabolic phenomena in fungi that have been related to manganese in a rational framework (Tables 1 and 2).

The manganese requirements of fungi for growth vary considerably (0.01 to 0.005 ppm). Early studies revealed species of the genus *Chaetomium* as the paradigm for fungi that exhibit negligible growth when incubated in manganese-free medium and improve growth with increasing Mn^{2+} concentrations for a range of different fungi (Barnett and Lilly, 1966; Zonneveld, 1976). Manganese limitation induces increased branching and swelling of hyphae during growth on solid and/or in liquid media in *Hypoxylum punctatum, Glomerella cingulata* (Barnett and Lilly, 1966), *Aspergillus niger* (Kubicek and Röhr, 1986), and *Penicillium urticae* (Scott et al., 1986a). In liquid media, increased branching yields pelletlike growth, which is often preferred in fungal biotechnology in order to improve handling of large-scale cultivations. The presence of manganese clearly suppresses pelletlike growth of *Aspergillus niger* (Kubicek and Röhr, 1986) and of *Penicillium urticae* (Scott et al., 1986a). In the view of the author, pelletlike growth appears to be a more general response of filamentous fungi to manganese depletion (Table 1), although its achievement by a diversity of special treatments has been claimed in the biotechnological literature (Milsom and Meers, 1985).

Table 1 Effects of Manganese Depletion in Filamentous Fungi

Organism	Observed effects	Hypothesis	References
Chaetomium globosum	No growth or crippled growth		Barnett and Lilly, 1966
Hypoxylum punctatum			
Glomerella cingulata			
Aspergillus parasiticus	Control of dimorphism	Mn-RRase[a]	Detroy and Ciegler, 1971
Phialophora verrucosa			Reiss and Nickerson, 1971
Penicillium marneffi			Garrison and Boyd, 1973
Aspergillus niger	Pellets in liquid media		Kubicek and Röhr, 1986
Penicillium urticae			Scott et al., 1986a
Aspergillus nidulans	No α-1,3-glucan in the cell wall		Zonneveld, 1975a
Aspergillus niger			Kisser et al., 1980
Aspergillus nidulans	No fruit bodies		Zonneveld, 1975b
Aspergillus niger	No conidiation		Ahrens, Auling (unpublished results)

[a] The hypothesis of Hockertz et al. (1987a) that the effects of manganese depletion in *Aspergillus niger* can be traced back to a Mn^{2+}-dependent ribonucleotide reductase (Mn-RRase) was extended.

Table 2 Effects by Supplementation with Low (Micromolar) and High (Millimolar) Mn^{2+} Concentrations in Fungi

Organism	Observed effects	References
Phanerochaete chrysosporium *Trametes versicolor and other white-rot fungi*[a]	Induction of Mn-peroxidases, repression of lignin peroxidases	Boominathan and Reddy, 1992 Brown et al., 1991 Paice et al., 1993
Pleurotus ostreatus[a] *Lentinus edodes*[a] *Pholiota nameko*[a] *Agrocybe aegerita*[a]	Improved growth and yield of fruit bodies on straw, improved resistance toward antagonistic fungi	Nies, 1990
Penicillium clavigerum[a] *Penicillium claviforme*[a]	Stimulation of conidiation	Tinnel et al., 1974 Tinnel et al., 1976
Candida albicans[a]	True yeastlike growth	Widra, 1964
Aspergillus spp.[a]	Stimulation of malformin synthesis	Steenbergen and Weinberg, 1968
Rhizopus arrhizus[a]	Squalene biosynthesis	Campbell and Weete, 1978
Neurospora crassa[a]	Triterpene biosynthesis	Bobowski et al., 1977
Penicillium urticae[a]	Patulin biosynthesis	Scott et al., 1984
Saccharomyces cerevisiae[b]	Mutagenic, growth inhibitory, impaired fidelity of DNA polymerase reaction	Putrament et al., 1977, Goodman et al., 1983
Aspergillus niger[b]	Inhibition of conidiation, perturbation of the phosphate/potassium ratio	Seehaus, 1990

[a] Low (micromolar) Mn^{2+} concentration.
[b] High (millimolar) Mn^{2+} concentration.

The biochemical basis of the shift from the hyphal filaments to the short, highly branched hyphae was studied in the genus *Aspergillus*. Manganese limitation alters in *A. nidulans* and *A. niger* the composition of the cell wall in favor of chitin while reducing the amount of α-1,3-glucan (Zonneveld, 1971; Kisser et al., 1980).

A shift to a yeast-like development as response to inoculation in a manganese-deficient synthetic medium was described for *Aspergillus parasiticus* (Detroy and Ciegler, 1971). However, ultrastructurally these altered forms display a thickened inner cell wall and probably resemble arthrospores (Garrison and Boyd, 1971) rather than a true yeast-like phase with a potential for budding as exhibited by dimorphic fungi. Due to another electron microscopical study from the same laboratory (Garrison and Boyd, 1973) manganese control of dimorphism may also exist in *Penicillium marneffi*.

In conclusion, manganese limitation causes reduced or crippled growth in a number of fungal genera (Table 1), sometimes inducing a shift from vegetative growth to a special status of differentiation. The question is whether there is a common manganese-requiring target (cf. Sec. V).

B. Mn^{2+} Requirement for Sexual and Asexual Differentiation

Sexual differentiation in *Aspergillus nidulans* is clearly manganese controlled. Manganese deficiency prevents cleistothecium formation, and this has been related to the absence of α-1,3-glucan (Zonneveld, 1975a). Under these conditions, phosphoglucomutase shows only 60% activity of the control cultures. Because glutamine synthetase is a manganese enzyme in *A. nidulans* (Pateman, 1969), lack of this trace element may also lead to accumulation of glutamate, which might explain the observed higher levels of NAD glutamate dehydrogenase activity. However, since addition of 0.125 μM Mn^{2+} allows the synthesis of sufficient amounts of α-1,3-glucan but not a complete development of cleistothecia, an extra manganese requirement for DNA synthesis during fructification was alternatively postulated (Zonneveld, 1975b).

The stimulation of conidiation in the presence of Mn^{2+} ions indicates manganese control of morphogenesis in *Penicillium claviforme* and *Penicillium clavigerum* (Tinnel et al., 1974, 1976). The formation of conidia is also clearly manganese dependent in *Aspergillus niger*. Consequently, the manganese content of conidia used as inoculum for citric acid production with *A. niger* influences the yield (see Sec. V). When exposed to rather high (1 mM) Mn^{2+} concentrations, *A. niger* preferably accumulates manganese in the conidia forming hyphae. As observed by energy-dispersive x-ray microanalysis (Seehaus, 1990), these bulbous structures display a threefold elevated manganese level over the regular thin hyphae. Mutants of industrial strains of *A. niger* have been isolated that have lost the ability to form conidia. However, because of their genetic instability, no further studies for elucidation of the defect were undertaken (W. Ahrens, unpublished results). The question arises whether the biochemical target of the manganese requirement for conidiation is the same in the two species of *Penicillium* and in *A. niger*. Since we may be dealing with a general feature of fungal development, a more extended survey of deuteromycetes and other fungi (including plant pathogenic species) for a possible manganese requirement of conidiation would be of interest.

C. Mn^{2+} Requirement for Secondary Metabolite Production

Manganese affects the production of diverse metabolites in fungi to a variable extent (Table 2). A very narrow range of manganese concentrations (1–10 μM Mn^{2+}) stimulates synthesis of malformin by strains of *Aspergillus* (Steenbergen and Weinberg, 1968). Manganese is also stimulatory for squalene biosynthesis in cell-free extracts of *Rhizopus arrhizus* (Campbell and Weete, 1978). An absolute requirement for manganese has been reported for conversion of mevalonic acid into triterpene by *Neurospora crassa* (Bobowski et al., 1977) and for patulin synthesis by *Penicillium urticae* (Scott et al., 1984). The antibiotic patulin is the best-studied manganese-dependent secondary metabolite (Scott and Gaucher, 1986; Scott et al., 1986a,b). Additional factors required for production of patulin in the presence of manganese are a carbon source and NADPH, acetyl-CoA, and malonyl-CoA, which are provided by the primary metabolism. Manganese deficient (1.5 μM *P. urticae* secretes only the patulin precursor 6-methylsalicylic acid. Timely addition of magnesium merely restores the primary metabolism, but not the production of patulin. Because production of patulin was successfully blocked by actinomycin D and cycloheximide it was concluded that transcription is the target of manganese depletion. The elevated manganese requirement of *P. urticae* during the iodophase is another example of the important roles trace elements play in transcriptional control of microbial secondary metabolism (review by Weinberg, 1990).

Differential levels of manganese exert even a diverse transcriptional control of individual protein families during secondary metabolism of the white-rot fungus *Phanerochaete chrysosporium,* a widely used model organism for studies on lignin degradation (Boominathan and Reddy, 1992). Low levels of Mn^{2+} in the growth medium allow production of both lignin peroxidase (LIPs) and manganese-dependent peroxidases (MNPs), which along with H_2O_2 producing enzymes are major components of the lignin-degrading system of this organism. Higher concentrations of Mn^{2+} enhance MNP production but repress LIP production. Brown et al. (1991) showed that Mn^{2+} is required for the transcription of MNP genes, but the mechanism of LIP gene repression by elevated levels of Mn^{2+} remains unclear. Because lignin degradation is clearly secondary metabolism, a current approach is by studying mutants that are deregulated for either carbon and/or nitrogen prior to unraveling the manganese effects. However, as indicated by studies with another fungus (*Trametes versicolor*), lignin peroxidase itself may not be important for biological bleaching and delignification of unbleached kraft pulp in order to replace Cl_2 (Archibald, 1992). According to the current knowledge, the MNPs preferentially oxidize Mn(II) to Mn(III), and the latter is chelated by organic acids. The Mn(III)—organic acid complex produced is considered to be primarily responsible for the oxidation of the phenylpropanoic matrix of lignin (Wariishi et al., 1989; Paice et al., 1993).

Another commercial aspect of white rot fungi is that some of them are used as edible mushrooms. Recently it has been reported that both the mycelial growth of *Pleurotus ostreatus, Lentinus edodes, Pholiota nameko,* and *Agrocybe aegerita* on straw and the yield of fruit bodies is distinctly improved when Mn^{2+} ions are supplemented. The risk of contamination by deleterious fungi is also reduced by supplementation of Mn^{2+} (Nies, 1990).

IV. Mn^{2+}/Mg^{2+} ANTAGONISM: MUTAGENICITY OF HIGH Mn^{2+} LEVELS

Although required as an essential trace element, manganese can be deleterious to fungi at high concentrations as well. In a series of studies on environmental factors that affect toxicity of heavy metals toward microorganisms it was found that toxicity of manganese to fungi is influenced by pH (Babich and Stotzky, 1981). For a detailed discussion of the fungal responses toward heavy metals the reader is referred to the review of Gadd (1986). For the scope of this article it is more important that high Mn^{2+} concentrations are known to be mutagenic and growth inhibitory for bacteria and yeasts (Silver and Jasper, 1977).

Remarkably, little apparent toxic effect on respiration and viability by Mn^{2+} concentrations up to 300 mM were reported for the yeast *Candida utilis* (Parkin and Ross, 1985). Exposition of the Mn^{2+}-requiring filamentous fungus *Aspergillus niger* to increasing external Mn^{2+} concentrations induces multiple effects (Seehaus, 1990). Whereas impairment of growth starts at 5 mM and above, respiration is only slightly inhibited even at 100 mM Mn^{2+}. Intracellular elemental analysis by EDXA revealed a reversal of the phosphate-to-potassium ratio upon raising the Mn^{2+} content of the medium from 1 to 10 mM. Furthermore, conidiation is inhibited at 10 mM and above. Due to the elevated external manganese levels, *Aspergillus niger* reduces the size and number of its vacuoles and displays a homogeneous cytoplasm as a more general stress response (R. Stelzer, personal communication) while growing only very slowly in the presence of 100 mM Mn^{2+}.

Addition of magnesium alleviates the impairment of growth exerted by high manganese concentrations in *Aspergillus niger* (Seehaus, 1990). The observation of a Mn^{2+}/Mg^{2+} antagonism is not surprising. Since the chemistry of the Mn^{2+} ion is very similar to that of the Mg^{2+} ion, each can substitute for the other in the Mn^{2+}/Mg^{2+}-activated proteins (McEuen, 1981). However, exchange of the chemically related magnesium by manganese can be deleterious to the organisms. For example, in yeasts the mutagenicity of excess manganese due to a proposed Mg^{2+}/Mn^{2+} antagonism in DNA metabolism (Putrament et al., 1977) has been exploited for isolation of mutants. This antagonism can be traced back to a decreased fidelity of DNA-polymerases in the presence of an excess of Mn^{2+} ions (Goodman et al., 1983). As will be discussed in Sec. VI, antagonism of both ions is also observed in certain Mn^{2+} transport systems.

V. PERTURBATIONS BY Mn^{2+} DEFICIENCY: ESSENTIAL FOR HIGH YIELDS IN CITRIC ACID PRODUCTION WITH *ASPERGILLUS NIGER*?

In a pioneering study in the early days of the surface process, Currie discovered that restricted growth increased the yield of citric acid with *A. niger* (cf. Röhr et al., 1992). Among the environmental conditions that influence the amount of citric acid, a stringent manganese deficiency is obligatory for the submerse process (Shu and Johnson, 1947), developed later, which nowadays provides 80% of the world market of 500,000 tons per annum. Care has to be taken on effects reported for other trace elements (cf. the controversy over inhibition of citric acid accumulation by higher Fe^{2+} concentrations, reviewed by Kubicek and Röhr, 1986). During studies with manganese-requiring bacteria (Auling and Plönzig, 1987), we repeatedly observed that manganese impurities of the Fe^{2+} salts accounted for a number of unexpected effects. The metabolic pathway for citric acid accumulation, which involves enhanced glycolysis and the reactions of the tricarboxylic acid cycle, has been elucidated, and the key regulatory enzymes have been identified (Kubicek and Röhr, 1986). However, although a model for the carbon flow has been developed by these authors, it does not consider the central role of manganese deprivation for restricted growth and citric acid accumulation (Bigelis and Arora, 1992). For abundant acid production in submerged fermentation, the mycelia must undergo certain physiological changes during the first day after germination of the inoculum spores. The question is which of the many phenomena associated with manganese deprivation in *A. niger* (Röhr et al., 1992) indicate key regulatory parameters and which represent only secondary events accompanying the shift from filamentous to restricted growth. At least the shifts in lipid metabolism (Meixner et al., 1985) remind us of similar late perturbations accompanying nucleotide fermentation with *Brevibacterium ammoniagenes* (Thaler and Diekmann, 1979), which is also strictly controlled by manganese deprivation. In this bacterial fermentation, manganese-dependent ribonucleotide reduction, which delivers the DNA-precursors, has been identified as the trigger for impairment of DNA formation and successive unbalanced growth (Auling et al., 1980; Auling and Follmann, 1993). A reversible inhibition of DNA formation in *A. niger* ATCC 11414 upon manganese deprivation was found by Hockertz et al. (1987a). This inhibition can also be achieved in the presence of manganese by the addition of hydroxyurea, an inhibitor of ribonucleotide reductase. A survey of the literature reveals that, although it is an essential reaction in primary metabolism, ribonucleotide reduction has been completely neglected in filamentous fungi. Recently we could establish an *in vitro* assay with ammonium sulfate fractionated protein fractions (Dutt et al., 1992). However, we have learned from our long-term study of the new type of bacterial manganese-ribonucleotide reductase (Auling and Follmann, 1993) that the final confirmation of the metal character of such enzymes requires persevering efforts.

VI. MANGANESE UPTAKE

A. General Considerations

Evidence for manganese as an important regulator of morphogenesis and metabolism in fungi has been presented, yet compared with bacterial systems (Silver and Lusk, 1987) the information regarding the character and specificity of Mn^{2+} uptake in filamentous fungi is either rather fragmented (Auling, 1989) or (sometimes) presented in the context of the toxic effects of heavy metals (Gadd, 1986). Metal uptake by fungi generally occurs in two separate independent steps, i.e., surface binding (often referred to as unspecific binding in the literature) is observed prior to energy-dependent intracellular influx, which may be regulated or not. Evidence has been accumulated in the last few years that more than one manganese transport system may be present in the same fungal organism (Table 3).

B. Binding to Surface

The initial rapid phase of wall binding, which requires no metabolic energy, has been broadly discussed previously (Borst-Pauwels, 1981; Gadd, 1986). Upon exposure to millimolar concentrations of Mn^{2+}, the yeast *Saccharomyces cerevisiae* binds a quarter of the total accumulated manganese to its surface (Okorokov et al., 1977). Less (14–18%) unspecific binding of manganese under these conditions was determined by three independent methods for the filamentous fungus *Aspergillus niger* (Seehaus and Auling, submitted). The bound manganese can be seen as an electron dense surface layer in electron micrographs from thin sections of *Aspergillus niger* grown in the presence of 1 mM Mn^{2+}, whereas hyphae from manganese-depleted cultures are devoid of this layer (Seehaus, 1990). However, the values reported for unspecific binding of manganese to the anionic species of cell surfaces may also reflect the efficacy of the washing procedure. For exam-

Table 3 Low- and High-Affinity Mn Uptake Systems in Fungi

Organism	K_m	V_{max} ($g^{-1} \times min^{-1}$)	Inhibitor	References
S. cerevisiae[a]			Mg and others	Rothstein et al., 1958 Nieuwenhuis et al., 1981
C. utilis[a]	65 μM		Mg, Co, Zn	Parkin and Ross, 1985
A. niger[a]	18 mM		Fe and others	Seehaus, 1990
P. notatum[a]	22 μM	0.50 nmol	Mg and others	Starling, 1990
C. utilis[b]	16 nM	1.00 nmol	Zn	Parkin and Ross, 1986a
A. niger[b]				
(mineral medium)	3 μM	6.00 nmol	Cu, Cd, Zn	Hockertz et al., 1987b
(MES-buffer)	3 μM	7.00 μmol		Seehaus, 1990
P. notatum[b]	4 nM	0.05 nmol	Cd?	Starling, 1990

[a] Low-affinity uptake.
[b] High-affinity uptake.

ple, reduction from 20 to merely 7% was observed when *Candida utilis* was washed with distilled water (Parkin and Ross, 1986b) instead of 2 mM $CaCl_2$ (Parkin and Ross, 1985).

C. High-Affinity Transport

As late as 1986 the influx systems in fungi other than yeasts were assumed to be of low affinity only (Gadd, 1986). It is obvious that the paucity of information relating to metal ion uptake of filamentous fungi partly arises from problems associated with mycelial growth behavior. However, the recent breakthrough in detection of energy-linked manganese transporters in filamentous fungi (Hockertz et al., 1987b; Starling and Ross, 1990), which exhibit both high affinity and high specificity, came from studies with cells grown at very low metal concentrations. Clearly both fungal high-affinity Mn^{2+} transporters operate independently of the broad divalent cation transport systems (discussed in the following section), as is noticeable from the observation that additions of higher Mg^{2+} concentrations are not inhibitory (a 10^5 molar excess of Mg^{2+} in *Aspergillus niger,* a 10^3 molar excess in *Penicillium notatum,* respectively). A third high-affinity Mn^{2+} transporter, which was unaffected by a 100-fold molar excess of Mg^{2+}, is known in the respiratory yeast *Candida utilis* (Parkin and Ross, 1986a). Ironically, as yet no such system has been described for *Saccharomyces cerevisiae,* although energy-dependent uptake of Mn^{2+} was first detected in this yeast (Rothstein et al., 1958). A survey of the above three high-affinity Mn^{2+} transporters (Table 3) reveals their high specificity, which nevertheless is not absolute. Of the related transition metal ions, Cu^{2+}, Cd^{2+}, and Zn^{2+} are inhibitors in *A. niger,* Zn^{2+} in *C. utilis,* and Cd^{2+} in *P. notatum,* some of them with inhibition constants in the range of the K_m of the Mn^{2+} uptake. Remarkably, the distinct inhibitory action of Cu^{2+} ions on the high-affinity Mn^{2+} uptake of *A. niger* has been previously observed (although not understood!) when addition of Cu^{2+} ions was empirically exploited (Kisser et al., 1980) for counteracting elevated levels of manganese contamination deleterious for citric acid production (cf. Sec. V). Mn-insensitive mutants of *A. niger* with a defect in the high-affinity Mn^{2+} uptake have been studied and were found to have altered kinetic characteristics (Seehaus, 1990).

Bacterial high-affinity Mn^{2+} transporters are known to be inducible (Silver and Lusk, 1987). Less evidence has been provided for the inducibility of the analogous fungal systems. One elegant approach, difficult to apply with filamentous fungi, is to assay the activity of the Mn^{2+} transporter with washed cells previously adapted to low or high Mn^{2+} concentrations in continuous culture. Parkin and Ross (1986b) have demonstrated that the initial uptake rate was reduced tenfold when *C. utilis* was continuously grown in a high-Mn^{2+} medium versus a low-Mn^{2+} medium. Inducibility of the *A. niger* high-affinity Mn^{2+} transporter was also indicated by the total repression of $^{54}Mn^{2+}$ uptake using washed mycelia in a standard buffer assay after harvest from previous growth in a high-Mn^{2+} medium (Seehaus, 1990). Regulation of the Mn^{2+} transporter gene(s) would be even more convincingly demonstrated by susceptibility to the translational inhibitor cycloheximide, which has been successfully applied in order to suppress the increasing derepression of a high-affinity Zn^{2+} transporter in *Sclerotium rolfsii* concomitantly with the decrease of the intracellular zinc content (Kluttig, 1992). In accordance with this view on regulation, the rather low velocity of the high-affinity Mn^{2+} transporter of *P. notatum* (Starling and Ross, 1990) could be explained as partial repression by the residual manganese contamination of the nondecationized glucose (cf. Sec. II) in the growth medium. That the inherent Mn^{2+} concentration (1.5 μM) of so-called manganese deficient media might be still too high for the demonstration of an energy-dependent high-affinity Mn^{2+} transporter is noticeable from experiments with *P. urticae,* which obviously increases Mn^{2+} uptake during transition from growth to secondary metabolism (cf. Sec. III.C). The study of Scott and Gaucher (1986) clearly points to the dilemma that sometimes the cation uptake of interest for a metabolic question occurs at a concentration range that is not at the optimum of the responsible transport system to be detected, i.e., on the one hand in that particular study the high-affinity seemed to be repressed by the background level of manganese contamination in the growth medium used, and on the other hand an alternative low-affinity Mn^{2+} transporter could not (and may often not!) be detected with isotopic methods instead of using AAS for detection of manganese accumulated within the biomass after previous exposure to a millimolar range of concentrations.

D. Low-Affinity Uptake

As an alternative route to the previously described inducible Mn^{2+} transporters with low capacity, constitutive transport systems are operating with high capacity when fungal cells are exposed to elevated Mn^{2+} concentrations in the environment. These alternative systems generally comprise Mg^{2+} transporters with a broad specificity for divalent cations including Mn^{2+} in yeasts and filamentous fungi (Table 3) and have been intensively studied in *S. cerevisiae* (cf. Borst-Pauwels, 1981; Gadd, 1986; and Chap. 1 in this volume. The low-affinity Mn^{2+} uptake of *A. niger* is exceptional because it has characteristics of an Fe^{2+} transporter (Seehaus and Auling, submitted). Clearly, increased influx (rates) can be measured in media with reduced magnesium content (Starling, 1990) or during transition to secondary metabolism as well (Scott and Gaucher, 1986). Whether the wide variation of the kinetic data of the few low-affinity Mn^2 transport systems (Table 3) reflects variations of experimental conditions or differences between yeasts or filamentous fungi cannot be decided yet. However, the sometimes unphysiologically high amounts of ions with which such low-affinity uptake systems can deal makes a discussion of their real function difficult.

E. Cotransport with Citrate

It is known that in prokaryotes divalent cations can be taken up together with citrate. The transport of citrate is coupled with Mg^{2+} in *Bacillus subtilis* (Willecke et al., 1973) and with Fe^{2+} in *Escherichia coli* (Silver and Walderhaug, 1992). The question whether an analogous cotransport of Mn^{2+} and citrate occurs in fungi can be answered more easily when fungal mutants are available that have lost either of the two energy-dependent Mn^{2+} transporters described in Sections VI.B and C. Mutants of *A. niger* with a defective high-affinity Mn^{2+} transporter show merely one tenth of the $^{54}Mn^{2+}$ uptake of the wild type (Seehaus, 1990). This low $^{54}Mn^{2+}$ uptake is distinctly stimulated by the addition of citrate and reaches a wild-type level in the presence of 10 mM citrate. Since the strength of this stimulation depends on the external citrate concentration, it may be that a metal ion–citrate sensor protein is involved, as has been postulated for the Fe^{3+}-dicitrate transport in *E. coli* (Van Hove et al., 1990). The citrate-stimulated Mn^{2+} uptake of *A. niger* is abolished by withdrawal of the carbohydrate source and by energy poisoning with uncouplers. Thus this cotransport of

Mn^{2+} and citrate is an energy-dependent process (Seehaus and Auling, 1989). Because no transport system for citrate has been described for *A. niger* as yet, further studies with radiolabeled citrate are required to determine its kinetic characteristics. However, recently two transport systems with different substrate specificities for citric acid were described for the yeast *Candida utilis* (Cassio and Leao, 1991). Both systems are inducible by growth on citrate and are repressed by glucose. They operate simultaneously at pH 3.5 with a K_m of 0.59 mM for the undissociated acid (low-affinity) and a K_m of 0.056 mM for citrate (high-affinity). However, at pH 5.0 or above the low-affinity citrate transport is either absent or cannot be observed. It would be of interest to analyze these citrate transport systems for any coupled transport of divalent cations.

F. Manganese Homeostasis

With regard to manganese, the filamentous fungus *Aspergillus niger* is a remarkable organism, because its requirement for growth is fully saturated at very low concentrations while its tolerance to high levels of manganese is extreme. Thus for the purpose of this review *A. niger* and *Saccharomyces cerevisiae* have been chosen as the paradigms for manganese homeostasis in filamentous fungi and yeasts. It is clear from the preceding paragraphs that up to three different routes can be used by fungi for energy-dependent Mn^{2+} uptake (Fig. 1), some of them inducible. This situation may be reflected by the protein status of the cytoplasmic membrane. A comparison of purified cytoplasmic membranes of *A. niger* strain ATCC 11414 (Seehaus, 1990) harvested from different growth conditions revealed that a 39

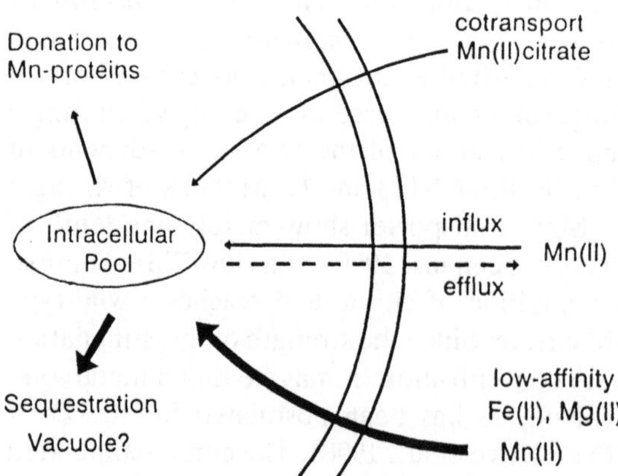

Figure 1 Scheme of manganese transport and homeostasis in fungi.

kD protein was absent under manganese deprivation. Wallrath et al. (1991) found a defective assembly of the respiratory chain NADH:ubiquinone reductase in the *A. niger* mutant strain B60 under citric acid-forming conditions (i.e., upon limited supply with manganese; cf. Sec. V). Although both findings should be brought to a broader basis they nevertheless indicate that Mn^{2+} controls the protein composition of the cytoplasmic membrane of filamentous fungi more effectively than Fe^{2+} (Huschka and Winkelmann, 1989) and Zn^{2+} (Kluttig, 1992).

Elevated levels of transition metals in fungal spores over mycelium have been reported. This was accounted for by greater binding to the melanized walls with chlamydospores of *Aureobasidium pullulans* (Gadd, 1986). However, using neutron activation analysis, translocation of manganese during differentiation from vegetative cells to conidia has also been shown, with the result of an increased accumulation of manganese within spores or more distinctly within coremia (spore-bearing structures) of *Penicillium* (Tinnel et al., 1974, 1976; Scott and Gaucher, 1986) and *Aspergillus* (Seehaus, 1990).

The yeast has been the model organism for studies on intracellular localization of divalent cations taken up by the low-affinity transport in vegetative fungal cells, and this uptake depends on the presence of polyphosphates located at the exterior side of the plasma membrane (Borst-Pauwels, 1981; Gadd, 1986). After phosphorylation, manganese is transported and then sequestered into the vacuole by binding as low molecular weight polyphosphates (Okorokov et al., 1980) and/or manganese orthophosphate, identified by ESR (Kihn et al., 1988). The observation by energy-dispersive x-ray microanalysis that increased accumulation of manganese in *Aspergillus niger* is accompanied by reduction of the potassium content (Seehaus and Auling, submitted) parallels previous findings in yeast (Lichko et al., 1980) and indicates at least one similar mechanism of manganese sequestration in both yeasts and filamentous fungi. Some manganese-resistant strains of *S. cerevisiae* that accumulated more manganese than the wild type have been isolated but were not deeply studied (Bianchi et al., 1981). Whereas the intracellular manganese concentration seems to be regulated by both vacuolar and plasma membrane transport systems in the yeast, Mn^{2+} efflux has been detected as an additional control mechanism in the filamentous fungus *Aspergillus niger.* Coupling of this efflux with the high-affinity Mn^{2+} transport (Fig. 1) was postulated (Seehaus, 1990) based on the fact that the mutants that were defective in the high-affinity Mn^{2+} uptake were also devoid of the energy-dependent Mn^{2+} efflux. Further support for this hypothesis comes from the observation that the intracellular manganese content increases only slightly upon raising the external Mn^{2+} level until a concentration of 1 mM is exceeded. Whether more than one protein is responsible for

Mn^{2+} efflux in *A. niger* may be elucidated by studies similar to those reported on the magnesium transport system in the bacterium *Salmonella typhimurium,* which requires at least three proteins for Mg^{2+} efflux (Gibson et al., 1991). Thus fungi have evolved a different homeostasis of the micronutrients Mn^{2+} and Zn^{2+} because no efflux of the latter has been detected so far in yeasts (Failla et al., 1976; White and Gadd, 1987) and filamentous fungi (Pilz et al., 1991).

Since *A. niger* is nevertheless able to accumulate more than 5 mg manganese per gram dry weight in response to external Mn^{2+} levels higher than 1 mM due to the high capacity of the alternative low-affinity Mn^{2+} transport (Seehaus and Auling, submitted), a search for manganese-sequestring proteins or peptides would be of interest. Mn-phytochelatins have not been looked for in *A. niger* and no phytochelatins were not found in *Aspergillus nidulans* (B. Tomsett, personal communication), whereas both Zn-binding metallothioneins and phytochelatins were recently detected in the filamentous fungus *Sclerotium rolfsii* (Kluttig, 1992). However, considering the detection of cadmium-mycophosphatins in mushrooms of the genus *Agaricus* (Meisch and Schmitt, 1986), a different type of low molecular weight proteins may perhaps sequester manganese in *Aspergillus* and other fungi by binding with phosphoserine groups. The model depicted in Fig. 1 summarizes the present knowledge from studies in yeast and *Aspergillus niger* on manganese transport and homeostasis in fungi.

VIII. CONCLUSIONS

When we shifted our manganese studies some years ago from bacteria to filamentous fungi, I expected many experimental difficulties. However, as discussed above, most of these can be overcome by increased experience in the handling of these microorganisms and by skillful adaptation of known techniques to their mycelial growth behavior. Comparing the wealth of published literature on the transport of transition metals in yeast to the limited information available within filamentous fungi, I am now convinced that, although we can learn much from studies with yeast, this organism is no longer the paradigm for the diversity of filamentous fungi. The availability of both mutants with defective Mn^{2+} uptake and purified plasma membranes makes *Aspergillus niger* an attractive organism for the study of manganese transport and related physiological processes in filamentous fungi.

There is no doubt that further progress on the exact function of this neglected transition metal in the primary and secondary metabolism of fungi is to be expected when the new genetic approach will aid the microbiological and biochemical studies reviewed here. With regard to the role of manganese in control of filamentous growth, it may be advisable to adapt experimental strategies from previous studies with coryneform bacteria (Auling and Follmann, 1993). On the other hand, the involvement of manganese in lignin degradation by white rot fungi is attracting growing efforts due both to its complexity and to the potential application in agricultural biotechnology or delignification of unbleached kraft pulp in the paper industry.

REFERENCES

Archibald, F. (1986). Manganese: Its acquisition and function in the lactic acid bacteria, *CRC Crit. Rev. Microbiol., 13:* 63.

Archibald, F. S. (1992). Lignin peroxidase activity is not important in biological bleaching and delignification of unbleached kraft pulp by *Trametes versicolor, Appl. Environ. Microbiol., 58:* 3101.

Ash, D. E., and Schramm, V. L. (1982). Determination of free and bound manganese (II) in hepatocytes from fed and fasted rats, *J. Biol. Chem., 257:* 9261.

Auling, G. (1989). Bacterial and fungal high affinity transport systems specific for manganese, *Metal Ion Homeostasis: Molecular Biology and Chemistry,* Alan R. Liss, New York, p. 459.

Auling, G. and Follmann, H. (1993). Manganese-dependent ribonucleotide reduction and overproduction of nucleotides in coryneform bacteria, *Metal Ions in Biological Systems,* vol. 30, *Metalloenzymes Involving Amino Acid Residue and Related Radicals,* Marcel Dekker, New York, p. 131.

Auling, G., Thaler, M., and Diekmann, H. (1980). Parameters of unbalanced growth and reversible inhibition of deoxyribonucleic acid synthesis in *Brevibacterium ammoniagenes* ATCC 6872 induced by depletion of Mn^{2+}. Inhibitor studies on the reversibility of deoxyribonucleic acid synthesis, *Arch. Microbiol., 127:* 105.

Babich, H., and Stotzky, G. (1981). Manganese toxicity to fungi: Influence of pH, *Bull. Environ. Contam. Toxicol., 27:* 474.

Barnett, H. L., and Lilly, V. G. (1966). Manganese requirements and deficiency symptoms of some fungi, *Mycologia, 58:* 585.

Bianchi, M. E., Carbone, M. L., and Lucchini, G. (1981). Mn^{2+} and Mg^{2+} uptake in Mn-sensitive and Mn-resistant yeast strains, *Plant Sci. Letters, 22:* 345.

Bigelis, R., and Arora, D. K. (1992). Organic acids of fungi, *Handbook of Applied Mycology,* vol. 4, *Fungal Biotechnology* (D. K. Arora, R. P. Elander, and K. G. Mukerji, eds.), Marcel Dekker, New York, p. 357.

Bobowski, G. S., Barker, W. G., and Subden, R. E. (1977). The conversion of (2-^{14}C)-mevalonic acid into triterpenes by cell-free extracts of *Neurospora crassa* albino mutant, *Can. J. Bot., 55:* 2137.

Boominathan, K., and Reddy, C. A. (1992). Fungal degradation of lignin: Biotechnological applications, *Handbook of Applied Mycology,* vol. 4, *Fungal Biotechnology* (D. K. Arora, R. P. Elander, and K. G. Mukerji, eds.), Marcel Dekker, p. 763.

Bonn, G., Reiffenstuhl, S., and Jandik, P. (1990). Ion chromatography of transition metals on an iminodiacetic acid bonded stationary phase, *J. Chromatogr., 499:*669.

Borst-Pauwels, G. W. F. H. (1981). Ion transport in yeast, *Biochim. Biophys. Acta, 650:* 88.

Brockaert, J. A. C. (1990). ICP-Massenspektrometrie, *Analytiker Taschenbuch,* Springer-Verlag, Berlin, p. 127.

Brown, J. A., Alic, M., and Gold, M. H. (1991). Manganese peroxidase gene transcription in *Phanerochaete chrysosporium:* Activation by manganese, *J. Bacteriol., 173:* 4101.

Campbell, O. A., and Weete, J. D. (1978). Squalene biosynthesis by cell-free extract of *Rhizopus arrhizus, Phytochem., 17:* 431.

Cassio, F., and Leao, C. (1991). Low- and high-affinity transport systems for citric acid in the yeast *Candida utilis, Appl. Environ. Microbiol., 57:* 3623.

Clark, D. S., Ito, K., and Horitsu, H. (1966). Effect of manganese and other heavy metals on submerged citric acid fermentation of molasses, *Biotechnol. Bioeng., 3:* 465.

Detroy, R. W., and Ciegler, A. (1971). Induction of yeastlike development in *Aspergillus parasiticus, J. Gen. Microbiol., 65:* 259.

Dutt, R., Blasczyk, K., and Auling, G. (1992). Manganese dependence of ribonucleotide reductase in *Aspergillus niger? BioEngineering, 2:* 47.

Failla, M. L., Benedict, C. D., and Weinberg, E. D. (1976). Accumulation and storage of Zn^{2+} by *Candida utilis, J. Gen. Microbiol., 94:* 23.

Follmann, H., Willing, A., Auling, G., and Plönzig, J. (1986). Manganese and ribonucleotide reduction in Gram-positive bacteria, *Thioredoxin and Glutaredoxin Systems: Structure and Function* (A. Holmgren, C.-I. Bränden, H. Jörnvall, and B.-M. Sjöberg, eds.), Raven Press, New York, p. 217.

Gadd, G. M. (1986). Fungal responses towards heavy metals, *Microbes in Extreme Environments* (R. A. Herbert and G. A. Codd, eds.), Academic Press, London, p. 83.

Gadd, G. M., Watt, F., Grime, G. W., and Brunton, A. H. (1988). Element localisation and distribution in the fungi *Aureobasidium pullulans* and *Aspergillus niger, Cytobios, 56:* 69.

Galpin, M. F. J., Jennings, D. H., Oates, K., and Hobot, J. (1978). Localization by X-ray microanalysis of soluble ions, particularly potassium and sodium in fungal hyphae, *Exp. Mycol., 12:* 258.

Garraway, M. O., and Evans, R. C. (1984). Fungal nutrition and physiology, John Wiley, New York.

Garrison, R. G., and Boyd, K. S. (1971). Ultrastructural studies of induced morphogenesis in *Aspergillus parasiticus, Sabouraudia, 12:* 179.

Garrison, R. G., and Boyd, K. S. (1973). Dimorphism of *Penicillium marneffei* as observed by electron microscopy, *Can. J. Microbiol., 19:* 1305.

Gibson, M. M., Bagga, D. A., Miller, C. G., and Maguire, M. E. (1991). Magnesium transport in *Salmonella typhimurium:* Influence of new mutations conferring Co^{2+} resistance on the CorA Mg^{2+} transport system, *Mol. Microbiol., 5:* 2753.

Good, N. E., Winget, G. D., Winter, W., Conally, T. N., Izawa, S., and Sing, R. M. M. (1966). Hydrogen ion buffers for biological research, *Biochemistry, 5:* 467.

Goodman, M. F., Keener, S., and Guidotti, S. (1983). On the enzymatic basis for mutagenesis by manganese, *J. Biol. Chem., 25:* 3469.

Harold, F. (1986). *The Vital Force: A Study of Bioenergetics,* W. H. Freeman, New York.

Hockertz S., Plönzig, J., and Auling, G. (1987a). Impairment of DNA formation is an early event in *Aspergillus niger* under manganese starvation, *Appl. Microbiol. Biotechnol., 25:* 590.

Hockertz, S., Schmid, J., and Auling, G. (1987b). A specific transport system for manganese in the filamentous fungus *Aspergillus niger, J. Gen. Microbiol., 133:* 3513.

Hughes, M. N., and Poole, R. K. (1991). Metal speciation and microbial growth—The hard (and soft) facts, *J. Gen. Microbiol., 137:* 725.

Huschka, H.-G., and Winkelmann, G. (1989). Iron limitation and its effect on membrane proteins and siderophore transport in *Neurospora crassa, Biol. Metals, 2:* 108.

Kihn, J. C., Dassargues, C. M., and Mestdagh, M. M. (1988). Preliminary ESR study of Mn(II) retention by the yeast *Saccharomyces, Can. J. Microbiol., 34:* 1230.

Kisser, M., Kubicek, C. P., and Röhr, M. (1980). Influence of manganese on morphology and cell wall composition of *Aspergillus niger* during citric acid fermentation, *Arch. Microbiol., 128:* 26.

Kluttig, S. (1992). Transport und physiologische Einflüsse von Zink bei *Sclerotium rolfsii,* Ph.D. thesis, Universität Hannover, Germany.

Kubicek, C. P., and Röhr, M. (1986). Citric acid fermentation, *CRC Crit. Rev. Biochem., 3:* 331.

Lichko, L. P., Okorokov, L. A., and Kulaev, I. S. (1980). Role of vacuolar ion pool in *Saccharomyces carlsbergensis:* Potassium efflux from vacuoles is coupled with manganese or magnesium influx, *J. Bacteriol., 144:* 666.

McEuen, A. R. (1981). Manganese metalloproteins and manganese activated enzymes, *Inorg. Biochem., 2:* 249.

Meisch, H.-U., and Schmitt, J. A. (1986). Characterization studies on cadmium-mycophosphatin from the mushroom *Agaricus macrosporus, Environ. Health Perspectives, 65:* 29.

Meixner, O., Mishak, H., Kubicek, C. P., and Röhr, M. (1985). Effect of manganese deficiency on plasma-membrane lipid composition and glucose uptake in *Aspergillus niger, FEMS Microbiol. Lett., 26:* 271.

Milsom, P. E., and Meers, J. L. (1985). Citric acid, *Comprehensive Biotechnology,* vol. 3 (M. Moo-Young, ed.), Pergamon Press, Oxford, p. 665.

Nieuwenhuis, B. J. W. M., Weijers, C. A. G. M., and Borst-Pauwels, G. W. F. H. (1981). Uptake and accumulation of Mn^{2+} and Sr^{2+} in *Saccharomyces cerevisiae, Biochim. Biophys. Acta, 649:* 83.

Nies, A. (1990). Untersuchungen zur verbesserten Substratbesiedelung für den

Anbau des Austernseitlings *Pleurotus ostreatus* (Jacq. ex Fr.) Kummer, Ph.D. thesis, Universität Hannover, Germany.

Okorokov, L. A., Lichko, L. P., Kadomtseva, V. M., Kholodenko, V. P., Titovsky, V. T., and Kulaev, I. S. (1977). Energy-dependent transport of manganese into yeast cells and distribution of accumulated ions, *Eur. J. Biochem., 75:* 373.

Okorokov, L. A., Lichko, L. P., and Kulaev, I. S. (1980). Vacuoles: Main compartments of potassium, magnesium, and phosphate ions in *Saccharomyces carlsbergensis* cells, *J. Bacteriol., 144:* 661.

Paice, M. G., Reid, I. D., Bourbonnais, R., Archibald, F. S., and Jurasek, L. (1993). Manganese peroxidase, produced by *Trametes versicolor* during pulp bleaching, demethylates and delignifies kraft pulp, *Appl. Environ. Microbiol., 59:* 260.

Parkin, M. J., and Ross, I. S. (1985). International Conference on Heavy Metals in the Environment, Athens, vol. 2 (T. D. Lekkas, ed.), CPC Consultants, Edinburgh, U.K., p. 289.

Parkin, M., and Ross, I. S. (1986a). The specific uptake of manganese in the yeast Candida utilis, *J. Gen. Microbiol., 132:* 2155.

Parkin, M. J., and Ross (1986b). The regulation of Mn^{2+} and Cu^{2+} uptake in cells of the yeast *Candida utilis* grown in continuous culture, *FEMS Microbiol. Lett., 37:* 59.

Pateman, J. A. (1969). Regulation of synthesis of glutamate dehydrogenase and glutamine synthetase in micro-organisms, *Biochem. J., 115:* 769.

Pilz, F., Auling, G., Stephan, D., Rau, U., and Wagner, F. (1991). A high-affinity Zn^{2+} uptake system controls growth and biosynthesis of an extracellular, branched β-1,3-β-1,6-glucan in *Sclerotium rolfsii* ATCC 15205, *Exp. Mycol., 15:* 181.

Plönzig, J., and Auling, G. (1987). Manganese deficiency impairs ribonucleotide reduction but not DNA replication in *Arthrobacter* species, *Arch. Microbiol., 146:* 396.

Putrament, A., Baranowska, H., Ejchart, A., and Jachymczyk, W. (1977). Manganese mutagenesis in yeast: VI. Manganese ion uptake, mitochondrial DNA replication and erythromycin resistance induction, comparison with other divalent cations, *Mol. Gen. Genet., 151:* 69.

Reiss, E., and Nickerson, W. J. (1971). Control of dimorphism in *Phialophora verrucosa, Sabouraudia, 12:* 202.

Röhr, M., Kubicek, C. P., and Kominek, J. (1992). Industrial acids and other small molecules, *Biology of Aspergillus* (J. Bennett, ed.), Butterworth, Stoneham, p. 91.

Rothstein, A., Hayes, A., Jennings, D., and Hooper, D. (1958). The active transport of Mg^{2+} and Mn^{2+} into the yeast cell, *J. Gen. Physiol., 41:* 585.

Scott, R. E., and Gaucher, G. M. (1986). Manganese and antibiotic biosynthesis, II. Cellular levels of manganese during the transition to patulin production in *Penicillium urticae, Can. J. Microbiol., 32:* 268.

Scott, R. E., Jones, A., and Gaucher, G. M. (1984). A manganese requirement for patulin biosynthesis by cultures of *Penicillium urticae, Biotechnol. Lett., 6:* 231.

Scott, R. E., Jones, A., Lam, K. S., and Gaucher , G. M. (1986a). Manganese and antibiotic biosynthesis, I. A specific manganese requirement for patulin production in *Penicillium urticae, Can. J. Microbiol., 32:* 259.

Scott, R. E., Jones, A., and Gaucher, G. M. (1986b). Manganese and antibiotic biosynthesis, III. The site of manganese control of patulin production in *Penicillium urticae, Can. J. Microbiol., 32:* 273.

Seehaus C. (1990). Charakterisierung von Transportsystemen für Mangan bei *Aspergillus niger* ATCC 11414, Ph.D. thesis, Universität Hannover, Germany.

Seehaus, C., and Auling, G. (1989). Effect of citrate on the uptake of Mn^{2+} by *Aspergillus niger, Forum Mikrobiologie, 1–2:* 58.

Seehaus, C., and Auling, G. Manganese homeostasis in *Aspergillus niger* ATCC 11414: Uptake and accumulation of manganese by an energy-dependent, low-affinity transport system (manuscript submitted).

Shu, P., and Johnson, M. J. (1947). Effect of the composition of the sporulation medium on citric acid production by *Aspergillus niger* in submerged culture, *J. Bacteriol., 54:* 161.

Silver, S., and Jasper, P. (1977). Manganese transport in microorganisms, *Microorganisms and Minerals* (E. D. Weinberg, ed.), Marcel Dekker, New York, Basel, p. 105.

Silver, S., and Lusk, J. E. (1987). Bacterial magnesium, manganese and zinc transport, *Ion Transport in Prokaryotes* (B. P. Rosen and S. Silver, eds.), Academic Press, New York, p. 165.

Silver, S., and Walderhaug, M. (1992). Gene regulation of plasmid- and chromosome-determined inorganic ion transport in bacteria, *Microbiol. Rev., 56:* 195.

Starling, A. P. (1990). The uptake of heavy metals by protoplasts and whole cells of filamentous fungi, Ph.D. thesis, University of Keele, U.K.

Starling, A. P., and Ross, I. (1990). Uptake of manganese by *Penicillium notatum, Microbios, 63:* 93.

Steenbergen, S. T., and Weinberg, E. D. (1968). Trace metal requirements for malformin biosynthesis, *Growth, 32:* 125.

Sulochana, C. B., and Lakshmanan, M. (1968). *Aspergillus niger* technique for the bioassay of manganese, *J. Gen. Microbiol., 50:* 285.

Thaler, M., and Diekmann, H. (1979). The effect of manganese deficiency on lipid content and composition in *Brevibacterium ammoniagenes, Eur. J. Appl. Microbiol., 6:* 379.

Tinnel, W. H., Jefferson, B. L., and Benoit, R. E. (1974). The organic nitrogen exigency of and effects of manganese on coremia production of *Penicillium clavigerum* and *Penicillium claviforme, Can. J. Microbiol., 20:* 91.

Tinnel, W. H., Jefferson, B. L., and Benoit, R. E. (1976). Manganese-mediated morphogenesis in *Penicillium claviforme* and *Penicillium clavigerum, Can. J. Microbiol., 23,* 209.

Van Hove, B., Staudenmaier, H., and Braun, V. (1990). Novel two-component transmembrane transcription control: Regulation of iron dicitrate transport in *Escherichia coli* K-12, *J. Bacteriol., 172:* 6749.

Wallrath, J., Schmidt, M., and Weiss, H. (1991) Concomitant loss of respiratory chain NADH:ubiquinone reductase (complexI) and citric acid accumulation in *Aspergillus niger, Appl. Microbiol. Biotechnol.,* 36: *76.*

Wariishi, H., Dunford, H. B., MacDonald, I. D., and Gold, M. H. (1989). Manganese peroxidase from the lignin-degrading basidiomycete *Phanerochaete chrysosporium:* Transient-state kinetics and reaction mechanism, *J. Biol. Chem., 264:* 3335.

Weinberg, E. D. (1990). Roles of trace metals in transcriptional control of microbial secondary metabolism, *Biol. Metals, 2:* 191.

White, C., and Gadd, G. M. (1987). The uptake and cellular distribution of zinc in *Saccharomyces cerevisiae, J. Gen. Microbiol., 133:* 727.

Widra, A. (1964). Phosphate directed Y-M variation in *Candida albicans, Mycopathol. Mycol. Appl., 23:* 197.

Willecke, K., Gries, E.-M., and Oehr, P. (1973). Coupled transport of citrate and magnesium in *Bacillus subtilis, J. Biol. Chem., 248:* 807.

Zonneveld, B. J. M. (1971). Biochemical analysis of the cell wall of *Aspergillus nidulans, Biochim. Biophys. Acta,* 249: 506.

Zonneveld, B. J. M. (1975a). Sexual differentiation in *Aspergillus nidulans.* The requirements for manganese and its effect on α-1,3-glucan synthesis and degradation, *Arch. Microbiol., 105:* 101.

Zonneveld, B. J. M. (1975b). Sexual differentiation in *Aspergillus nidulans.* The requirements for manganese and the correlation between phosphoglucomutase and the synthesis of reserve material, *Arch. Microbiol., 105:* 105.

Zonneveld, B. J. M. (1976). The effect of glucose and manganese on adenosine-3′-5′-monophosphate levels during growth and differentiation of *Aspergillus nidulans. Arch. Microbiol., 108:* 41.

8

Uptake of Zinc by Fungi

I. S. Ross
University of Keele, Keele, Staffordshire, England

I. INTRODUCTION

Zinc has long been known to be essential for the growth of microorganisms including fungi and is important in the structure and function of many enzymes, nucleic acid metabolism, and cell division (Failla, 1977). Zinc is also important in the production of secondary metabolites including aflatoxins, other polyketides, ergot alkaloids, penicillin (Weinberg, 1977; Failla and Niehaus, 1986), and extracellular polysaccharides such as scleroglucan from *Sclerotium rolfsii* (Pilz et al., 1991). It is therefore somewhat suprising that relatively little work on the uptake of zinc has been carried out in fungi, especially in mycelial species. This discussion will deal only with the uptake of zinc, but it should be noted that there are similarities between the uptake of zinc and that of a number of other divalent metals including manganese, cobalt, cadmium, and nickel, and also that toxic effects of and resistance to divalent metals may be relevant to uptake studies. There are a number of reviews that deal with broader aspects of divalent metal uptake, toxicity, and resistance (Gadd, 1986; Gadd, 1990a; Ross, 1993), and these aspects will not be considered here.

For clarity, yeasts and filamentous fungi will be discussed separately, but this should not be taken to indicate that there are important differences between the two groups, though there are special problems that can arise in handling mycelial cultures, which may influence the experimental approach. Attention will be drawn to similarities and possible differences between yeasts and filamentous fungi in the last section of this discussion.

II. UPTAKE BY CELLS AND ADSORPTION TO HYPHAL WALLS

In both yeasts and filamentous fungi, the first potential barrier to penetration of the cells by metal ions is the cell wall. There is much literature on metal uptake by fungal cell walls with emphasis on use as biosorbents. Langmuir or Freundlich isotherms have been used to describe binding. This work together with data on the chemical nature of metal binding has recently been reviewed by Gadd (1990b). However, relatively little of this work relates specifically to zinc, which has received less attention than other more toxic metals such as cadmium and copper. Also, the application of the Langmuir or Freundlich isotherm needs to be used with caution, as both assume that binding is to a uniform unimolecular layer with each ion adsorbed to an adsorption site, a situation unlikely to be found in fungal cell walls, where binding is likely to be of more than one type and may well be complex (Gadd, 1986; Gadd, 1990b). This discussion will concentrate on data obtained from zinc uptake studies aimed at understanding the transport of metal ions into cells, when more attention has generally been paid to cell viability.

A. Yeasts

Studies of the uptake of metals by yeasts, chiefly *Saccharomyces cerevisiae,* have shown that uptake is generally biphasic, with an initial rapid phase, independent of metabolic energy, followed by a slower period of uptake, dependent on metabolic energy. The initial phase is generally regarded as due to binding to the cell wall, while the second phase is due to the transport of metal ions into the cells across the plasma membrane. Such studies have largely been based on time-course experiments on nongrowing suspensions in the presence and absence of energy sources and/or metabolic inhibitors such as potassium cyanide or carbonylcyanide-*m*-chlorophenylhydrazone (CCCP). The initial surface binding is generally taken as the amount of metal bound within a few minutes in the absence of energy sources, at low temperature (usually about 4°C) or in the presence of metabolic inhibitors. These techniques do not distinguish between metal bound to the cell wall and that bound to the outer surface of the plasma membrane. A typical example is the uptake of Zn^{2+} by *Saccharomyces cerevisiae,* where relatively low amounts are bound in the absence of glucose or in the presence of metabolic inhibitors including cyanide, CCCP, *N,N'*-dicyclohexylcarbodiimide (DCCD), or diethylstilboestrol (DES) (Furmann and Rothstein, 1968; Ponta and Broda, 1970; Mowll and Gadd, 1983; White and Gadd, 1987). This pattern of uptake has also been seen in other yeast species including *Candida utilis* (Failla et al., 1976; Failla and Weinberg, 1977) and *Sporobolomyces roseus* (Mowll and Gadd, 1983).

Comparisons of published data on amounts of various metal ions such as Zn^{2+} bound by yeast or filamentous fungi (see below) are of little value even when external Zn^{2+} concentrations are apparently the same, as the actual figures will depend on the particular experimental conditions used, especially the composition and pH of the uptake medium. Metal ions such as Zn^{2+} may form complexes with various medium or buffer components or may be precipitated as complexes with phosphate or hydroxide, especially at pH near neutrality or above. This may lead to the presence of lower than expected concentrations of free Zn^{2+} in uptake media. The choice of pH is also known to influence surface binding of metal ions. This generally decreases as the pH falls (Gadd, 1986), probably due to competition for binding sites by hydrogen ions. Surface-binding of Zn^{2+} by cells of *S. cerevisiae* was markedly reduced by pH below 5 (Fuhrmann and Rothstein, 1968). The density of cell suspensions (usually expressed as mg dry weight mL^{-1} but sometimes expressed as cells mL^{-1}) varies between reports, and this may influence metal ion binding, especially when the external metal ion concentration is low. There is, however, some unequivocal data that show that the capacity of yeast cell walls to bind Zn^{2+} varies between species. For example, the amount of Zn^{2+} bound externally by *Sporobolomyces roseus* from a solution containing 100 μM Zn^{2+} was 28.9 nmol/mg dry wt, while under the same experimental conditions that bound by *Saccharomyces cerevisiae* was 3.8 nmol/mg dry wt (Mowll and Gadd, 1983). Differences in cell wall composition may account for this, since the wall of *Spor. roseus* consists largely of chitin and mannan, while that of *S. cerevisiae* is mainly glucan and mannan with traces of chitin (Mowll and Gadd, 1986). A table showing surface binding of Zn^{2+} by a range of yeasts cannot usefully be compiled from published data due to wide variations in external Zn^{2+} concentrations used by different workers (ranging from 1–500 μM) as well as the factors mentioned above. However, examination of data presented for *C. utilis* (Failla et al., 1976), *S. cerevisiae* (Fuhrmann and Rothstein, 1968; Mowll and Gadd, 1983; White and Gadd, 1987), and *Spor. roseus* (Mowll and Gadd, 1983) suggests that surface binding generally accounts for less than 30% of the Zn^{2+} taken up during 30 min incubation.

B. Filamentous Fungi

Studies of Zn^{2+} uptake in *Neocosmospora vasinfecta* (Paton and Budd, 1972) showed that about 84% of the metal taken up after 30 min was desorbed into phthalate buffer, indicating extensive binding to the exterior of the hyphae. Further investigation showed that this binding was unaffected by anaerobiosis, sodium azide, or low temperature and was complete within 2 min. Paton and Budd (1972) also observed that incubation of mycelium for 3 h in anaerobic conditions or in the presence of sodium azide prior to addition of Zn^{2+}resulted in a reduction in binding of 25–30%. They suggested that this indicated that some binding sites were subject to metabolic turnover, and that these might be located in the plasma membrane. If this is the case, some 25–30% of overall surface binding of Zn^{2+} represents binding to the plasma membrane in this organism.

Townsley (1985) found that large amounts of Zn^{2+} were bound by nongrowing suspensions of mycelium of *Penicillium spinulosum, Aspergillus niger,* and *Trichoderma viride* from an external Zn^{2+} concentration of 38 μM that was not toxic to the cells. This binding was rapid and independent of metabolic energy, and it largely masked any intracellular uptake. After 30 min incubation in the absence of glucose, Zn^{2+} binding reached 833 nmol/mg for *P. spinulosum,* 650 nmol/mg for *A. niger,* and 462 nmol/mg for *T. viride.* This indicates that as with yeasts, some variation in Zn^{2+} binding between species can be seen when experimental conditions are constant. The data presented above suggest that the surface binding of Zn^{2+} by filamentous fungi is much greater than that seen in yeasts, to the extent that transport across the plasma membrane may be difficult to demonstrate unless surface-bound Zn^{2+} is desorbed in some way. That this is not always the case has been demonstrated by studies of Zn^{2+} uptake in *Penicillium notatum* (Starling and Ross, 1991), where surface binding accounted for less than 30% of total Zn^{2+} uptake from 10 nM or 20 μM Zn^{2+}, and by investigation of Zn^{2+} uptake in *Sclerotium rolfsii,* where surface binding was reported to be very low (Pilz et al., 1991). In the case of *P. notatum,* the low surface binding may have resulted from the use of spore inoculation and harvesting of the mycelium after 19 h incubation. With *P. spinulosum,* the use of mycelial fragments as inoculum and harvesting after 48 h produced mycelium with a high surface-binding capacity (Townsley, 1985). The possibility that the difference in binding was due to differences in wall composition between *P. notatum* and *P. spinulosum* cannot be ruled out, but the age of the mycelium may be important. Pilz et al. (1991) used mycelial fragments allowed to regenerate for 24 h; and Hockertz et al. (1987) using young (24 h) spore-inoculated cultures of *A. niger* did not experience problems with high surface binding in studies of manganese uptake.

In the polymorphic fungus *Aureobasidium pullulans,* an organism commonly found in heavy-metal-polluted habitats, it has been noted that thick-walled, melanized chlamydospores bind much greater amounts of Zn^{2+} than do yeastlike cells and mycelium, and investigations using intact cells and protoplasts have shown that penetration of Zn^{2+} into chlamydospores is largely prevented by the capacity of the walls to bind the metal (Gadd et al., 1987). In intact yeastlike cells, about 60% of the Zn^{2+} taken up over 120 min was apparently surfacebound.

While there is no doubt that the cell walls of both yeasts and filamentous fungi may bind significant amounts of Zn^{2+}, there is now evidence that such binding is not a prerequisite for subsequent transport across the plasma membrane. As with other metal ions including copper, cadmium, and cobalt, Zn^{2+} is taken up by protoplasts of yeastlike and mycelial cells of *A. pullulans,* and the K_m for Zn^{2+} uptake in both protoplasts and intact yeastlike cells was about 11 μM. The V_{max} for Zn^{2+} uptake in protoplasts of yeastlike cells was lower than in intact cells, which may reflect damage to the plasma membrane during production of protoplasts (Gadd et al., 1987). Data obtained using protoplasts of other fungi including *S. cerevisiae* (Gadd et al. 1984), *Penicillium ochro-chloron* (Gadd and White, 1985), and *P. notatum* (Starling and Ross, 1990) also confirm that binding to the cell wall is not essential for uptake of metal ions, though Zn^{2+} was not used in these studies.

III. UPTAKE ACROSS THE PLASMA MEMBRANE

A. Yeasts

1. *Energy Dependence*

Early studies of Zn^{2+} uptake in the yeast *Saccharomyces cerevisiae* established that Zn^{2+} uptake was strongly stimulated by glucose, especially in cells that had been starved for some time (Fuhrmann and Rothstein, 1968; Ponta and Broda, 1970), and that this stimulation was reduced by low temperature (Ponta and Broda, 1970; Mowll and Gadd, 1983). This effect of glucose was abolished by metabolic inhibitors such as antimycin A (Mowll and Gadd, 1983), potassium cyanide, 2,4-dinitrophenol (DNP), DCCD, diethylstilboestrol (DES), and the glucose analog 2-deoxyglucose (White and Gadd, 1987). Similar results were obtained for *Candida utilis* using the inhibitors potassium cyanide, sodium arsenate, and CCCP by Failla et al. (1976), who also observed reduced uptake at low temperature (5°C). All these observations indicate that uptake of Zn^{2+} into the cells of yeasts is an energy-dependent process, and various suggestions have been made about the nature of energy coupling to the uptake process, generally based on the perceived effects of metabolic inhibitors on intact yeast cells. Thus it was stated that the inhibitory effect of DCCD and DES on Zn^{2+} uptake by *S. cerevisiae* implicated ATPase activity, especially that of the plasma membrane ATPase (White and Gadd, 1987), while the inhibitory effect of DNP indicated a requirement for a transmembrane proton gradient. While both of these statements may well be true, it should be noted that the specificity of these inhibitors is open to question. DCCD is not specific for the plasma membrane ATPase in fungi (Goffeau and Slayman, 1981) and can cause inhibition of the mitochondrial ATPase, with consequent effects on energy metabolism, while DCCD and DES can provoke an all-or-none loss of cellular potassium in *S. cerevisiae,* with some of these cells showing increased permeability for the dye bromophenol blue, indicating drastic changes in membrane permeability (Borst-Pauwells et al., 1983). Apart from this all-or-none effect, a more gradual loss of potassium from cells was also noted. In the light of these observations, Borst-Pauwels et al. (1983) warn against the use of these inhibitors in evaluating the role of the ATPase in the energy transduction of membrane transport processes. The work most usually cited in support of the use of inhibitors such as DCCD and DES is that of Serrano (1980), though it is stated that a respiratory-deficient strain of *S. cerevisiae* lacking functional mitochondria was specifically chosen to avoid the complicating effects of the inhibitors on oxidative phosphorylation that would occur in strains of *S. cerevisiae* with normal respiratory metabolism.

The complete inhibition of Zn^{2+} uptake in *S. cerevisiae* by uncouplers of oxidative phosphorylation such as DNP has been taken as evidence for the requirement of a transmembrane proton gradient (White and Gadd, 1987), but it should be noted that there are again potential nonspecific effects that can occur in intact cells. The optimal concentration of DNP for inducing maximum proton influx into yeast cells has been given as 0.5 mM (Gadd and Mowll, 1985) while White and Gadd (1987) used 2 mM DNP, and other effects may have occurred. DNP may exert a number of disruptive effects on cellular metabolism, and Jennings (1963) has warned that conclusions based on the assumption that DNP acts only as an uncoupler, at concentrations higher than the minimum necessary to uncouple phosphorylation, need careful interpretation. Other inhibitors such as CCCP, which is known to be an uncoupler (Heytler and Pritchard, 1962; Harold et al., 1974), are also effective in blocking divalent cation uptake; for example, Zn^{2+} uptake by *C. utilis* is inhibited by CCCP (Failla et al., 1976; Lawford *et al.,* 1980), but again, other effects cannot be ruled out. CCCP not only inhibited Zn^{2+} uptake in *C. utilis* but also induced a loss of recently accumulated Zn^{2+} from the cells (Failla et al., 1976). This could have resulted from altered membrane permeability caused by CCCP, though no other direct evidence for this, such as loss of cellular potassium, was reported.

There have been attempts to investigate the energy coupling of divalent cations other than Zn^{2+}, notably Mn^{2+} and Sr^{2+}. It was suggested that there was a direct involvement of the plasma membrane ATPase, as the order of affinity for uptake of several divalent cations was the same as that for the stimulation of the plasma membrane ATPase activity by the cations; the rate of divalent cation uptake in yeast plasma membrane vesicles was proportional to the initial ATP content, and membrane ATPase inhibitors completely inhibited cation uptake (Fuhrmann, 1973, 1974a,b). More recent work showed that there was no detectable selectivity in the initial uptake rates of Mn^{2+} or Sr^{2+}, while plasma membrane ATPase is specific for Mn^{2+} but not Sr^{2+} ions (Peters and Borst-Pauwels, 1979; Nieuwenhuis et al., 1981). A direct involvement of the plasma membrane ATPase can therefore be ruled out, but involvement of the ATPase in the generation of a transmembrane proton gradient is probable. For example, Ca^{2+}/H^+ antiport is driven by the transmembrane proton gradient in *S. cerevisiae* (Eilam et al., 1985), and evidence that Mn^{2+} uptake in yeast membrane vesicles is driven by a proton gradient (Borst-Pauwels, 1981) has been presented. It is likely that Zn^{2+} uptake in yeast is also driven by a transmembrane proton gradient, as suggested by White and Gadd (1987), but no other data not subject to the cautions mentioned above appears to be available.

2. *Uptake Kinetics*

It was rapidly established that Zn^{2+} uptake in yeasts such as *S. cerevisiae* followed Michaelis-Menten kinetics (Fuhrmann and Rothstein, 1968; Ponta and Broda, 1970), and in these and subsequent studies a range of half-saturation constants (K_m) have been determined. Some examples together with references are given in Table 1. It will be apparent that there is considerable variation in published K_m values even in the same yeast species. Similar variations have been seen in data for other divalent metals (Borst-Pauwels, 1981), and this was attributed to a number of factors including differences in techniques used to measure uptake, possible complexing of ions by buffers, which affects apparent K_m, and effects of the surface potential. It is the view of this author that comparison of K_m values may be useful only when experimental conditions are similar, and care is necessary in selecting concentration ranges to be used. An examination of the published data on the kinetics of uptake of a range of divalent cations gives the impression that the apparent K_m quoted may be a function of the concentration range used in its determination. Deviations of Zn^{2+} uptake from Michaelis-Menten kinetics may also occur, as has been observed by White and Gadd (1987) at Zn^{2+} concentrations about 80 μM in a strain of *S. cerevisiae.*

Table 1 Half-Saturation Constants K_m for Zinc Uptake by Yeasts

K_m (μM)	Conc. range (μM)	Reference
<10	[a]	Fuhrmann and Rothstein, 1968
1.3	0.6–10	Failla et al., 1976
2.0*	0.6–10	Failla and Weinberg, 1977
0.36*	[b]	Lawford et al., 1980
3.7	0–80	White and Gadd, 1987
90**	10–50	Mowll and Gadd, 1993
500	10–50	Mowll and Gadd, 1983
1150	0–8000	Ponta and Broda, 1970

Data for *S. cerevisiae* except where marked: **C. utilis;* ***Sporobolomyces roseus.*
[a]Not specified precisely but apparently in the range 0–5 mM.
[b]Estimated from figure to be in the range 0.2–20 μM.

There is obviously the possibility that two or more uptake systems for divalent cations may be present in yeasts, and there is unequivocal evidence that low- and high-affinity uptake systems for Mn^{2+}, with different specificity for divalent cations, exist in *C. utilis* (Parkin and Ross, 1985, 1986). Comparative data obtained under identical conditions are not available for Zn^{2+} uptake in yeasts.

3. Specificity of Uptake Systems

In *C. utilis,* uptake of Zn^{2+} from concentrations of 1.1 μM was not inhibited by 10 μM or 100 μM Na^+, Ca^{2+}, Mn^{2+}, Co^{2+}, Cu^{2+}, or Cr^{3+} ions, but Hg^{2+}, Ag^+, and Cd^{2+} were inhibitory. Of these, Hg^{2+} and Ag^+ killed the cells, while Cd^{2+} and the other ions tested were not toxic at 100 μM. Further investigation showed that inhibition of Zn^{2+} uptake by Cd^{2+} was competitive. Thus Zn^{2+} uptake in *C. utilis* involves a high-affinity, highly specific mechanism similar to the high-affinity system for Mn^{2+} in this organism (Parkin and Ross, 1986). Unfortunately, the effect of Mg^{2+} on Zn^{2+} uptake by *C. utilis* does not appear to have been investigated, and it is not clear whether Zn^{2+} uptake sensitive to inhibition by Mg^{2+}, analogous to the low-specificity uptake of Mn^{2+} reported by Parkin and Ross (1985), occurs in this yeast.

In *S. cerevisiae* the situation is more uncertain. Zn^{2+} uptake by a relatively high affinity system was inhibited by K^+ and Mg^{2+} but not by Na^+ or Ca^{2+}, while the effect of other divalent metal ions including Ni^{2+}, Mn^{2+}, Co^{2+}, and Cu^{2+} was complex (White and Gadd, 1987). Ni^{2+} inhibited Zn^{2+} uptake from 20 μM Zn^{2+} when present at equimolar concentration but stimulated Zn^{2+} uptake at 50 and 100 μM, while Mn^{2+} was increasingly inhibitory as the Mn^{2+} concentration was increased. Co^{2+} was slightly inhibitory at 20 and 100 μM, while Cu^{2+} stimulated uptake slightly at 10 μM but inhibited it at 50 μM. It is difficult from this data to make comparisons with the specificity of Zn^{2+} uptake in *C. utilis,* and little other data appears to be available. An affinity series of Mg^{2+}, Co^{2+}, $Zn^{2+} > Mn^{2+} > Ni^{2+} > Ca^{2+} > Sr^{2+}$ was proposed for divalent cation uptake in *S. cerevisiae* by Fuhrmann and Rothstein (1968) that assumes a common uptake system, and this is rather similar to the nonspecific system that takes up Mn^{2+} in *C. utilis* (Parkin and Ross, 1985) and in bacteria (Silver and Lusk, 1987). The kinetic data available for *S. cerevisiae* (Table 1) indicate that both high-affinity, K_m 3.7 μM (White and Gadd, 1987), and low-affinity, K_m 500 μM (Mowll and Gadd, 1983), uptake systems exist, but the specificity of the high-affinity system is uncertain. A characteristic of high-affinity Mn^{2+} uptake in both *C. utilis* and bacteria is its insensitivity to inhibition by Mg^{2+} (Parkin and Ross, 1986; Silver and Lusk, 1987). In *S. cerevisiae* the high-affinity uptake of Zn^{2+} is inhibited by Mg^{2+}. It is possible that uptake of divalent cations by *S. cerevisiae* is less specific than in *C. utilis,* but because no data on the effect of Mg^{2+} on Zn^{2+} uptake by *C. utilis,* and no data on *S. cerevisiae* directly comparable to that on high- and low-affinity Mn^{2+} uptake in *C. utilis,* is available, this remains speculation.

4. Relationship Between Uptake and Potassium Efflux

Efflux of K^+ during uptake of Zn^{2+} was first reported by Fuhrmann and Rothstein (1968), who observed an exchange of K^+ for Zn^{2+} in the ratio $2K^+$:Zn^{2+}. It was proposed that uptake of Zn^{2+} was coupled to efflux of K^+ to maintain charge balance. Gadd and Mowll (1983) did not detect loss of K^+ from viable cells of *S. cerevisiae* or *Sporobolomyces roseus* during Zn^{2+} uptake, but a sudden loss of viability of *S. cerevisiae* at Zn^{2+} concentrations over 100 μM was accompanied by release of K^+. In the case of *Spor. roseus,* a progressive loss of viability occurred with increasing Zn^{2+} concentration but was not accompanied by loss of K^+. There is considerable evidence that K^+ loss is linked to the toxic effects of heavy metals on yeasts and other fungal cells, especially membrane disruption (Kuypers and Roomans, 1979; Borst-Pauwels, 1981; Gadd and Mowll, 1985); this has been further investigated for Zn^{2+} and *S. cerevisiae* (White and Gadd, 1987). At concentrations below 25 μM Zn^{2+}, the rate of K^+ efflux was constant, indicating no toxic effect of Zn^{2+} on K^+ efflux, and total efflux increased. Efflux was followed by a period of K^+ uptake that restored the cellular potassium content to its initial level. Neither total efflux nor efflux rate showed any constant stoichiometric relationship to Zn^{2+} uptake, and the time courses of Zn^{2+} uptake and K^+ efflux were different, which indicates that Zn^{2+} uptake is not tightly coupled to K^+ efflux. At concentrations of Zn^{2+} above 50 μM, the initial rate of K^+ efflux increased with increasing Zn^{2+} concentration, and no subsequent period of uptake of K^+ was noted. There was evidence that K^+ uptake was totally inhibited by Zn^{2+} concentrations above about 45 μM. This concentration of Zn^{2+} also inhibited K^+ uptake in potassium-starved cells. The initial rate of K^+ efflux continued to increase at higher Zn^{2+} concentrations, and this was assumed to be due to continuing damage to membranes (White and Gadd, 1987).

Stoichiometric relationships between K^+ efflux and uptake of some divalent metal ions has been observed (Gadd, 1986), but not in all cases, especially where toxic effects of the metal ion have been noted (Norris and Kelly, 1977; Gadd and Mowll, 1983; Mowll and Gadd, 1983), and as can be seen from the data of White and Gadd (1987), the relationship of K^+ efflux to metal ion uptake can be complex. While no stoichiometric relationship between Zn^{2+} uptake and K^+ loss has been demonstrated in yeast, the complexity of interactions between divalent metal ion uptake and potassium efflux suggests that such a relationship cannot be conclusively discounted; more data are needed.

5. *Effect of pH*

Early work established that uptake of Zn^{2+} by *S. cerevisiae* was dependent on pH, with maximal uptake between pH 5 and 6, and was reduced sharply by pH below 5 (Fuhrmann and Rothstein, 1968). Other studies with *S. cerevisiae* have used pH 5 (Ponta and Broda, 1970) or pH 6.5 (Mowll and Gadd, 1983; White and Gadd, 1987), though no data were presented on the effect of changes in pH in these studies. The effect of pH on Zn^{2+} uptake in *C. utilis* was marked, with increasing reduction in uptake as pH was increased from 4.8 to 8.2. At pH 6.6, uptake was reduced by about 66% compared with pH 4.8, and it was reduced by 83% at pH 8.2 (Failla et al., 1976). This effect of increasing pH was ascribed to the relative insolubility of Zn^{2+} at pH above about 6.8 due to formation of complexes with polyphosphates, carbonates, and hydroxides. These data are in accord with observations on the pH dependence of the uptake of other heavy metal ions by yeasts and other fungi (Gadd, 1986). From a practical point of view, it is clearly desirable to avoid pH below about 5 or above about 6 in uptake studies with metal ions such as Zn^{2+}.

6. *Intracellular Distribution*

The main compartment of cells of Zn^{2+}-loaded *S. cerevisiae* containing Zn^{2+} was the vacuole, which contained a little under 60% of the cellular Zn^{2+}, with about 40% firmly bound in a fraction composed of membrane and organelles with little (apparently less than 10%) in the cytosol (Gadd and White, 1987). Uptake of Zn^{2+} into the vacuole was apparently an energy-dependent process requiring both ATPase activity and a proton gradient across the vacuolar membrane (White and Gadd, 1987). There is little other data available for yeast, though it has been suggested that a metallothioneinlike protein may be involved in Zn^{2+} storage in *C. utilis* (Failla et al., 1976).

7. *Efflux*

There is to date no evidence that an efflux system for Zn^{2+} exists in yeasts. No efflux could be detected in Zn^{2+}-loaded *C. utilis* (Failla et al., 1976; Lawford et al., 1980) or in *S. cerevisiae* (White and Gadd, 1987).

8. *Uptake by Growing Cultures*

Growing cultures of *C. utilis* were found to accumulate Zn^{2+} during lag phase and during late exponential phase, but not during early exponential phase or stationary phase (Failla and Weinberg, 1977). This appears to represent a pattern of uptake during lag phase, dilution of Zn^{2+} content due to cell growth during early exponential phase, and further uptake during late exponential phase. There was no evidence of Zn^{2+} efflux, and it was suggested that dilution into daughter cells represents the only means of reducing the Zn^{2+} content of cells in this organism (Failla and Weinberg, 1977).

9. *Influence of Medium Zinc Concentration on Uptake*

Growth of *C. utilis* in medium with a low Zn^{2+} concentration did not affect growth rate or final cell yield, but Zn^{2+} uptake was markedly stimulated (Failla and Weinberg, 1977). Cells grown in low Zn^{2+} (regarded as zinc-deficient) took up 12 times as much Zn^{2+} as those from zinc-supplemented medium. Zn^{2+} uptake of *C. utilis* cells harvested from medium containing 0.1–1.1 μM Zn^{2+} was identical, but as the Zn^{2+} concentration was increased, uptake was reduced, and it ceased altogether at Zn^{2+} concentrations of 5.1 μM and above. The increase in Zn^{2+} uptake was reflected in an increase of 17 times in the maximum uptake rate, but K_m was unchanged (Failla and Weinberg, 1977). A similar stimulation of Zn^{2+} uptake in cells grown under zinc-limiting conditions in a chemostat was observed by Lawford et al. (1980). Zn^{2+} accumulation was investigated in *C. utilis* cells maintained in a chemostat under carbon- and energy-limited conditions with a dilution rate of 0.4 h^{-1}, which represents a specific growth rate close to the maximum exhibited by exponential phase batch cultures. Cells harvested from cultures grown in chemostats with excess Zn^{+} in the medium showed little uptake of Zn^{2+}, while those grown with Zn^{2+} at limiting concentration took up 14 times as much Zn^{2+} after 60 min incubation. When cells were grown under conditions where growth limitation was poised between carbon (energy) limitation and Zn^{2+} limitation, Zn^{2+} uptake was about 3.5 times that of cells grown in Zn^{2+}-sufficient medium (Lawford et al., 1980). The ability of zinc-deficient cells to take up large amounts of Zn^{2+} was independent of growth rate (Lawford et al., 1980) and not directly related to energy status (Failla and Weinberg, 1977) but was determined by the zinc content of the cells (Failla and Weinberg, 1977; Lawford et al., 1980). Lawford et al. (1980) state that an internal zinc content of less than 1 nmol mg^{-1} was necessary for enhancement of Zn^{2+} uptake. Normal zinc content of *C. utilis* from zinc-sufficient cultures was between 4 nmol mg^{-1} (Lawford et al., 1980) and 6 nmol mg^{-1} (Failla and Weinberg, 1977).

10. Role of Zinc-Binding Proteins in Uptake

Cycloheximide (100 μM) had no effect on uptake of Zn^{2+} by zinc-deficient cells that had not been starved (Failla et al., 1976; Lawford et al., 1980) but did prevent previously starved cells from taking up Zn^{2+} above that taken up by such cells in the absence of glucose (Failla et al., 1976). This suggests that *de novo* protein synthesis is required for Zn^{2+} uptake, at least in starved cells. This may reflect the synthesis of membrane transport proteins, intracellular zinc-binding proteins, or possibly proteins analogous to siderophores involved in iron uptake. No evidence for siderophore analogs was found, but a protein-containing fraction that also contained 70% of supplied ^{65}Zn was detected by column chromatography using Sephadex G-75 (Failla et al., 1976). This material was not further characterized but may be a zinc-binding protein of the metallothionein type.

B. Filamentous Fungi

1. Energy Dependence

Inhibition of Zn^{2+} uptake by metabolic inhibitors has been observed in several filamentous fungi. In *Neocosmospora vasinfecta,* uptake from 100 μM Zn^{2+} was inhibited by sodium azide, low temperature (3°C), or anaerobiosis by more than 70% (Paton and Budd, 1972) or from 10 μM Zn^{2+} by CCCP (85%), anaerobiosis (78%), and less effectively by DES (42%) and DCCD (38%) (Budd, 1988). In *Penicillium notatum,* uptake from 10 nM and 20 μM Zn^{2+} was abolished by carbonylcyanide-*p*-trifluoromethoxyphenylhydrazone (FCCP) (Starling and Ross, 1991), while in *Sclerotium rolfsii,* uptake from 0.58 nM Zn^{2+} was inhibited by anaerobiosis, incubation at 4°C, or DNP by more than 90%; and uptake from 2 μM Zn^{2+} was inhibited by CCCP, DES, sodium azide, and iodoacetic acid by more than 90%, but by only 18% by DCCD (Pilz et al., 1991). In *Aspergillus parasiticus,* uptake of Zn^{2+} was inhibited by 99% by incubation at 0°C or addition of antimycin, DNP, rotenone, or 2-deoxyglucose. All these data support the view that Zn^{2+} uptake is energy-dependent in these fungi. It should be noted that glucose has been reported either to inhibit Zn^{2+} uptake (Paton and Budd, 1972) or to have no effect (Budd, 1988), unlike yeasts, where uptake depends on the presence of glucose (see above). If mycelium is starved prior to uptake experiments, however, a requirement for glucose can be demonstrated (Starling and Ross, 1991), which suggests that energy reserves of unstarved mycelium may be adequate to support Zn^{2+} uptake. The inhibitory effects of glucose reported by Paton and Budd (1972) have not been observed in other studies and remain unexplained.

The mechanism of energy coupling to Zn^{2+} transport in filamentous fungi has been little studied, but the role of the membrane potential has been investigated in *N. vasinfecta* (Budd, 1989). The membrane potential was estimated using the distribution of the tetraphenylphosphonium ion (TPP^+), and a minimum value of −68 mV (interior negative) under aerobic conditions at pH 6.5 in the absence of other cations was calculated. The membrane was depolarized by CCCP and anaerobiosis, which depress Zn^{2+} uptake, but was hyperpolarized by gramicidin, which produces a slight increase in Zn^{2+} uptake (Budd, 1988), by trifluoperazine (TFP), by and prior treatment with acetate at pH 4.6. TFP stimulated Zn^{2+} uptake by up to 200 times compared to control at concentrations of 30 μM or less, but it caused inhibition at 100 μM. This inhibition was accompanied by foaming of the mycelial suspension (Budd, 1989). Presumably TFP caused cell damage at this concentration. Incubation of mycelium in acetate buffer prior to determination of Zn^{2+} uptake produced a 50% increase compared to control. These data point strongly to Zn^{2+} uptake being driven by the membrane potential in *N. vasinfecta* (Budd, 1989). Comparable studies with other fungi have apparently not been carried out.

2. *Uptake Kinetics*

A summary of K_m values for Zn^{2+} uptake obtained for filamentous fungi is given in Table 2, and as with yeasts (see Table 1), considerable variation in reported values exists in the literature. As mentioned in the discussion of uptake kinetics in yeasts earlier, care is necessary in making comparisons, as experimental conditions may affect the values determined. It is clear, however, that in *N. vasinfecta* there were two uptake systems for Zn^{2+} revealed by a Hofstee plot over a wide Zn^{2+} concentration range (Budd, 1988), one with high affinity for Zn^{2+} and K_m 1.1 μM, and a lower-affinity system with K_m 770 μM. High and low affinity uptake systems also apparently exist in *P. notatum* (Starling and Ross, 1991), with K_m of 6.7 nM and 5.9 μM, respectively. It should be noted, however, that determined K_m values may be affected by the range of metal ion concentrations used to determine them, so caution is necessary in interpreting such data. The use

Table 2 Half-Saturation Constants K_m for Zinc Uptake in Some Filamentous Fungi

Fungus	K_m (μM)	Conc. range (μM)	Reference
N. vasinfecta	200	0–1000	Paton and Budd, 1972
N. vasinfecta	1.1, 770[a]	0.29–984	Budd, 1988
P. notatum	0.0066	0.005–0.1	Starling and Ross, 1991
P. notatum	5.9	10–100	Starling and Ross, 1991
A. parasiticus	20	5–50	Failla and Niehaus, 1986
S. rolfsii	3.8	0.1–10	Pilz et al., 1991
A. pullulans	11[a]	0–20	Gadd et al., 1987

[a]See text.

of terms such as high and low affinity also needs care. For example, K_m values around 6 μM for the low-affinity system of *P. notatum* could be regarded as high-affinity values by comparison with 770 μM for the low-affinity uptake system of *N. vasinfecta.* In the dimorphic fungus *Aureobasidium pullulans,* the affinity of the Zn^{2+} uptake system of yeastlike cells was not significantly altered by the removal of the cell wall by enzymic digestion, with K_m 11.1 for whole cells and 11.8 for protoplasts, though V_{max} was reduced from 0.29 nmol Zn^{2+} min^{-1} $(10^7$ cells$)^{-1}$ to 0.08 nmol Zn^{2+} min^{-1} $(10^7$ cells$)^{-1}$ in protoplasts, indicating that wall removal may have caused some damage to Zn^{2+} uptake sites in the cell membrane (Gadd et al., 1987). The K_m for protoplasts of mycelial cells was 10.8 μM, while that for chlamydospores was 66.7 μM. There was thus little difference between the affinity of yeastlike and mycelial cells for Zn^{2+}, but chlamydospores had reduced affinity for Zn^{2+}. This might be expected in dormant structures, but it may contribute to the high resistance to metals exhibited by these structures (Gadd et al., 1987).

3. *Specificity of Uptake Systems*

There is relatively little data on the specificity of Zn^{2+} uptake in filamentous fungi. Paton and Budd (1972) found that Mn^{2+} at 500 μM was a competitive inhibitor of Zn^{2+} uptake in *N. vasinfecta* over a range of Zn^{2+} concentrations up to 1 mM, while Mg^{2+} was stated to increase the K_m for Zn^{2+} and to reduce the rate of uptake at all but the highest Zn^{2+} concentrations, when the uptake rate increased; but no supporting data were presented. A more detailed study in the same fungus (Budd, 1988) showed that with a Zn^{2+} concentration of 10 μM, Mg^{2+} and Ca^{2+} at 1 and 2 mM respectively had no effect on Zn^{2+} uptake, while Mn^{2+}, Cd^{2+}, Co^{2+}, and Ni^{2+} supplied at 10 μM reduced uptake by no more than 20%, though inhibition was more severe when these cations were supplied at 500 μM, with uptake ranging from about 30% of control with Mn^{2+} and Cd^{2+} to 27% of control for Ni^{2+} and 11.9% of control for Co^{2+}. It was nevertheless concluded that with 10 μM Zn^{2+}, which represented uptake by the high-affinity uptake system, uptake of Zn^{2+} was selective for Zn^{2+} (Budd, 1988).

In *S. rolfsii,* Zn^{2+} uptake from a concentration of 0.58 nM was found to be insensitive to inhibition by Ca^{2+} (1 mM) or Mg^{2+} (2 mM), and uptake from 2 μM was not significantly inhibited by Mn^{2+}, Co^{2+}, Ni^{2+}, Cd^{2+}, or Fe^{2+} supplied at 10 or 100 μM; Cu^{2+} had no effect at 10 μM (Pilz et al., 1991). Copper was not tested at 100 μM, presumably because of the possibility of toxicity to *S. rolfsii* at this concentration. Thus it appears that the high-affinity Zn^{2+} uptake system in *S. rolfsii* has a very high specificity. In *P. notatum,* uptake from 10 nM Zn^{2+} was also very specific (Table 3), being inhibited only by Cd^{2+} and Cu^{2+}, while Mg^{2+}, Mn^{2+}, Ni^{2+}, and Co^{2+} had no effect (Starling and Ross, 1991). Further investigation of the effects of Zn^{2+} uptake by Cd^{2+} showed inhibition to be competitive, while that of Cu^{2+} was noncompetitive and presumably due to toxic effects of Cu^{2+} on the fungus. However, when the effect of these metal ions (concentration 100 μM) was tested on Zn^{2+} uptake from an external concentration of 20 μM, all were inhibitory. Magnesium ions were the most effective, while Mn^{2+} and Co^{2+} were least effective (Table 3). This suggests that the high-affinity uptake

Table 3 Effect of Competing Metal Ions on Uptake of Zinc

Zn^{2+} conc.	Competing metal ion and conc.	Zn^{2+} uptake, % control
10 nM	Mg^{2+} 10 μM	110
	Mn^{2+}	92
	Ni^{2+}	91
	Co^{2+}	89
	Cu^{2+}	49
	Cd^{2+}	2
20 μM	Mg^{2+} 100 μM	26
	Mn^{2+}	64
	Ni^{2+}	40
	Co^{2+}	61
	Cu^{2+}	41
	Cd^{2+}	39

Data represent the uptake of zinc in the presence of a competing cation expressed as a percentage of uptake of zinc alone after 20 min incubation at 30° calculated from mean data from five replicate experiments. (From Starling and Ross, 1991, with permission.)

system active with low Zn^{2+} concentrations was very specific, while a second, lower-affinity uptake system, active at concentrations that presumably saturate the high-affinity system, has a much lower specificity (Starling and Ross, 1991). It is possible that the low affinity uptake of Zn^{2+} represents uptake by the Mg^{2+} uptake system, as it was strongly inhibited by Mg^{2+}. There is, however, no direct evidence for this.

4. *Effect of pH*

There appears to be almost no information on the effect of pH on Zn^{2+} uptake in filamentous fungi. Paton and Budd (1972) observed a bimodal effect with peaks of uptake at pH 4.5 and pH 6.5, the latter being greater than the former. However, the data presented consist of two determinations per pH value, and the reliability of the data is therefore limited. The bimodal effect may not in fact be real. No other data appear to be available. Various pH values of 4.5 (Pilz et al., 1991), 5.5 (Starling and Ross, 1991), and 6.5 (Paton and Budd, 1972; Gadd et al., 1987; Budd, 1988) have been used in Zn^{2+} uptake studies.

5. *Efflux*

The loss of $^{65}Zn^{2+}$ from preloaded mycelia of *Sclerotium rolfsii* could not be detected in the presence of excess nonradioactive Zn^{2+}, though addition of 1% toluene caused 95% of the $^{65}Zn^{2+}$ to be released. This indicates that there is no efflux of accumulated Zn^{2+} from intact mycelium of *S. rolfsii* (Pilz et al., 1991). No other information seems to be available for filamentous fungi.

6. *Influence of Medium Zinc Concentration on Uptake*

Growth of *Aspergillus parasiticus* in medium described as zinc-deficient (0.8 μM Zn) caused enhanced uptake of Zn^{2+} by washed mycelium by comparison with mycelium from cultures grown in normal medium, though uptake was also influenced by mycelial age, with maximal uptake by mycelium from 60 h cultures (Failla and Niehaus, 1986). The data also show that the uptake of Zn^{2+} was inversely proportional to mycelial zinc content with a threshold of 7.6 nmol mg/dry wt for inhibition of Zn^{2+} uptake by mycelium from 72 h cultures in this fungus. In *Penicillium notatum,* mycelium grown in medium with a normal zinc content (26.8 μM Zn) took up relatively little Zn^{2+} from 10 nM Zn^{2+}, but this uptake was doubled in mycelium grown in medium with no added zinc. Growth in medium with a zinc content of 1 mM resulted in mycelium with a negligible capacity for uptake of Zn^{2+} (Starling and Ross, 1991). These data show that the activity of the high-affinity specific uptake system for Zn^{2+} was influenced by the zinc content of the original growth medium and by the initial mycelial zinc content, but comparable data for the low-affinity nonspecific system are not available. Some as yet unpublished data (G. Auling, personal communication) shows that the high-affinity Zn^{2+} uptake system in *S. rolfsii* is activated by a drop in the internal zinc concentration of the mycelium below a threshold value and that *de novo* protein synthesis is involved, as the process is blocked by cycloheximide. This suggests that the high-affinity uptake system in *S. rolfsii* depends on proteins whose synthesis is induced by low mycelial zinc content.

In *Neocosmospora vasinfecta,* Zn^{2+} uptake was also inversely related to the zinc content of the growth medium. Mycelium from cultures grown in low zinc medium contained no measurable zinc and showed enhanced Zn^{2+} uptake by comparison with normal mycelium, which contained 10.7 nmol mg dry wt^{-1} (Budd, 1989). It was also noted that mycelium from medium with a low magnesium content showed marked stimulation of Zn^{2+} uptake of up to 300-fold, while zinc content of such mycelium was near normal (12.2 nmol mg dry wt^{-1}). This observation was somewhat unexpected, and it was suggested (Budd, 1989) that this result reflected Zn^{2+} uptake by an activated Mg^{2+} uptake system. A similar stimulation of the low-affinity Zn^{2+} of *P. notatum* by growth in medium low in magnesium was observed by Starling (1990), though this might be expected, as the low-affinity Zn^{2+} uptake system was strongly inhibited by Mg^{2+} and may in fact be an Mg^{2+} uptake system. There does not seem to be any other data on the effect of magnesium content of growth media on Zn^{2+} uptake in filamentous fungi.

IV. CONCLUSIONS

It will be clear from the above discussion that more information on Zn^{2+} uptake is available for yeasts than for filamentous fungi; this probably reflects the greater technical difficulty of working with mycelial fungi. However, as might be expected, there are many similarities in observed uptake systems between the two groups. There appears to be two kinetically distinct energy-dependent systems, one with high affinity for Zn^{2+} and high specificity, and another with low affinity and low specificity. The difference between these systems is not always clearly defined as is the case in *S. cerevisiae.* A characteristic of high-affinity, specific uptake systems appears to be their insensitivity to inhibition by Mg^{2+}, yet the high-affinity uptake system of *S. cerevisiae* is inhibited by Mg^{2+}. The status of the low-affinity uptake system for Zn^{2+} is also less than certain. This system is generally of low specificity and strongly inhibited by Mg^{2+}, leading to the suggestion that it may primarily be an Mg^{2+} uptake system.

Mutants with altered Zn^{2+} uptake systems would be of value in determining the number and specificity of Zn^{2+} uptake systems in yeasts such as *S. cerevisiae,* but the recognition of such mutants poses problems. Detection of mutations in high-affinity uptake systems might be possible by comparison of growth on zinc-deficient medium with that of wild-type cultures, when an inability to grow on such medium could indicate a reduced capacity for uptake, though ultimate confirmation would only be possible by kinetic studies of Zn^{2+} uptake. Mutations in low-affinity uptake systems might be detected by increased resistance to zinc toxicity, though again, confirmation by kinetic studies of Zn^{2+} uptake would be necessary to establish that resistance was due to altered uptake and not to other mechanisms. In this connection, it is worth noting that a nickel-resistant strain of *Candida utilis,* recently isolated in my laboratory, has a reduced uptake of nickel, and though further characterization of this mutant is needed, this does support the view that mutants with altered low-affinity uptake systems for divalent ions may be obtained in this way.

Another noteworthy aspect of uptake of Zn^{2+} by both yeasts and filamentous fungi is its relationship to cellular zinc content. The uptake of Zn^{2+} at least by high-affinity systems is inversely related to the zinc content of the cells, and uptake may be undetectable in cells grown in medium with a high zinc content. This suggests that there is regulation of the zinc uptake process in response to the internal zinc concentration at least over a limited range. There are no data to show whether such regulation of Zn^{2+} uptake is limited to uptake by high-affinity, specific uptake systems, nor is there any information on possible mechanisms other than that discussed for *S. rolfsii* above. At high external Zn^{2+} concentrations when toxicity can be observed, unregulated Zn^{2+} uptake may occur, possibly via the low-affinity uptake system.

The loss of K^+ during divalent metal ion uptake has been widely reported in yeasts (Gadd, 1986), and this also occurs during uptake of Zn^{2+}. However, as discussed above, the relationship between K^+ loss and Zn^{2+} uptake may be complex, and it may be associated with toxicity of Zn^{2+}. There is no comparable data for filamentous fungi, so it cannot be ascertained whether K^+ loss is a feature of Zn^{2+} uptake in these organisms or is peculiar to yeasts. Similarly, there is no data in filamentous fungi to indicate whether compartmentalization of Zn^{2+} into vacuoles, comparable with that which occurs in yeast, occurs in these organisms.

Finally, in both yeasts and filamentous fungi, there is no evidence of an efflux system for zinc, which implies that tolerance of elevated zinc concentrations in growth media must depend either on reduced uptake or on the production of substances capable of detoxifying zinc. There is some evidence of production of metallothioneinlike zinc-binding proteins in *C. utilis* (see above), and recent unpublished data on *S. rolfsii* indicates the production of two zinc-binding proteins/peptides (G. Auling, personal communication), one of which is a metallothionein and the other a phytochelatin; but information on other fungi does not seem to be available.

REFERENCES

Borst-Pauwels, G. W. F. H. (1981). Ion transport in yeast, *Biochim. Biophys. Acta, 650:* 88.

Budd, K. (1988). A high-affinity system for the transport of zinc in *Neocosmospora vasinfecta, Exp. Mycol., 12:* 195.

Budd, K. (1989). Role of the membrane potential in the transport of zinc by *Neocosmospora vasinfecta, Exp. Mycol., 13:* 356.

Eilam, Y., Lavi, H., and Grossowicz, N. (1985). Cytoplasmic Ca^{2+} homeostasis maintained by a vacuolar Ca^{2+} transport system in the yeast *Saccharomyces cerevisiae, J. Gen. Microbiol., 131:* 623.

Failla, M. L. (1977). Zinc: Function and transport in microorganisms, *Microorganisms and Minerals* (E. D. Weinberg, ed.), Marcel Dekker, New York, p. 151.

Failla, L. J., and Niehaus, W. G. (1986). Regulation of Zn^{2+} uptake and versicolorin A synthesis in a mutant strain of *Aspergillus parasiticus, Exp. Mycol., 10:* 35.

Failla, M. L., and Weinberg, E. D. (1977). Cyclic accumulation of zinc by *Candida utilis* during growth in batch culture, *J. Gen. Microbiol., 99:* 85.

Failla, M. L., Benedict, C. D., and Weinberg, E. D. (1976). Accumulation and storage of Zn^{2+} by *Candida utilis, J. Gen. Microbiol., 94:* 23.

Fuhrmann, G. F. (1973). Dependence of divalent cation transport on ATP in yeast, *Experientia, 29:* 742.

Fuhrmann, G. F. (1974a). Transport of divalent cations in *Saccharomyces cerevisiae, 4th International Symposium on Yeasts,* Part 1, H1, Vienna.

Fuhrmann, G. F. (1974b). Relation between divalent cation transport and divalent cation activated ATPase in yeast plasma membranes, *Experientia, 30:* 686.

Fuhrmann, G., and Rothstein, A. (1968). The transport of Zn^{2+}, Co^{2+} and Ni^{2+} into yeast cells, *Biochim. Biophys. Acta, 163:* 325.

Gadd, G. M. (1986). Fungal responses towards heavy metals, *Microbes in Extreme Environments* (R. A. Herbert and G. A. Codd, eds.), Academic Press, London, p. 85.

Gadd, G. M. (1990a). Metal tolerance, *Microbiology of Extreme Environments* (C. Edwards, ed.), Open University Press, Milton Keynes, p. 178.

Gadd, G. M. (1990b). Fungi and yeasts for metal accumulation, *Microbial Mineral Recovery* (H. L. Ehrlich and C. L. Brierley, eds.), McGraw-Hill, New York, p. 249.

Gadd, G. M., and Mowll, J. L. (1983). The relationship between cadmium uptake, potassium release and viability in *Saccharomyces cerevisiae, FEMS Microbiol. Lett., 16:* 45.

Gadd, G. M., and Mowll, J. L. (1985). Copper uptake by yeast-like cells, hyphae and chlamydospores of *Aureobasidium pullulans, Exp. Mycol., 9:* 230.

Gadd, G. M., and White, C. (1985). Copper uptake by *Penicillium ochro-chloron:* Influence of pH on toxicity and demonstration of energy-dependent copper influx using protoplasts, *J. Gen. Microbiol., 131:* 1875.

Gadd, G. M., White, C., and Mowll, J. L. (1987). Heavy metal uptake by intact cells and protoplasts of *Aureobasidium pullulans, FEMS Microbiol. Ecol., 45:* 261.

Goffeau, A., and Slayman, C. W. (1981). The proton-translocating ATPase of the fungal plasma membrane, *Biochim. Biophys. Acta, 639:* 197.

Harold, F. M., Altendorf, K. H., and Hirata, H. (1974). Probing membrane transport mechanisms with ionophores, *Ann. N.Y. Acad. Sci., 235:* 149.

Heytler, R. G., and Pritchard, W. W. (1962). The new class of uncoupling agents—Carbonyl cyanide phenyl hydrazones, *Biochem. Biophys. Res. Commun., 7:* 272.

Hockertz, S., Schmid, J., and Auling, G. (1987). A specific transport system for manganese in the filamentous fungus *Aspergillus niger, J. Gen. Microbiol., 133:* 3513.

Jennings, D. H. (1963). *The absorption of solutes by plant cells,* Oliver and Boyd, Edinburgh.

Kuypers, G. A. J., and Roomans, G. M. (1979). Mercury-induced loss of K^+ from yeast cells investigated by electron probe X-ray microanalysis, *J. Gen. Microbiol., 115:* 13.

Lawford, H. G., Pik, J. R., Lawford, G. R., Williams, T., and Kligerman, A. (1980). Hyperaccumulation of zinc by zinc-depleted *Candida utilis* grown in chemostat culture, *Can. J. Microbiol., 26:* 71.

Mowll, J. L., and Gadd, G. M. (1983). Zinc uptake and toxicity in the yeasts *Sporobolomyces roseus* and *Saccharomyces cerevisiae, J. Gen. Microbiol., 129:* 3421.

Nieuwenhuis, B. J. W. M., Weijers, C. A. G. M., and Borst-Pauwels, G. W. F. H. (1981). Uptake and accumulation of Mn^{2+} and Sr^{2+} in *Saccharomyces cerevisiae, Biochim. Biophys. Acta, 649:* 83.

Norris, P. R., and Kelly, D. P. (1977). Accumulation of cadmium and cobalt by *Saccharomyces cerevisiae, J. Gen. Microbiol., 99:* 317.

Parkin, M. J., and Ross, I. S. (1985). Uptake of copper and manganese by the yeast *Candida utilis, Microbios Lett. 29:* 115.

Parkin, M. J., and Ross, I. S. (1986). The specific uptake of manganese in the yeast *Candida utilis, J. Gen. Microbiol., 132:* 2155.

Paton, W. H. N., and Budd, K. (1972). Zinc uptake in *Neocosmospora vasinfecta, J. Gen. Microbiol., 72:* 173.

Peters, P. H. J., and Borst-Pauwels, G. W. F. H. (1979). Properties of plasmamembrane ATPase and mitochondrial ATPase of *Saccharomyces cerevisiae, Physiol. Plant., 46:* 330.

Pilz, F., Auling, G., Stephan, D., Rau, U., and Wagner, F. (1991). A high-affinity Zn^{2+} uptake system controls growth and biosynthesis of an extracellular branched β-1,3-β-1,6-glucan in *Sclerotium rolfsii* ATCC 15205, *Exp. Mycol., 15:* 181.

Ponta, H., and Broda, E. (1970). Mechanismen der Aufnahme von Zink durch Bäckerhefe, *Planta, 95:* 18.

Ross, I. S. (1993). Membrane transport processes and response to exposure to heavy metals, *Stress Tolerance in Fungi,* Marcel Dekker, New York.

Serrano, R. (1980). Effect of ATPase inhibitors on the proton pump of respiratory-deficient yeast, *Eur. J. Biochem., 105:* 419.

Silver, S., and Lusk, J. E. (1987). Bacterial magnesium, manganese and zinc trans-

port, *Ion Transport in Prokaryotes* (B. P. Rosen and S. Silver, eds.), Academic Press, New York, p. 165.
Starling, A. P. (1990). The uptake of heavy metals by protoplasts and whole cells of filamentous fungi, Ph.D. thesis, University of Keele.
Starling, A. P., and Ross, I. S. (1990). Uptake of manganese by *Penicillium notatum, Microbios, 63:* 93.
Starling, A. P., and Ross, I. S. (1991). Uptake of zinc by *Penicillium notatum, Mycol. Res., 95:* 712.
Townsley, C. C. (1985). Heavy metal accumulation in filamentous fungi, Ph.D. thesis, University of Keele.
Weinberg, E. D. (1977). Mineral element control of microbial secondary metabolism, *Microorganisms and Minerals* (E. D. Weinberg, ed.), Marcel Dekker, New York, p. 289.
White, C., and Gadd, G. M. (1987). The uptake and cellular distribution of zinc in *Saccharomyces cerevisiae, J. Gen. Microbiol., 133:* 727.

9

Accumulation of Radionuclides in Fungi

Kurt Haselwandter and Michael Berreck
University of Innsbruck, Innsbruck, Austria

I. INTRODUCTION

Radionuclides either occur naturally or are man-made. Naturally occurring are radionuclides such as ^{40}K and ^{87}Rb, or members of the three decay chains (uranium-radium series, thorium series, actinium series). Man-made radionuclides can be derived from all the various elements, either as fission products (e.g., ^{85}Kr, ^{89}Sr, ^{90}Sr, ^{95}Zr, ^{99}Mo, ^{106}Ru, ^{137}Cs, ^{140}Ba, ^{144}Ce, ^{147}Nd) or as activation products (e.g., ^{14}C, ^{3}H, ^{54}Mn, ^{55}Fe).

On a global basis, for most of the toxic metals the natural fluxes are small compared with emissions from industrial activities (Nriagu, 1989). Similarly, mankind appears to have become the key agent in the global atmospheric distribution of radionuclides, e.g., through the release of radiocesium (Cambray et al., 1987; UNSCEAR, 1988).

Fungi have been shown to take up naturally occurring as well as man-made radionuclides. This paper reviews the literature on the species-specific accumulation pattern as revealed by analyses of fungal fruit-bodies collected at different sampling sites, and the use of fungi as bioindicators of biosphere contamination. Furthermore, it summarizes the role of fungi in biogenous migration of radionuclides in soils and in the transfer along the food chain. In addition, the biochemical basis for the uptake mechanism is described, prior to a brief survey of the potential application of fungi as biosorbents of radionuclides.

II. RADIONUCLIDE ACCUMULATION PATTERN

A. Pre-Chernobyl

The fallout radionuclide ^{137}Cs is ubiquitous in the environment as a result of atmospheric weapons testing in the 1950s and 1960s. Investigating the accumulation of fallout in vegetation, Grüter (1971) found an uncommonly high and selective enrichment of the fission product ^{137}Cs in fruit-bodies of different basidiomycete fungi. The radioactive contamination of fungal fruit-bodies culminated in 1965 (Kiefer and Maushart, 1965; Rohleder, 1967). Before the accident at Chernobyl, the highest concentration of ^{137}Cs (up to 25,200 Bq/kg dry weight) was found in fruit-bodies of *Cortinarius armillatus* (Haselwandter, 1977) (see Table 1). The ^{137}Cs accumulation seems to be species-specific rather than site-specific. *Paxillus filamentosus,* for example, did not accumulate ^{137}Cs, whereas *Paxillus involutus* did (Haselwandter, 1978).

Beside the species-specific accumulation behavior of fungi, Eckl et al. (1986) found radiocesium uptake to be substrate dependent. The ^{137}Cs transfer from soil to fungi was correlated with the soil pH. Like other mineral elements the solubility and mobility of Cs (including ^{137}Cs) increases with decreasing pH, because Cs^+ bound by clay minerals can be exchanged for H^+. With increasing pH there is less ion exchange; the Cs remains bound and is therefore not available for the fungus.

B. Post-Chernobyl

On April 26th, 1986, the worldwide inventory of ^{137}Cs was increased by 5% following the release of about 3.8×10^{16} Bq of ^{137}Cs. Deposition of ^{137}Cs was substantial in some parts of Europe (Zifferero, 1988; Cambray et al., 1987). The deposition of fallout from Chernobyl showed extreme geographical variability over short distances, depending on rainfall patterns prevailing when the radioactive clouds passed (Hohenemser et al., 1986).

Analysis of fungal fruit-bodies, collected in summer 1986, revealed a significant increase in the radiocesium content in comparison to the situation before. Basidiocarps contained ^{134}Cs in addition to ^{137}Cs (Haselwandter et al., 1988).

In comparison to the fallout from the nuclear bomb tests, the Chernobyl fallout differs in composition. One of the marked differences is the presence of ^{134}Cs, another long-lived isotope. ^{134}Cs builds up in nuclear fuel in a reactor and hence is absent from the fallout of atmospheric weapon tests

(De Meijer et al., 1988). At the time of release the $^{137}Cs/^{134}Cs$ ratio was about 2.0. This is the basis for differentiating between Chernobyl-derived ^{137}Cs and "old bomb ^{137}Cs" (Rückert et al., 1990).

In addition to radiocesium, fungi take up ^{7}Be, ^{60}Co, ^{90}Sr, ^{95}Zr, ^{95}Nb, ^{110m}Ag, ^{125}Sb, ^{144}Ce, ^{226}Ra, and ^{238}U (Haselwandter and Irlweck, 1977; Eckl et al., 1986; Seeger et al., 1982; Gentili et al., 1990). Fungi contain between 0.15 and 11.7% dry weight potassium (Seeger, 1978), including its natural radioactive isotope ^{40}K. The Cs content covers with 0.1 to 308 ppm (Seeger and Schweinshaut, 1981), an even wider range than K.

Within fruit-bodies the different tissues can contain different amounts of cesium and potassium. In the cap, the K and Cs contents are highest, whereas the lowest concentration is found in the gills and the stem (Seeger, 1978; Seeger and Schweinshaut, 1981; Haselwandter et al., 1988). For the radiocesium, this can also be demonstrated by autoradiography of fungal fruit-bodies (Fig. 1).

No correlation between the concentration of ^{40}K and that of the Cs nuclides was found by Rückert and Diehl (1987). In all mushrooms except *Xerocomus badius,* Elstner et al. (1987) report that the ^{40}K activity was generally higher than that of ^{137}Cs. The authors claim that the analyzed mushrooms (except *X. badius*) do not actively take up cesium from soil, in contrast to potassium. These results could not be confirmed by other authors, e.g., Dighton and Horrill (1988), Horyna et al. (1988), Byrne (1988), and Heinrich (1992), who found higher ^{137}Cs activities and transfer factors for ^{137}Cs than for ^{40}K, suggesting preferential uptake of the radionuclide ^{137}Cs. Both saprophytic as well as ectomycorrhiza-forming species show a considerable variation in their level of Cs uptake (cf. Haselwandter et al., 1988; Hofmann et al., 1988; Bakken and Olson, 1990; Heinrich, 1992). Some authors assume that mycorrhizal fungi are more efficient in accumulating radionuclides than saprophytic fungi (e.g., Oolbekkink and Kuyper, 1989; Bakken and Olson, 1990), while others indicate that mycorrhizal and saprophytic fungi do not differ in Cs uptake (cf. De Meijer et al., 1989). Gerzabek et al. (1988) relate the accumulation pattern (e.g., *Agaricus campestris:* 222 Bq ^{137}Cs/kg; *X. badius:* 49,543 Bq ^{137}Cs/kg dry weight) to the typical habitat of the different fungi. Considering the number of fungi that have been investigated with regard to radionuclide accumulation, it appears to be impossible to answer this question precisely, especially as the range of species investigated so far cannot be considered to be representative for either category of fungi.

Since 1966 the ^{137}Cs activity in the most important species of mushrooms has been monitored regularly by official food control stations in order to assess the possible radiation burden for humans from the consumption of edible fungi. In general, cultivated mushrooms such as *Agaricus bisporus* and *Pleurotus ostreatus* contain comparatively lower radiocesium concen-

Table 1 ^{137}Cs **Contamination of Selected Species of Basidiomycetes Collected in Different European Countries and in Japan.**

Fungal species	Bq ^{137}Cs kg^{-1} dry weight	Sampling sites	Reference
Collected before the Chernobyl accident			
Cortinarius armillatus	25160[a]	Austria	Haselwandter (1977)
Russula emetica	21275[c]	Austria	Eckl et al. (1986)
Amanita fulva	14837[c]		
Suillus variegatus	9213[c]		
Rozites caperatus	8621[c]		
Paxillus involutus	2141[a]	Austria	Heinrich (1992)
Suillus grevillei	333[a]		
Lactarius rufus	266[a]		
Collected after the Chernobyl accident			
Cantharellus lutescens	1124–35000[b]	Italy	Govi and Innocenti (1987)
Boletus edulis	1069[c]		
Rozites caperatus	2100–62000[b]	Slovenia	Byrne (1988)
Cortinarius armillatus	21000–96000[b]		
Laccaria amethystina	12000–117000[b]		
Lactarius rufus	1780–7297[b]	UK (upland forests)	Dighton and Horrill (1988)
Inocybe longicystis	8736–14060[b]		

Boletus edulis	440–1200[b]	Czechia (Prague)	Horyna and Randa (1988)
Leccinum scabrum	33300[c]		
Lactarius rufus	5400[c]		
Cortinarius armillatus	18119–48100[b]	Austria	Haselwandter et al. (1988)
Rozites caperatus	8384–46298[b]		
Lactarius rufus	984–16668[b]		
Boletus edulis	95–477[b]	Austria and Northern Italy	Battiston et al. (1989)
Cantharellus lutescens	1147–27626[b]		
Boletus edulis	292–437[b]	Poland	Bem et al. (1990)
Xerocomus badius	1498–8855[b]		
Agrocybe erebia	1520[c]	Japan	Muramatsu et al. (1991)
Lactarius rufus	44904[a]	Ukraine	Wasser and Grodzinskaya (1993)
Xerocomus badius	19547[a]		
Boletus edulis	6489–21400[b]		
Cortinarius armillatus	69863–123761*[b]	Finland Chernobyl-contaminated (*) and uncontaminated (·) areas	Berreck et al. (1992)
Cortinarius armillatus	3715–5968·[b]		
Lactarius rufus	1283–19229 *[b]		
Lactarius rufus	163–703·[b]		
Gomphidius glutinosus	3774999[a]	Ukraine (30 km zone of Chernobyl nuclear power station)	Grodzinskaya et al. (1994)
Suillus luteus	83485–947400[b]		
Boletus edulis	155305[c]		

[a]Maximum; [b]minimum–maximum; [c]single measurement.

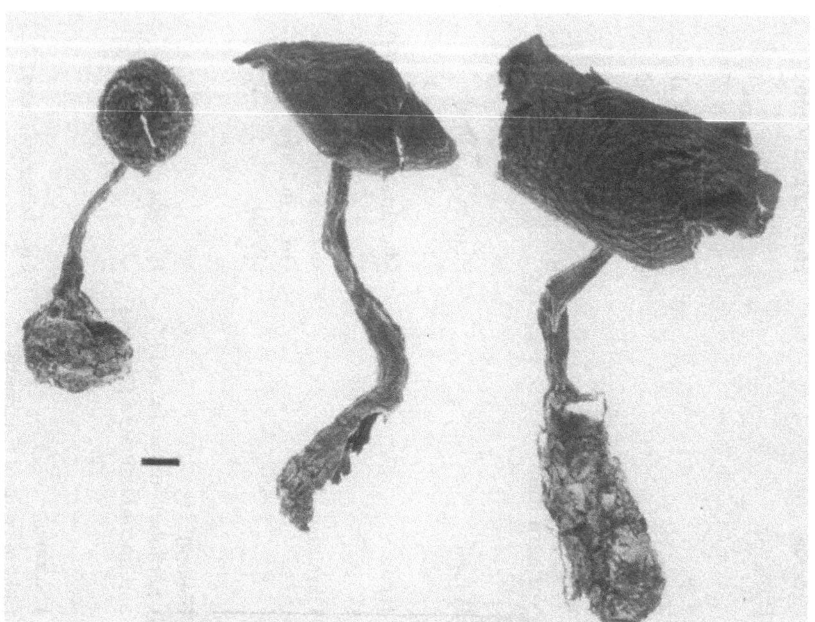
(a)

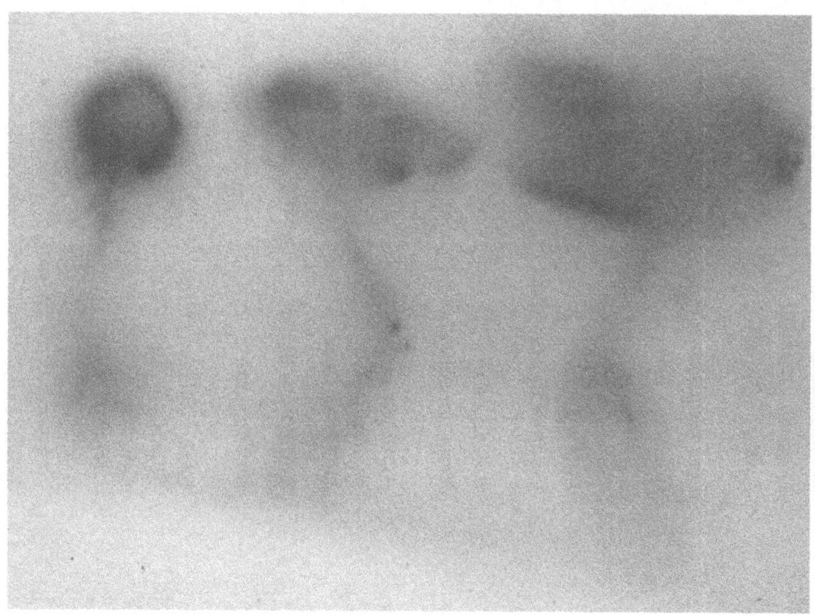
(b)

trations than those collected from the field (Haselwandter and Berreck, 1989). While single collections of *Boletus edulis* and *Cantharellus cibarius* may exceed the highest permissible radionuclide concentration for vegetables, including fungi, this can be observed regularly in the case of *Rozites caperatus* (Berreck and Haselwandter, 1989). Hence *R. caperatus* should not be considered an edible fungus anymore.

III. FUNGI AS BIOINDICATORS OF RADIOACTIVE CONTAMINATION OF THE BIOSPHERE

It has been postulated that fungi can be used as bioindicators of radioactive contamination of the environment (Mihok et al., 1989; Van Tran and Le Duy, 1991; Wasser et al., 1991). Fungi accumulate radionuclides such as ^{137}Cs in a species-specific manner (Haselwandter, 1978). Despite considerable variation, this feature of higher fungi in particular can be used to monitor the radioactivity of the environment. On the basis of a set of seven species, the mean ^{137}Cs content of the basidiomycete fruit-bodies was 3.0 to 4.8 times higher in 1986 after the Chernobyl accident than in 1974 (Haselwandter et al., 1988). A study carried out in Finland revealed a close correlation of the radiocesium (^{137}Cs, ^{134}Cs) content, and the $^{137}Cs/^{134}Cs$ ratio, of fungal fruit-bodies (*C. armillatus, Lactarius rufus*) with the deposition of radiocesium in the area where the fruit bodies were collected (Berreck et al., 1992). In the heavily contaminated areas, the level of radiocesium content in fungal fruit bodies was high while the $^{137}Cs/^{134}Cs$ ratio was low, whereas in samples from uncontaminated areas not only was the radiocesium content low but also the $^{137}Cs/^{134}Cs$ ratio was high due to the lack of ^{134}Cs in the radiocesium derived from nuclear bomb testing. This emphasizes the potential use of fungi as bioindicators for radioactive contamination of the biosphere.

IV. SIGNIFICANCE OF FUNGI IN BIOGENOUS IMMOBILIZATION OF RADIONUCLIDES IN SOILS

Fungi play a key role in the biogenous migration of radionuclides in soil (O'Donnell and Johnson, 1989; Boháč et al., 1991). Johnson et al. (1991)

Figure 1 (a) Dried specimens of *Cortinarius armillatus* collected in Finland in 1987 (bar = 10 mm); (b) autoradiography of the same specimens of *C. armillatus* (time of exposure 92 d).

developed a simple autoradiographic technique for the selective isolation of ^{137}Cs-sorbing soil microorganisms. Under pure culture conditions, uptake of Cs by fungal mycelium of grassland soil fungi ranged from 44 (*Fusarium* sp.) to 236 (*Epicoccum nigrum*) nmol Cs g^{-1} dry weight h^{-1} (Dighton et al., 1991). More than 40% of the Cs taken up was bound within the hyphae. This suggests that the fungal soil biomass could immobilize substantial quantities of radiocesium for a yet unknown period of time.

Concentration factors were determined for the transfer of Cs from soil into fungal fruit-bodies. The range of the concentration factors may extend over four orders of magnitude (Klan et al., 1988). Interestingly, even within one species (e.g., *X. badius*) it may range from 7 to 99 (Horyna and Randa, 1989). It may well be that the microbial composition, the fungal component in particular, of the soil determines to a great extent the soil radioactivity, and hence has a strong effect upon the concentration factor, which is calculated as the ratio of the radioactivity of the fungal fruit-bodies to that of the corresponding soil. Indirect evidence for this is provided by Bunzl and Schimmack (1988), who irradiated six different soils with 40 and 80 kGy from a ^{60}Co source. Irradiation led not only to a decrease in the microbial biomass but also to a change in the radionuclide sorption properties of the soils.

V. ACCUMULATION MECHANISMS

Living and dead microbial cells are capable of radionuclide accumulation; they differ, however, in the mechanisms involved in these processes. The main difference is based upon the metabolic dependence of the accumulation.

Living and dead cells can be capable of generally rapid, metabolism-independent binding of radionuclides to cell walls, extracellular polysaccharides, or other materials, which is now frequently referred to as biosorption. Fungi vary in the chemical composition of their cell walls (Farkas, 1985) and the extracellular material that they release (Burnett, 1976; Berry, 1988); hence, variation in the accumulation capacity of different fungi is bound to be considerable.

In uranium biosorption by *Rhizopus arrhizus*, at least the following three processes are involved: rapid and simultaneous coordination of uranium to the amine nitrogen of chitin; adsorption to cell wall chitin; and rather slow precipitation of uranyl hydroxide ($UO_2(OH)_2$; Tsezos and Volesky, 1982). Uranium oxide (UO_2^{2+}) seems to bind to phosphate and carboxyl groups of the cell wall of *S. cerevisiae* prior to the deposition of the uranium as needlelike fibrils; this process can lead to uranium accumulation of as much as 15% of yeast dry weight (Strandberg et al., 1981).

Biosorption can be related to the ionic radius of an element and to the pH of the environment, which affects the solubility of metals. In general, low external pH decreases the rate and extent of metal biosorption. Between approximately 4 and 30°C fungal biosorption is relatively unaffected by temperature, whereas the presence of other cations and anions can exert a strong influence upon metabolism-independent metal accumulation (Gadd, 1988).

Albeit normally slower than biosorption, metabolism-dependent intracellular uptake can lead to the accumulation of greater amounts of metal, e.g., in yeasts, whereas in filamentous fungi active transport may be less significant than metabolism-independent accumulation (Gadd, 1990). Metabolism-dependent intracellular accumulation is affected by temperature, metabolic inhibitors, and the availability of energy sources. Thus the metabolic activity determines the uptake rate. In fungi, the membrane potential seems to be responsible for the electrophoretic mono- and divalent cation transport, although other gradients, e.g., K^+, may also be involved (Borst-Pauwels, 1981). In cases where toxic effects change the membrane permeability, intracellular metal uptake may be based upon diffusion (Gadd, 1990).

De Rome and Gadd (1991) have shown that in *S. cerevisiae* the uptake of cesium, strontium, and uranium is biphasic, surface biosorption being followed by energy-dependent influx. It seems that three sites are involved in the translocation of Cs^+ across the yeast (*S. cerevisiae*) cell membrane (Derks and Borst-Pauwels, 1979). In *Candida utilis* neither Li, Na, Cs nor ammonium ions could functionally substitute for K; K was effectively replaced only by Rb, which gave, on a molar basis, and under conditions where cation availability limited growth, the same cell yield as did K (Aiking and Tempest, 1977). On the contrary, in the case of the filamentous fungus *Fusarium solani,* Rb or Cs could replace K, whereas Li, Na, and the divalent cations could not (Das, 1991).

The uptake of Sr^{2+} and Ca^{2+} by *S. cerevisiae* is energy dependent and shows a deviation from simple Michaelis-Menten kinetics (Roomans et al., 1979). Plasma membrane ATPase does not seem to be involved in Sr^{2+} uptake (Nieuwenhuis et al., 1981). The selectivity of Sr^{2+} versus Ca^{2+} uptake by *S. cerevisiae* is probably related to differences in their affinity for the negative groups on the cell membrane determining the surface potential, rather than to differences in their affinity for a transport system (Borst-Pauwels and Theuvenet, 1984). A specific extrusion pump, presumably a Ca^{2+} pump, seems to enable *S. cerevisiae* cells to regulate accumulation of Sr^{2+} and Mn^{2+} to different levels (Theuvenet et al., 1986). It is important to note that different strains of *S. cerevisiae* differ considerably in the mechanisms involved in both metabolism-independent Sr^{2+} adsorption and metabolism-dependent intracellular Sr^{2+} uptake (Avery and Tobin, 1992).

Voltage-dependent and Ca^{2+}-activated cation-selective channels were recently found in the vacuolar membrane of *S. cerevisiae* (Sato et al., 1989; Bertl and Slayman, 1990). In yeasts, a majority of metals such as K^+, Co^{2+}, Mn^{2+}, Zn^{2+}, and Mg^{2+} are located in the vacuole, where they may be bound to low molecular weight polyphosphates (Gadd and White, 1989). However, in phosphate-rich *S. cerevisiae* cells the major part of divalent cations is sequestered in cytoplasmic polyphosphate granules that may serve as an important store for these cations (Roomans, 1980).

In the basidiomycete fungus *X. badius,* ^{137}Cs is complexed by the cap pigments badione A and norbadione A, which are pulvinic acid derivatives. Aumann et al. (1989) postulate that this mechanism is responsible for the radiocesium accumulation in *Boletus erythropus, B. mirabilis,* and *X. badius.*

Short-term influx of ^{137}Cs into fungal hyphae ranged from 85 (*Cenococcum geophilum*) to 276 (*Mycena polygramma*) nmol Cs g^{-1} dry weight h^{-1} (Clint et al., 1991). This wide range of influx values obtained for fungal hyphae of different species may help to explain the large species-specific differences in radiocesium accumulation of fruit-bodies of Basidiomycetes.

VI. EFFECT OF MYCORRHIZAL SYMBIOSIS UPON RADIONUCLIDE UPTAKE BY PLANTS

Evans and Dekker (1966) have demonstrated that ^{137}Cs uptake by plants from soils can be appreciable. The ^{137}Cs concentrations found in vegetable crops were generally higher than those of forage crops, which in turn were generally higher than those found for cereals (Evans and Dekker, 1968). These authors also indicated that the $^{137}Cs/K$ ratios were not constant, and only in cereal crops was there a significant correlation between ^{137}Cs and K concentrations. Thus the relationship between ^{137}Cs and K was not as close as that usually found between ^{90}Sr and Ca.

Mycorrhizal fungal infection, in general, has a significant effect on the mineral nutrition of plants (Harley and Smith, 1983; Moser and Haselwandter, 1983). Hence it can be anticipated that this also applies to the uptake of radionuclides such as ^{137}Cs.

McGraw et al. (1979) have first reported a possible interaction between arbuscular mycorrhizal infection and the uptake of ^{137}Cs by plants. Infection of Bahia grass (*Paspalum notatum*) roots by two out of ten arbuscular mycorrhizal fungi resulted in a two-fold increase of the ^{137}Cs content of leaf tissue 48 h after its injection into soil, indicating that some mycorrhizal species may enhance the uptake of radiocesium.

Rogers and Williams (1986) have studied in more detail the influence of three inoculum types with different arbuscular mycorrhizal fungi predominating on the uptake of ^{137}Cs and ^{60}Co by yellow blossom sweet clover (*Melilotus officinalis*) and Sudan grass (*Sorghum sudanense*). While the ^{137}Cs content of mycorrhizal clover increased in comparison to nonmycorrhizal plants, the ^{60}Co content was not significantly different at the two harvests (65 and 93 days). The mycorrhizal grass, harvested at 85 and 119 days, showed greater, albeit not significantly different, mean ^{137}Cs contents than the nonmycorrhizal controls. The ^{60}Co content was increased over the control at the first but not at the second harvest of the grass. This study indicates that the harvest time seems to be crucial for what is being observed with regard to Cs or Co accumulation in plant shoots, in addition to specific effects of different fungus and plant species.

On the other hand, arbuscular mycorrhizal infection can lead to a decrease in radiocesium content of *Festuca ovina* (Fig. 2). At the first and second harvest, the difference in radiocesium content of mycorrhizal and nonmycorrhizal plants is highly significant. Shoot tissue radioactivity of

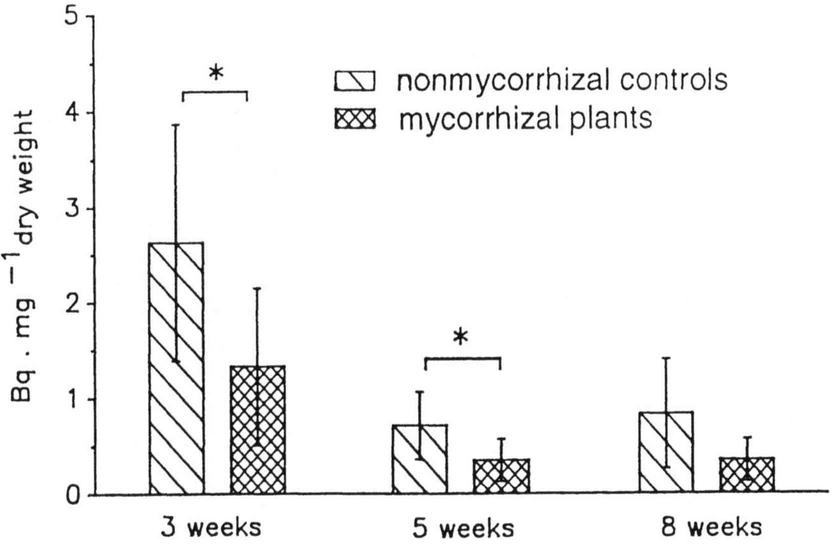

Figure 2 **^{137}Cs content (means of *n* = 7, ± SD) of the shoot system of arbuscular mycorrhizal and nonmycorrhizal grass (*Festuca ovina*). Harvests: 3, 5, and 8 weeks after application of the radiocesium. * indicates that mycorrhizal and control treatment means are significantly different at the 5% level.**

mycorrhizal plants drops to about half that of the nonmycorrhizal controls (Haselwandter and Berreck, 1994).

Jackson et al. (1973) provided evidence that the arbuscular mycorrhizal infection of soybean (*Glycine max* var. Harasoy 63) by *Glomus mosseae* can lead to an increase in root absorption of ^{90}Sr. This enhanced absorption occurred in soil with normal amounts of Ca and Mg, which have biological and chemical properties similar to those of Sr.

The results of studies on the effect of mycorrhiza on plant uptake of radionuclides are controversial; at least they are inadequate for drawing any general conclusions.

VII. ROLE OF FUNGI IN RADIONUCLIDE ACCUMULATION ALONG THE FOOD CHAIN

Renewed interest in the behavior of ^{137}Cs in natural food chains has been generated by the recent release of fission products from the accident at Chernobyl. It is well documented that fungi accumulate radionuclides such as ^{137}Cs. Therefore particular attention is paid to food chains leading from fungi to man, e.g., the lichen–reindeer/caribou or the mushroom–roe-deer food chain, or food chains that include fungi-grazing ruminants (Hanson, 1967; Liden and Gustafsson, 1966; Johnson and Nayfield, 1970; Hove et al., 1989; Karlen and Johanson, 1991).

The transport of ^{137}Cs fallout through the various components of the food chain lichen–reindeer–man has been studied since 1961. For about 6 months lichens are the main fodder for reindeer during the winter period. In the lichen symbiosis, the fungus appears to be responsible for the accumulation of radioactive nuclides like ^{137}Cs, ^{90}Sr, ^{239}Pu, and ^{210}Pb (Eckl et al., 1986). As a result of the grazing habits of the reindeer and its rapid cesium metabolism (biological half-time $T_{1/2} \sim 30$ d), there is a considerable seasonal variation of the ^{137}Cs content of reindeer, with a maximum at the end of winter in March and a minimum in autumn in September. Lapps, as the final link in this food chain, show a pronounced seasonal variation, with a maximum body burden in June from consumption of reindeer meat (Liden and Gustavson, 1966).

After the accident of Chernobyl, the whole-body counts of sheep, goat, and reindeer increased rapidly in contaminated parts of northern Europe. This coincided with an increased level of radioactivity in fungal fruit-bodies. Fungal radiocesium was found to be highly available in a digestibility study carried out with goats. Milk radioactivity levels could be accounted for by consumption of as little as 20–100 g of fungal dry matter per day (Hove et al., 1990).

VIII. FUNGAL BIOSORBENTS FOR RADIONUCLIDES

Microbes, such as the fungus *Penicillium,* can be used to recover metals like U and Sr from radioactive wastes (Zajic and Chiu, 1972). The potential use of fungi (e.g., *Paecilomyces marquandii*) as microbial collectors for radioisotopes such as $^{137}Cs^{+}$, $^{45}Ca^{2+}$, and $^{204}Tl^{+}$ has been described by Fisel et al. (1976). White and Gadd (1990) demonstrated that fungal biomass could remove approximately 90–95% of the thorium supplied in an air-lift bioreactor at an inflow concentration of 3 mM. With loading capacities of 116 mg (0.5 mmol) and 139 mg (0.6 mmol) per g dry weight of fungal biomass, *R. arrhizus* and *Aspergillus niger* were more effective biosorbents than *Penicillium italicum* and *P. chrysogenum.* Interestingly, thorium biosorption by *R. arrhizus* was relatively unaffected by the presence of other inorganic solutes, whereas thorium biosorption by *A. niger* was reduced under the same conditions.

Uptake of Cs by mycelial biomass of *P. chrysogenum* led to a 50% removal, while *R. arrhizus* achieved only 41%. In the case of uranium, *R. arrhizus* was most efficient, able to remove over 90% of the metal from the solution. With 44 and 39%, Sr removal was similar in both *R. arrhizus* and *P. chrysogenum* (De Rome and Gadd, 1991). In the presence of 50 mM glucose, immobilized cells of *S. cerevisiae* removed 83% of the uranium, 45% of the strontium, and 43% of the cesium from the metal solution. In the absence of glucose, strontium removal dropped to 46% and cesium removal to 76% of that achieved in the presence of glucose; this indicates the ratio between energy-dependent and energy-independent biosorption of Sr and Cs.

Uranium uptake displayed similar kinetics in the presence or absence of glucose; hence, the accumulation mechanism for U appears to be energy-independent (De Rome and Gadd, 1991). This was also demonstrated by Strandberg et al. (1981) for *S. cerevisiae* and *Pseudomonas aeruginosa.* Dependent on environmental parameters (e.g., pH, temperature, interference by certain anions and cations), *S. cerevisiae* accumulated U extracellularly on the cell surface. Without any response to environmental parameters, *P. aeruginosa* accumulated U extremely rapidly intracellularly. Cell-bound U reached 10–15% of the cell dry weight, but only 32% of the *S. cerevisiae* cells and 44% of the *P. aeruginosa* cells contained U deposits that were visible in the EM. U could be removed chemically from the cells of *S. cerevisiae,* which then could be reused as a biosorbent (e.g., for removal of radionuclides from waste water streams of nuclear fuel processing plants). The biotechnological application of radionuclide biosorption for nuclear waste treatment is reviewed by Ashley and Roach (1990).

IX. EPILOGUE

The main objective of this review is to highlight our current knowledge on the accumulation of radionuclides by fungi. In addition, it is feasible to identify gaps in our knowledge, a few of which are described below.

With the few exceptions mentioned above, remarkably little research has been done on the effect of mycorrhizal infection on plant uptake of radionuclides; only a very limited number of species of both mycorrhizal fungi and host plants has been investigated in this respect. Comprehensive understanding of the mechanisms involved in radionuclide uptake is bound to have practical implications. Knowledge of its regulation at the molecular, cellular, and organismic level could help us to control the transfer of radionuclides from the soil into plants and hence through the food chain. Such knowledge is essential if we have to design or to evaluate any preventive measures in order to minimize radionuclide transfer along the food chain.

So far, the biochemistry of radionuclide uptake has been studied in a very limited number of fungi, most of them yeasts. Because fungi, especially higher fungi, Basidiomycetes in particular, show a very distinct radionuclide accumulation pattern, a detailed study of the uptake mechanisms involved seems appropriate. Precise information on the molecular biology of radionuclide uptake and its regulation could provide the basis for the transfer of the corresponding genes into fungi that are easier to handle as biosorbents in biotechnological processes than Basidiomycetes. Both living and dead fungal biomass as well as derived or excreted products can be used to remove radionuclides from solution. With continued pollution of the biosphere with radionuclides, microbe-based technologies may play an important role in environmental protection in the future (Gadd, 1990).

ACKNOWLEDGMENT

We thank the Bundesministerium für Wissenschaft und Forschung, Vienna, and the Amt der Tiroler Landesregierung, Innsbruck, for financial support of part of this work.

REFERENCES

Aiking, H., and Tempest, D. W. (1977). Rubidium as a probe for function and transport of potassium in the yeast *Candida utilis* NCYC-321 grown in chemostat culture, *Arch. Microbiol., 115:* 215.

Ashley, N. V., and Roach, D. J. W. (1990). Review of biotechnology applications to nuclear waste treatment, *J. Chem. Technol. Biotechnol., 49:* 381.

Aumann, D. C., Clooth, G., Steffan, B., and Steglich, W. (1989). Komplexierung von Caesium-137 durch die Hutfarbstoffe des Maronenröhrlings (*Xerocomus badius*), *Angew. Chemie, 101:* 495.

Avery, S. V., and Tobin, J. M. (1992). Mechanisms of strontium uptake by laboratory and brewing strains of *Saccharomyces cerevisiae, Appl. Environ. Microbiol., 58:* 3883.

Bakken L. R., and Olson, R. A. (1990). Accumulation of radiocaesium by fungi, *Can. J. Microbiol., 36:* 704.

Battiston, G. A., Degetto, S., Gerbasi, R., and Sbrignadello, G. (1989). Radioactivity in mushrooms in northeast Italy following the Chernobyl accident, *J. Environ. Radioactivity, 9:* 53.

Bem, H., Lasota, W., Kuśmierek, E., and Witusik, M. (1990). Accumulation of ^{137}Cs by mushrooms from Rogozno area of Poland over a period 1984–1988, *J. Radioanal. Nucl. Chem., Lett., 145:* 39.

Berreck, M., and Haselwandter, K. (1989). Belastung wildwachsender Pilze durch Cs-137 und Cs-134, *Österreichische Forstzeitung, 89:* 57.

Berreck, M., Ohenoja, E., and Haselwandter, K. (1992). Mycorrhizal fungi as bioindicators of radioactivity, *Responses of Forest Ecosystems to Environmental Changes* (A. Teller, P. Mathy, and J. N. R. Jeffers, eds.), Elsevier, London, p. 800.

Berry, D. R. (1988). *Physiology of Industrial Fungi,* Blackwell, Oxford.

Bertl, A., and Slayman, C. L. (1990). Cation-selective channels in the vacuolar membrane of *Saccharomyces:* Dependence on calcium, redox state, and voltage, *Proc. Natl. Acad. Sci. USA, 87:* 7824.

Boháč, J., Krivolutskii, D. A., and Antonova, T. B. (1990). The role of fungi in the biogenous migration of elements and in the accumulation of radionuclides, *Agric. Ecosyst. Environ., 28:* 31.

Borst-Pauwels, G. W. F. H. (1981). Ion transport in yeast, *Biochim. Biophys. Acta, 650:* 88.

Borst-Pauwels, G. W. F. H., and Theuvenet, A. P. R. (1984). Apparent saturation kinetics of divalent cation uptake in yeast (*Saccharomyces cerevisiae*) caused by a reduction in the surface potential, *Biochim. Biophys. Acta, 771:* 171.

Bunzl, K., and Schimmack, W. (1988). Effect of microbial biomass reduction by gamma-irradiation on the sorption of ^{137}Cs, ^{85}Sr, ^{139}Ce, ^{57}Co, ^{109}Cd, ^{65}Zn, ^{103}Ru, ^{95m}Tc and ^{131}I by soils, *Radiat. Environ. Biophys., 27:* 165.

Burnett, J. H. (1976). *Fundamentals of Mycology,* Arnold, London.

Byrne, A. R. (1988). Radioactivity in fungi in Slovenia, Yugoslavia, following the Chernobyl accident, *J. Environ. Radioactivity, 6:* 177.

Cambray, R. S., Cawse, P. A., Garland, J. A., Gibson, J. A. B., Johnson, P., Lewis, G. N. J., Newton, D., Salmon, L., and Wade, B. O. (1987). Observations on radioactivity from the Chernobyl accident, *Nucl. Energy, 26:* 77.

Clint, G. M., Dighton, J., and Rees, S. (1991). Influx of 137 Cs into hyphae of basidiomycete fungi, *Mycol. Res., 95:* 1047.

Das, J. (1991). Influence of potassium in the agar medium on the growth pattern of the filamentous fungus *Fusarium solani, Appl. Environ. Microbiol., 57:* 3033.

De Meijer, R. J., Aldenkamp, F. J., and Jansen, A. E. (1988). Resorption of cesium radionuclides by various fungi, *Oecologia, 77:* 268.

Derks, W. J. G., and Borst-Pauwels, G. W. F. H. (1979). Apparent 3 site kinetics of cesium ion uptake by yeast, *Physiol. Plant., 46:* 241.

De Rome, L., and Gadd, G. M. (1991) Use of pelleted and immobilized yeast and fungal biomass for heavy metal and radionuclide recovery, *J. Industrial Microbiol., 7:* 97.

Dighton, J., and Horrill, A. D. (1988). Radiocaesium accumulation in the mycorrhizal fungi *Lactarius rufus* and *Inocybe longicystis,* in upland Britain following the Chernobyl accident, *Trans. Br. Mycol. Soc., 91:* 335.

Dighton, J., Clint, G. M., and Poskitt, J. (1991). Uptake and accumulation of 137 cesium by upland grassland soil fungi: A potential pool of cesium immobilization, *Mycol. Res., 95:* 1052.

Eckl, P., Hofmann, W., and Türk, R. (1986). Uptake of natural and man-made radionuclides by lichens and mushrooms, *Radiat. Environ. Biophys.,* 25: 43.

Elstner, E. F., Fink, R., Höll, W., Lengenfelder, E., and Ziegler, R. (1987). Natural and Chernobyl-caused radioactivity in mushrooms, mosses and soil-samples of defined biotops in SW Bavaria, *Oecologia, 73:* 553.

Evans, E. J., and Dekker, A. J. (1966). Plant uptake of Cs-137 from nine Canadian soils, *Can. J. Soil Sci., 46:* 167.

Evans, E. J., and Dekker, A. J. (1968). Comparative Cs-137 content of agricultural crops grown in a contaminated soil, *Can. J. Plant Sci., 48:* 183.

Farkas, V. (1985). The fungal cell wall, *Fungal Protoplasts* (J. F. Peberdy and L. Ferency, eds.), Marcel Dekker, New York, p. 3.

Fisel, S., Dulman, V., and Cecal, A. (1976). Enrichment of cesium ions, calcium ions and thallium ions with microbiological collectors, *J. Radioanal. Chem., 34:* 285.

Gadd, G. M. (1988). Accumulation of metals by microorganisms and algae, *Biotechnology—A Comprehensive Treatise,* vol. 6 (H.-J. Rehm and G. Reed, eds.), VCH, Weinheim, p. 401.

Gadd, G. M. (1990). Heavy metal accumulation by bacteria and other microorganisms, *Experientia, 46:* 834.

Gadd, G. M., and White, C. (1989). Heavy metal and radionuclide accumulation and toxicity in fungi and yeasts, *Metal-Microbe Interactions* (R. K. Poole and G. M. Gadd, eds.), IRL Press, Oxford, p. 19.

Gentili, A., Gremigni, G., and Sabbatini, V. (1990) Ag-110m in fungi in central Italy after the Chernobyl accident, *J. Environ. Radioactivity, 13:* 75.

Gerzabek, M., Haunold, E., and Horak, O. (1988). Radioaktivität in Pilzen, *Die Bodenkultur, 39:* 37.

Govi, G., and Innocenti, G. (1987). Presenza di cesio 137 e cesio 134 nei fungi, *Micologia italiana, 16:* 123.

Grodzinskaya, A. A., Berreck, M., Wasser, S. P., and Haselwandter, K. (1994). In preparation.

Grüter, H. (1971). Radioactive fission product ^{137}Cs in mushrooms in W. Germany during 1963–1970, *Health Physics, 20:* 655.

Hanson, W. C. (1967). ^{137}Cs in Alaskan lichens, caribou and eskimos, *Health Phys., 13:* 383.

Harley, J. L., and Smith, S. E. (1983). *Mycorrhizal Symbiosis,* Academic Press, London.

Haselwandter, K. (1977). Radioaktives Cäsium (Cs137) in Fruchtkörpern verschiedener Basidiomycetes, *Zeitschr. f. Pilzkunde, 43:* 323.

Haselwandter, K. (1978). Accumulation of the radioactive nuclide ^{137}Cs in fruitbodies of Basidiomycetes, *Health Phys., 34:* 713.

Haselwandter, K., and Berreck, M. (1989). Accumulation of ^{137}Cs in fruit-bodies of edible fungi—A comparison between wild and cultivated mushrooms, *Mushroom Science, 12:* 587.

Haselwandter, K., and Berreck, M. (1994). Effect of arbuscular mycorrhizae upon ^{137}Cs uptake by plants, in preparation.

Haselwandter, K., Berreck, M., and Brunner, P. (1988). Fungi as bioindicators of radiocaesium contamination: Pre- and post-Chernobyl activities, *Trans. Br. Mycol. Soc., 90:* 171.

Haselwandter, K., and Irlweck, K. (1976). Uran in Fruchtkörpern von Basidiomyceten, *Anzeiger der österr. Akad. d. Wiss., math.-naturwiss. Klasse, 10:* 165.

Heinrich, G. (1992). Uptake and transfer factors of 137Cs by mushrooms, *Radiat. Environ. Biophys., 31:* 39.

Hofmann, W., Attarpour, N., and Türk, R. (1988). Verteilung von Caesium-137 in Wald-Ökosystemen im Bundesland Salzburg (Österreich), *Waldsterben in Österreich: Theorien, Tendenzen, Therapien* (E. Führer and F. Neuhuber, eds.), Bundesministerium f. Wissenschaft und Forschung, Wien, p. 269.

Hohenemser, C., Deicher, M., Hofsäss, H., Lindner, G., Recknagel, E., and Budnick, J. I. (1986). Agricultural impact of Chernobyl: A warning, *Nature, 321:* 817.

Horyna, J., and Randa, Z. (1988). Uptake of radiocesium and alkali metals by mushrooms, *J. Radioanal. Nucl. Chem., 127:* 107.

Horyna, J., Randa, Z., Benada, J., and Klan, J. (1988). Beitrag zum Problem der Akkumulation von Cäsium und Radiocäsium durch höhere Pilze, *Zeitschrift für Mykologie, 54:* 179.

Hove, K., Pedersen, Ø., Garmo, T. H., Hansen, H. S., and Staaland, H. (1990). Fungi: A major source of radiocesium contamination of grazing ruminants in Norway, *Health Physics, 59:* 189.

Jackson, N. E., Miller, R. H., and Franklin, R. E. (1973). The influence of vesicular-arbuscular mycorrhizae on uptake of 90 Sr from soil by soybeans, *Soil Biol. Biochem., 5:* 205.

Johnson, W., and Nayfield, C. L. (1970). Elevated levels of ^{137}Cs in common mushrooms (Agaricaceae) with possible relationship to high levels of ^{137}Cs in whitetail deer 1968–1969, USA, *Radiol. Health Data Rep., 11:* 527.

Johnson, E. E., O'Donnell, A. G., and Ineson, P. (1991). An autoradiographic technique for selecting cesium-137-sorbing microorganisms from soil, *J. Microbiol. Methods, 13:* 293.

Karlen, G., and Johanson, K. (1991). Seasonal variation in the activity concentration of ^{137}Cs in Swedish roe-deer and in their daily intake, *J. Environ. Radioactivity, 14:* 91.

Kiefer, H., and Maushart, R. (1965). Erhöhter Cs-137-Gehalt im menschlichen Körper nach Pilzgenuß, *Atompraxis: Direct Information, 15.*

Klán, J., Řanda, Z., Benada, J., and Horyna, J. (1988). Investigation of nonradioactive Rb, Cs, and radiocaesium in higher fungi, *Česká Mykologie, 42:* 158.

Liden, K., and Gustafsson, M. (1966). Relationships and seasonal variations of ^{137}Cs in lichen, reindeer and man in northern Sweden 1961–65, *Proc. Int. Symp. Radioecological Concentration Processes, Stockholm, Sweden, 25–29 April 1966:* 193.

McGraw, A. C., Gamble, J. F., and Schenck, N. C. (1979). Vesicular-arbuscular mycorrhizal uptake of cesium-134 in two tropical pasture grass species, *Phytopathology, 69:* 1038.

Mihok, S., Schwartz, B., and Wiewel, A. M. (1989). Bioconcentration of fallout 137 cesium by fungi and red-backed voles (*Clethrionomys gapperi*), *Health Phys., 57:* 959.

Moser, M., and Haselwandter, K. (1983). Ecophysiology of mycorrhizal symbioses, *Encyclopedia of Plant Physiology, New Series, Volume 12 C* (= *Physiological Plant Ecology III,* O. L. Lange, P. S. Nobel, C. B. Osmond, and H. Ziegler, eds.), Springer-Verlag, Berlin, p. 391.

Muramatsu, Y., Yoshida, S., and Sumiya, M. (1991). Concentrations of radiocesium and potassium in basidiomycetes collected in Japan, *The Science of the Total Environment, 105:* 29.

Nieuwenhuis, B. J. W. M., Weijers, C. A. G. M., and Borst-Pauwels, G. W. F. H. (1981). Uptake and accumulation of manganese and strontium in *S. cerevisiae, Biochim. Biophys. Acta, 649:* 83.

Nriagu, J. O. (1989). A global assessment of natural sources of atmospheric trace metals, *Nature, 338:* 47.

O'Donnell, A. G., and Johnson, E. E. (1989). Influence of microorganisms on the retention and migration of radionuclides in soils, *J. Sci. Food Agric., 49:* 128.

Oolbekkink, G. T., and Kuyper, T. W. (1989). Radioactive caesium from Chernobyl in fungi, *The Mycologist, 3:* 3.

Rogers, R. D., and Williams, S. E. (1986). Vesicular-arbuscular mycorrhiza: Influence on plant uptake of cesium and cobalt, *Soil Biol. Biochem., 18:* 371.

Rohleder, K. (1967). Zur radioaktiven Kontamination von Speisepilzen, *Dtsch. Lebensm. Rundsch., 63:* 135.

Roomans, G. M. (1980). Localization of divalent cations in phosphate-rich cytoplasmic granules in yeast (*Saccharomyces cerevisiae*), *Physiol. Plant., 48:* 47.

Roomans, G. M., Theuvenet, A. P. R., Van den Berg, T. P. R., and Borst-Pauwels, G. W. F. H. (1979). Kinetics of calcium ion and strontium ion uptake by yeast: Effects of pH, cations and phosphate, *Biochim. Biophys. Acta, 551:* 187.

Rückert, G., and Diehl, J. F. (1987). Anreicherung von Cäsium-137 und Cäsium-

134 in 34 Pilzarten nach dem Reaktorunglück von Tschernobyl, *Z. Lebensm. Unters. Forsch., 185:* 91.

Rückert, G., Diehl, J. F., and Heilgeist, M. (1990). Radioaktivitätsgehalte von 1987 und 1988 im Raum Karlsruhe gesammelten Pilzen, *Z. Lebensm. Unters. Forsch., 190:* 496.

Sato, M., Tanifuji, M., and Kasai, M. (1989). Further characterization of the cation channel of a yeast vacuolar membrane in a planar lipid bilayer, *Cell Struct. Funct., 14:* 659.

Seeger, R. (1978). Kaliumgehalt höherer Pilze, *Z. Lebensm. Unters. Forsch., 167:* 23.

Seeger, R., and Schweinshaut, P. (1981). Vorkommen von Caesium in höheren Pilzen, *The Science of the Total Environment, 19:* 253.

Seeger, R., Orth, H., and Schweinshaut, P. (1982). Strontiumvorkommen in Pilzen. *Z. Lebensm. Unters. Forsch., 174:* 381.

Strandberg, G. W., Shumate, S. E., and Parrott, J. R. (1981). Microbial cells as biosorbents for heavy metals: Accumulation of uranium by *S. cerevisiae* and *Pseudomonas aeruginosa, Appl. Envir. Microbiol., 41:* 237.

Theuvenet, A. P. R., Nieuwenhuis, B. J. W. M., Van de Mortel, J., and Borst-Pauwels, G. W. F. H. (1986). Effect of ethidium bromide and DEAE-dextran on divalent cation accumulation in yeast (*Saccharomyces cerevisiae*): Evidence for an ion-selective extrusion pump for divalent cations, *Biochim. Biophys. Acta, 855:* 383.

Tsezos, M., and Volesky, B. (1982). The mechanism of uranium biosorption by *Rhizopus arrhizus, Biotechnol. Bioeng., 24:* 385.

UNSCEAR. (1988). Sources, effects and risks of ionizing radiation. *United Nations Scientific Committee on the Effects of Atomic Radiation,* Report to the General Assembly, United Nations, New York.

Van Tran, L., and Le Duy, T. (1991). Linhchi mushrooms as biological monitors for cesium-137 pollution, *J. Radioanal. Nucl. Chem., 155:* 451.

Wasser, S. P., and Grodzinskaya, A.A. (1993). Content of radionuclides in macromycetes of the Ukraine in 1990–1991, *Fungi of Europe: Investigation, Recording and Conservation* (D.N. Pegler, L. Boddy, B. Ing, and P.M. Kirk, eds.), Royal Botanic Gardens, Kew, p. 189.

White, C., and Gadd, G. M. (1990). Biosorption of radionuclides by fungal biomass, *J. Chem. Tech. Biotechnol., 49:* 331.

Zajic, J. E., and Chiu, Y. S. (1972). Recovery of heavy metals by microbes, *Developments in Industrial Microbiology, Vol. 13* (E. D. Murray, ed.), American Institute of Biological Sciences, Washington, D.C., p. 91.

Zifferero, M. (1988). A post-Chernobyl view, *Radionuclides in the Food Chain* (M. W. Carter, J. H. Harley, G. D. Schmidt, and G. Silini, eds.), Springer-Verlag, Berlin, p. 3.

10

Metal Ion Resistance and the Role of Metallothionein in Yeast

Ian G. Macreadie
Biomolecular Research Institute, Parkville, Victoria, Australia

Andrew K. Sewell and Dennis R. Winge
University of Utah Medical Center, Salt Lake City, Utah

I. INTRODUCTION

All cells are presented with the challenge of living with metal ions. Cobalt, copper, iron, molybdenum, nickel, and zinc ions are essential for biological functions, while ions of cadmium, lead, and mercury are nonessential yet are absorbed by cells. Cells must be able to regulate the intracellular concentration of metal ions, since excess concentrations of both essential and nonessential metal ions cause toxicity and death of the organism. Cells need to maintain a balance between levels of metal ions that are nutritious and those that are toxic.

Copper ions are essential but extremely toxic at higher concentrations. Copper is therefore a good example of the dilemma that faces an organism in keeping the concentration of an element within strict physiological limits. Copper ions are essential for a number of enzymes including Cu,Zn-superoxide dismutase, lysyl oxidase, dopamine β-hydroxylase, galactose oxidase, ascorbate oxidase, and cytochrome *c* oxidase (Mason, 1979; Adman, 1991). Copper ion deficiency results in impaired physiology, whereas excessive accumulation of copper ions can lead to toxicosis (Mason, 1979; Winge and Mehra, 1990). The cytotoxicity of copper is illustrated by the use of copper salts in fungicides, molluscides, and algicides (Scheinberg

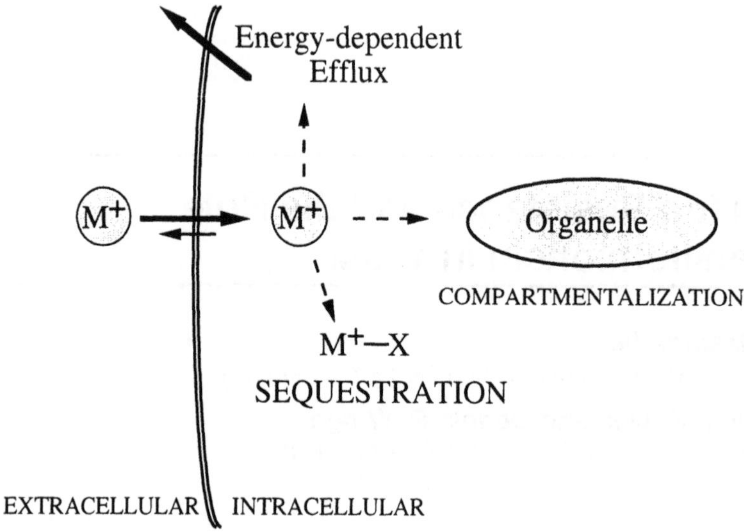

Figure 1 Metal ion detoxification mechanisms in yeast.

and Sternlieb, 1976). Excess copper accumulation may affect the stability of membranes, the cellular redox state, the synthesis and function of proteins, and the replication and transcription of DNA (Agarwal et al., 1989).

The cellular copper levels separating copper balance from toxicosis are dictated largely by the available homeostatic mechanisms in a given species. Copper homeostasis in microorganisms can involve regulation of absorption and/or efflux, compartmentalization in subcellular organelles, and sequestration in macromolecular complexes (Fig. 1). This regulation must result in the delivery of adequate levels of copper ions for critical copper metalloenzymes and also provide adequate copper buffer capacity to prevent copper-induced toxicosis.

The focus of this chapter is on intracellular copper sequestration by metallothionein (MT), although glutathione (GSH)-related isopeptides and sulfide also act in copper sequestration in certain species. Regulation of sequestering molecules is important for homeostatic control of the intracellular copper concentration. The mechanism of regulation of the cellular concentration of metallothionein will be discussed in detail.

II. *CUP1*, A GENE ENCODING COPPER RESISTANCE IN *Saccharomyces cerevisiae*

The identification of metallothionein as an important molecule in copper buffering emerged from early studies on the adaptation of yeast to overcome the toxic effects of copper (Yanagishima et al., 1949; Minagawa et al., 1951). It was observed that subculturing of yeast on media containing copper ions led to the selection of strains that were resistant, a significant observation since repeated many times.

Genetic analysis showed copper resistance was due to a dominant genetic locus (Brenes-Pomales et al., 1955) now known as *CUP1*. Later analyses showed that *CUP1* was located on chromosome VIII and that copper resistance was a consequence of tandem amplification of *CUP1* (Fogel and Welch, 1982). Restriction endonuclease digests showed that the *CUP1* locus comprised a 2 kb sequence that was repeated up to fifteen times in a copper-resistant strain, while only one copy was present in the sensitive strain (Fogel and Welch, 1982).

The cloning and sequencing of the *CUP1* gene was independently performed by two groups. Fogel and Welch (1982) cloned *CUP1* by transforming a copper-sensitive strain to resistance with a library of yeast DNA fragments cloned into a yeast replicating plasmid. The second group identified a *CUP1* clone from a library by hybridization with a probe enriched for *CUP1* sequences (Butt et al., 1984a). The probe was total mRNA made from a copper-induced yeast strain with the main RNA species being *CUP1* transcripts, highlighting the predominance of *CUP1* transcription under such conditions.

The sequence of the *CUP1* locus (Butt et al., 1984b; Karin et al., 1984) revealed one open reading frame (ORF) encoding a product exhibiting features in common with known MT molecules. Specifically, the 61 amino acid product contained twelve cysteinyl residues with multiple Cys-X-Cys sequence motifs. In addition, the sequence had a limited number of hydrophobic residues. The abundance of Cys residues in repeated Cys-X-Cys motifs and a paucity of hydrophobic residues are hallmarks of all known metallothioneins (Kagi and Schaffer, 1988). Although the ORF showed no homology in primary structure to animal MTs, these common salient features justify the ORF product being designated as a yeast MT.

That *CUP1* is the gene responsible for copper resistance was demonstrated by its genetic disruption. This was not lethal to *S. cerevisiae* cells, but it did confer copper hypersensitivity (Gorman et al., 1986; Hamer et al., 1986). It is now generally believed that the major function of yeast MT is in copper homeostasis.

The predicted translation product of *CUP1* was eight amino acids longer than the mature metallothionein isolated from cells (Winge et al., 1985). The N-terminal extension contains five hydrophobic residues, whereas the mature metallothionein is devoid of hydrophobic residues. It was proposed that *CUP1* encoded a primary translation product with a presequence of eight amino acids that was posttranslationally cleaved from the precursor (Winge et al., 1985). A presequence is rare among known metallothioneins, and its significance is still unknown. A mutation in the *CUP1* locus that resulted in N-terminal truncations exhibited slightly reduced copper resistance, but a mutation that resulted in the retention of the N-terminal extension on the metallothionein molecule was without effect on copper resistance (Wright et al., 1987). Likewise, insertions of nine amino acids have been produced at the N-terminus without affecting the copper resistance afforded by the altered product (Macreadie et al., 1989).

The *CUP1* locus was shown to contain a second ORF upstream of the MT gene (Karin et al., 1984). The function of this gene, designated gene X, is unresolved, but the gene disruption showed that it is not essential for copper resistance (Karin et al., 1984).

Industrial strains of *Saccharomyces* sp. have also been found to exhibit various amounts of tandem gene amplification at the *CUP1* locus and to have occasionally an additional locus (Welch et al., 1983; Fogel and Welch, 1983). This locus has been found to exist in repeat lengths of 1.9, 1.8, 1.7, 1.6, 1.1, and 0.9 kb, resulting in an incomplete gene X (Welch et al., 1983; Naumov et al., 1992). In a variety of natural strains of *S. cerevisiae,* Naumov et al. (1992) showed that this second locus was on chromosome XVI.

III. THE COPPER METALLOTHIONEIN COMPLEX IN *Saccharomyces cerevisiae*

Biochemical studies of *S. cerevisiae* showed that most intracellular copper ions are sequestered by MT (Prinz and Weser, 1975a,b; Weser et al., 1977). *S. cerevisiae* MT was reported to bind 7–8 Cu(I) ions through cysteinyl thiolates (Winge et al., 1985; George et al., 1988). Subsequently, heteronuclear multiple quantum correlation nuclear magnetic resonance (NMR) analysis of MT with Ag(I) ions, isoelectronic to Cu(I), revealed seven metal sites ligated exclusively by thiolate ligands (Narula et al., 1991). Nuclear Overhauser enhancement spectroscopy (NOESY) NMR studies revealed similar long range NOEs for AgMT and CuMT, suggesting that both complexes have a similar tertiary fold (Narula et al., 1993).

The metal ions are sequestered within a single polymetallic cluster (George et al., 1988; Narula et al., 1991). The cluster has structural features analogous to those of synthetic copper-thiolate (CuS) model clusters (Dance, 1986). From such structures, a number of generalizations have been formulated (Dance, 1986).

1. The structure of the synthetic clusters is maintained by doubly bridged thiolate coordinate bonds.
2. The Cu atoms are present as Cu(I) ions.

3. Coordination numbers of two (digonal) or three (trigonal) are common, and both coordination numbers can exist within the same cluster.
4. The actual Cu-S bond distance is dependent on the coordination number.
5. Short Cu-Cu interactions are common.

Mixed Cu(I) coordination geometries appear to exist in yeast CuMT (Narula et al., 1991). The metal-Cys connectivities shown in Fig. 2 are based on the AgMT NMR studies of Narula et al. (1991). Heteronuclear NMR studies suggested that in addition to five trigonal coordinate Cu(I) ions, two Cu(I) ions with digonal coordination geometry may be present (Narula et al., 1991). A detailed extended x-ray absorption fine structure (EXAFS) study of CuMT and a series of model CuS clusters (Pickering et al., 1993) demonstrated a correlation between the mean Cu-S bond distance and the fraction of digonal Cu(I) ions in a polymetallic cluster. The mean Cu-S bond distance determined in yeast CuMT by EXAFS analysis is also consistent with digonal Cu(I) ions. Yeast CuMT appears similar to synthetic CuS cage clusters with mixed Cu(I) coordination. It should be noted that not all known CuMTs will have mixed digonal and trigonal Cu(I) coordination.

As mentioned, the *S. cerevisiae* MT contains 12 cysteinyl residues in the 53 residue polypeptide. The arrangement of cysteines in the protein is similar to other MTs in the abundance of Cys-X-Cys and Cys-X-X-Cys sequence motifs (Winge et al., 1993). The only detailed structure of a metallothionein is the mammalian Cd,Zn-MT (Robbins et al., 1991). In this structure, each Cys-X-Cys and Cys-X-X-Cys motif exists within a reverse turn with a novel NH-S hydrogen bond stabilizing the turn (Robbins et al., 1991). This configuration juxtaposes a pair of cysteinyl sulfurs in van der Waals contact, thereby favoring metal ligation. Each cysteine pair can serve as ligands for one or two metal ions. Eight of the twelve cysteines in

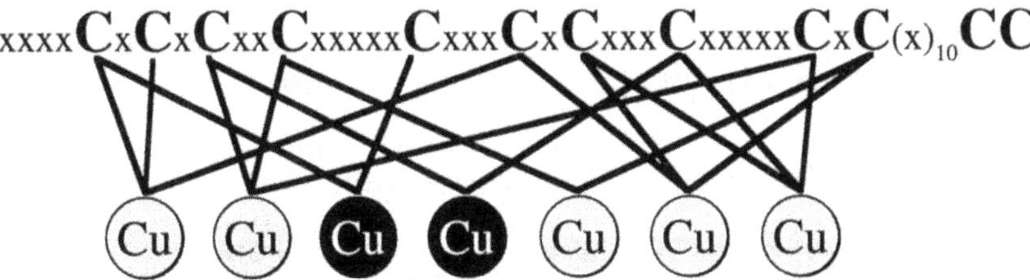

Figure 2 Cu-S connectivities in *S. cerevisiae* MT. The connectivities were elucidated as described by Narula et al. (1991).

S. cerevisiae MT are present in such sequence motifs. Two other cysteines, the C-terminal Cys-Cys pair, are not involved as ligands (Narula et al., 1991). MT mutants that remove this pair of cysteines are as effective as the wild-type protein in mediating copper resistance (Byrd et al., 1988; Thrower et al., 1988). The mutant proteins form native-like CuS polynuclear clusters, and cells harboring these C-terminal mutant MT genes are as effective in resisting copper poisoning as cells harboring the wild-type MT gene (Thrower et al., 1988; Narula et al., 1991). It therefore appears that the C-terminal Cys pair has no role in the structure and function of MT.

The CuS polynuclear cluster in CuMT thus consists of 7 metal ions and 10 cysteinyl sulfurs. As with synthetic CuS clusters, short range Cu-Cu distances are observed. Cu(I) ions are separated from each other by a mean distance of 2.7 Å in CuMT as well as in Cu_4S_6 cage clusters (George et al., 1988; Winge et al., 1993; Dance, 1986). Little Cu-Cu bonding is expected at this distance; the energetics of the cluster is likely to be dominated by Cu-S bonding and doubly bridging thiolates, as in the synthetic clusters.

The CuS cluster exists as the core of the tertiary fold, with the polypeptide chain enfolding the CuS cage. This conclusion is based on two observations. From luminescence measurements, the Cu(I) ions are clearly shielded from solvent interaction and therefore must be within a compact tertiary fold (Winge et al., 1993). Secondly, the mature *S. cerevisiae* MT contains no hydrophobic residues. Most globular proteins have a hydrophobic core formed by hydrophobic residues. In the absence of a hydrophobic core, the hydrophilic polymetallic cage is the likely candidate as the structural framework. The CuS cluster provides the stabilization energy for the tertiary fold. The protein is devoid of ordered structure in the absence of bound metal. The Cu(I) ions within the cluster are reactive toward Cu(I) chelators. It was found that a limited number of Cu(I) ions reacted with bathocuproinesulfonate and cuproine (Weser and Hartmann, 1988). It is not clear whether the reactive Cu(I) ions are the digonally coordinated ions.

As one might expect, other MTs can functionally replace the MT in *S. cerevisiae.* The ORFs from both monkey and *Drosophila* MT genes were shown to provide copper resistance in yeast when placed under the *CUP1* promoter (Thiele et al., 1986; Silar and Wegenz, 1990).

IV. METALLOTHIONEINS IN *Candida glabrata*

A yeast that can exhibit very high levels of copper and cadmium resistance is *Candida* (formerly *Torulopsis*) *glabrata.* Biochemical analyses showed that this yeast also produced metallothionein in a copper-specific response (Fig. 3; Mehra et al., 1988). Two distinct MTs denoted MTI and MTII were

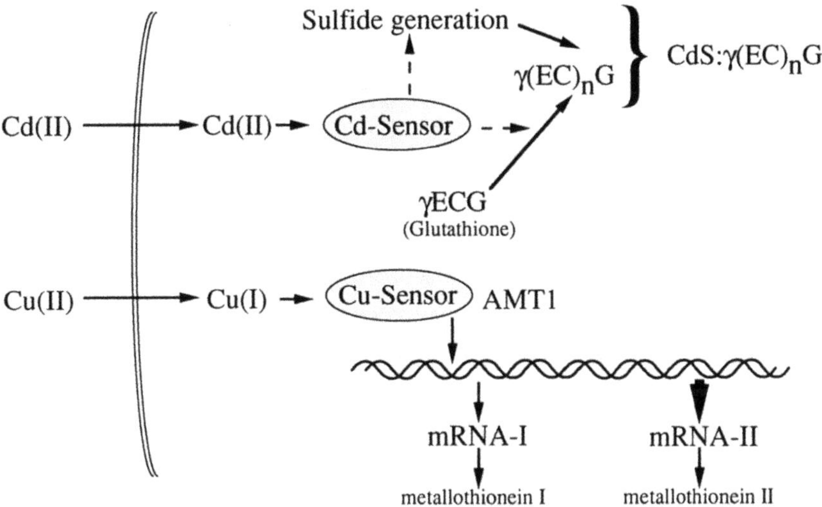

Figure 3 Metal responsive pathways in *C. glabrata*. The one-letter code for amino acids is used to designate the γ-Glu-Cys isopeptides of general structure, $(\gamma\text{-Glu-Cys})_n$Gly.

isolated from *C. glabrata* (Mehra et al., 1988). Analysis of their composition and sequence indicated that both MTs had a cysteine content of about 30 mol % with multiple Cys-X-Cys sequence motifs (Mehra et al., 1988, 1989). MTI and MTII consist of 62 and 51 amino acids, respectively (Mehra et al., 1989). Fourteen of the 18 Cys residues in MTI are present in Cys-X-Cys motifs, and two additional Cys residues are present as Cys-X-X-Cys. In MTII, 14 of the 16 cysteines are present as Cys-X-Cys. As with most metallothioneins, the *C. glabrata* MTs contain only a limited number of hydrophobic residues. The tertiary folds are likely to be dictated by Cu(I) binding. Processing is occasionally observed in the MTII protein: one such event is an N-terminal truncation of MTII resulting in Gln_7 becoming the N-terminus (Mehra et al., 1989).

Both *C. glabrata* MTs bind copper as Cu(I). Cu(I)-sulfur coordination was indicated in both proteins by spectroscopic techniques. The CuMTs share structural similarities to CuMT from *S. cerevisiae.* The copper centers are luminescent, indicative of Cu(I) binding within the protein core. MTI and MTII bind 11–12 and 10 Cu atoms, respectively. The Cu(I) ions are distributed within two domains in both MTI and MTII (Mehra and Winge, unpublished observation). This is analogous to mammalian MTs in which two distinct domains enfold separate metal:thiolate polymetallic clusters (Winge and Miklossy, 1982).

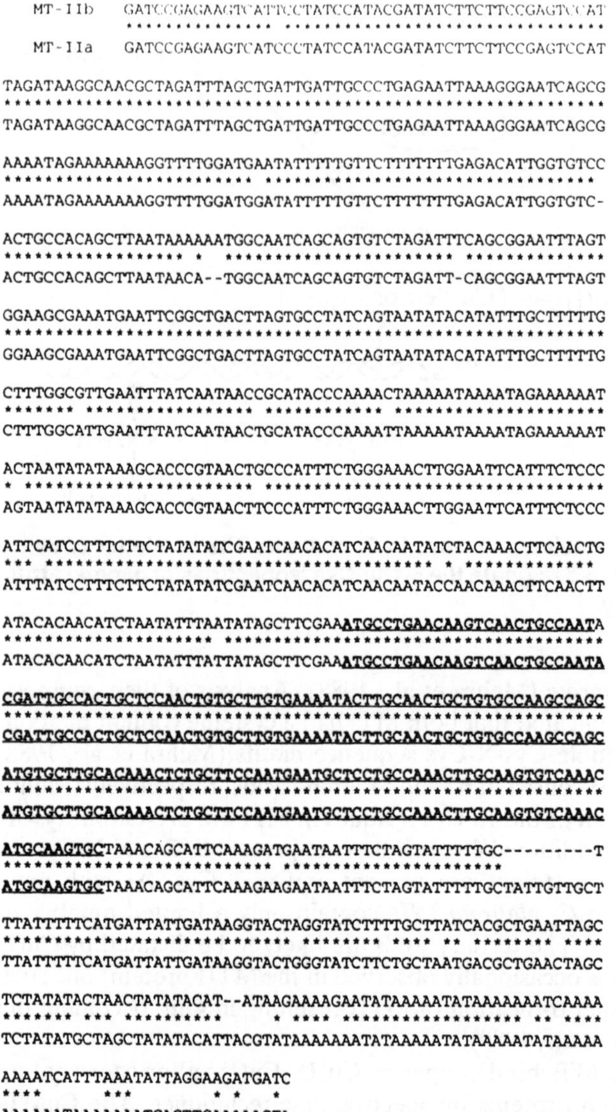

Figure 4 Sequence comparison of *MTIIa* and *MTIIb* genes from *C. glabrata*. The ORFs are shown in bold and underlined. The alignment shown is in contrast to that published previously (Mehra et al., 1992b) in that significant homology is observed in the 3′ flanking region.

Oligonucleotide probes based on the amino acid sequence were used to identify clones containing the MT genes (Mehra et al., 1989). The MTI gene was always found to be present in a single copy and never amplified in any of the strains tested. MTII was encoded by the *MTIIa* locus that comprised a gene tandemly duplicated three to nine times in wild-type isolates (Mehra et al., 1990), although in one clinical isolate the *MTIIa* locus was present as a single copy gene. Southern blots indicated the existence of an additional sequence related to *MTIIa.* This sequence was cloned by inverse polymerase chain reaction (PCR) and conventional methods (Mehra et al., 1992b). Analysis of the sequence showed that it was identical in the coding sequence to *MTIIa* and extremely similar elsewhere (Fig. 4). This sequence, now denoted *MTIIb,* was shown by gene disruption experiments to be a functional gene further contributing to the high level of copper resistance seen in this yeast (Mehra et al., 1992b). The *MTIIb* gene is always present as a single copy, unlike the amplified *MTIIa* locus.

The relative importance of these MT genes in yeast has been assessed by gene disruption of *MTIIa* and *MTIIb*. The results showed a reduction in the copper concentrations which inhibits growth by 50% from 7 mM to 1 mM when the amplified *MTIIa* locus was deleted, and a reduction to 0.1 mM when *MTIIb* was further deleted. The remaining resistance was attributed to MTI, since there was no evidence of $(\gamma\text{Glu-Cys})_n\text{Gly}$ peptide synthesis under the growth conditions of the experiment. MTI contributes little to the Cu(I) buffering in *C. glabrata.* Extraction of CuMTs from cells grown in medium containing copper salts revealed predominantly CuMTII even in cells containing only a single *MTIIa* gene (Mehra et al., 1992b).

V. REGULATION OF METALLOTHIONEIN BIOSYNTHESIS GENE

A. Components or the *CUP1* System Essential for Metalloregulation

Unlike their mammalian counterparts, the *S. cerevisiae* and *C. glabrata* MTs are specifically regulated by Cu(I) and Ag(I) ions and not other metals (see Thiele, 1992). Thus MT usually only affords copper resistance *in vivo,* although it binds many metals, such as Cd(II), *in vitro* (Winge et al., 1985; Berka et al., 1988). However, when the *CUP1* MT gene is placed under a constitutive promoter, the resulting cells become highly resistant to other metal ions such as Cd(II) (Ecker et al., 1986). An exception to this is a selected strain of *S. cerevisiae,* 301N, that exhibits Cd-induced *CUP1* expression (Inouhe et al., 1989,1991; Tohoyama et al., 1992a). CdMT accumulates in these cells when cultured in medium containing cadmium salts. This is discussed in greater detail in Sec. VII.

Both trans acting and cis acting elements are important for metalloregulation (Fig. 5). Metalloregulation of MT gene expression is mediated through a trans acting factor. The *S. cerevisiae* factor was identified from the isolation of copper-sensitive (Cu^s) mutants after mutagenesis of a multicopy *CUP1* strain (Thiele, 1988; Welch et al., 1989). Transformation of these Cu^s cells with a yeast DNA bank led to a Cu^r phenotype in certain clones. A single gene controlling the regulation of *CUP1* in *S. cerevisiae* was identified (Thiele, 1988; Furst et al., 1988; Welch et al., 1989). The new locus, denoted *ACE1* (activator of *CUP1* expression; Thiele, 1988) or *CUP2* (Welch et al., 1989), was proposed to encode a positive transcriptional regulator for the *CUP1* gene, since the mutants were defective for the transcription of *CUP1* mRNA following copper induction (Thiele, 1988; Buchman et al., 1989; Welch et al., 1989; Fig. 5). Since *ACE1* and *CUP2* are identical, we shall use the name *ACE1* in this review.

Analysis of the DNA sequences in the 5′ flanking region of *CUP1* revealed the presence of metal-responsive elements that are sites of ACE1 binding (Thiele and Hamer, 1986; Furst et al., 1988; Culotta et al., 1989; Buchman et al., 1990; Hu et al., 1990; Fig. 6). Fungal transcriptional elements are typically designated upstream activation sequences (UAS).

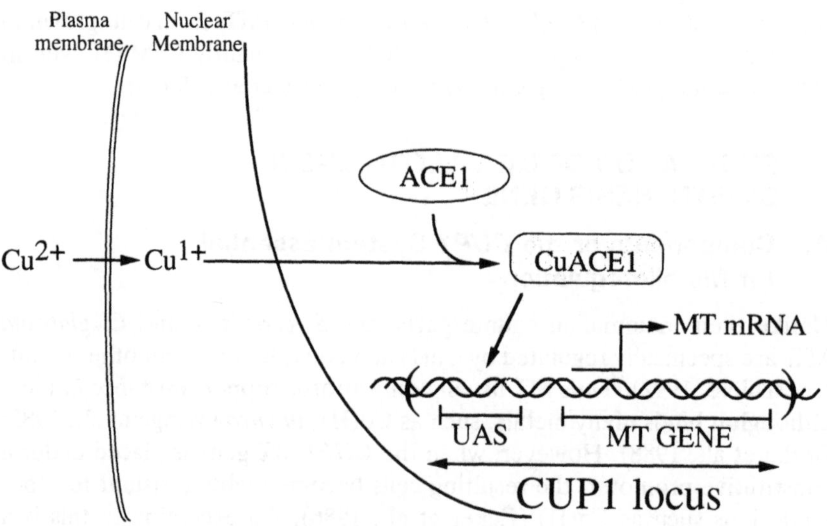

Figure 5 Cu activation pathway in *S. cerevisiae*.

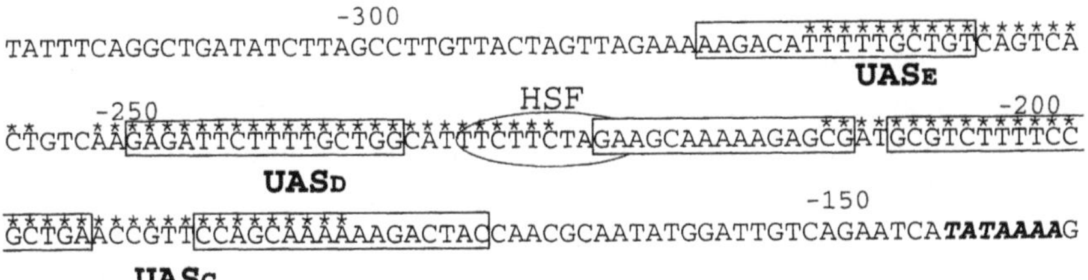

ACE1 Binding Sites

SOD1	AGCGGCATTTGCGCTGTCA
CUP1	
UASC(L)	TGCGTCTTTTCCGCTGAAC
UASC(R)	GTAGTCTTTTTTGCTGGAA
UASD(L)	AGAGATTCTTTTGCTGGCA
UASD(R)	ATCGCTCTTTTTGCTTCTA
UASE	AAAGACATTTTTGCTGTCA
MTI	
UAS1	AAAGCTATTATTGCTGTCA
UAS2	GAAGCTATTATTGCTGATT

Consensus

DMGNYHYTWNYGCTGD

D - A,G or T
H - A,C or T
M - C or A
N - A,C,G or T
W - A or T
Y - C or T

Figure 6 Regulatory elements in the *CUP1* promoter. The sequence upstream of the *CUP1* TATA box (shown in boldface italics) is shown with nucleotide numbering from the *CUP1* start codon. Sequences homologous to UAS_C are outlined in boxes; those sequences that exist within regions footprinted by CuACE1 are shown by filled boxes. Sequences that exhibit a CuACE1 footprint have a * above the bases. A candidate HSF binding site is outlined in the filled oval. A comparison of the five homologous core sequences in the *CUP1* promoter are shown below, along with the candidate ACE1 binding sites in the *SOD1* gene and the *MTI* gene from *C. glabrata.* From the four candidate UAS elements (those homologous core sequences within regions footprinted by CuACE1) in the *CUP1* promoter and the homologous element within the *SOD1* and *MTI* genes, a consensus sequence is derived for ACE1 binding and is shown at the right. Those conserved sequences that do not conform to the consensus sequence are listed with a dot above the nonconserved bases.

Two palindromic UAS elements were initially mapped in the 5′ flanking region of the MT gene (Thiele and Hamer, 1986; Huibregtse et al., 1989; Evans et al., 1990). The effectiveness of one UAS palindrome (originally UAS_P, but designated UAS_C more recently) to confer Cu-inducibility on downstream genes was a strong indication that this element was the binding site of ACE1 (Thiele and Hamer, 1986; Furst et al., 1988). The ability of UAS_C to mediate Cu-induced regulation on downstream genes was shown to be dependent on the orientation of the DNA element and copy number (Thiele and Hamer, 1986). Subsequently, Hu et al. (1990) demonstrated that the 5′ half of the UAS_C palindrome was effective in mediating limited metalloregulation on downstream genes. A number of mutations in the 5′ half of UAS_C abolish Cu-inducibility (Furst et al., 1988). Although this DNA element appears to be the major cis acting element in metalloregulation, maximal Cu-inducibility requires sequences outside of UAS_C (Thiele and Hamer, 1986; Culotta et al., 1989; Fig.6).

In the *CUP1* promoter region there are four sequences sharing homology with the 5′ half site of UAS_C (Fig. 6). Footprinting analyses of the *CUP1* 5′ sequences have revealed three regions of ACE1 binding encompassing the four common sequences (Huibregtse et al., 1989; Evans et al., 1990). Comparison of the DNA sequences within the regions of ACE1 binding suggests a candidate consensus sequence of at least 16 bp. The consensus as given (see Fig. 6 for description of letters) is

$$D^1M^2G^3N^4Y^5H^6Y^7T^8W^9N^{10}Y^{11}G^{12}C^{13}T^{14}G^{15}D^{16}$$

UAS_D was formerly proposed to be a palindromic site much like UAS_C (Thiele and Hamer 1986); however, the right half of the UAS_D candidate palindrome was not protected in DNaseI footprinting studies (Huibregtse et al., 1989; Evans et al., 1990). This is not surprising, since the right half site of UAS_D does not conform to the predicted consensus in that it lacks a G at position 15. This G has been shown to be essential for the binding of ACE1 to UAS_C(L) by methylation interference (Furst and Hamer, 1989; Buchman et al., 1990). The *CUP1* promoter therefore appears to consist of four ACE1 binding sites, one palindromic UAS element, and two elements lacking any dyad symmetry. Since there is no evidence to suggest that ACE1 binds to a palindrome at UAS_D, we suggest that UAS_D be viewed as a single site. Furthermore, as the region −274 to −265 (relative to translation start) in the *CUP1* promoter contains the correct consensus for ACE1 binding and is also protected in footprint analyses, we propose that it is a UAS element and have termed it UAS_E to conform to the already established nomenclature. Studies of these UAS elements in UAS-less promoter vectors will be necessary to establish if they are indeed capable of independent metalloregulation.

CuACE1 is also known to mediate the Cu-induced expression of *SOD1,* the gene encoding Cu,Zn-superoxide dismutase (Carri et al., 1991; Gralla et al., 1991). The 5′ promoter sequence of the *SOD1* gene contains a single sequence element homologous to the consensus ACE1 binding site (Fig. 6). This region is within a CuACE1 footprint (Gralla et al., 1991).

Within the *CUP1* promoter the W at position 9 is always a T. However, Thorvaldsen et al. (1993) showed ACE1 to give limited metalloregulation of the *C. glabrata* MTI promoter, where the consensus binding sites have an A in this position (see Fig. 9).

Footprinting of UAS_C was dependent on the copper binding to ACE1 (Huibregtse et al., 1989; Buchman et al., 1990). ACE1 expressed in *E. coli* binds to *CUP1* UAS_C in a copper-dependent manner (Evans et al., 1990). *In vitro* transcription studies showed that transcription of a reporter gene downstream of TATA and UAS_C elements required the presence of ACE1 and copper (Culotta et al., 1989; Buchman et al., 1990).

Sequence analysis of the *ACE1* gene indicated that the 225 amino acid translation product contained a C-terminal half that was highly acidic in nature (Szczypka and Thiele, 1989; Furst et al., 1988), a common property of many yeast transcription factors (Fig. 7). In addition, an *ACE1-lacZ* fusion product was localized within the nucleus (Szczypka and Thiele, 1989). The N-terminal half was rich in positively charged amino acids and cysteines occurring in three Cys-X-Cys and two Cys-X-X-Cys sequence motifs. This led to the suggestion that these motifs may constitute a metal binding domain (Furst et al., 1988; Szczypka and Thiele, 1989).

Further studies clearly showed that Cu ions bind to the N-terminal domain in ACE1 (Furst et al., 1988; Hu et al., 1990; Dameron et al., 1991). Eleven of the twelve cysteinyl residues in the N-terminal half of ACE1 are critical for the functioning of CuACE1 in *S. cerevisiae* (Hu et al., 1990). Physical studies with the N-terminal half of the protein expressed in *E. coli* showed that DNA binding required the presence of bound Cu(I) or Ag(I) ions (Furst et al., 1988; Dameron et al., 1991). Cu(I) binding appears

Figure 7 The sequence of ACE1 showing the Cys-X-Cys and Cys-X-X-Cys sequence motifs and richness of positively charged amino acids in the N-terminal DNA-binding domain and the highly acidic C-terminal half common to many yeast transcription factors.

maximal between 6–7 mol eq. Cu(I) (Dameron et al., 1991). It is unclear whether the Cu(I) ions are clustered in a single or multiple centers. The Cu-protein exhibited charge transfer bands in the ultraviolet as expected if cysteinyl thiolates are ligands. CuACE1 is luminescent, with emission occurring between 580–620 nm (Dameron et al., 1991; Casas-Finet et al., 1991). The emission was indicative of bound Cu(I) ions within a compact protein structure in which the Cu(I) ions are shielded from solvent contact.

X-ray absorption spectroscopy has been performed on *E. coli* isolates of CuACE1 as well as CuACE1 prepared by *in vitro* reconstitution protocols. EXAFS of CuACE1 revealed a major scatterer in the first coordination shell that fits well with sulfur atoms. A similar mean Cu-S bond distance of 2.26 Å was observed by two groups independently (Dameron et al., 1991; Nakagawa et al., 1991). The mean Cu-S distance of 2.26 Å is consistent with predominant trigonal Cu(I) coordination. From the synthetic CuS cluster data one may conclude that CuACE1 may contain a single two coordinate Cu(I) ion at most (Pickering et al., 1993). Cluster formation in CuACE1 was indicated by the observed Cu-Cu interactions apparent in the EXAFS (Dameron et al., 1991). These Cu-Cu interactions, occurring at 2.72 Å, are equivalent to the short Cu-Cu distances observed in CuMTs and synthetic trigonally coordinated CuS cages.

B. Mechanism of Metalloregulation

Cu(I) binding to ACE1 activates it for high-affinity binding to the UAS elements of the *CUP1* locus. Binding of CuACE1 to UAS_C may stabilize CuACE1 as a dimer, whereas CuACE1 binding to UAS_D and UAS_E is likely to be monomeric CuACE1. The binding of CuACE1 to UAS elements brings the C-terminal transcriptional activation domain into position for assembly of the transcriptional machinery.

It is not known how ACE1 contacts DNA within the UAS sequences. The proposed consensus sequence would spread over one and one-half turns of the B-form DNA helix. It is possible that CuACE1 makes base-specific contacts at the ends of the consensus sequence and backbone contacts in the A/T-rich middle section as it crosses the minor groove (Buchman et al., 1990). According to this model, CuACE1 contacts the major groove in two different sites and crosses over the minor groove. One implication of this model is that the CuACE1 protein may contain two protruding structural elements appropriately spaced for insertion within the major groove spanning one and one-half turns of the DNA helix. Mutagenesis studies have shown that $G_{37}E$, $K_{53}E$, and $R_{94}E$ amino acid substitutions in ACE1 result in molecules incapable of trans activation but containing a CuS cluster (Buchman et al., 1989; Hu et al., 1990). One model is that sequences around Gly_{37}, Arg_{94}, or Lys_{53} form either separate helices that enter the major grooves at the periphery of the UAS elements or a segment important in spanning the minor groove. It will be of interest to determine whether each of the two putative DNA binding loops is stabilized by separate or a single metal center(s).

The Cu-activation process involves formation of a polymetallic CuS cluster that mimics CuS cages in CuMT and certain synthetic CuS complexes. Similarities between CuACE1 and CuMT lead to the postulate that MT is a good structural model of the core structure of the active CuACE1. The polypeptide fold in both CuMT and CuACE1 may be dictated in part by the CuS cluster. This prediction is supported by the lack of an aromatic core in each protein. The tertiary folds in CuACE1 and CuMT contain limited periodic secondary structure elements indicating that protein stabilization is conferred by the CuS cluster. The data do not distinguish between a single polymetallic cluster or two separate CuS clusters. If CuMT is a structural model of the CuS cluster in CuACE1, then a single polymetallic cluster will exist.

ACE1 and CuACE1 adopt distinct conformers. The metal-free ACE1 must exist as a conformer lacking the juxtaposition of critical residues to form contacts with bases in the major groove of the UAS sites. Likewise, ACE1 binds Cu(I) and Cd(II) ions in different conformers, as is the case with MT (Nielson et al., 1985). CdACE1 can readily be formed *in vitro,* but the CdACE1 like apoACE1 does not form a high-affinity complex with DNA containing the specific UAS (Dameron et al., 1993). Maximal Cd(II) binding occurs between 3–4 mol eq. with tetrahedral coordination, as is usually seen in CdS clusters (Dameron et al., 1993). Similar Cd(II) binding stoichiometry and coordination geometry are seen in yeast CdMT. The tetrahedral coordination of Cd(II) and Zn(II) ions in ACE1 and MT necessitates a distinct protein fold from the Cu(I) conformers. The ability of ACE1 to form distinct conformers with Cu(I) and Cd(II) may form the basis for the exquisite metal ion specificity in the function of ACE1. The only metal ions that yield an active conformer, Cu(I) and Ag(I), are ions that yield homologous metal:thiolate cage clusters. The use of a multinuclear cluster as the structural core of the molecule ensures a unique conformer with Cu(I). Distinct conformers may also contribute to the activation of ACE1 through Cu(I)/Zn(II) exchange (Dameron et al., 1993).

In vitro studies have demonstrated that ACE1 binds to the DNA UAS_C element only in the presence of Cu(I). ApoACE1 or CdACE1 forms only a low-affinity protein:DNA complex, and this complex is competed with nonspecific DNA. ACE1 is effective in sensing the intracellular copper concentration and mediating MT biosynthesis in proportion to the magnitude of the copper pool. No information is available on how Cu(I) is presented to ACE1 *in vivo.* Glutathione may be involved in the Cu(I)-activation of ACE1 by virtue of its high intracellular concentration and tendency to form CuS complexes. The Cu(I):GSH complex is kinetically labile and will donate Cu(I) ions to either MT or ACE1 (Dameron and Winge, unpublished observation). As ACE1 is constitutively expressed under steady state conditions, it may exist as ZnACE1 or the metal-free molecule.

The trans acting factor that mediates Cu-induced expression of the three *C. glabrata* MT genes is AMT1 (Zhou and Thiele, 1991). AMT1 was identified as the gene responsible for Cu-specific induction of the MTI promoter fused to a β-galactosidase reporter gene in *S. cerevisiae* (Zhou and Thiele, 1991). The N-terminal half of AMT1 is 50% identical to the N-terminal, DNA-binding domain of ACE1 with complete conservation of the sequence positions of the critical cysteine residues (Zhou and Thiele, 1991; Zhou et al., 1992a; Fig. 8). DNA methylation interference assays have indicated that AMT1 has specific binding sites in regions upstream of sequences encoding MTI and MTIIa (Zhou et al., 1992a). Although the protected sequence extends over some 12 to 22 nucleotides, there is a consensus sequence, shown in Fig. 9, in the core of the AMT1 binding sites. This AMT1 consensus is conserved in *MTIIb* and is part of the ACE1 binding site in *CUP1* and in *SOD1*. We demonstrated that AMT1 is able to trans activate these promoters in a copper-dependent manner (Thorvald-

```
       1                                      40
ACE1:  MVVINGVKYACETCIRGHRAAQCTHTDGPLQMIRRKGRPS
       :::::::::::..::. :.:::: : : ::  .. .::: 
AMT1:  MVVINGVKYACDSCIKSHKAAQCEHNDRPLKILKPRGRPP

      41                                      80
       TTCGHCKELRRTKNFNPSGGCMCAS-AR-RPAVGSK-EDE
       ::: :::. :.::: :::: : :    . :   :   :..
       TTCDHCKDMRKTKNVNPSGSCNCSKLEKIRQEKGITIEED

      81                                     120
       -------TRCRCDEGEPCKCHTKRKSSRKSKGGSCHRRAN
                : :  ::::.:: .:: . ::
       MLMSGNMDMCLCVRGEPCRCHARRKRTQKS
```

Figure 8 Comparison of the N-terminal DNA-binding domains of ACE1 and AMT1 showing 50% identity at the amino acid level and a conservation of 11 critical cysteines.

Saccharomyces cerevisae

Gene	Start	Sequence	End
CUP1	-274	T**T**TTT**GCTG**T	-265
	-245	C**T**TTT**GCTG**G	-236
	-201	T**T**TCC**GCTG**A	-192
	-176	T**T**TTT**GCTG**G	-185
SOD1	-233	T**T**TGC**GCTG**T	-242

Neurospora crassa

Gene	Start	Sequence	End
MT	-208	T**T**CCT**GCTG**T	-217

Candida glabrata

Gene	Start	Sequence	End
MTI	-203	T**T**ATT**GCTG**A	-212
	-165	T**T**ATT**GCTG**T	-174
MTIIa/b	-432/-436	A**T**TTA**GCTG**A	-423/-427
	-387/-391	T**T**TTC**GCTG**A	-396/-400
	-316/-319	A**T**TAA**GCTG**T	-325/-328
	-280	A**T**TCC**GCTG**A	-289
	-261	A**T**TCG**GCTG**A	-252
AMT1	-250	A**T**TTG**GCTG**A	-241

Consensus H**T**HNN**GCTG**D

D is A, G or T
H is A, C or T
N is A, C, G, or T

Figure 9 Conserved motifs found in copper-regulated genes. These motifs are in the ACE1 binding sites in *CUP1* (Thiele and Hamer, 1986; Huibregtse et al., 1989; Evans et al., 1990) and *SOD1* (Carri et al., 1991; Gralla et al., 1991) and the AMT1 binding sites in *MTI, MTIIa* (Zhou et al., 1992a) and *AMT1* (Zhou et al., 1992b). No data is yet available for *MTIIb* or the *N. crassa* MT gene. Nucleotide numbering is from the start codon of each gene. Sequences are from: *CUP1* (Karin et al., 1984; Butt et al., 1984b); *SOD1* (Bermingham-McDonogh et al., 1988); *MTI* (Mehra et al., 1989); *MTIIa* (Mehra et al., 1990); *MTIIb* (Mehra et al., 1992b); *AMT1* (Zhou and Thiele, 1991); *N. crassa* MT gene (Munger et al., 1985).

sen et al., 1993). The core consensus motif may also exist in copper-responsive genes of nonyeast organisms as well. For example, the MT gene of the fungus *Neurospora crassa* (Munger et al., 1985) contains one copy of this motif. There is also an AMT1 binding site containing the motif upstream of the sequences encoding AMT1. *AMT1,* unlike *ACE1,* exhibits Cu-inducible expression (Zhou et al., 1992b). The proposed consensus may be a critical sequence for binding, but sequences outside this consensus may be important for full Cu-inducibility, as is the case with CuACE1 (Furst et al., 1988).

The similarities between the MT genes of *C. glabrata* and *S. cerevisiae* suggest that components may be interchanged with a conservation of the expression. We have demonstrated that *AMT1* can replace *ACE1* and mediate the Cu(I)-induced expression of *CUP1* MT in *S. cerevisiae* (Thorvaldsen et al., 1993). Likewise, *ACE1* is functional in the Cu-mediated expression of the *C. glabrata* MTI gene (Thorvaldsen et al., 1993). ACE1 and AMT1 therefore share both structural and functional homology.

C. Other Factors Affecting MT Gene Expression

MT genes display a high basal level of expression, and it is reasonable to expect that other factors may promote transcription. Thus there have been several searches for such factors. Selecting copper resistance in *acel* deletion strains, Yang et al. (1991) and Silar et al. (1991) independently found a suppressor with an alanine-to-valine change in the DNA binding domain of the heat shock transcription factor (HSF). This mutation enables the HSF to bind to the *CUP1* promotor. The putative binding sequence, TTCTAGAA (Yang et al., 1991; shown in Fig. 6) appears to be a truncated portion of the heat shock element (HSE) that is found upstream of heat shock genes in all organisms. With wild-type HSF there was binding to the *CUP1* HSE as well; however, the level of binding was greatly reduced when compared to the mutant form. Nonetheless, it would appear that this binding may contribute to at least some of the basal level of expression of *CUP1* in the normal situation. The *CUP1* locus is only marginally responsive to heat shock in wild-type cells (Yang et al., 1991).

Another copper-resistant mutant mapped to a locus designated *ACE2* (Butler and Thiele, 1991). ACE2 is homologous to SWI5, a transcription factor involved in switching of yeast mating type. Both proteins contain zinc finger motifs, and both are effective in enhancing the basal expression of *CUP1* (Butler and Thiele, 1991; Thorvaldsen et al., 1993).

It has also been determined that the depletion of nucleosomes from *CUP1* leads to a transcription level that is equivalent to the fully induced level (Durrin et al., 1992). Rapid responsiveness of *CUP1* may necessitate the absence of nucleosome structures in the *CUP1* region.

Methylation has been noted to be important in the expression of MT genes in mammalian cells (Hamer, 1986). The fungus *Mucor* has also been shown to have a developmentally regulated *CUP* gene (Cano-Canchola et al., 1992); however, since yeasts do not appreciably methylate their DNA, this cannot be a mechanism for MT regulation in yeast.

VI. GENE AMPLIFICATION AND COPPER RESISTANCE

In addition to Cu-induced synthesis of MT, amplification of MT genes is another response to copper stress that leads to enhanced copper resistance in *S. cerevisiae* and *C. glabrata* (Fig. 10). Generally, copper-resistant strains contain three or more MT genes, while copper-sensitive strains only contain one copy. Strains with one copy have never been demonstrated to undergo amplification. In a detailed study of meiotic progeny from crosses between strains having different *CUP1* copy number, Fogel et al. (1983, 1984) showed that strains increase their *CUP1* copy number by the process of gene conversion.

Exposure of *C. glabrata* cells containing an amplified *MTIIa* locus to copper stress led to the further amplification of the *MTIIa* locus to upwards of 30 copies or to duplication of the chromosome on which it is contained

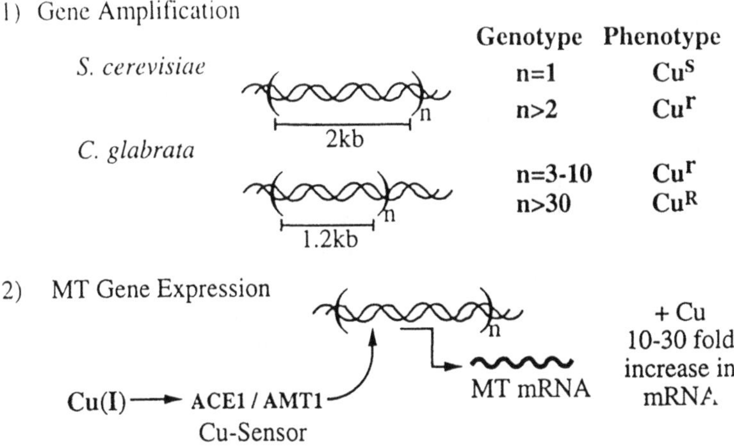

Figure 10 Responses of MT loci to copper stress in the yeasts *S. cerevisae* and *C. glabrata*. An increase in MT product is achieved by both gene amplification and regulation of expression by ACE1 or AMT1, respectively.

(Mehra et al., 1990). No amplification of the *MTIIa* locus was observed in copper-resistant isolates of the strain containing only a single *MTIIa* gene. The *MTIIa* locus contains an autonomously replicating sequence (Mehra et al., 1992a) suggesting a means by which amplification of the *MTIIa* gene occurs.

Another stress that has led to amplification is exposure to carcinogens, where copper-resistant colonies increased by several hundred fold (Aladjem et al., 1988). In strains containing multiple *CUP1* repeats, this resistance arose from gene amplification, but in strains containing only one *CUP1* repeat, there was no gene amplification.

Amplification during copper stress also occurs in genetically engineered strains, where a foreign sequence is embedded into the *CUP1* locus (Butt and Ecker, 1987). Thus copper stress leads to the amplification of this segment, resulting in an increase in copy number for the foreign gene as well. Such modifications have great utility in leading to the overproduction of foreign proteins that can be stably expressed in the absence of selection.

VII. CADMIUM METALLOTHIONEIN

Resistance to cadmium salts is mediated by binding to glutathione and γ-Glu-Cys isopeptides (Kneer et al., 1992) and by the effects of a number of nuclear gene products (Tohoyama et al., 1990). Searches for other mechanisms of cadmium resistance have uncovered a diverse array of possible mechanisms. For example, a protein with protective properties was found in *S. cerevisiae* strain 301N (Inouhe et al., 1989, 1991). Cadmium resistance in 301N was found to occur by constitutive expression of *CUP1* (Tohoyama et al., 1992b) that is further induced by copper or cadmium ions, resulting in the production of cadmium or copper metallothionein. The induction characteristics suggest that strain 301N could be a candidate for having an altered transcription factor.

Jeyaprakash et al. (1991) also isolated a cadmium-resistant mutant after transforming a YEp351-library into a cadmium-sensitive strain. They found that the DNA sequence conferring resistance consisted of two repeated copies of the *CUP1* gene. While the sequence was insufficient to confer copper resistance at low copy number, it was capable of conferring resistance to cadmium at high copy number. Curiously, both copper and cadmium induced expression of *CUP1,* but copper led to higher levels of transcription. An implication of this work is that higher product levels are required to confer cadmium resistance. Similarly, it has been found that a *S. cerevisiae* transformant carrying an episomal copy of the *C. glabrata MTIIa* gene could give rise to cadmium-resistant colonies by *in vivo* rearrangement of the promoter elements or by tandem duplication of the *MTIIa* genes (Yu et al., unpublished observation).

Metallothionein may also function to sequester metal ions in cells where MT does not occur. For example, the placement of *CUP1* in *Kluyveromyces lactis,* where the promoter is constitutively expressed, results in *K. lactis* transformants that are found to be both cadmium resistant and copper resistant (Macreadie et al., 1991b). It has been determined that in the case of cadmium this resistance is due to sequestering of intracellular cadmium by the *S. cerevisiae* metallothionein (Macreadie et al., unpublished observation).

VIII. EXPLOITATION OF METALLOTHIONEIN AND ITS GENE

The characteristics of metallothionein, the structure of its gene, and its expression have all offered intriguing possibilities for yeast biotechnologists to exploit. The metallothionein protein itself has been exploited for several uses. *CUP1* has served as a dominant selectable marker for copper resistance, making possible the transformation and selection of brewing strains without the need to acquire mutants and develop the genetics of such strains (Henderson et al., 1985; Hinchliffe and Daubney, 1986; Penttila et al., 1987; Villanueba et al., 1990). Resnick et al. (1990) have shown in laboratory strains that *CUP1* on centromere plasmids can be used as a copper-resistance marker to quantitate stability and plasmid copy number directly. Butt and Ecker (1987) have also proposed that metallothionein in copper-resistant yeast could be used as an aid to metal ion recovery in mining.

The amplification of the *CUP1* gene can also be exploited in a number of ways. First, determining the extent of *CUP1* amplification can be useful for the fingerprinting of brewer's yeast strains (Meaden, 1990). In addition, examination of other *CUP* loci (Fogel and Welch, 1983; Welch et al., 1983; Naumov et al., 1992) can provide additional characterization, especially of wild strains of yeast. The amplification of foreign genes inserted into the chromosomal *CUP1* locus appears the most generally useful application of the locus (Butt and Ecker, 1987). This results in a recombinant yeast that can retain foreign sequences in the absence of selection.

Metallothionein promoters have been popular in mammalian cell expression, and they are just as useful for driving expression of foreign genes in yeast, because of their convenient and rapid response to copper ions. Strains can be grown in the absence of added copper until the desired biomass is obtained; then copper can be added, though the precise induction protocol can vary and can depend on the genotype of the *CUP1* strain. Induction protocols have been described by Etcheverry (1990). A great attraction of the copper-inducible expression system is that first copper is conveniently added to very large cultures, and second the prior expression can be sufficiently low that even products toxic to yeast can be produced following induction.

A considerable number of proteins have been produced in *S. cerevisiae* with the aid of the *CUP1* promoter. Given here are some examples: *E. coli* galactokinase (Butt et al., 1984a; Gorman et al., 1986); human serum albumin (Etcheverry et al., 1986); phosphoglycerate kinase (Etcheverry et al., 1986); chicken oviduct progesterone receptor (Mak et al., 1989); *E. coli* β-galactosidase (Macreadie et al., 1989); infectious bursal disease virus antigens (Macreadie et al., 1990; Jagadish et al., 1990); rat liver c-*erbA* thyroid hormone receptor (Lu et al., 1990); cytochrome P450 (Fujita et al., 1990); human vitamin D receptor (Sone et al., 1990); helminth antigen (Macreadie et al., 1991a); blue tongue virus protein (Martyn et al., 1991); house dust mite allergen (Chua et al., 1992); human immunodeficiency virus Vpu protein (Macreadie et al., 1992); human estrogen receptor (Lyttle et al., 1992); human immunodeficiency virus Nef proteins (Macreadie et al., 1993).

The principles established for the copper-inducible expression of foreign proteins in *S. cerevisiae* may be applicable to other yeast or fungal expression systems. For example a *Neurospora crassa* copper-inducible system is now being used for the production of valuable proteins (Schilling et al., 1992).

IX. OTHER BIOMOLECULES INVOLVED IN PROTECTION

A. Glutathione-Related Peptides

Metal-isopeptide complexes, referred to as phytochelatin, cadystin, or γ-Glu-Cys isopeptides, have been reported from yeasts and plants (Rauser, 1990). These isopeptides have the structure $(\gamma\text{Glu-Cys})_n\text{Gly}$, where the value of n typically ranges from 2 to 5 in *Schizosaccharomyces pombe* and plants. The isopeptides from *C. glabrata* consist of predominantly n_2 and a desGly variant of n_2. DesGly sequence variants have also been shown to occur in *S. pombe* (Reese et al., 1988). In *S. cerevisiae,* only the n_2 isopeptide has been observed (Kneer et al., 1992; Winge et al., unpublished observation). The yeast *S. pombe* is unusual among yeasts in that it has a life cycle more typical of higher eukaryotes and divides by fission; however, like many yeasts it has no metallothionein and appears to derive limited protection against metal poisoning by the isopeptides.

Cd(II)-peptide complexes form a matrix that can incorporate sulfide ions to yield inorganic crystalline lattices coated with a mixture of isopeptides (Dameron et al., 1989). A Cu(I)-isopeptide complex has been characterized from *S. pombe.* This complex is a Cu(I)-thiolate polymetallic complex. The mean Cu-S bond distance and x-ray absorption edge spectrum is consistent with Cu(I) coordination in trigonal geometry (Pickering et al., 1993).

Biosynthesis of the peptide from its constituents, glutathione or the γ-Glu-Cys dipeptide unit of glutathione, is thought to be aided by a transpeptidase (Grill et al., 1989). In contrast to MT synthesis, the transpeptidase is thought to be constitutively produced, since *in vitro* activity is found in *S. pombe* (Grill et al., 1989). This topic is dealt with extensively in Chap. 11 and will not be discussed further.

B. Other Mechanisms of Protection

Metallothioneins or the γ-Glu-Cys isopeptides are not the only determinants of copper resistance in *S. cerevisiae.* There are over 600 species of yeast, and metal resistance determinants have been studied in relatively few of them. It is likely that in many, such as *K. lactis,* there is neither MT or peptide. Despite this, these yeast exhibit various degrees of resistance to metal ions, and hyper-resistant mutants can be readily isolated.

Naiki et al. (1961) and Yanagishima et al. (1949) reported that in certain copper-resistant strains most of the copper was precipitated as copper sulfide within or on the cell wall. Surface precipitation of copper by cells producing high quantities of sulfide has been observed in other yeasts and bacteria (Minney and Quirk, 1985; Erardi et al., 1987). In yet other cases, intracellular metal ions may be sequestered into organelles such as the vacuole (reviewed in Gadd and White, 1989).

The gene *ZRC1,* which confers resistance to both zinc and cadmium ions, was isolated by Kamizono et al. (1989). The sequence of the gene revealed that it encoded a 48 kDa protein with six potential membrane-spanning domains. It seems plausible that the protein could serve as a pump for zinc and cadmium ions. Another 48 kDa protein of 60% identity to the *ZRC1* product has recently been identified as a protein involved in cobalt transport (Conklin et al., 1992). It appears to be involved in compartmentalization or sequestration of cobalt ions into mitochondria.

While there are some further examples emerging of proteins involved in metal resistance, it is clear that some of these proteins are not primarily involved in metal ion resistance; they have other critical functions. The procedures used to search out proteins involved in metal ion interactions are sometimes extremely powerful and prone to lead to artifacts. For example, in a search for the gene encoding the transpeptidase that is thought to catalyze the synthesis of the γ-Glu-Cys isopeptide, Yu et al. (1991) searched an *S. pombe* cDNA library for the ability to confer cadmium resistance to *S. cerevisiae.* In analyzing the unexpectedly high frequency of cadmium-resistant transformants, plasmids conferring the resistance were found to contain the gene for alcohol dehydrogenase (ADH). Further analysis of these yeasts showed cadmium to be bound to ADH. Therefore, in this unnatural instance the enzyme ADH was able to sequester cadmium ions and protect the yeast from being poisoned.

X. CONCLUDING REMARKS

Studies on proteinaceous protection from toxicosis by metal ions have been dominated by copper resistance and the protection provided by MT. *S. cerevisiae* and *C. glabrata* have MTs that appear quite unrelated in protein sequence. Even the MTs within *C. glabrata* are unrelated in sequence. However, there are strong analogies in expression of yeast MTs at the level of gene duplication and copper-regulated transcription.

In both yeasts the rapid expression of MT genes appears to be mediated by a copper sensor that is a nuclear localized polypeptide with two domains. The N-terminal cysteine-rich domain is the metal switch. Cu(I) binding activates ACE1/AMT1 for specific binding to recognition motifs upstream of MT genes. The C-terminal domain is the transcription activator. Therefore, binding of the CuACE1 leads to expression of the downstream MT gene. Expression is regulated by the availability of copper ions to bind the sensor. The expression product, MT, sequesters available copper ions leading to the downregulation of the sensor.

At the DNA level, copper resistance levels are often directly related to the MT gene copy number. There is tandem gene amplification of *CUP1* and *MTIIa,* and further amplification can be selected. The extent of amplification may be restricted by limiting concentrations of the trans acting factors.

Copper homeostasis is an important role for MTs, but it is tempting to speculate, by comparison to the expression of mammalian MTs, that there may be other roles. On the other hand, it seems that metallothionein is not an obligatory requirement of yeast in general. Why then do species such as *S. cerevisiae* have a metallothionein, and why does a human pathogen, *C. glabrata,* have three loci encoding metallothionein? It could be argued that in the laboratory situation we do not simulate the natural demands put on a strain in the wild. For example, in *S. cerevisiae* the fermentation of beer in copper vats may have led to selection for copper resistance. It is possible that metallothionein may also have a significant role in the wild that is yet to be discovered.

REFERENCES

Adman, E. T. (1991). Structure and function of copper-containing proteins, *Curr. Opin. Struct. Biol., 1:* 895.

Aladjem, M. I., Koltin, Y., and Lavi, S. (1988). Enhancement of copper resistance and *Cup*I amplification in carcinogen-treated yeast cells, *Mol. Gen. Genet., 211:* 88.

Agarwal, K., Sharma, A., and Talukder, G. (1989). Effects of copper on mammalian cell components, *Chem. Biol. Interact., 69:* 1.

Berka, T., Shatzman, A., Zimmerman, J., Strickler, J., and Rosenberg, M. (1988). Efficient expression of the yeast metallothionein gene in *Escherichia coli, J. Bact., 170:* 21.

Bermingham-McDonogh, O., Gralla, E. G., and Valentine, J. S. (1988). The copper, zinc-superoxide dismutase gene of *Saccharomyces cerevisiae:* cloning, sequencing, and biological activity, *Proc. Natl. Acad. Sci. USA, 85:* 4789.

Brenes-Pomales, A., Lindegren, G., and Lindegren, C. C. (1955). Gene control of copper sensitivity in Saccharomyces, *Nature* (London), *136:* 841.

Buchman, C., Skroch, P., Welch, J., Fogel, S., and Karin, M. (1989). The CUP2 gene product, regulator of yeast metallothionein expression, is a copper-activated DNA-binding protein, *Mol. Cell. Biol., 9:* 4091.

Buchman, C., Skroch, P., Dixon, W., Tullius, T. D., and Karin, M. (1990). A single amino acid change in CUP2 alters its mode of DNA binding, *Mol. Cell. Biol., 10:* 4778.

Butler, G., and Thiele, D. J. (1991). *ACE2,* an activator of yeast metallothionein expression which is homologous to *SWI5, Mol. Cell. Biol., 11:* 476.

Butt, T. R., and Ecker, D. J. (1987). Yeast metallothionein and applications in biotechnology, *Micro. Rev., 51:* 351.

Butt, T. R., Sternberg, E. J., Herd, J., and Crooke, S. T. (1984a). Cloning and expression of a yeast copper metallothionein, *Gene, 27:* 23.

Butt, T. R., Sternberg, E., Gorman, J. A., Clarke, P., Hamer, D., Rosenberg, M., and Crooke, S. T. (1984b). Copper metallothionein of yeast, structure of the gene, and regulation of expression, *Proc. Natl. Acad. Sci. USA, 81:* 3332.

Byrd, J., Berger, R. M., McMillin, D. R., Wright, C. F., Hamer, D., and Winge, D. R. (1988). Characterization of the copper-thiolate cluster in yeast metallothionein and two truncated mutants, *J. Biol. Chem., 263:* 6688.

Cano-Canchola, C., Sosa, L., Fonzi, W., Sypherd, P., and Ruiz-Herrera, J. (1992). Developmental regulation of *CUP* gene expression through DNA methylation in *Mucor* spp., *J. Bact., 174:* 362.

Carri, M. T., Galiazzo, F., Ciriolo, M. R., and Rotilio, G. (1991). Evidence for co-regulation of Cu,Zn superoxide dismutase and metallothionein gene expression in yeast through transcriptional control by copper via the ACE 1 factor, *FEBS Lett., 278:* 263.

Casas-Finet, J. R., Hu, S., Hamer, D., and Karpel, R. L. (1991). Spectroscopic characterization of the copper(I)-thiolate cluster in the DNA-binding domain of yeast ACE1 transcription factor, *FEBS Lett., 281:* 205.

Chua, K.-W., Kehal, P. K., Thomas, W. R., Vaughan, P. R., and Macreadie, I. G. (1992). High-frequency binding of IgE to the *Der p* allergen expressed in yeast, *J. Allergy Clin. Immunol., 89:* 95.

Conklin, D. S., McMaster, J. A., Culbertson, M. R., and Kung, C. (1992). *COT1,* a gene involved in cobalt accumulation in *Saccharomyces cerevisiae, Mol. Cell. Biol., 12:* 3678.

Culotta, V. C., Hsu, T., Hu, S., Furst, P., and Hamer, D. (1989). Copper and the ACE1 regulatory protein reversibly induce yeast metallothionein gene transcription in a mouse extract, *Proc. Natl. Acad. Sci. USA, 86:* 8377.

Dameron, C. T., Reese, R. N., Mehra, R. K., Kortan, A. R., Carroll, P. J., Steigerwald, M. L., Brus, L. E., and Winge, D. R. (1989). Biosynthesis of cadmium sulfide semiconductor crystallites, *Nature, 338:* 596.

Dameron, C. T., Winge, D. R., George, G. N., Sansone, M., Hu, S., and Hamer, D. (1991). A copper-thiolate polynuclear cluster in the ACE1 transcription factor, *Proc. Natl. Acad. Sci. USA, 88:* 6127.

Dameron, C. T., George, G. N., Arnold, P., Santhanagoyalan, V., and Winge, D. R. (1993). Distinct metal binding configurations in ACEI, *Biochemistry, 32:* 7294.

Dance, I. G. (1986). The structural chemistry of metal thiolate complexes, *Polyhedron, 5:* 1037.

Durrin, L. K., Mann, R. K., and Grunstein, M. (1992). Nucleosome loss activates *CUP1* and *HIS3* promoters to fully induced levels in the yeast *Saccharomyces cerevisiae, Mol. Cell. Biol., 12:* 1621.

Ecker, D. J., Butt, T. R., Sternberg, E. J., Neeper, M. P., Debouck, C., Gorman, J. A., and Crooke, S. T. (1986). Yeast metallothionein function in metal ion detoxification, *J. Biol. Chem., 261:* 16895.

Eradi, F. X., Failla, M. L., and Falkinham, J. O. (1988). Plasmid-encoded copper resistance and precipitation by *Mycobacterium scrofulaceum, Appl. Environ. Micro., 53:* 1951.

Etcheverry, T. (1990). Induced expression using yeast copper metallothionein promoter, *Methods Enzymol., 185:* 319.

Etcheverry, T., Forrester, W., and Hitzeman, R. (1986). Regulation of the chelatin promoter during the expression of human serum albumin or yeast phosphoglycerate kinase in yeast, *Biotech., 4:* 726.

Evans, C. F., Engelke, D. R., and Thiele, D. J. (1990). ACE1 transcription factor produced in *Escherichia coli* binds multiple regions within yeast metallothionein upstream activation sequences, *Mol. Cell. Biol., 10:* 426.

Fogel, S., and Welch., J. W. (1982). Tandem gene amplification mediates copper resistance in yeast, *Proc. Natl. Acad. Sci. USA, 79:* 5342.

Fogel, S., and Welch, J. W. (1983). A recombinant DNA strategy for characterizing industrial yeast strains, *Genetics: New Frontiers: Proceeding of the XV International Congress of Genetics* (V. L. Chopra et al., eds.), p. 133.

Fogel, S., Welch, J. W., Cathala, G., and Karin, M. (1983). Gene amplification in yeast: *CUP1* copy number regulates copper resistance, *Curr. Genet., 7:* 347.

Fogel, S., Welch, J. W., and Louis, E. J. (1984). Meiotic gene conversion mediates gene amplification in yeast, *Cold Spring Harbor Symp. Quant. Biol., XLIX:* 55.

Fujita, V. S., Thiele, D. J., and Coon, M. J. (1990). Expression of alcohol-inducible rabbit liver cytochrome P-450 3a (P-450IIE1) in *Saccharomyces cerevisiae* with the copper-inducible *CUP1* promoter, *DNA Cell. Biol., 9:* 111.

Furst, P., Hu, S., Hackett, R., and Hamer, D. (1988). Copper activates metallothionein gene transcription by altering the conformation of a specific DNA binding protein, *Cell, 55:* 705.

Gadd, G. M., and White, C. (1989). Heavy metal and radionuclide accumulation and toxicity in fungi and yeasts, *Metal-Microbe Interactions* (R. K. Poole and G. M. Gadd, eds.), IRL Press, Oxford, p. 19.

George, G. N., Byrd, J., and Winge, D. R. (1988). X-ray absorption studies of yeast copper metallothionein, *J. Biol. Chem., 263:* 8199.

Gralla, E. B., Thiele, D. J., Silar, P., and Valentine, J. S. (1991). ACE1, a copper-dependent transcription factor, activates expression of the yeast copper, zinc superoxide dismutase gene, *Proc. Natl. Acad. Sci. USA, 88:* 8558.

Grill, E., Loeffler, S., Winnacker, E. L., and Zenk, M. H. (1989). Phytochelatins, the heavy metal-binding peptides of plants, are synthesized from glutathione by a specific γ-glutamylcysteine dipeptidyl transpeptidase (phytochelatin synthase), *Proc. Natl. Acad. Sci. USA, 86:* 6838.

Gorman, J. A., Clark, P. E., Lee, M. C., Debouck, C., and Rosenberg, M. (1986). Regulation of the yeast metallothionein gene, *Gene, 48:* 13.

Hamer, D. H. (1986). Metallothionein, *Ann. Rev. Biochem., 55:* 913.

Hamer, D. H., Thiele, D. J., and Lemontt, J. E. (1986). Function and autoregulation of yeast copper thionein, *Science, 228:* 685.

Henderson, R. C. A., Cox, B. S., and Tubb, R. (1985). The transformation of brewing yeasts with a plasmid containing the gene for copper resistance, *Curr. Genet., 9:* 133.

Hinchliffe, E., and Daubney, C. J. (1986). The genetic modification of brewing yeast with recombinant DNA, *J. Am. Soc. Brew. Chem., 44:* 98.

Hu, S., Furst, P., and Hamer, D. (1990). The DNA and Cu binding functions of ACE1 are interdigitated within a single domain, *New Biol., 2:* 544.

Huibregtse, J. M., Engelke, D. R., and Thiele, D. J. (1989). Copper-induced binding of cellular factors to yeast metallothionein upstream activation sequences, *Proc. Natl. Acad. Sci. USA, 86:* 65.

Inouhe, M., Hiyama, M., Tohoyama, H., Joho, M., and Murayama, T. (1989). Cadmium-binding protein in a cadmium-resistant strain of *Saccharomyces cerevisiae, Biochim. Biophys. Acta, 993:* 51.

Inouhe, M., Inagawa, A., Morita, M., Tohoyama, H., Joho, M., and Murayama, T. (1991). Native cadmium-metallothionein from the yeast *Saccharomyces cerevisiae:* Its primary structure and function in heavy-metal resistance, *Plant Cell Physiol., 32:* 475.

Jagadish, M. N., Vaughan, P. R., Irving, R. A., Azad, A. A., and Macreadie, I. G.

(1990). Expression and characterization of the infectious bursal disease virus polyprotein in yeast, *Gene, 95:* 179.

Jeyaprakash, A., Welch, J. W., and Fogel, S. (1991). Multicopy *CUP1* plasmids enhance cadmium and copper resistance levels in yeast, *Mol. Gen. Genet., 225:* 363.

Kagi, J. H. R., and Schaffer, A. (1988). Biochemistry of metallothionein, *Biochemistry, 27:* 8509.

Kamizono, A., Nishizawa, M., Teranishi, Y., Murata, K., and Kimura, A. (1989). Identification of a gene conferring resistance to zinc and cadmium ions in the yeast *Saccharomyces cerevisiae, Mol. Gen. Genet., 219:* 161.

Karin, M., Najarian, R., Haslinger, A., Valenzuela, P., Welch, J., and Fogel, S. (1984). Primary structure and transcription of an amplified genetic locus: The *CUP1* locus of yeast, *Proc. Natl. Acad. Sci. USA, 81:* 337.

Kneer, R., Kutchan, T. M., Hochberger, A., and Zenk, M. H. (1992). *Saccharomyces cerevisiae* and *Neurospora crassa* contain heavy metal sequestering phytochelatin, *Arch. Microbiol., 157:* 305.

Lu, C., Yang, Y. F., Ohashi, H., and Walfish, P. G. (1990). *In vivo* expression of rat liver c-*erbA* β thyroid hormone receptor in yeast (*Saccharomyces cerevisiae*), *Biochem. Biophys. Res. Commun., 171:* 138.

Lyttle, C. R., Damian-Matsumura, P., Juul, H., and Butt, T. R. (1992). Human estrogen receptor regulation in a yeast model system and studies on receptor agonists and antagonists, *J. Steroid Biochem. Mol. Biol., 42:* 677.

Macreadie, I. G., Jagadish, M. N., Azad, A. A., and Vaughan, P. R. (1989). Versatile cassettes designed for the copper inducible expression of proteins in yeast, *Plasmid, 21:* 147.

Macreadie, I. G., Vaughan, P. R., Chapman, A. J., McKern, N. M., Jagadish, M. N., Heine, H.-G., Ward, C. W., Fahey, K. J., and Azad, A. A. (1990). Passive protection against infectious bursal disease virus by viral VP2 expressed in yeast, *Vaccine, 14:* 549.

Macreadie, I. G., Horaitis, O., Verkuylen, A. J., and Savin, K. W. (1991a). Improved shuttle vectors for cloning and high level Cu^{2+}-mediated expression of foreign genes in yeast, *Gene, 104:* 107.

Macreadie, I. G., Horaitis, O., Vaughan, P. R., and Clark-Walker, G. D. (1991b). Constitutive expression of the *Saccharomyces cerevisiae CUP1* gene in *Kluyveromyces lactis, Yeast, 7:* 127.

Macreadie, I. G., Failla, P., Horaitis, O., and Azad, A. A. (1992). Production of HIV-1 Vpu with pYEULCBX, a convenient vector for the production of non-fused proteins in yeast, *Biotech. Lett., 14:* 639.

Macreadie, I. G., Ward, A. C., Failla, P., Grgacic, E., McPhee, D., and Azad, A. A. (1993). Expression of HIV-1 *nef* in yeast: The 27 kDa Nef protein is myristylated and fractionates with the nucleus, *Yeast, 9:* 565.

Mak, P., McDonnell, D. P., Weigel, N. L., Schrader, W. T., and O'Malley, B. W. (1989). Expression of functional chicken oviduct progesterone receptors in yeast (*Saccharomyces cerevisiae*), *J. Biol. Chem., 264:* 21613.

Martyn, J. C., Gould, A. R., and Eaton, B. T. (1991). High level expression of the major core protein VP7 and the non-structural protein NS3 of bluetongue virus in yeast: Use of expressed VP7 as a diagnostic, group-reactive antigen in a blocking ELISA, *Virus Res., 18:* 165.

Mason, K. E. (1979). A conspectus of research on copper metabolism and requirements of man, *J. Nutr., 109:* 1979.

Meaden, P. (1990). DNA fingerprinting of brewer's yeast: Current perspectives, *J. Inst. Brew., 96:* 195.

Mehra, R. K., Target, E. B., Gray, W. R., and Winge, D. R. (1988). Metal-specific synthesis of two metallothioneins and γ-glutamyl peptides in *Candida glabrata, Proc. Natl. Acad. Sci. USA, 85:* 8815.

Mehra, R. K., Garey, J. R., Butt, T. R., Gray, W. R., and Winge, D. R. (1989). *Candida glabrata* metallothioneins. Cloning and sequence of the genes and characterization of proteins, *J. Biol. Chem., 264:* 19747.

Mehra, R. K., Garey, J. R., and Winge, D. R. (1990). Selective and tandem amplification of a member of the metallothionein gene family in *Candida glabrata, J. Biol. Chem., 265:* 6369.

Mehra, R. K., Thorvaldsen, J. L., Macreadie, I. G., and Winge, D. R. (1992a). Cloning system for *Candida glabrata* using elements from the metallothionein-IIa-encoding gene that confer autonomous replication, *Gene, 113:* 119.

Mehra, R. K., Thorvaldsen, J. L., Macreadie, I. G., and Winge, D. R. (1992b). Disruption analysis of metallothionein-encoding genes in *Candida glabrata, Gene, 114:* 75.

Minagawa, T., Yanagishima, N., Arakatsu, Y., Nagasaki, S., and Ashida, J. (1951). The adaption of yeast to copper, II. Studies on the yeast nucleic acid fraction counteracting the inhibition of yeast growth by copper, *Botan. Mag., 64:* 65.

Minney, S. R., and Quirk, A. V. (1985). Growth and adaptation of *Saccharomyces cerevisiae* at different cadmium concentrations, *Microbios, 42:* 37.

Munger, K., Germann, U. A., and Lerch, K. (1985). Isolation and structural organzation of the *Neurospora crassa* copper metallothionein gene, *EMBO J. 4:* 2665.

Naiki, N., and Yamagata, S. (1981). Isolation and some properties of copper-binding protein in copper-resistant strains of yeast, *Plant Cell. Physiol., 17:* 1281.

Nakagawa, K. H., Inouye, C., Hedman, B., Karin, M., Tullius, T. D., Hodgson, K. O. (1991). Evidence from EXAFS for a copper cluster in the metalloregulatory protein CUP2 from yeast, *J. Am. Chem. Soc., 113:* 3621.

Narula, S. S., Mehra, R. K., Winge, D. R., and Armitage, I. M (1991). Establishment of the metal-to-cysteine connectivities in silver-substituted yeast metallothionein, *J. Am. Chem. Soc., 113:* 9354.

Narula, S. S., Winge, D. R., and Armitage, I. M. (1993). Copper and silver substituted yeast metallothionein: NMR sequential assignments, *J. Am. Chem. Soc., 32:* 6773.

Naumov, G. I., Naumova, E. S., Turakainen, H., and Korhola, M. (1992). A new family of polymorphic metallothionein-encoding genes *MTH1 (CUP1)* and *MTH2* in *Saccharomyces cerevisiae, Gene, 119:* 65.

Nielson, R. B., Atkin, C. L., and Winge, D. R. (1985). Distinct metal-binding configurations in metallothionein, *J. Biol. Chem., 260:* 5324.

Penttila, M. E., Suihko, M.-L., Lehtinen, U., Nikkola, M., and Knowles, J. K. C. (1987). Construction of brewer's yeasts secreting fungal endo-β-glucanase, *Curr. Genet., 12:* 413.

Pickering, I. J., George, G. N., Dameron, C. T., Kurz, B., Winge, D. R., and Dance, I. e. (1993). X-ray absorption spectroscopy of cuprous-thiolate multinuclear clusters, *J. Ann. Chem. Soc.,* in press.

Prinz, R., and Weser, U. (1975a). Cuprodoxin, *FEBS Lett., 54:* 224.

Prinz, R., and Weser, U. (1975b). Naturally occurring Cu-thionein in *Saccharomyces cerevisiae, J. Physiol. Chem., 356:* 767.

Rauser, W. E. (1990). Phytochelatin, *Ann. Rev. Biochem., 59:* 61.

Reese, R. N., and Winge, D. R. (1988). Sulfide stabilization of the cadmium-γ-glutamyl peptide complex of *Schizosaccharomyces pombe, J. Biol. Chem., 263:* 12832.

Resnick, M. A., Westmoreland, J., and Bloom, K. (1990). Heterogeneity and maintenance of centromere plasmid copy number in *Saccharomyces cerevisiae, Chromosoma, 99:* 281

Robbins, A. H. McRee, D. E., Williamson, M., Collett, S. A., Xuong, N. H., Furey, W. F., Wang, B. C., and Stout, C. D. (1991). Refined crystal structure of Cd,Zn metallothionein at 2.0 Å resolution, *J. Mol. Biol., 221:* 1269.

Scheinberg, I. H., and Sternlieb, I. (1976). *Trace Elements in Human Health and Disease* (A. S. Prasad and D. Oberlease, eds.), Vol. 1, Academic Press, New York, p. 415.

Schilling, B., Linden, K. M., Kupper, U., and Lerch, K. (1992). Expression of *Neurospora crassa* laccase under the control of the copper-inducible metallothionein-promoter, *Curr. Genet., 22:* 197.

Silar, P., and Wegnez, M. (1990). Expression of the *Drosophila melanogaster* metallothionein genes in yeast, *FEBS Lett., 269:* 273.

Silar, P., Butler, G., and Thiele, D. (1991). Heat shock transcription factor activates transcription of the yeast metallothionein gene, *Mol. Cell. Biol., 11:* 1232.

Sone, T., McDonnell, D. P., O'Malley, B. W., and Pike, J. W. (1990). Expression of human vitamin D receptor in *Saccharomyces cerevisiae.* Purification, properties, and generation of polyclonal antibodies, *J. Biol. Chem., 265:* 21997.

Szczypka, M. S., and Thiele, D. J. (1989). A cysteine-rich nuclear protein activates yeast metallothionein gene transcription, *Mol. Cell. Biol., 9:* 421.

Thiele, D. J. (1988). *ACE1* regulates expression of the *Sacharomyces cerevisiae* metallothionein gene, *Mol. Cell. Biol., 9:* 2745.

Thiele, D. J. (1992). Metal-regulated transcription in eukaryotes, *Nucl. Acids Res., 20:* 1183.

Thiele, D. J., and Hamer, D. H. (1986) Tandemly duplicated upstream control sequences mediate copper-induced transcription of the *Saccharomyces cerevisiae* copper-metallothionein gene, *Mol. Cell. Biol., 6:* 1158.

Thiele, D. J., Walling, M. J., and Hamer, D. H. (1986). Mammalian metallothionein is functional in yeast, *Science, 231:* 854.

Thorvaldsen, J. L., Sewell, A. K., McCowen, C., and Winge, D. R. (1993). Regulation of metallothionein genes by ACE1 and AMT1 transcription factors, *J. Biol. Chem., 268:* 12512.

Thrower, A. R., Byrd, J., Tarbet, E. B., Mehra, R. K., Hamer, D. H., and Winge, D. R. (1988). Effect of mutation of cysteinyl residues in yeast Cu-metallothionein, *J. Biol. Chem., 263:* 7037.

Tohoyama, H., Inouhe, M., Joho, M., and Murayama, T. (1990). Resistance to cadmium is under control of the *CAD2* gene in the yeast *Saccharomyces cerevisiae, Curr. Genet., 18:* 181.

Tohoyama, H., Tomoyasu, T., Inouhe, M., Joho, M., and Murayama, T. (1992a). The gene for cadmium metallothionein from a cadmium-resistant yeast appears to be identical to *CUP1* in a copper-resistant strain, *Curr. Genet., 21:* 275.

Tohoyama, H., Inagawa, A., Koike, H., Inouhe, M., Joho, M., and Murayama, T. (1992b). Constitutive transcription of the gene for metallothionein in a cadmium-resistant yeast, *FEMS Microbiol. Lett., 95:* 81.

Villanueba, K. D., Goossens, E., and Masschelein, C. A. (1990). Subthreshold vicinal diketone levels in lager brewing yeast fermentations by means of *ILV5* gene amplification, *J. Am. Soc. Brew. Chem., 48:* 111.

Welch, J. W., Fogel, S., Cathala, G., and Karin, M. (1983). Industrial yeasts display tandem gene reiteration at the *CUP1* region, *Mol. Cell. Biol., 3:* 1353.

Welch, J., Fogel, S., Buchman, C., and Karin, M. (1989). The *CUP2* gene product regulates the expression of the *CUP1* gene, coding for yeast metallothionein, *EMBO J., 8:* 255.

Weser, U., and Hartmann, H. J. (1988). Differently bound copper(I) in yeast Cu_8-thionein, *Biochem. Biophys. Acta, 953:* 1.

Weser, U., Hartmann, H.-J., Fretzdorff, A., and Strobel, G.-J. (1977). Homologous copper(I)-$(thiolate)_2$-chromophores in yeast copper thionein, *Biochim. Biophys. Acta, 493:* 465.

Winge, D. R., and Mehra, R. K. (1990). Host defenses against copper toxicity, *Int. Rev. Exp. Pathol., 31:* 47.

Winge, D. R., and Miklossy, K. A. (1982). Domain nature of metallothionein, *J. Biol. Chem., 257:* 3471.

Winge, D. R., Nielson, R. B., Gray, W. R., and Hamer, D. H. (1985). Yeast metallothionein: Sequence and metal binding properties, *J. Biol. Chem., 260:* 14464.

Winge, D. R., Dameron, C. T., and George, G. N. (1993). The metallothionein structural motif in gene expression, *Advances Inorg. Biochem., 10,* in press.

Wright, C. F., McKenney, K., Hamer, D. H., Byrd, J., and Winge, D. R. (1987). Structure and functional studies of the amino terminus of yeast metallothionein, *J. Biol. Chem., 262:* 12912.

Yanagishima, N., Minagawa, T., and Sasaki, T. (1949). Studies on the adaption of yeast to copper, I. Changes in cells and in colonies associated with the increase of copper resistance, *Physiol. Ecol., 3:* 79.

Yang, W. M., Gahl, W., and Hamer, D. (1991). Role of heat shock transcription factor in yeast metallothionein gene expression, *Mol. Cell. Biol., 11:* 3676.

Yu, W., Macreadie, I. G., and Winge, D. R. (1991). Protection against cadmium toxicity in yeast by alcohol dehydrogenase, *J. Inorg. Biochem., 44:* 155.

Zhou, P., and Thiele, D. J. (1991). Isolation of a metal-activated transcription factor gene from *Candida glabrata* by complementation in *Saccharomyces cerevisiae, Proc. Natl. Acad. Sci. USA, 88:* 6112.

Zhou, P., Szczypka, M. S., Sosinowski, T., and Thiele, D. J. (1992a). Expression of a yeast metallothionein gene family is activated by a single metalloregulatory transcription factor, *Mol. Cell. Biol., 12:* 3766.

Zhou, P., Szczypka, M., Sosinowski, T., and Thiele, D. J. (1992b). Metal-regulated gene transcription in the yeast *Candida glabrata, Yeast, 8:* S160.

11

Cadystin (Phytochelatin) in Fungi

Yukimasa Hayashi and Norihiro Mutoh
Institute for Developmental Research, Aichi Colony, Kasugai, Aichi, Japan

I. INTRODUCTION

Metallothioneins are a class of low molecular weight, cysteine-rich proteins binding heavy metal ions. They were found initially in horse kidney (Margoshes and Vallee, 1957) and subsequently in various higher organisms. The protective effects of the thioneins against acute effects of heavy metal toxicity on the activity of certain SH-containing enzymes have been observed. Upon exposure of eukaryotic organisms to heavy metals, such as Cd, Zn, and Cu, the metallothioneins are inductively synthesized and result in detoxification and storage of these heavy metals (Hamer, 1986; Shaw et al., 1992).

The synthesis of heavy metal–binding peptides is induced in fission yeast by the administration of heavy metal ions (Murasugi et al., 1981a). These peptides exist in the cytosol of yeast clls in complexes with heavy metal ions. The metal-peptide complexes are detected by the analysis of the cytosol using the gel filtration by Sephadex G50 column. Since the molecular weights of these complexes are rather small in comparison with metallothionein complexes, the metal-peptide complexes of the fission yeast are eluted in the middle position between the excluded proteins and the small molecular weight materials within the inner volume. Two kinds of metal-peptide complexes are observed in the cytosol of fission yeast upon cadmium ion induction. The larger Cd-peptide complex eluted earlier is designated as the cadmium-binding peptide 1 (Cd-BP1), and the smaller Cd-peptide complex eluted later is designated as the cadmium-binding peptide 2 (Cd-BP2). The absorbance ratio of 280/250 nm differs clearly between the two metal-peptide complexes, indicating the difference of the components. These were the initial observations of the metal-binding peptide complexes (Murasugi et al., 1981a, b).

The physicochemical properties of these Cd-peptide complexes were determined (Murasugi et al., 1981b), and acid-labile sulfur was found in Cd-BP1 but not in Cd-BP2 (Murasugi et al., 1983). It was also apparent that the accumulated amounts of Cd-BP allomorphs (Cd-BP1 and Cd-BP2) varied depending on the concentration of cadmium in the induction medium and the duration of the metal administration (Murasugi et al., 1984).

The chemical structures of cadystins were determined to be homologous to glutathione, as expressed in the formula $(\gamma\text{-Glu-Cys})_n$-Gly; this formula can be abbreviated as $(\gamma\text{-EC})_n$G. We have given the trivial name *cadystin* to these molecules (Kondo et al., 1983, 1984), since these peptides are induced by cadmium ions at a 10 times higher efficiency than by zinc, copper, and other heavy metal ions; the content of the cysteine residue in the peptides is quite high, indicating that the characteristic properties of these peptides are mostly derived from the cysteine residues of the peptides. Later, Grill et al. (1985) found that these peptides were also induced in many kinds of plants by the administration of heavy metals, and the name of *phytochelatin* was given to these peptides by them.

The accumulation of cadystin species $(\gamma\text{-EC})_n$G (n=2, 3 or 4) varies during induction, in that the larger species of cadystin is synthesized at the later stage of induction (Hayashi et al., 1988). The *in vitro* synthesis of cadystins has been studied by using the crude cell extract of fission yeast. The main pathway of biosynthesis involves transfer of γ-Glu-Cys from glutathione to another glutathione or a $(\gamma\text{-EC})_n$G homolog. In some cases the polymerization of γ-Glu-Cys occurs first, and then glycine is added to the C-terminal of poly(γ-Glu-Cys) (Hayashi et al., 1991). In Zenk's group, a dipeptidyl transpeptidase was purified from cultured plant cells and named *Silene cucubalus* (Grill et al., 1989; Loeffler et al., 1989).

Cadystin synthesis is also induced by stresses other than heavy metal administration, such as cell-wall wounds or the addition of antifungal agents or chitosan or polysaccharides to the incubation medium. In these cases, the uptake of zinc and copper from the medium is increased 2- to 3-fold over those of untreated cells, but these increased concentrations of heavy metals in the cell were not enough to explain the more than 20 times increase of cadystins in the treated cells (Hayashi et al., 1992).

The above-mentioned phenomena or properties of cadystins of fission yeast are mostly our findings. In the following discussion these phenomena will be presented in detail. In addition, the properties of the isolated mutants and genes of fission yeast concerning the synthesis of cadystins will be discussed.

II. INDUCTION OF CADYSTIN SYNTHESIS IN FISSION YEAST BY CD IONS

The fission yeast *Schizosaccharomyces pombe L972h*$^-$ was grown at 30°C in YPD (1% yeast extract, 2% polypeptone, and 2% dextrose), and at the early logarithmic phase of growth (3×10^6 cells/mL) 1mM cadmium chloride was added to the medium. The growth rate of the fission yeast decreased somewhat by the addition of $CdCl_2$, but growth continued to the stationary phase at 1.6×10^8 cells/mL. Cellular cadmium uptake during the culture was determined after washing the cells with distilled water and digestion with sulfuric acid by using an atomic absorption spectrophotometer (Perkin-Elmer 403 or Hitachi Z-7000). The chromatographic fraction and aqueous solution of Cd-BPs were analyzed without prior digestion. Cellular cadmium uptake increased with time to a plateau at about 5 mg/10^8 cells. During this period, the change of Cd ion concentration in the medium was less than a few percent in the case of 1 mM $CdCl_2$ induction. After appropriate incubation time a volume of the culture was harvested, washed twice with distilled water and once with 50 mM Tris-HCl (pH7.6) 0.1 M KCl (E-buffer), and stored frozen at −25°C until use. Frozen cells were homogenized with a mortar and pestle at 4°C with two to three times the weight of acid-washed quartz sand and extracted with a small volume of E-buffer. The extract was centrifuged at 15,000 g for 20 min at 4°C to remove the membrane fraction, nuclei and mitochondria, and the supernatant was analyzed on a Sephadex G50 SF (Pharmacia) column (1.6 × 65 cm) equilibrated with E-buffer. In the cytosol sample of 5-h induced cells, two Cd peaks in the low molecular weight region were detected by monitoring the absorbance at 250 nm, which originated from the mercaptide bonds with Cd ions (Cd-BP1 and Cd-BP2), as shown in Fig. 1. The molecular weights of Cd-BP1 and Cd-BP2 were approximately 4×10^3 and 1.8×10^3, respectively, by gel filtration at an ionic strength of 0.2 (E-buffer). With time after the induction, the amounts of Cd-BP1 gradually increased, and concomitantly the fractional amount of Cd-BP2 decreased (Figs. 1c,d and Fig. 2). Total Cd uptake in the total cell and Cd-BPs in the total cell extract increased steadily as long as cell growth continued, indicating accumulation of these peptides. The amount of Cd-BPs per cell increased for about 10 h under these conditions; then it remained constant

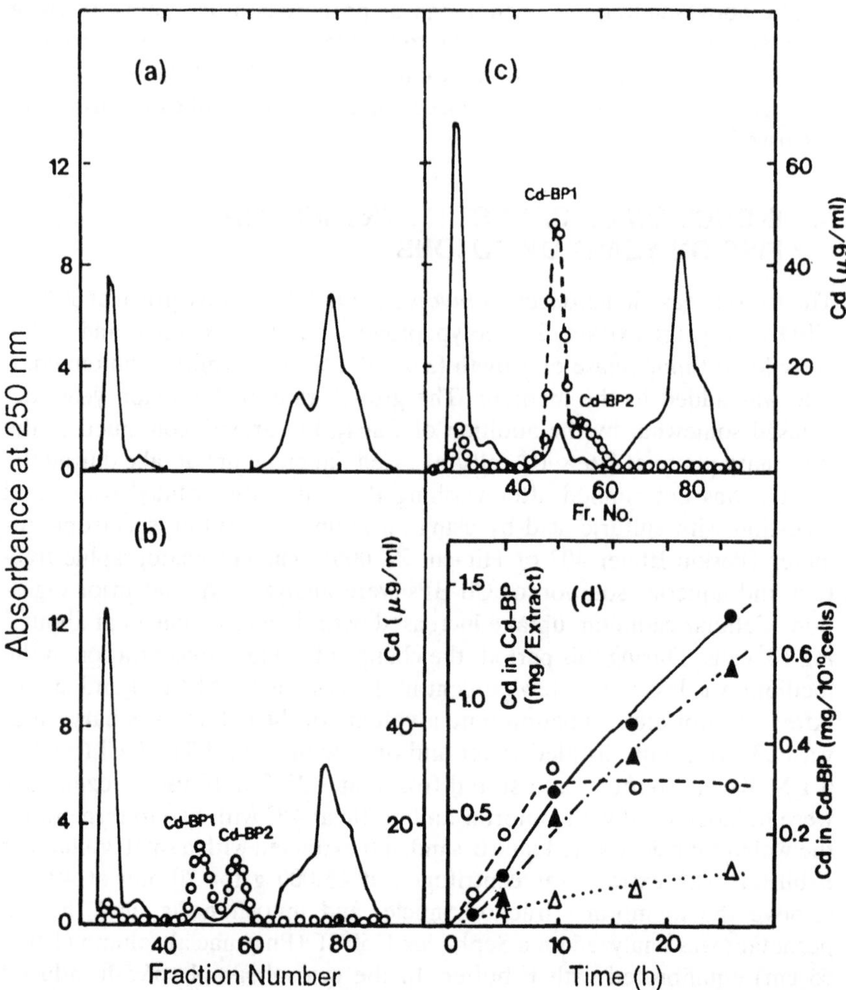

Figure 1 Induction of Cd-BPs analyzed by Sephadex G50 gel filtration. Each fraction contained 1.5 mL of eluate. Extracts were obtained from cells of 500-mL cultures (a) without Cd ions; (b) with Cd ions for 5 h; and (c) with Cd ions for 9.5 h. Absorbances at 250 nm (solid line) and Cd concentration (broken line with open circles) in each fraction were determined and expressed as normalized values per 10^{10} cells. Increases of Cd-BPs in extracts with time (d) were expressed as their Cd amounts. Cd-BPs in the total cell extract (closed circles); Cd-BPs per 10^{10} cells (open circles); Cd-BP1 in the total cell extract (closed triangles); Cd-BP2 in the total cell extract (open triangles).

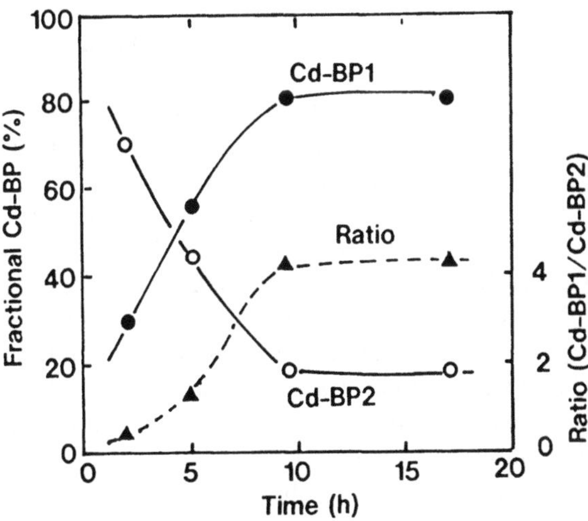

Figure 2 The ratio of Cd-BP1 to Cd-BP2 varies with time after Cd induction. Cd-BPs were induced in the cells on exposure to 1 mM $CdCl_2$. The molar fractional amounts of Cd-BP1 (closed circles) and Cd-BP2 (open circles), and the ratio of Cd-BP1 to Cd-BP2 (triangles), are plotted against the induction time.

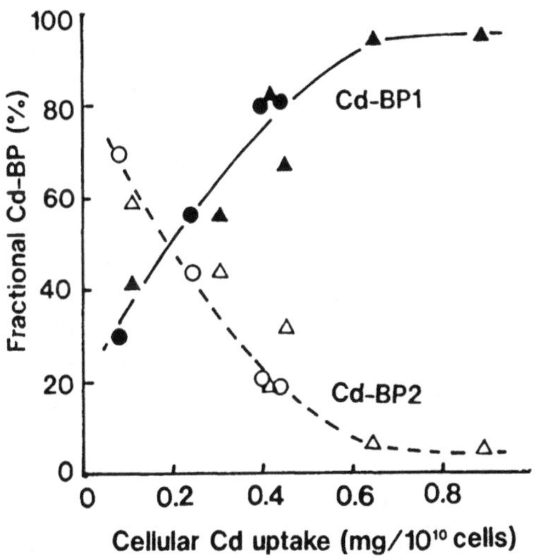

Figure 3 The effect of the cellular Cd uptake on the relative amounts of Cd-BP1 and Cd-BP2. The molar fractional amounts of Cd-BP1 (closed circles) and Cd-BP2 (open circles) from the time course experiment and the amounts of Cd-BP1 (closed triangles) and Cd-BP2 (open triangles), which are dependent on $CdCl_2$ concentration in the medium, are plotted against the cellular Cd uptake.

(Fig. 1d). This is consistent with the time to get the maximum Cd uptake per cell, implying cellular Cd uptake is allowed as long as cadystin synthesis continues (Fig. 1d). The amount of Cd-BP2 per cell increased at lower Cd concentrations in the medium and decreased over 1 mM Cd ion, indicating a complex relationship between Cd concentration in the medium and Cd-BP2 formation. The ratio of Cd-BP1 to Cd-BP2 increased with increasing Cd concentration up to 2 mM in the medium. When the relative amounts of Cd-BP1 and Cd-BP2 were plotted against the cellular Cd uptake, the fractional amount of the two Cd-BPs changed inversely as a function of cellular Cd uptake (Fig. 3), indicating that cellular Cd concentration plays a role in the regulation of the relative amount of each kind of Cd-BP.

The isolated Cd-peptides are purified by the column of DEAE-cellulose (DE-52, Whatman, Kent, England) or DEAE-Toyopearl 650 (Tosoh, Tokyo, Japan) by elution with a linear gradient of KCl in the buffer containing 20 mM $CdCl_2$. Cd was included in the eluting buffer of the ion exchange column to stabilize Cd-BPs, especially Cd-BP2.

III. COMPONENTS OF Cd-PEPTIDE COMPLEXES

When the purified Cd-BP1 is acidified, we are able to detect hydrogen sulfide by the characteristic smell. On the other hand, by the acidification of Cd-BP2, we cannot detect hydrogen sulfide. By the analyses for the contents of the labile sulfide in Cd-BP1 and in Cd-BP2, the presence of the labile sulfide is shown in Cd-BP1 but not in Cd-BP2 (Murasugi et al., 1983). The molar amount of the sulfide per mole of Cd-BP1 or cadmium differed among the respective Cd-BP1 samples. At first we estimated the amount of the labile sulfide in Cd-BP1 to be one mole per mole of Cd-BP1, since we had mixed Cd-BP1 samples obtained from cells cultured in the various conditions. Later, the heterogeneity of Cd-BP1 was noticed, and the content of the sulfide was reexamined on the respective Cd-BP1 species (Hayashi and Winge, 1992). The overall molar ratio of sulfide per Cd increased with the increase of the concentration of Cd in the medium and the increase of the time of incubation. It is in accordance with the previous observation that both Cd uptake per cell and sulfide production per cell are proportional to the Cd concentration in the medium (Fig. 4). By the analyses of the molar ratio of the sulfide in Cd-BP1 it was shown that the larger the molecular weight, the higher the molar ratio of sulfide per Cd in the complex. The peptide length variation was determined by the analyses of isolated Cd-BP1 in a DEAE-Toyopearl column, as shown in Fig. 5. Some properties of Cd-BP1 species concerning its heterogeneity are shown in Table 1. A gradient of components varying in cadystin $(\gamma EC)_nG$ compo-

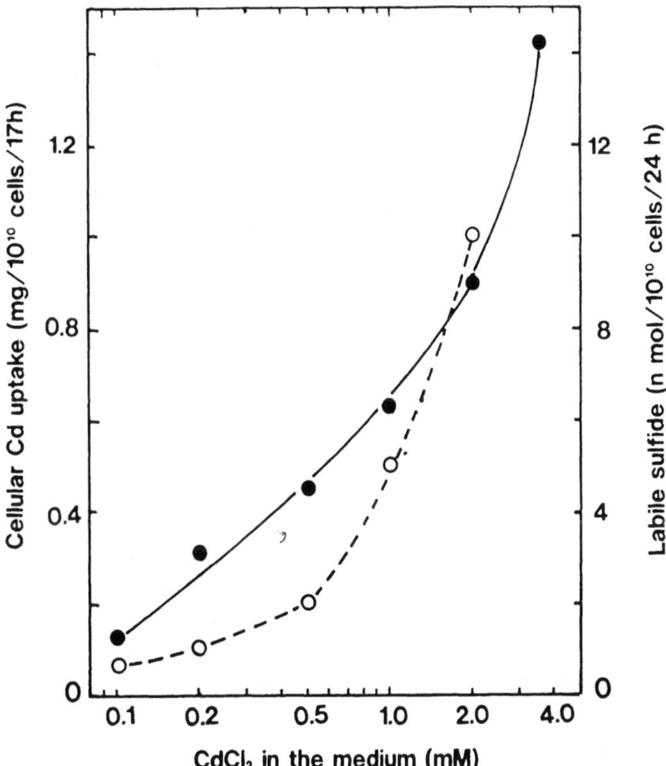

Figure 4 Cellular Cd uptake and labile sulfide production are correlated with $CdCl_2$ concentration in the medium. Cellular Cd uptake in 17 h (closed circles) and labile sulfide production in 24 h (open circles) were determined and plotted against $CdCl_2$ concentration in the medium.

Table 1 Some Properties of the Isolated Cd-BP1 spe

Cd-BP1 species[a]	Labile S:Cd	Cd:Cys	n=3/n=2[b]	Mr
0.13 M	0.10	0.74	0.58	3,500
0.16 M	0.16	0.88	0.74	3,600
0.18 M	0.22	0.97	1.21	5,400
0.20 M	0.51	0.87	1.75	6,000
0.22 M	0.62	1.09	2.00	6,400
0.25 M	0.58	1.38	2.86	6,000

[a]Cd-BP1 species were isolated by DEAE-Toyopearl 650S column chromatography.
[b]The ratio of the amounts between n=2 and 3 in cadystin $(\gamma EC)_nG$.

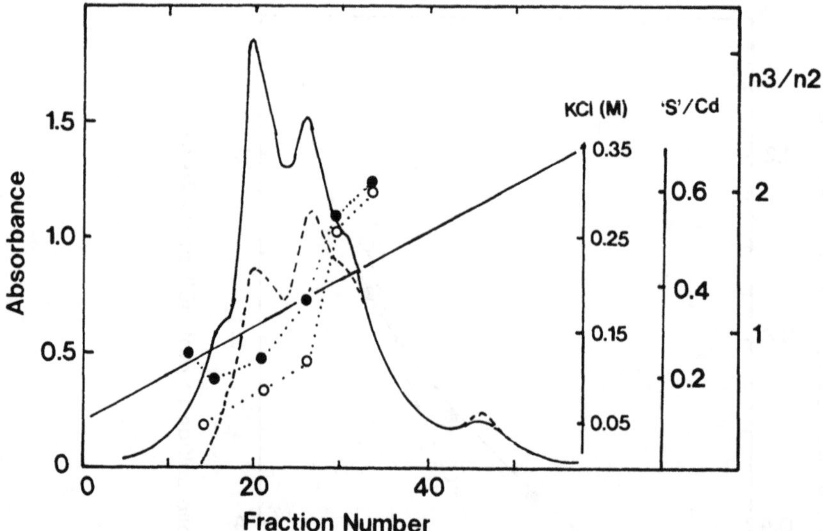

Figure 5 **Heterogeniety of Cd-BP1 was analyzed by anionic exchanger. The Cd-BP1 fraction from Sephadex G50 SF was chromatographed on DEAE-Toyopearl 650S, and absorbances at 250 nm (solid line) and 280 nm (broken line) were determined. The molar ratios of $(\gamma EC)_3G$:$(\gamma EC)_2G$ (closed circles) and sulfide:cadmium (open circles) were determined.**

nents was observed. The n3/n2 ratio of cadystin $(\gamma EC)_nG$ increased with the anionic nature of the complexes. In the high salt eluting complexes the ratio exceeded 2. The heterogeneity in sulfide content in Cd-BP1 was also apparent by ultraviolet absorption spectroscopy (Fig. 6). Fractions across the elution profile of a DEAE-Toyopearl column exhibited distinct absorption spectra. The near ultraviolet optical transition varied with the sulfide per Cd ratio of fractions; as the sulfide per Cd ratio increased, the transition became more red-shifted. The physical basis of the optical properties relates to the quantum nature of the CdS crystallite that forms in these cadystin $(\gamma EC)_nG$ complexes (Dameron et al., 1989).

The ratio of Cd to cysteinyl thiolate concentrations increased in direct proportion to the sulfide per Cd ratio, indicating some kind of direct interaction of the labile sulfide with Cd, as suggested previously (Murasugi et al., 1983).

Cd-BP2 was also resolved by a DEAE-Toyopearl 650 column into multiple components, as shown in Fig. 7. The UV absorption spectra of the elution fractions revealed Cd complexes with characteristic thiolate to Cd charge transfer transitions (Fig. 8). These components were devoid of sul-

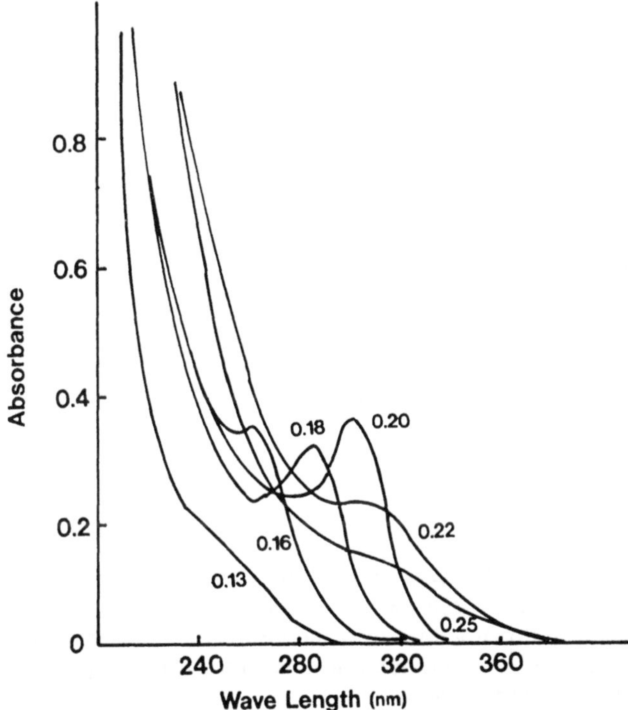

Figure 6 Ultraviolet absorbance spectra of Cd-BP1 fractions. Spectra in 25 mM Tris-HC1 (pH 7.5), 10 μM $CdCl_2$ are shown for elution fractions from DEAE-Toyopearl 650S, and the numbers by the curves indicate the salt concentration of the elution fractions.

fide ions and did not show transitions between 270 and 320 nm due to CdS. Analysis of the elution fractions by HPLC indicated that cadystin $(\gamma EC)_2G$ was the predominant peptide component in fractions eluting below 0.14 M salt, and cadystin $(\gamma EC)_3G$ was the most abundant peptide species in fractions eluting above 0.16 M salt. Elution fractions of Cd-BP2 were pooled into six fractions for subsequent rechromatography in the DEAE-Toyopearl column. Homogeneous cadystin $(\gamma EC)nG$ species in Cd-BP2 were observed after repeated rechromatography. Complexes eluting at low salt concentrations contained exclusively $(\gamma EC)_2G$ peptides, and those eluting above 0.18 M salt had only $(\gamma EC)_3G$ peptides. The Cd content varied for the different complexes; the Cd stoichiometry in cadystin $(\gamma EC)_2G$ complexes was 1.25 mol equivalent. The stoichiometry in Cd:$(\gamma EC)_3G$ complexes was around 2 mol equivalent or more. The lability of the sulfide-free complexes in the salt

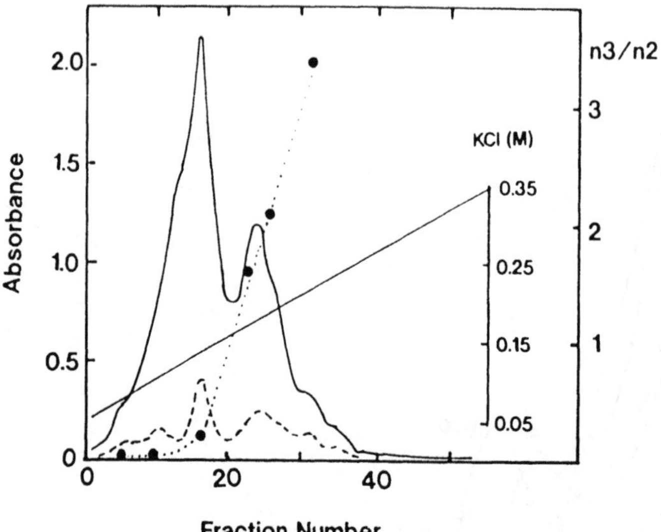

Figure 7 Heterogeniety of Cd-BP2 was analyzed by anionic exchanger. The Cd-BP2 fraction from Sephadex G50 SF was chromatographed on DEAE-Toyopearl 650S, and the elution fractions were monitored for absorbances at 250 nm (solid line) and 280 nm (broken line). The mole ratio of $(\gamma EC)_3G$:$(\gamma EC)_2G$ was determined for the fractions shown.

solution makes cadmium stoichiometry of the complexes inaccurate and may somehow induce the rearrangement of complexes. This lability of Cd-BP2 also made difficult the determination of the molecular weight by gel filtration at the ionic strength of 0.2.

IV. INDUCTION OF CADYSTIN SYNTHESIS BY OTHER HEAVY METALS

Metallothioneins in mammals are induced by other heavy metals than Cd, such as Zn, Cu, Ag, Hg, Co, and Pb. Cadystin synthesis is also induced by heavy metals other than Cd, though the induced amounts are usually less than 10% of that by Cd. In Table 2, the induced amounts of cadystins by various heavy metals are shown. The synthesized cadystins exist as complexes with the administered heavy metals. Examples of the analysis by gel filtration of the cytosol are depicted in Fig. 9 (Hayashi et al., 1986). With 2.5 mM $CuCl_2$ in YPD medium, a copper-cadystin complex is observed in the cytosol of fission yeast. By Sephadex G50 gel filtration, the Cu-cadystin

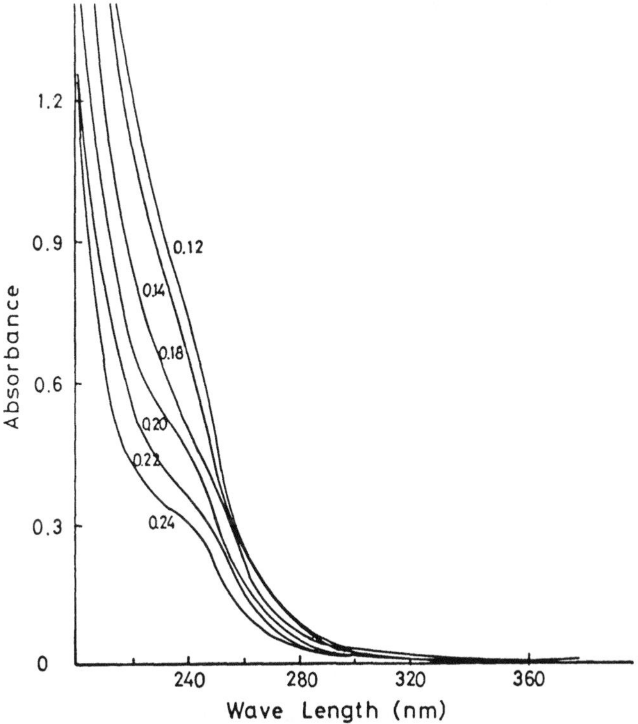

Figure 8 Ultraviolet absorbance spectra of Cd-BP2 elution fractions. Spectra in 25 mM Tris-HC1 (pH 7.5), 10 μM $CdCl_2$ are shown for elution fractions from DEAE-Toyopearl 650S, and the numbers by the curves indicate the salt concentration of the elution fractions.

complex eluted in a volume corresponding to Mr 3000, which is an intermediate molecular weight between two Cd-BPs (Fig. 9). Induction of cadystin synthesis by Zn ions was also enhanced by the simultaneous administration of aluminum ions as shown in Table 2 and in Fig. 10. By analysis of gel filtration, zinc peaks were observed in the small molecular regions centered at Mr 1200 and 800. The higher molecular weight peak (fr. 30) consisted of zinc and cadystins $(\gamma EC)_nG$, n=2 and 3, and the lower molecular weight peak (fr. 33) consisted of zinc and glutathione. The Al peak coincident with the Zn peak may indicate the cooperative complex formation by both metals. In any event, the presence of Al or coinduction with Al increased the inductive synthesis of cadystins. The addition of Al ion to the medium somewhat induced cadystin synthesis by itself, but the significance of alumi-

Table 2 Induction of Cadystin Synthesis by Various Metals

Metal	(mM)	Time[a] (h)	GSH[b]	Cadystin[b]
None		8	105.3	3.0
—		20	80.2	4.6
$CdCl_2$	0.02	6	156.9	160.4
—	0.10	6	152.5	200.1
—	0.50	6	131.7	164.3
—	1.00	6	120.7	185.5
$CuSO_4$	2.00	20	95.3	11.0
$ZnCl_2$	2.00	8	179.4	16.5
—	5.00	8	131.8	18.6
—	10.00	8	122.2	36.5
$ZnCl_2$ 2.00 + $Al_2(SO_4)_3$	1.00	8	165.7	52.0
— 2.00 + —	2.00	8	123.6	36.2
— 2.00 + —	3.00	8	116.8	25.9

[a]Cultures of the fission yeast at 30°C in YPD started at the early log-phase (3.1×10^6 cells/mL).
[b]SH nmol/mg protein.

num ions in the medium was the cell-wall hardening of the fission yeast. This may lead to the change of the cell membrane architecture and could induce or stimulate cadystin synthesis.

The UV spectra of the Cu-cadystin complex and the zinc-cadystin complexes are shown in Figs. 11 and 12. From the comparison of the UV absorption spectrum of the native complex with that of Cu(I)- or Cu(II)-replaced complexes, the native Cu-cadystin complex is composed of Cu(I), as shown in Fig. 11. The UV spectrum of the complex induced by zinc and aluminum is the same as those of zinc-cadystin complexes, suggesting that the metals of formed complexes were Zn (Fig. 12). The zinc-cadystin complexes with different cadystin species are resolved into the zinc-cadystin complex by the DEAE-Toyopearl 650M column by KCl gradient elution. The zinc-cadystin complex shows a homologous spectrum as shown in Fig. 12. The labile sulfide is not detected in these complexes with Cu or Zn, as far as we have analyzed.

V. INDUCTION OF CADYSTIN SYNTHESIS OTHER THAN BY HEAVY METALS

Since various stresses to introduce tissue injury and inflammation induce metallothionein synthesis in mammals, the effect of wounds to the cell sur-

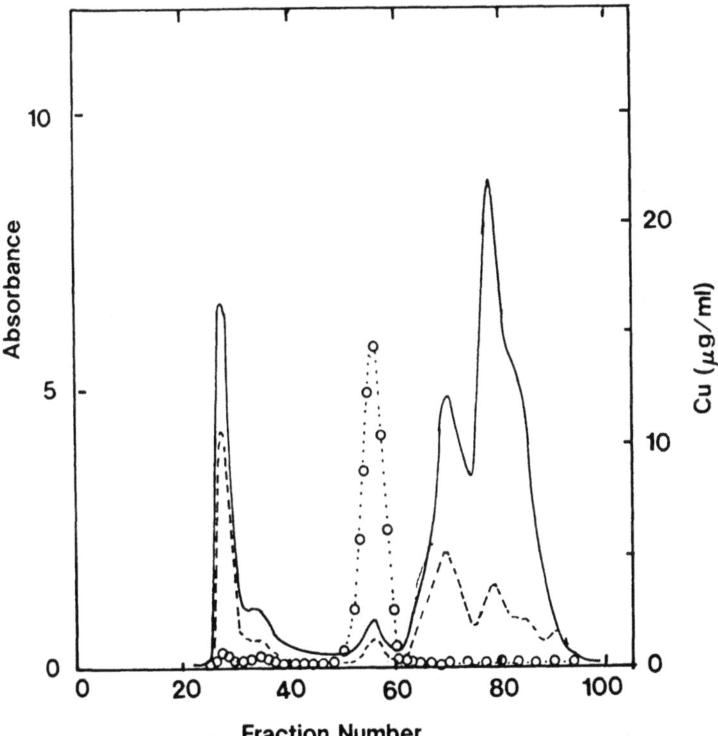

Figure 9 Induction of the Cu-cadystin complex. $CuCl_2$ was added at 2.5 mM to the logarithmic phase culture of the fission yeast in YPD medium, and culture continued for 17 h. One third of the cell extract from a 500-mL culture was applied to a Sephadex G50 SF column (1.6 × 54 cm). Absorbances at 250 nm (solid line) and 280 nm (broken line), and Cu concentration (open circles) in each fraction (1.5 mL) were determined. (Hayashi et al., 1986.)

face of fission yeast was examined on the induction of cadystin synthesis (Hayashi et al., 1992). To introduce wounds to the cell wall and/or to the membrane of the fission yeast, acid-washed glass beads (0.5 mm diameter) were added to the flask of rotation culture. After various incubation times, cadystin amounts in the cytosol were determined by HPLC analysis (Fig. 13). After 12 h incubation with the glass beads, the induced cadystin amount was about 15 nmol/mg protein with 10 mg/mL glass beads. On the other hand, chitosan is a deacetylated product of chitin and has the activity of changing the cell surface structure to let the fission yeast be permeable to high molecular weight compounds. Addition of chitosan to the culture medium of fission yeast also induced cadystin synthesis (Table 2 and Fig. 14). It

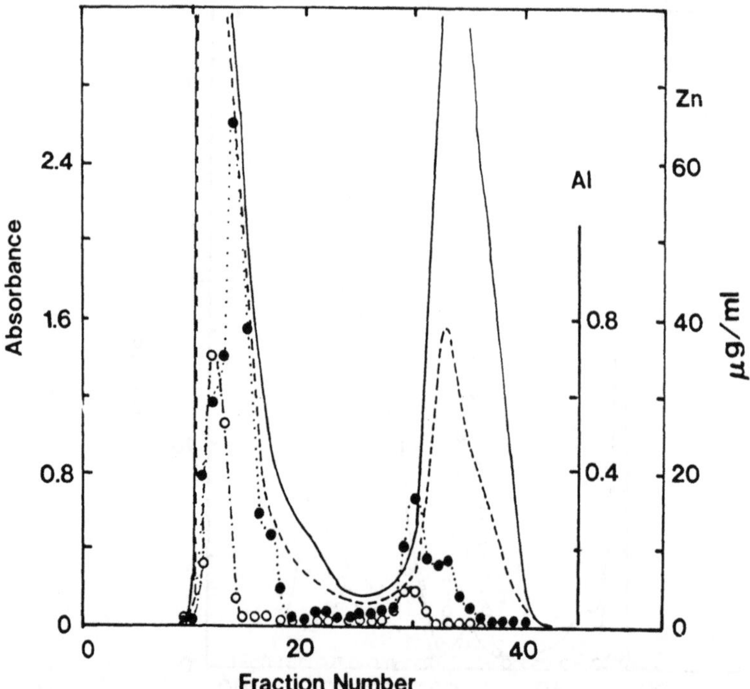

Figure 10 Induction of metal-cadystin complexes by the simultaneous exposure to zinc and aluminum ions. $ZnCl_2$ and $Al_2(SO_4)_3$ were added at 2 mM and 1 mM respectively to 500-mL culture of the fission yeast in YPD medium (A_{660} = 0.5), and incubation continued for 10 h. Absorbances at 250 nm (solid line) and at 280 nm (broken line), and Zn (closed circles), Al (open circles) were determined in each fraction (4 mL/fraction).

has been suggested by these facts that a certain signal transduction pathway from the cell surface may relate to the induction of cadystin synthesis. A highly hydrophobic compound, tetramethylthiuram disulfide (TMTD), and its reduced form, dimethyldithiocarbamate (DMDTC), are antifungal reagents. These compounds are also excellent inducers of cadystin synthesis in fission yeast (Mutoh et al., 1991), suggesting that some alteration of the cell surface structure is a positive induction signal for caydstin synthesis.

VI. CHEMICAL STRUCTURE OF CADYSTINS

The metal-cadystin complexes from the Sephadex G50 SF column fractions are purified by an anion-exchange column (DE52; Whatman, or DEAE-Toyopearl 650M; Tosoh). The purified metal-cadystin complex is concen-

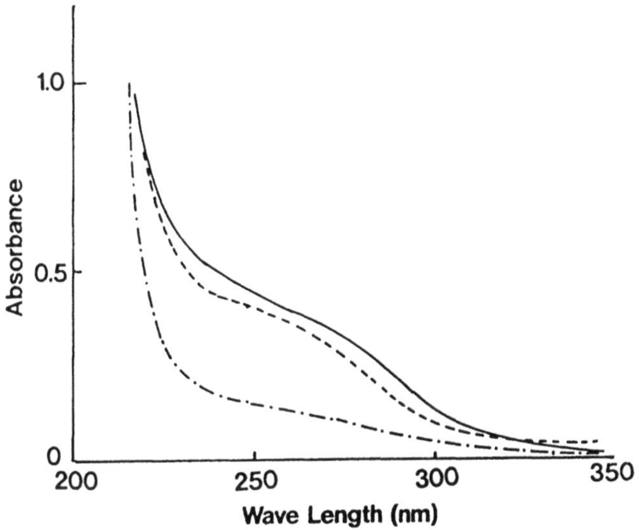

Figure 11 UV absorption spectra of copper-cadystin complexes. UV spectra were recorded for the native Cu-cadystin complex (solid line), Cu(I)-replaced complex (broken line), and Cu(II)-replaced complex (broken line with dots). Cu-replaced complexes were prepared by incubation of Cd-cadystin complex with an excess amount of Cu(I) or Cu(II) ion, and the replaced complexes were isolated by the use of Bio-Gel P2 or Sephadex G10 column. Cadystin amounts in complexes were about 100 μg/mL in 5 mM Tris-HCl (pH 7.5), 10 mM KCl.

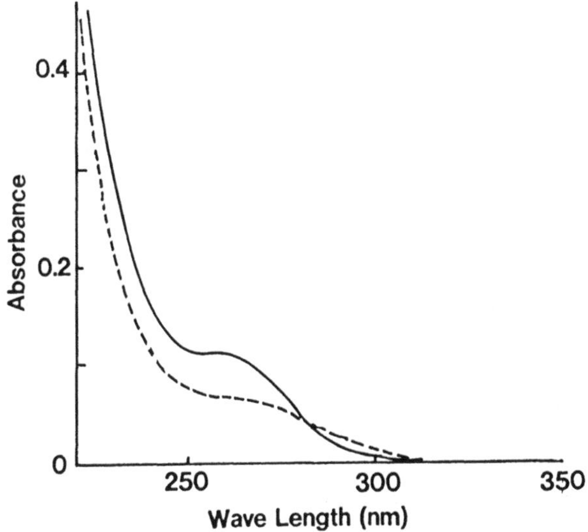

Figure 12 UV absorption spectra of zinc-cadystin complexes. UV spectra were recorded for Zn-cadystin $(\gamma EC)_3G$ complex (solid line) and for Zn-cadystin $(\gamma EC)_2G$ complex (broken line) in 25 mM Tris-HCl (pH 7.5), 0.1 M KCl. The amounts of cadystin $(\gamma EC)_2G$ and $(\gamma EC)_3G$ were 10.2 and 17.3 nmol/mL, respectively.

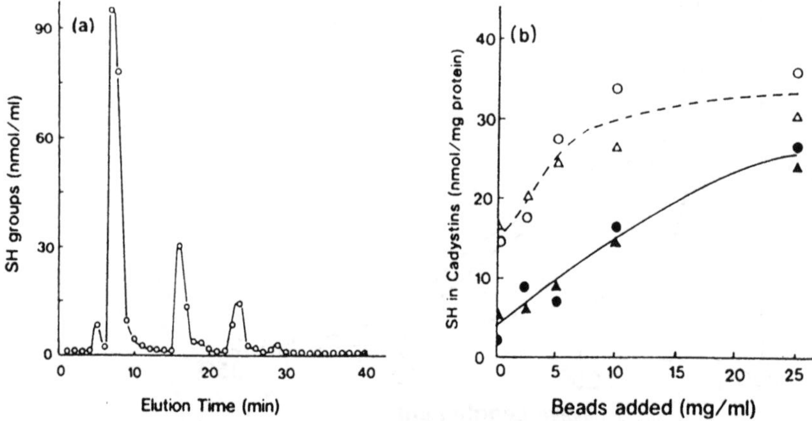

Figure 13 Induction of cadystin synthesis in the fission yeast by adding glass beads to the medium. The fission yeast growing at the log-phase was added with the various amounts of glass beads and incubated in a rotary shaker for 12 h. (a) The cell extract was analyzed for the content of cadystins by HPLC as described previously. (b) Cadystin synthesis as a function of added amounts of glass beads, without addition of zinc ion (closed circles and triangles) or with simultaneous addition of the zinc ion at 2 mM (open circles and triangles). Circles and triangles indicate the independent series of experiments.

trated by ultrafiltration using SPECTRA/POR membrane MW-cutoff 1000 (Spectrum, Los Angeles, CA). At first, the peptide component in the purified Cd-peptides complex was assumed to be homogeneous from the symmetrical elution pattern of the complex in gel filtration. Later, the peptide components were determined by reverse phase HPLC, and it was observed that the metal complexes were heterogeneous in the peptide component. In the analyses of the peptide component, the metal complex is acidified by the addition of trifluoroacetic acid (TFA) to 5%, centrifuged at 15,000 g, filtered through Chromatodisc 13A, 0.45 μm pore size (Kurabou, Tokyo), and applied to an HPLC column (TSK-ODS80T; Tosoh, Tokyo). Cadystin species are eluted by a linear gradient of 0–20% acetonitrile in 0.05% TFA in 40 min. Depending on the column size, 1- or 2-mL fractions are collected, and the content of SH groups is determined in an aliquot of each fraction to ascertain the position of the respective cadystin species and to estimate amounts of cadystins. Purified cadystin species are lyophilized and oxidized by performic acid and digested in 5.7 N HCl 0.02% phenol for 24 h at 120°C. Samples are lyophilized again to remove HCl completely and coupled to isothiocyanate (Tarr, 1981; Heinrikson and Meredith, 1984). Phenylthiocar-

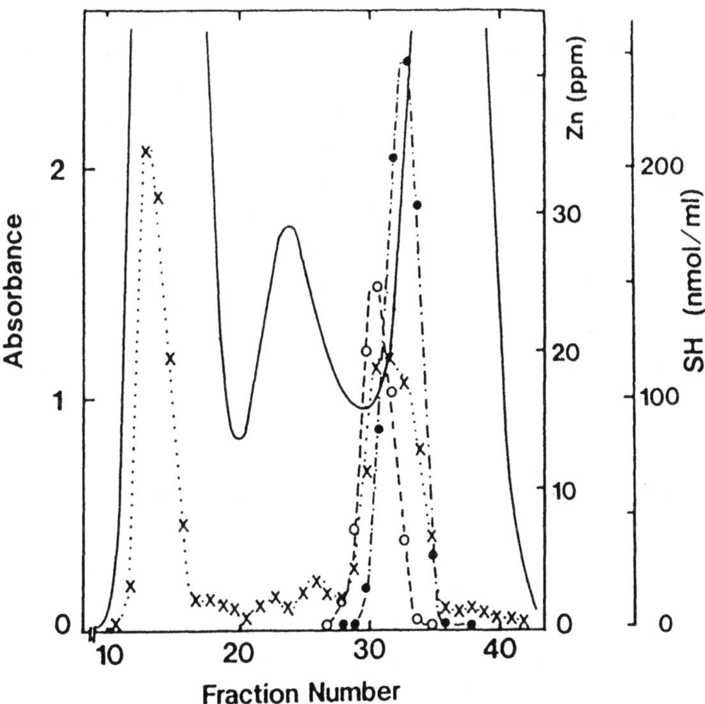

Figure 14 Analysis of the metal cadystin complexes. Cell extract obtained from the fission yeast treated for 15 h by hydrolyzed chitosan (0.05 mg/mL) was applied to sephadex G50 SF column. Four-mL fractions were collected, absorbance at 250 nm (solid line) was read, Zn content (dotted line with x) of each fraction was determined in an atomic absorption spectrophotometer, and an aliquot of each fraction was analyzed by HPLC for the content of GSH (closed circles) and cadystins (open circles).

bamyl-amino acids are analyzed in a TSK-ODS80T column by a linear gradient of A solution [0.14 M sodium acetate (pH 6.3), 0.05% triethylamine/acetonitrile (940:60)] and B solution [acetonitrile/H_2O (60:40)] in 50 min from 100% A to 100% B. The results of quantitative amino acid analyses give equimolar amounts of glutamic acid and cysteinic acid and the respective amount of glycine according to cadystin species.

To determine the chemical structure of cadystins (Kondo et al., 1983, 1984), the respective cadystin species is desulfurized with Raney nickel W-2 at 50°C for 12 h in 50 mM Tris-HCl (pH 7.6). The solvent is evaporated *in vacuo,* and the catalyst is dissolved in a small volume of 6 N HCl below 30°C; then the mixture is lyophilyzed. The residue is desalted in Bio-Gel P2 column (Bio-Rad, Richmond, CA) with 3 M acetic acid. Dethiocadystin thus obtained is separated by HPLC (ODS column) as previously described. Amino acid analysis of dethiocadystins shows equimolar amounts of glutamate and alanine and the respective amount of glycine depending on cadystin species. The smallest dethiocadystin (n=2) is digested with carboxypeptidase P for 96 h at 35°C in 0.1 M sodium acetate (pH 5.2). It yields glycine and two molar equivalents of dipeptides containing glutamate and alanine in 1:1 ratio. The dried dipeptide is esterified in 10 mL of a mixture of methanol, trimethyl orthophosphate, and thionyl chloride (80:20:5 w/w) for 4 h at 40°C and dried *in vacuo.* The residue is reduced in 10 mL of 0.3 M LiBH4 in tetrahydrofolic acid (THF) for 12 h at 80°C in a sealed tube. After evaporation to dryness, the residue is hydrolyzed with 50 mL of 6 N HCl for 12 h at 105°C. The hydrolysate gives 4-amino-5 hydroxyvaleric acid and 2-amino propanol in nearly quantitative yield without formation of alanine or 2-amino-5-hydroxyvaleric acid. Thus the sequence of the dipeptide should be γ-Glu-Ala. The deduced structure of this smallest dethio-cadystin is $(\gamma\text{-Glu-Ala})_2$Gly. The structure of other cadystin species is similarly determined. Then the general structure of cadystins is deduced to be $(\text{H-}\gamma\text{-Glu-Cys})_n$-Gly-OH. Absolute configuration of all the constituent amino acids was assigned to be L through circular dichroism (CD) spectroscopy. CD spectra of natural and synthetic (all L) cadystins were measured in the presence of Cd ion (II) in 5 mM Tris-HCl (pH 7.6) to find the same curves with identical signs (Kondo et al., 1984).

VII. BIOSYNTHESIS OF CADYSTIN

At the first finding in 1981, we have confirmed *de novo* synthesis of peptide by pulse-labelling with [^{3}H]cystine in synthetic medium (Murasugi et al., 1981a). In that experiment, we have already noticed that the peptides are soluble in 10% trichloroacetic acid, suggesting that the lengths of the peptides are short. After a while, the chemical structures of the peptides were determined in our group (Kondo et al., 1983, 1984). From the chemical structure having γ-glutamic acid and the short lengths of peptides, it has been assumed that synthesis does not involve ribosomal synthesis but rather an enzymatic pathway in the biosynthesis of cadystins *in vivo.*

We have found several Cd-hypersensitive mutants of fission yeast that cannot grow in the presence of 0.5 mM $CdCl_2$ in the medium (Mutoh and Hayashi, 1988). The analyses of these mutants revealed the following facts. (1) Absence of glutathione due to the deletion in either one of the two enzymes in the biosynthesis of glutathione, that is, γ-Glu-Cys synthetase (EC 6.3.2.2) or glutathione synthetase (EC 6.3.2.3), blocks the biosynthesis of cadystins $(\gamma EC)_nG$. (2) There is a mutant that is deficient in cadystins but has a normal content of glutathione. These facts indicate that both enzymes in glutathione synthesis are associated with the biosynthesis of cadystins or that glutathione itself relates to cadystin biosynthesis, and that another enzymatic step is necessary for cadystin biosynthesis in addition to two enzymatic steps in glutathione biosynthesis. We can think of several pathways for the enzymatic synthesis of the cadystin $(\gamma EC)_nG$ as follows.

Scheme 1. Elongation of $(\gamma EC)_nG$ with γEC addition

(1-a) γEC condensation with $(\gamma EC)_nG$

$$(\gamma EC)_nG + \gamma EC \rightarrow (\gamma EC)_{n+1}G$$

(1-b) $(\gamma EC)_nG$ condensation and glycine released in a certain case (i=n)

$$(\gamma EC)_mG + (\gamma EC)_nG \rightarrow (\gamma EC)_{m+i}G + (\gamma EC)_{n-i}G$$

Scheme 2. Having the step of γEC polymerization

Step 1.

(2-1a) γEC polymerization from γEC

$$n\ \gamma EC \rightarrow (\gamma EC)_n$$

(2-1b) γEC polymerization by the transfer of γEC from $(\gamma EC)_nG$

$$(\gamma EC)_m + (\gamma EC)_nG \rightarrow (\gamma EC)_{m+1} + (\gamma EC)_{n-1}G$$

Step 2.

(2-2a) Glycine addition to $(\gamma EC)_n$

$$(\gamma EC)_n + G \rightarrow (\gamma EC)_nG$$

(2-2b) Glutathione addition to $(\gamma EC)_n$

$$(\gamma EC)_n + \gamma ECG \rightarrow (\gamma EC)_nG + \gamma EC \text{ or } (\gamma EC)_{n+1}G$$

Here we describe studies for the determination of the pathway of cadystin biosynthesis, using the cell-free system of fission yeast (Hayashi et al., 1991).

The cell extract was prepared from the fission yeast *S. pombe* L972 h^-, grown to the late logarithmic phase in YPD medium. The cell extract was centrifuged at 120,000 g to remove the ribosome fraction, the protein was precipitated by 75% saturation of ammonium sulfate, and the resulting protein was desalted by Sephadex G25 column filtration. This crude enzyme preparation could be stored in 50% glycerol at −75°C and was used without further purification in the study of the cadystin biosynthesis pathway. The incubation mixture to synthesize cadystin $(\gamma EC)_nG$ or poly-γ-glutamyl-cysteine $(\gamma EC)_n$ in a cell-free system contained 50 mM Tri-acetate (pH 7.9), 10 mM Mg acetate, 1 mM dithiothreitol, 1 mM ATP, 0.25 mM GTP, ATP generating system, and radioisotope-labeled or nonlabeled substrates, such as glycine, $(\gamma$-Glu-Cys$)_n$, $(\gamma EC)_n$ (n=1–3), glutathione, or cadystin, $(\gamma EC)_nG$ (n=1–4). The reaction at 30°C was started by addition of the enzyme. After the reaction, TFA was added to 5% and the mixture was chilled at 0°C for 10 min. Protein was removed by centrifugation, and the supernatant was applied to a reverse-phase HPLC column (0.46 × 25 cm TSK-ODS 80T or 1.0 × 25 cm Beckman Ultrasphere-ODS). Peptides were eluted by a linear gradient of 0–20% acetonitrile in 0.05% TFA for 40 min at a flow rate of 1 or 2 mL/min. Fractions were collected (1 min/fraction), and radioactivity and thiol group content were determined (Ellman, 1959). The amino acid analyses of synthesized peptides were performed by HPLC after the digestion in 6 N HCl at 120°C for 20 h and coupling with phenyl isothiocyanate as described previously.

Cadystin synthesis by the dipeptide transfer from a glutathione molecule to another glutathione or to a cadystin molecule was detected by the use of $[^{35}S]$ or [glycine-^{3}H]glutathione as shown in Figs. 15 and 16. The time course study (Fig. 16) indicated that cadystin $(\gamma EC)_2G$ synthesis was most abundant with the substrate glutathione, though the other cadystin species were also synthesized. This meant that the γEC moiety from glutathione (γECG) was transferred to the synthesized cadystin $(\gamma EC)_2G$ or $(\gamma EC)_3G$.

The polymerization of γEC was also detected by the transfer of γEC moiety from $[^{35}S]$glutathione to $(\gamma EC)_n$ (Fig. 17). In the glutathione synthetase deficient mutant (*S. pombe* MN101), we observed the accumulation of great amounts of γEC. At the same time, $(\gamma EC)_n$ of n=2 or more were also observed in the cell extract. Although the amounts of the respective species were far less than γEC, the presence of $(\gamma EC)_n$ could be detected easily by culture of the cells with $[^3H]$cysteine or by thiol group detection in HPLC fractions. Even in the cell extract of the wild-type *S. pombe,* small amounts of $(\gamma EC)_n$ could be observed by thiol group detection after HPLC fractionation. With the use of $[^3H]\gamma EC$, *in vitro* polymerization to $(\gamma EC)_n$ was also observed (Fig. 18a). Addition of glycine to the carboxyl terminal of $(\gamma EC)_n$ yielded cadystins $(\gamma EC)_nG$ by the reaction of glutathione synthetase (Fig. 18b and 19). The predominant reaction in wild-type cells exposed to heavy metal may be the γEC transfer from glutathione or cadystins to other glutathione or cadystin molecules, though the polymerization of γEC is also performed at lesser degrees. This polymerization of γEC may also be catalyzed by the same enzyme in the dipeptide transfer from glutathione or cadystin, although the energy requirement is different from that of

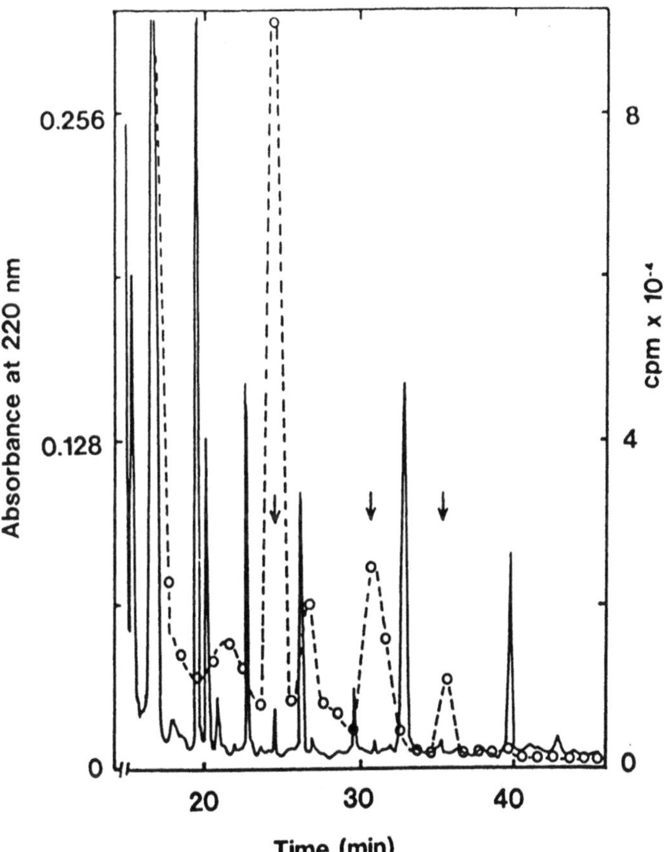

Figure 15 HPLC analysis of cadystin synthesis in the cell-free system of the fission yeast. Cold GSH was added to 1 mM together with 1 μCi [^{35}S]GSH/100 μL SRM. Incubation was for 2 h at 30°C with AS75. The HPLC column was Beckman ODS-Ultrasphere (1.0 $\times$ 25 cm). Fractions (1 min/fraction) were collected at the flow rate of 2 mL/min, reading absorbance at 220 nm (solid line). One-tenth volume of each fraction was used for the determination of radioactivity (broken line with open circles), and the rest of the fraction was used for SH determination and amino acid analyses. The percentages of radioactivity in the identified products to the input radioactivity were 1.30 for $(\gamma EC)_2G$, 0.23 for $(\gamma EC)_2$, 0.48 for $(\gamma EC)_3G$, and 0.13 for $(\gamma EC)_4G$. Arrows indicate the positions of marker cadystins $(\gamma EC)_2G$, $(\gamma EC)_3G$, and $(\gamma EC)_4G$, from the left.

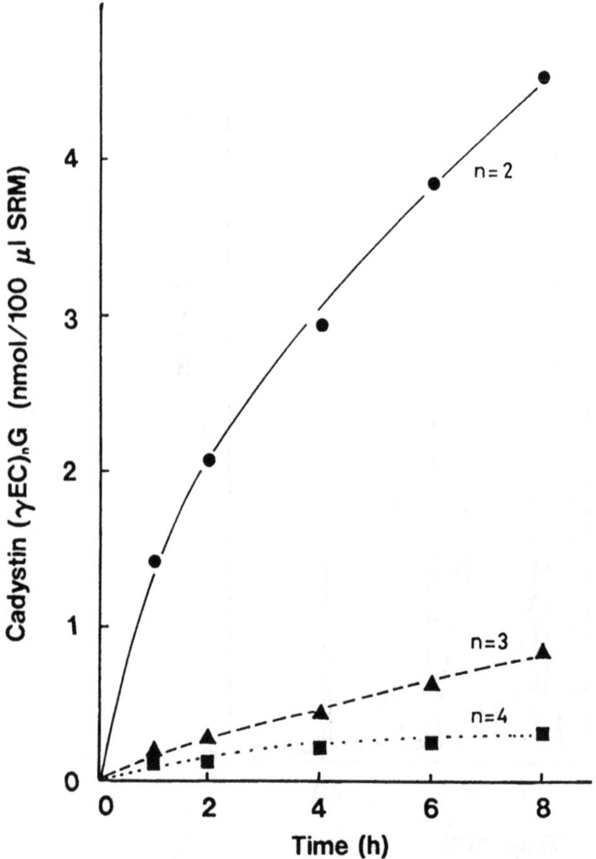

Figure 16 **Time-course of cadystin ($(\gamma EC)_nG$) synthesis with [glycine-^{3}H]GSH. [Glycine-^{3}H]GSH was used at 2 μCi/100 μL SRM together with cold 1 nM GSH. Synthesized molar amounts of the respective cadystin $(\gamma EC)_nG$ were expressed according to n of $(\gamma EC)_nG$ in 100 μL SRM.**

dipeptide transfer. In the transfer of dipeptides from glutathione energy is not required, while in γEC polymerization ATP is required. In both reactions, heavy metal ions, such as Cd, Zn, etc., are not required, but the presence of Mg ions greatly stimulates the reaction of the dipeptide transfer. This nonrequirement of heavy metal ions in the reaction is a clear difference from the transpeptidase of *Silene cucubalus* (Grill et al., 1989; Loeffler et al., 1989). Recently, the dipeptidyl transpeptidase from fission yeast was purified in our laboratory and showed the same properties with the crude enzyme preparation used previously (unpublished data).

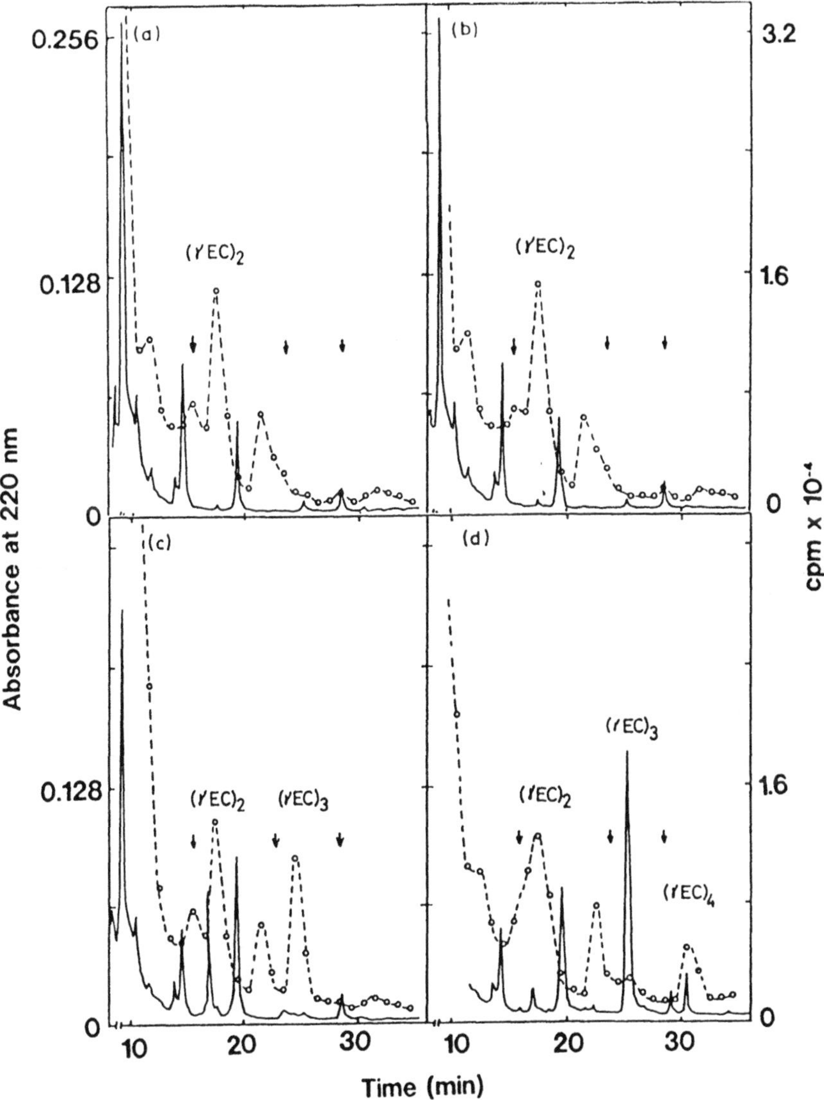

Figure 17 Polymerization of γEC from the substrate of [^{35}S]GSH in a cell-free system with AS75. Incubation was for 2 h at 30°C with 1.2 μCi [^{35}S]GSH / 1.2 nmol in 100 μL SRM, and the incubation mixture was applied to a column of TSK-ODS 80T (0.46 × 25 cm). Fractions (1 min/fraction) were collected at the flow rate of 1 mL/min. (a) With [^{35}S]GSH only. (b) With [^{35}S]GSH and 1 mM γEC. (c) With [^{35}S]GSH and 1 mM $(\gamma EC)_2$. (d) With [^{35}S]GSH and 1 mM $(\gamma EC)_3$. Arrows and the dotted line with open circles indicate the same as in Fig. 15.

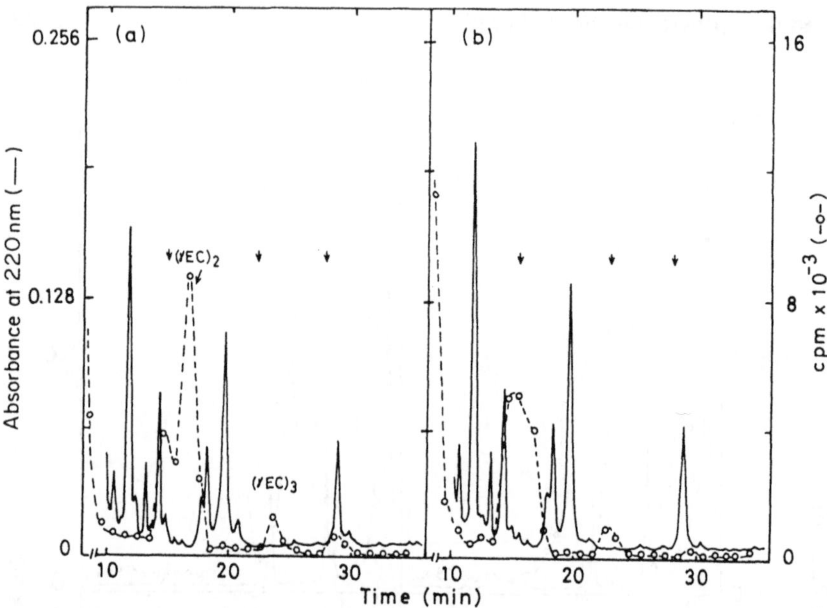

Figure 18 **Polymerization of γEC in a cell-free system with γE[^{3}H]. Incubation was 4 h and the product was analyzed by HPLC as in Fig. 17. (a) With [^{3}H]γEC (1.39 × 105 cpm/2.8 nmol in 100 μL SRM); (b) With [^{3}H]γEC and 1 mM glycine.**

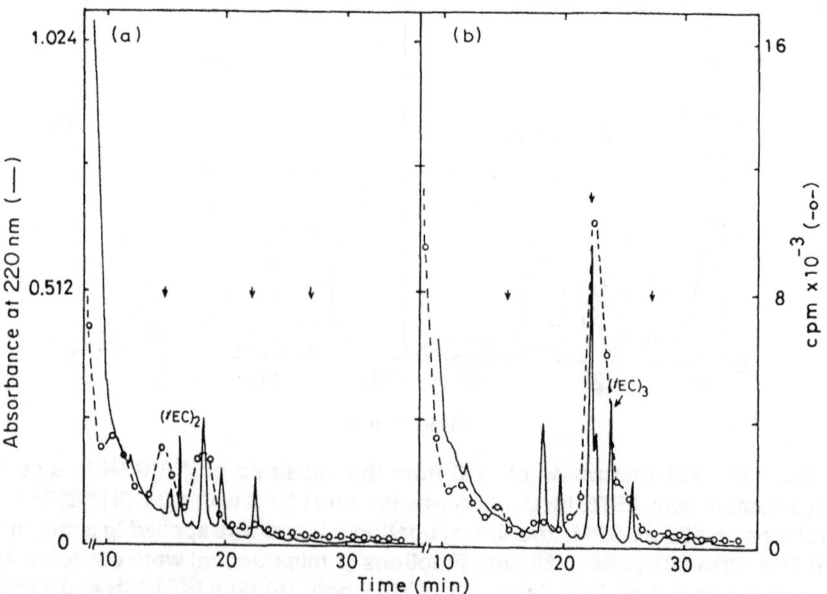

Figure 19 **[^{3}H]glycine incorporation into cadystin $(\gamma EC)_nG$ in a cell-free system with partially purified GSH synthetase. (a) With 2 μCi [^{3}H]glycine and 1 mM $(\gamma EC)_2$ in 100 μL SRM; (b) With [^{3}H]glycine and 1 mM $(\gamma EC)_3$. The other conditions were the same as with Fig. 18.**

VIII. THE PRESUMED FUNCTION IN THE CELL

The functional significance of metallothionein (MT) remains a topic of discussion 30 years after its discovery. The sequestration of nonessential and toxic metal is one role. In addition, MT may have a role in the homeostasis of physiologically essential metals such as Zn and Cu. As a homeostatic mediator, MT could distribute metal ions in the biosynthesis and activation of Zn- and Cu-containing metalloproteins. Conversely, when Zn and Cu accumulate intracellularly, the ions can be sequestered in a chemically inactive form by binding to newly synthesized apoMT. The biological role of MT, other than the roles mentioned above, is also suggested in that MT is induced by various forms of chemical, physical, and biological stress (Bremner, 1987; Kägi and Schäffer, 1988). The biological stresses include hormones (glucocorticoids, progesterone, estrogen, etc.), interferon, starvation, inflammation, and endotoxin. The direct effect of these biological stresses may relate to the disturbances of Zn metabolism that occur in stressed animals, namely increased hepatic uptake and decreased metal level in plasma. However, it is still unclear how MT production, which resembles an acute phase response in many ways, confers advantage on an animal subjected to the stress under conditions of exposure to electrophilic agents, such as oxigen, free radicals, and alkylating agents. The production of MT could provide the neutralizing nucleophilic equivalents. The free radical scavenger activity certainly relates to glutathione concentration in the cell. The level of glutathione decreases with the conjugation of glutathione to the alkylating agent or its metabolites or scavenging free radicals produced by x-ray irradiation.

The function of cadystin is analogous to that of MT and even more directly relates to glutathione. Cadystins are induced on exposure to heavy metals and bind the metals, detoxifying them (Murasugi, 1982). The exposure to chitosan or wounds in the cell wall influence the permeability of the cell membrane, and the release of electrolytes was observed before the inductive synthesis of cadystins. These treatments may correspond to the stresses in MT induction and may also be related to the production of free radicals, which should be scavenged by cadystins in turn. The other possible function of cadystins is to activate the influx of Zn, which may be used for the activation of Zn-requiring enzymes, as the result of the defense reaction. However, glutathione-deficient mutants and cadystin-deficient mutants of fission yeast are still alive with normal growth rate in the absence of heavy metal loading. Therefore, glutathione and cadystins in fission yeast are not essential for growth, though cadystin minus mutants is hypersensitive on exposure to hydrogen peroxide or UV irradiation as well as on exposure to the Cd ion. The recent finding that the cadystin-deficient mutant is also hypersensitive to starvation, while the glutathione-deficient mutant is not, may give us a new clue in considering the biological function of cadystins.

REFERENCES

Bremner, I. (1987). Nutritional and physiological significance of metallothionein, *Experientia Suppl.*, *52:* 81.

Dameron, C. T., Reese, R. N., Mehra, R. K., Kortan, A. R., Carroll, P. J., Steigerwald, M. L., Brus, L. E., and Winge, D. R. (1989). Biosynthesis of cadmium sulfide quantum semiconductor crystallites, *Nature*, *388:* 596.

Ellman, G. L. (1959). Tissue sulfhydryl groups, *Arch. Biochem. Biophys.*, *82:* 70.

Grill, E., Winnacker, E.-L., and Zenk, M. H. (1985). Phytochelatins: The principal heavy-metal complexing peptides of higher plants, *Science*, *230:* 674.

Grill, E., Löffler, S., Winnacker, E.-L., and Zenk, M. H. (1989). Phytochelatins, the heavy-metalbinding peptides of plants, are synthesized from glutathione by a specific γ-glutamylcysteine dipeptidyl transpeptidase (phytochelatin syntase), *Proc. Natl. Acad. Sci. USA*, *86:* 6838.

Hamer, D. H. (1986). Metallothionein, *Ann. Rev. Biochem.*, *55:* 913.

Hayashi, Y., and Winge, D. R. (1992). $(\gamma\text{-EC})_n$G peptides, *Metallothioneins: Synthesis, Structure and Properties of Metallothioneins, Phytochelatins and Metal-Thiolate Complexes* (M. J. Stillman, C. F. Shaw III, and K. T. Suzuki, eds.), VCH, New York, p. 271.

Hayashi, Y., Nakagawa, C. W., and Murasugi, A. (1986). Unique properties of Cd-binding peptides induced in fission yeast, *Schizosaccharomyces pombe, Environ. Health Perspect.*, *65:* 13.

Hayashi, Y., Nakagawa, C. W., Mutoh, N., Isobe, M., and Goto, T. (1991). Two pathways in the biosynthesis of cadystins $(\gamma\text{EC})_n$G in the cell-free system of the fission yeast, *Biochem. Cell Biol.*, *69:* 115.

Hayashi, Y., Nakagawa, C. W., Uyakul, D., Imai, K., Isobe, M., and Goto, T. (1988). The change of cadystin components in Cd-binding peptides from the fission yeast during their induction by cadmium, *Biochem. Cell Biol.*, *66:*288.

Hayashi, Y., Morikawa, S., Kawabata, M., and Hotta, Y. (1992). The synthesis of cadystins, heavy metal chelating peptides, is induced in the fission yeast by wounds of the cell wall or by incubation with chitosan, *Biochem. Biophys. Res. Comm.*, *188:* 388.

Heinrikson, R. L., and Meredith, S. C. (1984). Amino acid analysis by reverse-phase high-performance liquid chromatography: Precolumn derivatization with phenylisothiocyanate, *Anal. Biochem.*, *136:* 65.

Kägi, J. H. R., and Schäffer, A. (1988). Biochemistry of metallothionein, *Biochem.*, *27:* 8509.

Kondo, N., Isobe, M., Imai, K., Goto, T., Murasugi, A., and Hayashi, Y. (1983). Structure of cadystin, the unit-peptide of cadmium-binding peptides induced in a fission yeast, *Schizosaccharomyces pombe, Tetrahedron Lett.*, *24:* 925.

Kondo, N., Imai, K., Isobe, M., Goto, T., Murasugi, A., Nakagawa, C. W., and Hayashi, Y. (1984). Cadystin A and B, major unit peptides comprising cadmium binding peptides induced in a fission yeast—Separation, revision of structure and synthesis, *Tetrahedron Lett., 25:* 3869.

Löffler, S., Hochberger, A., Grill, E., Winnacker, E.-L., and Zenk, M. H. (1989). Termination of the phytochelatin synthase reaction through sequestration of heavy metals by the reaction product, *FEBS Lett., 258:* 42.

Margoshes, M., and Vallee, B. L. (1957). A cadmium protein from equine kidney cortex, *J. Amer. Chem. Soc., 79:* 4813.

Murasugi, A. (1982). Cd-binding peptides induced in fission yeast: Synthesis and properties, Ph.D. thesis, Nagoya University, Nagoya, Japan.

Murasugi, A., Wada, C., and Hayashi, Y. (1981a). Cadmium-binding peptide induced in fission yeast, *Schizosaccharomyces pombe, J. Biochem., 90:* 1561.

Murasugi, A., Wada, C., and Hayashi, Y. (1981b). Purification and unique properties in UV and CD spectra of Cd-binding peptide 1 from *Schizosaccharomyces pombe, Biochem. Biophys. Res. Comm., 103:* 1021.

Murasugi, A., Wada, C., and Hayashi, Y. (1983). Occurrence of acid-labile sulfide in cadmium-binding peptide 1 from fission yeast, *J. Biochem., 93:* 661.

Murasugi, A., Nakagawa, W. C., and Hayashi, Y. (1984). Formation of cadmium-binding peptide allomorphs in fission yeast, *J. Biochem., 96:* 1375.

Mutoh, N., and Hayashi, Y. (1988). Isolation of mutants of *Schizosaccharomyces pombe* unable to synthesize cadystin, small cadmium-binding peptides, *Biochem. Biophys. Res. Comm., 151:* 32.

Mutoh, N., Kawabata, M., and Hayashi, Y. (1991). Tetramethylthiuram disulfide or dimethyldithiocarbamate induces the synthesis of cadystins, heavy metal chelating peptides, in *Schizosaccharomyces pombe, Biochem. Biophys. Res. Comm., 176:* 1068.

Shaw, C. F., III, Stillman, M. J., and Suzuki, T. S. (1992). Metallothioneins: An overview of metal-thiolate complex formation in metallothioneins, *Metallothioneins: Synthesis, Structure and Properties of Metallothioneins, Phytochelatins and Metal-Thiolate Complexes* (M. J. Stillman, C. F. Shaw III, and K. T. Suzuki, eds.), VCH, New York, p. 1.

Tarr, G. E. (1981). Rapid separation of amino acid phenylthiohydantoins by isocratic high-performance liquid chromatography, *Anal. Biochem., 111:* 27.

12

Molecular Genetic Analysis of Cadmium Tolerance in *Schizosaccharomyces pombe*

David W. Ow, Daniel F. Ortiz, David M. Speiser, and Kent F. McCue
United States Department of Agriculture and University of California—Berkeley, Albany, California

I. INTRODUCTION

One cellular response to toxic levels of metals is the induced synthesis of intracellular chelators. To that effect, animals and certain fungi produce small cysteine-rich proteins through the transcriptional activation of metallothionein genes (for reviews, see Hamer, 1986; Winge, this volume). In contrast, plants and some fungi respond by the metal-activated enzymatic synthesis of peptides derived from glutathione (for reviews, see Tomsett and Thurman, 1988; Rauser, 1990; Robinson, 1990; Steffens, 1990; Hayashi, this volume). These peptides have the general structure of $(\gamma\text{Glu-Cys})_n$-Gly (Kondo et al., 1984; Grill et al., 1985), where the γGlu-Cys repeating unit extends up to 11 (Grill et al., 1987). In members of the *Fabales* thus far examined, these peptides contain β-alanine in place of glycine, consistent with the same carboxyl-end residue found in homoglutathione (Grill et al., 1986). First discovered in *Schizosaccharomyces pombe* (Murasugi et al., 1981), a variety of names have been used for these peptides: cadystins (Kondo et al., 1983), phytochelatins (Grill et al., 1985), poly(γ-glutamylcysteinyl)glycine (Jackson et al., 1987), γ-glutamyl peptides (Reese et al., 1988), and $(\gamma\text{EC})_n\text{G}$ (Delhaize et al., 1989). Although the chemical structure $(\gamma\text{EC})_n\text{G}$ is a most accurate description, the ubiquity of these peptides in the plant kingdom (Gekeler et al., 1989) has popularized the term phytochelatins proposed by Zenk and colleagues; as in several recent reviews, we will use the term phytochelatins (PCs).

In plants, enzymatic synthesis of PCs from glutathione by PC synthase requires metal cofactors (Grill et al., 1989). Subsequent chelation of the metal ions by the newly synthesized peptides deactivates enzyme activity (Loeffler et al., 1989). In the fission yeast *S. pombe,* Hayashi et al. (1991) found an additional route of PC synthesis *in vitro.* This alternative pathway consists of the polymerization of γGlu-Cys and glutathione to form a $(\gamma\text{Glu-Cys})_n$ polymer followed by the carboxyl-terminal addition of glycine. Since this pathway shows no regulation by metal ions, it is not yet clear what its physiological role might be.

Induction of PC synthesis has been reported with a wide variety of metal cations (Ag^+, Au^+, Cd^{2+}, Cu^{2+}, Hg^{2+}, Ni^{2+}, Pb^{2+}, Sn^{2+}, Zn^{2+}, Bi^{3+}, Sb^{3+}, Te^{4+}, W^{6+}) as well as some multiatomic anions (SeO_3^{2-} and AsO_4^{3-}) (Grill et al., 1987). Metal binding, however, has been shown for only a few of the metals (such as Cd^{2+} and Cu^{2+}, reviewed by Rauser, 1990). Based on the ability of Hg^{2+} and Pb^{2+} to displace bound Cd^{2+} ions from PCs, it has been suggested that PCs can also bind these metals (Abrahamson et al., 1992). With regard to the chelation of Cd, two types of PC complexes separable by gel filtration chromatography have been described from *S. pombe* and *Candida glabrata* (Murasugi et al., 1983; Mehra et al., 1988). The low molecular weight (LMW) complex consists primarily of PCs and Cd, whereas a higher molecular weight (HMW) complex contains a high proportion of acid labile sulfide (Murasugi et al., 1983). The PC-Cd-S^{2-} chelate has greater stability and higher Cd binding capacity than the PC-Cd complex and can be formed *in vitro* upon addition of sulfide to the LMW PC-Cd complex (Reese and Winge, 1988). The structure of the PC-Cd-S^{2-} complex has been proposed to consist of a CdS crystallite core with quantum semiconductor characteristics and an outer layer of PC peptides (Dameron et al., 1989).

A physiological effect of the sulfide-containing complex is its apparent ability to confer greater tolerance to Cd. Mutants of *S. pombe* failing to form the HMW complex show reduced tolerance to Cd (Mutoh and Hayashi, 1988; Ortiz et al., 1992; Speiser et al., 1992b). In plants, it is not clear whether sulfide accumulation is necessarily associated with the Cd-induced response. However, the presence of acid labile sulfide in the PC-Cd complex has been reported in several plants including *Silene vulgaris* (Verkleij et al., 1990), *Lycopersicum esculentum* (Reese et al., 1992), and *Brassica juncea* (Speiser et al., 1992a). In the case of the selenium-tolerant cultivar of *B. juncea,* the high proportion of HMW complex produced during Cd stress suggests that there might be an association between greater metal tolerance and production of the PC-Cd-S^{2-} complex.

The interest of this laboratory lies in the molecular genetics of metal detoxification, with the long-term goal of modifying plants. A lower accumulation of toxic metals in consumable tissues would be desirable for reducing dietary intake. Conversely, the specific engineering of high-level uptake of metals in noncrop plants might prove useful for phytoremediation of metal-contaminated soil and water systems. Although a few higher plants could be attractive models for metal research, none are comparable to the unicellular fission yeast as an organism for molecular genetic analysis. Anticipating that, at the cellular level, *S. pombe* can serve as a model system in defining the genes involved in metal detoxification, we initiated a search for Cd-sensitive fission yeast mutants. In this chapter, we describe findings derived from the investigation of two such mutants.

II. GENE ISOLATION

Our experimental approach began with identifying mutants that failed to grow in the presence of Cd. To exclude mutants that acquired a genetic defect nonspecific to the PC-mediated response, Cd-sensitive strains were assayed for the formation of PC-Cd complexes by gel filtration chromatography. Over 50 lines derived from progenitor strain Sp223 (*ade6*$^-$, *leu1*$^-$, *ura4*$^-$) were analyzed by Sephadex G-50 chromatography. Not surprisingly, most Cd-sensitive mutants did not produce a Cd-binding profile substantially different from that of Sp223. Two mutants, LK69 and LK100, however, exhibited a significant reduction in the accumulation of the HMW PC-Cd-S^{2-} complex. Figure 1 shows a typical Cd elution profile obtained from wild type and mutant cell extracts. A genomic library was used in transformation experiments to restore a Cd-tolerant phenotype to these mutants. This effort resulted in the isolation of pGS1 and pGS3 as the plasmid clones that complement LK69 and LK100, respectively. A set of deletion derivatives from each plasmid clone was generated by the subcloning of DNA fragments produced by restriction endonuclease or exonuclease treatment.

With pGS1, a 2-kilobase fragment was assigned as the minimal complementing domain. To insure that this fragment was not an extragenic suppressor of Cd sensitivity, gene disruption of the wild type allele was performed. A construct was made substituting a portion of the defined complementing domain of pGS1 with a selectable marker. Upon transfer of the construct to the progenitor Sp223 genome via homologous recombination, the newly created mutant strain, DS1, was examined for sensitivity to Cd. Indeed, DS1 was sensitive to Cd and showed reduced accumulation of the PC-Cd-S^{2-} complex. This showed that the defined complementing domain corresponded to the mutant locus of LK69.

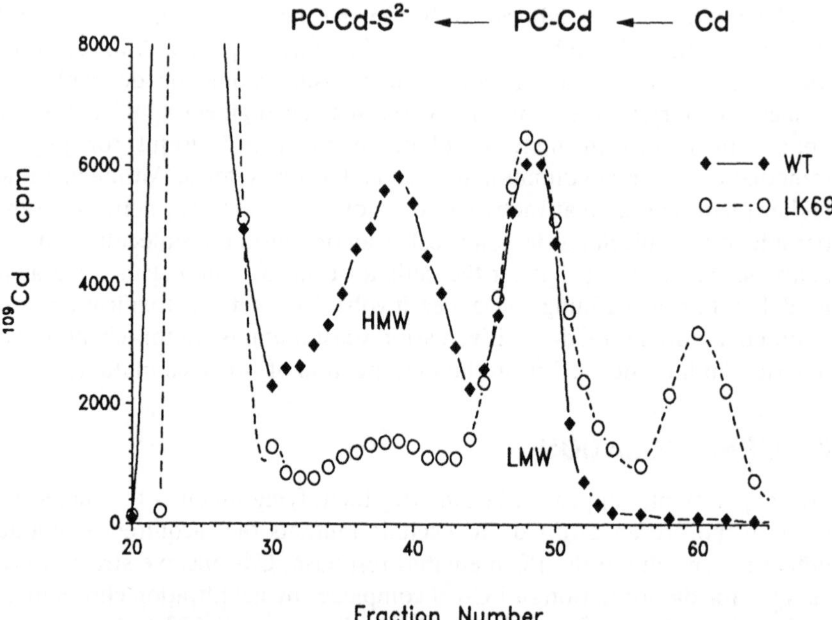

Figure 1 Sephadex G-50 chromatography of Cd-binding species from cell extracts of wild type and a mutant strain, LK69, that produces the PC-Cd complex but fails to accumulate a wild type level of the HMW PC-Cd-S^{2-} complex. Cd peaks, labeled by inclusion of ^{109}Cd, are, from right to left, free Cd, the LMW and HMW PC complexes, and Cd bound nonspecifically to macromolecules in the extracts.

In the case of pGS3, a similar analysis delimited the complementing region to a 3.5 kilobase fragment. Gene disruption of this region, however, was unsuccessful despite several attempts. Although the gene disruption construct was observed to integrate into the corresponding locus in the Sp223 genome, a wild type copy of the locus was observed in every case, as if gene duplication had occurred. To examine the possibility that the target gene might be indispensable, a diploid strain could have been used for gene disruption followed by assessing the viability of sporulating tetrads. However, stable diploids of *S. pombe* that could sporulate are not commonly available. In the absence of a more amenable alternative, we chose to clone the corresponding fragment of DNA from the mutant strain. Indeed, the corresponding DNA derived from LK100 was incapable of restoring Cd tolerance to LK100. Hence the minimal complementing region within pGS3 corresponded to the DNA lesion of LK100.

III. GENE IDENTITY

Both complementing loci are nuclearly encoded as pGS1 and pGS3 DNA hybridized exclusively to the DNA of chromosomes I and III, respectively. The complementing DNA also hybridized to polyadenylated RNA, revealing the presence of a mRNA species of approximately 1.5 or 2.75 kilobases in length for the loci derived from pGS1 or pGS3, respectively. In each case, the presence of Cd failed to induce accumulation of the transcripts. Over-production of the mRNA species was observed for both loci when present on a multicopy plasmid. For the LK100 complementing plasmid, over-expression of the 2.75 kilobase transcript enhanced Cd tolerance in Sp223. In contrast, an enhanced level of tolerance was not observed with over-expression of the 1.5 kilobase transcript from the LK69 complementing plasmid.

The cDNAs of the two mRNA species were isolated by DNA hybridization from a cDNA library. The longest cDNA of each gene was inserted into a plasmid vector such that its expression was driven by a plasmid promoter. In each case, restoration of Cd tolerance was observed in the corresponding mutant host, suggesting that near full length cDNAs were isolated. The sequences of these putative full length clones were determined and the deduced amino acid (aa) sequences compared with those in the data banks. The deduced product that complemented LK100 is a 90.5 kDa protein (830 aa) that has not been previously identified. Since this product is involved in heavy metal tolerance, the gene was named *hmtl* (Ortiz et al., 1992). As described earlier, the mutant *hmtl* allele was isolated from LK100. Through the exchange of DNA fragments between wild type and mutant alleles, the mutation was found to reside within a 1.1 kilobase fragment. The sequence of this small fragment of the mutant allele was determined, which revealed that a single G to A transition had altered the TGG triplet encoding for Trp to a TAG termination codon (Fig. 2a). Translational termination at this newly created stop codon would produce a truncated polypeptide only 253 aa in length. Thus the Cd-sensitive phenotype of LK100 is most likely the result of a severely reduced synthesis of the full length HMT1 protein.

As for the LK69 complementing product, the deduced aa sequence revealed a 48 kilodalton protein (434 aa) with substantial sequence similarity to adenylosuccinate synthetases (ASS, EC 6.3.4.4) from a variety of organisms. ASS converts IMP to AMP-S in a reaction leading to the formation of AMP (see Fig. 3; abbreviations for relevant intermediates and enzymes of purine biosynthesis are defined in the legend). This suggested that LK69, derived from the *ade6*$^-$ Sp223, harbored mutations in two genes of purine biosynthesis: *ade6*, encoding AIR carboxylase (ARC, EC 4.1.1.21), and

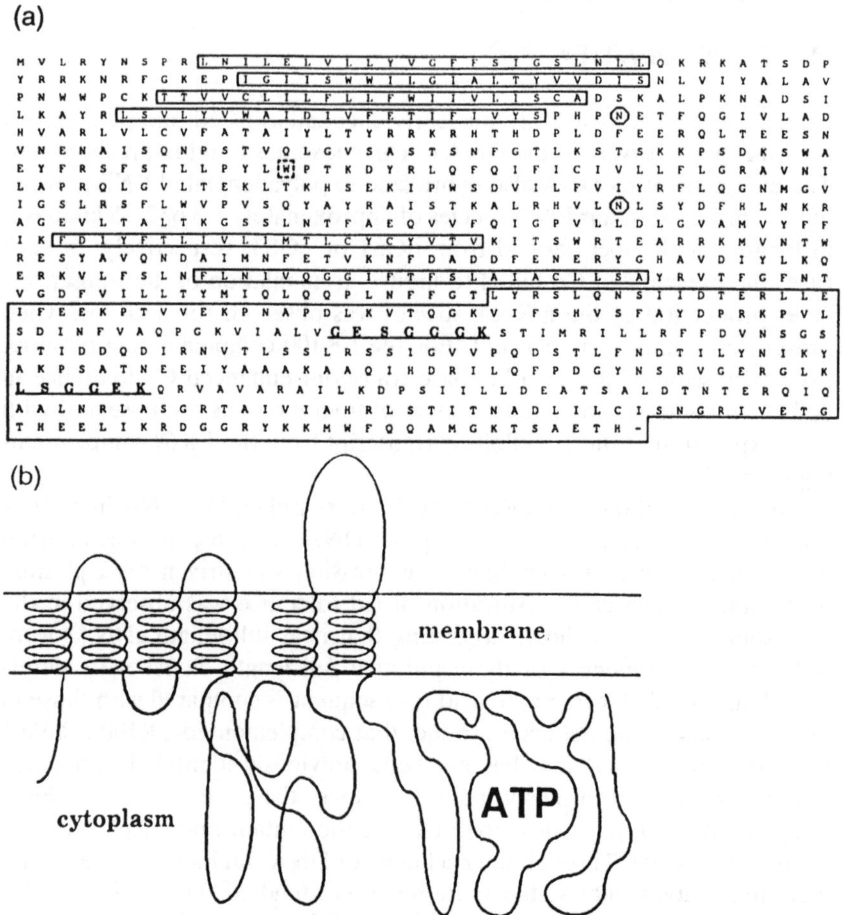

Figure 2 Amino acid sequence and putative structure of the product deduced from the *hmt1* cDNA. (a) In keeping with structures proposed for other ABC membrane proteins, six membrane-spanning domains are indicated by narrow boxes. The two Asp residues in sites matching the consensus motif susceptible to *N*-linked glycosylation (encircled) might be modified during passage through the vesicular transport pathway. The large box encompassing the carboxyl-terminal region contains the bipartite NBS (underlined bold characters) and exhibits sequence similarity with ABC-type membrane transport proteins. The TGG Trp codon (dashed box) was found mutated to a translation termination TAG triplet in the LK100 mutant allele. (b) The diagram represents a putative topology for the HMT1 protein based on aa sequence analysis and structural similarities with other members of the ABC-type family of transport proteins. This is only one of many possible structures. Other possible structures include those with a cleaved signal peptide, more membrane-spanning domains, or inverted direction of the diagrammed transmembrane regions following insertion of the putative signal peptide in the opposite orientation.

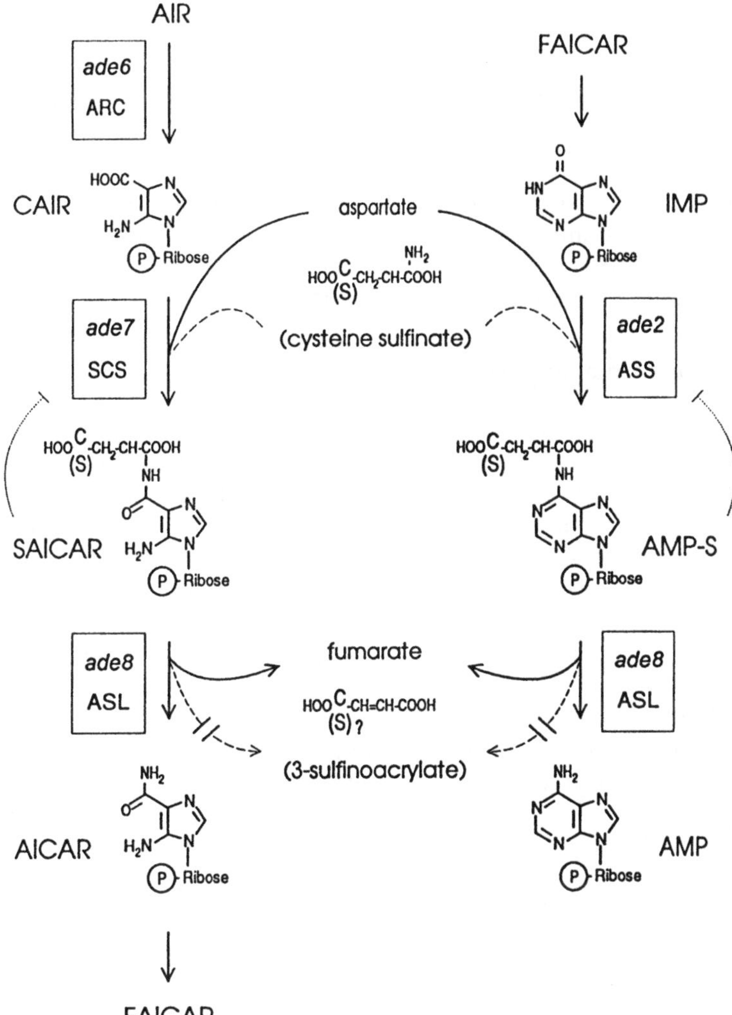

Figure 3 Purine biosynthesis pathway relevant to *ade* mutations and the proposed biosynthesis of cysteine sulfinate derived products. The pathway from AIR to AMP is shown along with hypothetical sulfur analog products. The abbreviations used are: AIR, aminoimidazole ribonucleotide; ARC, AIR carboxylase; CAIR, carboxyaminoimidazole ribonucleotide; SAICAR, succinoaminoimidazole carboxamide ribonucleotide; SCS, SAICAR synthetase; AICAR, aminoimidazolecarboxamide ribonucleotide; FAICAR, formamidoimidazolecarboxamide ribonucleotide; IMP, inosine monophosphate; AMP-S, adenylosuccinate; ASS, adenylosuccinate synthetase; ASL, adenylosuccinate lyase; AMP, adenosine monophosphate. As described in the text, formation of S-derivatives by SCS and ASS have been demonstrated *in vitro*. However, ASL does not react with these S-derivatives. Hence it is more likely that ASL activity is needed to prevent product inhibition (dotted lines) of SCS and ASS, rather than for the formation of 3-sulfinoacrylate.

ade2, encoding ASS. The identity of *ade2* was consistent with the following observations: (1) the cDNA was capable of restoring adenine prototrophy to ASS-deficient mutants of *Escherichia coli* and *S. pombe;* (2) both pGS1 DNA and *ade2* map to chromosome I; and (3) disruption of the DNA corresponding to the minimal complementing domain of pGS1 in an ade^{+} *S. pombe* strain, Sp806 ($leu1^{-}$, $ura4^{-}$), produced adenine auxotrophy that cannot be restored by supplementation with IMP.

IV. THE HMT1 PRODUCT

Analysis of the deduced aa sequence from the *hmtl* cDNA suggests that the HMT1 protein is a polytopic integral membrane protein. The amino terminal region contains a series of putative transmembrane domains that could span the membrane from 6 to 10 times (Fig. 2). The first of these transmembrane helices and the flanking aa sequences match the loose consensus sequence of the eukaryotic signal peptide implicated in cotranslational insertion in the endoplasmic reticulum. The carboxyl-terminal region is more hydrophilic and contains a bipartite nucleotide binding site (NBS) of the type present in a number of proteins involved in active transport (Walker et al., 1982).

A search of protein sequence banks with the HMT1 aa sequence indicated that the carboxyl-terminal region of about 250 aa, containing the NBS, displays a high level of sequence identity with the family of ATP-binding-cassette [ABC]-type membrane transport proteins (Juranka et al., 1989). Members of this group have been identified in organisms ranging from bacteria to mammals. In all instances, an NBS-containing domain, which presumably hydrolyzes ATP or GTP during substrate translocation across the membrane, is associated with a region containing six or more membrane spanning segments.

Amino acid sequence identity among ABC-type proteins is primarily restricted to the region surrounding the NBS; the hydrophobic domains containing the transmembrane helices display little or no similarity with each other. The different domains may be encoded by different genes, as for some of the bacterial transport systems, or by a single gene, as for all eukaryotic members of the family. Some bacterial genes encode two tandemly fused NBS domains or two sets of membrane-spanning regions. Almost all eukaryotic proteins, on the other hand, have two polytopic membrane-spanning domains and two NBS domains. The structure of these genes suggests that the arrangement of two membrane domains and two NBS domains, or a multiple thereof, reflects the native structure of the active ABC type transporter in the membrane. This would imply that HMT1, with one polytopic transmembrane region and a single NBS might function as a multimer. HMT1 is not the only eukaryotic protein that exhibits this arrangement. A few mammalian gene products display this same structure; interestingly, these proteins are sorted to intracellular membranes: PMP7 (Kamijo et al., 1990) to the peroxisome and RING4 to the endoplasmic reticulum (Trowsdale et al., 1990). Based on computer-aided sequence analysis and structural similarity with members of the ABC-type transport protein family, a putative topology of the HMT1 protein with six transmembrane domains is shown in Fig. 2b.

ABC-type proteins are implicated in membrane translocation of a wide variety of substrates that seem unrelated in size or chemical nature. For example, the cystic fibrosis transmembrane regulator functions as a chloride channel in mammalian epithelial cells (Riordan et al., 1990); the *E. coli hlyB* gene product transports hemolysin, a 105 kDa protein, across the bacterial membrane (Hess et al., 1986); the *Agrobacterium chvA* product transports β-1,2-glucan (Cangelosi et al., 1989); and the P-glycoprotein encoded by the mammalian multiple drug resistance gene exports small lipophilic drugs from the cell (Juranka et al., 1989). Thus while the high degree of similarity to ABC-type proteins in both structure and sequence allows us to propose that HMT1 is involved in active membrane transport, it sheds little light on the nature of the substrate mobilized by this protein. One possible function for HMT1 is that of a Cd efflux pump, since ABC-type transporters are known to mediate cellular export of a number of toxins, e.g., cytoplasmic drugs in mammalian cells and arsenate in bacteria (Chen et al., 1986). However, strains over-expressing *hmt1* from a plasmid accumulate more Cd than cells carrying the vector alone, suggesting that HMT1 may instead be involved in intracellular sequestration of the metal.

Cytoplasmic homeostasis of a number of amino acids and inorganic ions, such as Ca^{2+}, Zn^{2+}, and polyphosphate, is mediated in part by storage of these compounds in the vacuole (Klionski et al., 1990). Tobacco seedlings exposed to Cd accumulate most of the metal taken up, as well as PCs, in the vacuoles (Vögeli-Lange and Wagner, 1990). It is therefore possible that the vacuole plays a central role in PC-mediated heavy metal detoxification. In fungi, the vacuolar Ca^{2+} and arginine pools are maintained by specific transporters residing in the vacuolar membrane (for reviews, see Davis, 1986; Klionski et al., 1990). It is thus conceivable that HMT1 may be a vacuolar membrane protein responsible for PC and/or Cd^{2+} sequestration.

To test this hypothesis, it was necessary to ascertain if HMT1 is sorted to the vacuolar membrane. Addressing this question required the purification of fission yeast vacuoles. Various *S. pombe* mutants deficient in adenine biosynthesis, including the *ade6*$^-$ mutant, accumulate a red pigment in the vacuoles when grown in media low on adenine (Fig. 4). Lysis of these cells under specific conditions liberates vesicles that still retain this pigment,

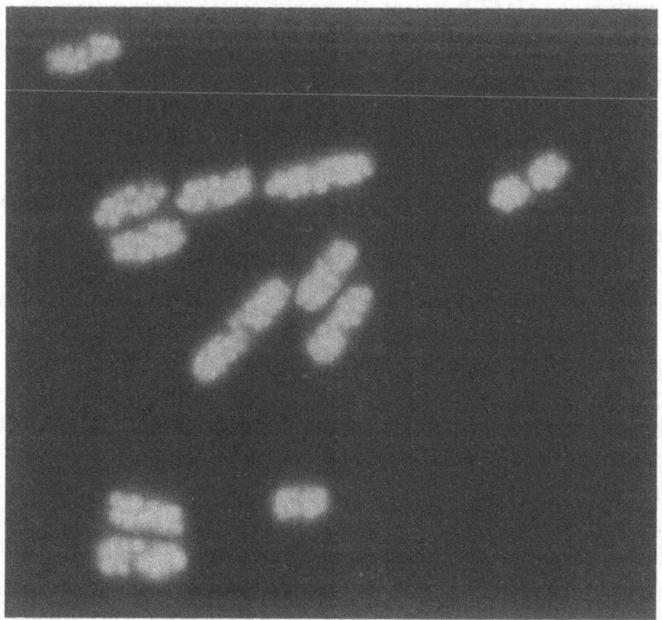

Figure 4 Vacuoles of *ade6⁻* strains of *Schizosaccharomyces pombe* shown under epifluorescence microscopy. These vesicles match the description of vacuoles derived from electron and light microscopy of fixed *S. pombe* cells (Robinow and Hyams, 1989).

which fluoresces in ultraviolet light. A fractionation procedure was developed that purified these fluorescent vacuoles, and the final fraction obtained exhibited a 50- to 200-fold increase in the specific activities of fungal vacuolar enzymes such as α-mannosidase and carboxypeptidase Y. Conversely, the cytoplasmic enzyme glucose 6-phosphate dehydrogenase displayed a 50-fold decrease in activity per mg protein. These results indicated that the fractionation procedure achieved at least a 50-fold enrichment of intact vacuoles.

Initially, antibodies directed against HMT1 were unavailable. To circumvent this problem, we constructed a gene fusion that over-produces an HMT1-β-galactosidase fusion protein. The plasmid containing this gene fusion complemented LK100, suggesting that the fusion protein functions correctly and is likely to be in its proper intracellular location. Cells harboring this gene fusion were fractionated as before, and the various fractions were examined by immunoblot analysis using monoclonal antibodies directed against β-galactosidase. The analysis revealed that the HMT1-β-

galactosidase chimeric protein is found predominantly associated with vacuolar membranes (Ortiz et al., 1992). Recently, we were able to over-express and purify from *E. coli* a partial HMT1 protein fragment that was used to raise antibodies against HMT1. Immunoblots of proteins derived from fractionated cellular components confirmed that the native HMT1 protein is sorted primarily to the vacuolar membrane of *S. pombe* (Fig. 5).

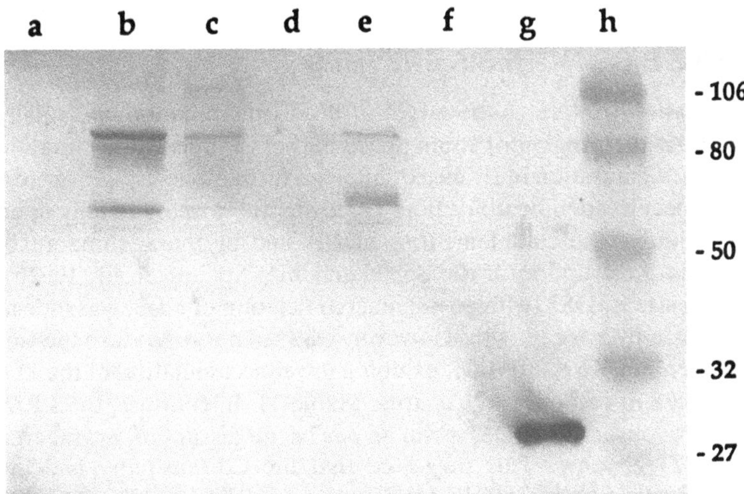

Figure 5 **Localization of HMT1 to vacuolar membranes. Immunoblot detection of HMT1 in protein extracts derived from *S. pombe* subcellular fractions: (a) P100 (total cellular membranes precipitated by centrifugation at 100,000 g) from LK100; (b) P100 from LK100 harboring a plasmid expressing *hmt1;* (c) P100 from Sp223; (d) S100 (supernatant from centrifugation at 100,000 g) from Sp223; (e) vacuolar membranes from Sp223; (f) P100 from a vacuole-depleted Sp223 fraction; (g) HMT1 protein fragment purified from *E. coli;* (h) prestained size standards (in kilodaltons). LK100 contains undetectable levels of HMT1 (lane a). This same strain expressing an *hmt1* cDNA (lane b) exhibits higher levels of HMT1 than does Sp223 (lane c). The *hmt1* gene product is found exclusively in the membrane fraction (lanes c versus d) and primarily in membranes derived from purified vacuoles (lane e versus lane f). It is not clear what generates the two bands at 85 and 70 kDa. They may be due to post-translational processing or the highly hydrophobic nature predicted for HMT1, which might impair denaturation and coating with SDS. The primary antiserum used was obtained from rabbits immunized with an HMT1 protein fragment purified from *E. coli* (lane g). The cross-reacting protein bands were visualized using mouse anti-rabbit IgG antibodies conjugated with alkaline phosphatase.**

The immunoblots also confirm that strains that express *hmtl* from a plasmid contain higher levels of HMT1 protein than does the progenitor strain Sp223; moreover, this protein is not detected in extracts derived from the mutant LK100 (Fig. 5). This implies that there is a direct correlation among the amount of HMT1 protein, the production of the HMW PC-Cd-S^{2-} complex, and Cd tolerance. The function of HMT1 has not yet been established; however, preliminary *in vitro* analysis with isolated vacuolar membrane vesicles suggessts that it may represent a transporter of PCs.

V. PURINE BIOSYNTHESIS ENZYMES

The identification of ASS as the target of the LK69 mutation was totally unexpected. As the progenitor strain Sp223 harbors a preexisting mutation in *ade6*, the *ade2* mutation in LK69 exhibited no further phenotypic expression with respect to adenine utilization. Gene identity was found only upon characterization of the complementing cDNA and by the adenine auxotrophy seen in a disruption of the Sp806 genome (Speiser et al., 1992b). The resulting strain, DS3, with an engineered deletion of *ade2*, was indeed unable to utilize IMP for growth. However, DS3 did not reproduce the Cd-sensitivity seen in LK69, nor did it exhibit a lower accumulation of the PC-Cd-S^{2-} complex in response to Cd stress (Table 1). In contrast, the LK69 phenotype was observed in DS1, with an *ade2* gene disruption engineered into the Sp223 genome. This suggested that the Cd-sensitivity and the deficient production of the HMW complex found in LK69 and DS1 must be attributed to a combination of the *ade6* and *ade2* mutant alleles.

Table 1 Effects of *ade* Mutations on Tolerance to Cd and Accumulation of the PC-Cd-S^{2-} Complex

ade genotype	Cd tolerance	PC-Cd-S^{2-}
ade$^+$	+	+
ade6$^-$	+	+
ade2$^-$ 6$^-$	sensitive	no
ade2$^-$	+	+
ade7$^-$	+	+
ade2$^-$ 7$^-$	sensitive	no
ade8$^-$	sensitive	slow

+ denotes a wild type phenotype.
Slow refers to the rate of formation.

It may seem peculiar that a double lesion in a linear pathway would produce a phenotype dissimilar from those of each mutation alone. ARC catalyzes a reaction that leads to the formation of IMP, whereas ASS converts IMP in a reaction leading to the formation of AMP (see Fig. 3). Although ARC acts at a step upstream of the ASS reaction, a mutation blocking IMP production via this linear pathway does not exert an epistatic effect on ASS. This is because adenine can be converted to IMP via a salvage pathway. Hence, the IMP-to-AMP reactions are operational in an *ade6* mutant grown in the requisite supplement adenine, just as the reactions leading to IMP production catalyzed by ARC are operational in an *ade2* mutant background. The operational independence of each segment of this pathway could account for the distinction between *ade6 ade2* double mutants as compared with either the *ade6* or *ade2* single lesion.

As for a biochemical basis for why two genetic lesions are needed to produce Cd sensitivity, a possibility would be that each segment of the pathway catalyzes a reaction that can be complemented by the other. The reactions performed by the two segments are indeed similar (see Fig. 3). The enzymatic step immediately following the action of the ARC is the conversion of CAIR to SAICAR, catalyzed by SAICAR synthetase (SCS, EC 6.3.2.6), product of the *ade7* gene. This reaction is analogous to the IMP-to-AMP-S reaction catalyzed by ASS, as both enzymes add aspartate to a nucleotide substrate. In the step immediately following each of these reactions, fumarate is cleaved from SAICAR and AMP-S by the same enzyme, adenylosuccinate lyase (ASL, EC 4.3.2.3), product of the *ade8* gene.

To test the hypothesis that ASS and SCS each perform a reaction that can be mutually complementary, PC-mediated Cd-tolerance was compared among *ade7*, *ade2*, and *ade7 ade2* mutant strains. Analogous to the construction of DS1, the isogenic *ade7 ade2* double mutant was made by disruption of the *ade2* gene in the *ade7* mutant B1048. The resulting double mutant, DS5, was indeed sensitive to Cd as well as deficient in accumulation of the PC-Cd-S^{2-} complex (Table 1). This phenotype is identical to that of the *ade2 ade6* mutant DS1, but it is in contrast to the Cd-tolerant phenotype of the *ade7* mutant B1048 and the *ade2* mutant DS3.

Since ASL participates in both segments of the *de novo* purine biosynthesis pathway, the effects of a mutation in the *ade8* gene was examined. The *ade8* mutant strain B23 was sensitive to Cd, but the accumulation of the HMW PC-Cd-S^{2-} complex was not entirely abolished. Compared to the *ade*$^{+}$ strain Sp806, B23 accumulated the HMW complex at slower rate when grown in rich medium in the presence of Cd. When grown in minimal medium, accumulation of the HMW complex was not observed. Since strains are generally more sensitive to the same concentration of Cd in minimal than in rich media, a greater proportion of cell death could account for the lack of detectable HMW complex. As for the sensitivity to Cd seen in B23, it could be attributed to the slower rate of HMW complex formation. The point in question is why a lesion in *ade8* would affect the rate of formation but not the final accumulated level of the PC-Cd-S^{2-} complex.

If ASL performs a function essential to the biogenesis of the PC-Cd-S^{2-} complex, then the slow rate of formation of the HMW complex could be due to "leaky" low-level synthesis of active enzyme. On the other hand, a lack of ASL could lead to a buildup of SAICAR and AMP-S. Product feedback inhibition by these intermediates might inhibit the activities of ASS and SCS. If this were possible, then a lesion in *ade8* would mimic a double lesion of *ade2* and *ade7*. Consistent with this latter proposal, feedback inhibition of ASL by AMP-S has been reported in bacteria (Stayton et al., 1983); however, it remains to be shown if this is also the case with SCS and SAICAR. If product inhibition of enzyme function is less effective than genetic blockage of enzyme synthesis, this would account for the "leaky" accumulation of the HMW complex seen in the ASL-deficient mutant.

VI. SULFIDE SOURCE AND THE CYSTEINE SULFINATE HYPOTHESIS

It has been demonstrated that the LMW PC-Cd complex can be converted to the HMW form *in vitro* by the addition of S^{2-} (Reese and Winge, 1988). Measurement of S^{2-} production during Cd stress by the strains described above showed that S^{2-} production was significantly lower in strains unable to produce the HMW PC-Cd-S^{2-} complex. However, the data could not address whether the loss of sulfide production was a cause or an effect of the loss of the HMW PC-Cd-S^{2-} complex. Nonetheless, this correlation is consistent with a possibility that these purine biosynthetic enzymes may be affecting the production of sulfide that is incorporated into the Cd chelate.

The source of the S^{2-} ion used to form the PC-Cd-S^{2-} complex is not known; the unstated assumption has been to attribute S^{2-} production to the assimilatory sulfate reduction pathway, and specifically to sulfite reductase or thiosulfate reductase. However, mutants LK69, DS1, DS5, and B23 have no apparent defect in assimilatory sulfate reduction as they required neither cysteine nor methionine. In the absence of Cd, basal levels of sulfide production in these mutants are also comparable to that of the wild type strain, which shows a sulfide level some 40-fold lower than during Cd stress. There are other known enzymatic reactions which have been shown to generate S^{2-} *in vitro*, including those catalyzed by cysteine desulfhydrase and thiosulfate cyanide sulfurtransferase, but there is no data to suggest that these might participate in generating sulfide *in vivo* for the PC-Cd complex.

A possible explanation for a deficient accumulation of the PC-Cd-S^{2-} complex seen in some of these *ade*$^-$ strains is that the activity of either ASS or SCS is needed to generate a product leading to the incorporation of sulfide into the PC-Cd complex. Porter et al. (1983) has shown indirectly by the hydrolysis of GTP that cysteine sulfinate, an oxidative product of cysteine, can replace aspartate in an *in vitro* reaction catalyzed by the ASS from *Azotobacter vinlandii.* If this same analog substrate is also utilized by SCS, then it may be that the sulfur-analog intermediates from both segments of the pathway serve as sulfide carriers or as sulfide donors upon further reduction. Additionally, ASL might catalyze the release of the S-analog of fumarate, 3-sulfinoacrylate, as an intermediate in the formation of sulfide and the HMW PC-Cd-S^{2-} complex. This scheme is depicted in Fig. 3. At the present, we have no data to verify that cysteine sulfinate is involved in the production of acid-labile sulfide *in vivo*, and it remains possible that the reduction in sulfide accumulation we observe is not a cause, but a result, of loss of the HMW PC-Cd-S^{2-} complex. In support for this hypothesis, however, we have found that cysteine sulfinate can be a substrate *in vitro* for both *S. pombe* ASS and SCS.

VII. *IN VITRO* UTILIZATION OF CYSTEINE SULFINATE BY ASS AND SCS

In testing the above hypothesis, we examined the activities of ASS and SCS, partially purified from the ASL-deficient mutant B23, and of ASL, partially purified from the ASS-deficient strain DS3 (Juang et al., 1993). Figure 6 shows the reactions of ASS with either Asp or cysteine sulfinate in a sequential reaction with ASL. Monitored spectrophotometrically, ASS catalyzed the formation of both AMP-S and sulfinylpropanoyladenylate (SPA) (S-analog of AMP-S), although the rate of the reaction with cysteine sulfinate is only about 25% of the rate observed with Asp. When ASL was added to the product of the ASS reaction, it was able to convert AMP-S to AMP but was unable to react with the putative sulfur analog. With removal of ASS prior to addition of ASL, a steady absorbance was observed, indicating a lack of further reaction of ASL with SPA. To check whether the enzymatic conversion of SPA by ASL required modification through exposure to cadmium or addition of factor(s), ASL and crude extracts of Cd-induced cultures were examined. Neither partially purified ASL nor crude extracts from Cd-induced cultures was able to alter the levels of SPA in the subsequent reaction.

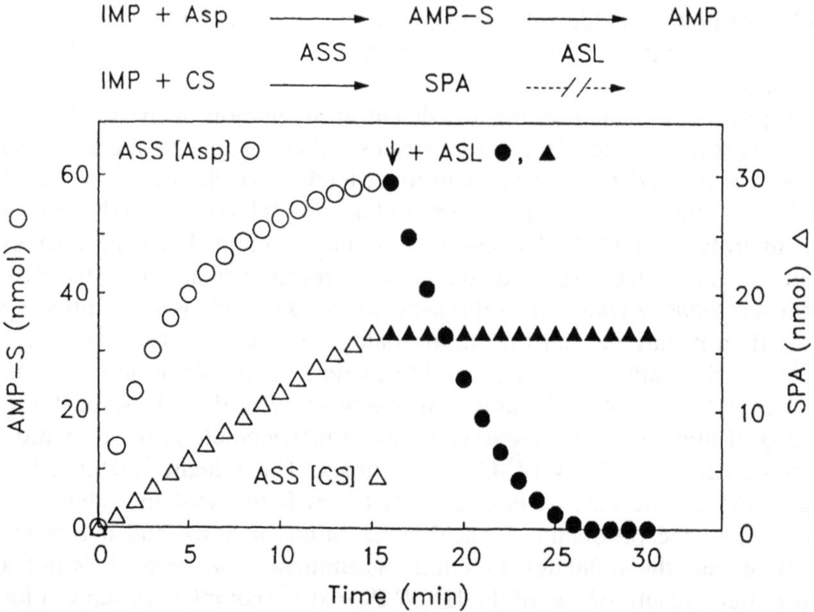

Figure 6 Utilization of cysteine sulfinate by adenylosuccinate synthetase. For the reaction with Asp, it was not necessary to remove ASS from the subsequent ASL reaction. In the reaction of ASS with cysteine sulfinate (CS), however, removal of ASS by ultrafiltration was required to prevent a steady increase in absorbance due to continued SPA formation.

Although the reaction of SCS with Asp could be monitored spectrophotometrically by the Bratton-Marshall reaction (Laikind et al., 1986), this was not possible with the reaction of SCS with cysteine sulfinate due to interference of cysteine sulfinate with the color reaction. To circumvent this problem, [^{35}S]-labeled cysteine sulfinate was synthesized and utilized in reactions with both SCS and ASS, and the products of the reactions were examined by thin layer chromatography (Juang et al., 1993). The reactions of SCS with [^{35}S]-cysteine sulfinate generated a novel [^{35}S]-labeled compound. Subsequent addition of ASL or crude extracts to the reactions failed to produce a change in the new spot. Neither were any additional spots seen that would indicate the release of [^{35}S]-sulfinoacrylate. As with SCS, ASS reacted with [^{35}S]-cysteine sulfinate to form a novel radiolabeled compound that also failed to react further upon addition of ASL or crude extracts. In both cases, the synthetases catalyzed the production of a novel radioactive compound with reduced mobilities as compared to the authentic carbon-succinyl derivatives. Despite the ability of the synthetases to generate sulfur analogs *in vitro*, the existence of the compounds and their significance *in vivo* require further study.

VIII. CONCLUSIONS

The realization that diverse cellular functions are required for accumulation of the HMW PC-Cd-S^{2-} complex in *S. pombe* has led to a more complicated model of the response of this organism to Cd. While the transformation of LMW PC-Cd complex to the HMW complex *in vitro* was accomplished by the simple addition of Na_2S, the same process *in vivo* depends on a number of different gene products including a putative vacuolar transport protein and the enzymes of purine biosynthesis. A model of the response of *S. pombe* to Cd exposure leading to the accumulation of the HMW PC-Cd-S^{2-} complex is presented in Fig. 7.

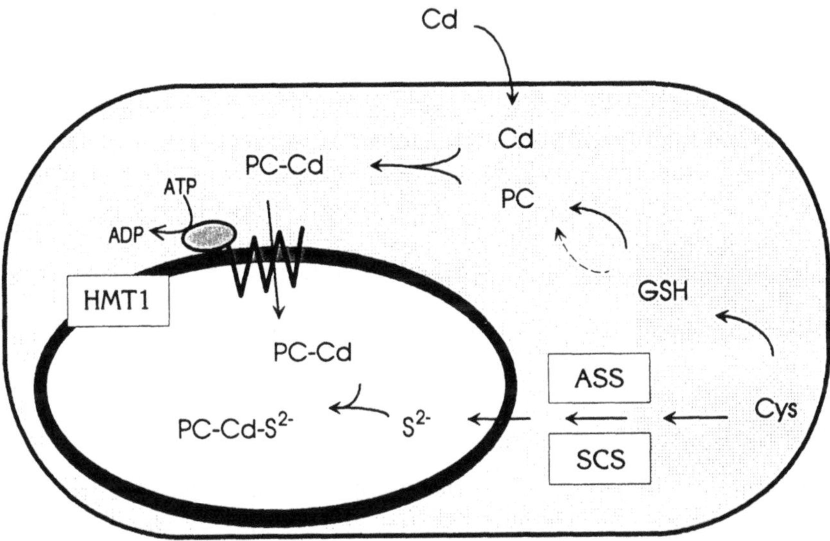

Figure 7 **Model of the *in vivo* maturation of the PC-Cd-S^{2-} complex. PC peptides, derived enzymatically from glutathione (GSH) via PC synthase or an alternate biosynthetic pathway (dashed lines), binds Cd to form a PC-Cd complex. As proposed: (1) This LMW complex is transported into the vacuole by the membrane protein HMT1; (2) addition of S^{2-} to the PC-Cd complex produces a storage form, the more stable sulfide-containing chelate; (3) the source of S^{2-} originates via the oxidation of Cys to produce cysteine sulfinate, which is then incorporated into sulfur-analog purine intermediates catalyzed by ASS and SCS. The sulfur-analog products are then further converted by as yet unidentified steps to yield the sulfide ion for the PC-Cd-S^{2-} complex.**

In this model, uptake of Cd^{2+} from the media is accomplished by unknown means, possibly by the action of a transport system used by other divalent cations such as Zn^{2+}. Preliminary data indicate that *S. pombe* can accumulate Cd^{2+} against a concentration gradient. Cytoplasmic Cd^{2+} results in activation of PC synthase and synthesis of PCs, which then bind Cd^{2+} ions to form the LMW PC-Cd complex. Given vacuolar accumulation of PCs, the LMW PC-Cd complex formed in the cytoplasm must be transported across the vacuolar membrane. In this model, this function is performed by HMT1 via hydrolysis of nucleoside triphosphates.

The LMW PC-Cd complex in the vacuole must be combined with S^{2-} to form the HMW complex. One possible source of the S^{2-} in this model is cysteine sulfinate, derived from cysteine oxidation, incorporated into purine intermediate analogs by ASS and SCS, and then subjected to reduction. Alternatively, the cysteine sulfinate–derived purine intermediates might serve as carriers of another sulfur compound that is the precursor of sulfide in the HMW complex. As for the role of ASL, the lack of an observable *in vitro* reaction of ASL with either sulfur analog suggests that these are not substrates for the enzyme, and that ASL activity is necessary merely to prevent product inhibition of ASS and SCS by AMP-S and SAICAR, respectively (see Fig. 3).

In this model, the HMW complex can be envisioned as a storage form of the metal chelate, consistent with its increased acid stability and metal-binding capacity, whereas the LMW complex could be the cytoplasmic scavenger and carrier of Cd. Much has been learned from the analysis of the two mutants; however, it is apparent that there are a number of functions that remain unspecified and could be identified by further biochemical and genetic analyses.

ACKNOWLEDGMENTS

Funding for completion of this manuscript was made possible by ARS project 5335-2300-005-00D (DWO) and USDA competitive grants 9101964 (DMS) and 9202488 (DFO). We also thank Jennifer VandeWeghe for reading the manuscript.

REFERENCES

Abrahamson, S. L., Speiser, D. M., and Ow, D. W. (1992). A gel electrophoresis assay for phytochelatins, *Anal. Biochem., 200:* 239–243.

Cangelosi, G. A., Martinetti, G., Leigh, J. A., Lee, C. C., Theines, C., and Nester, E. W. (1989). Role of *Agrobaceterium tumefaciens* ChvA protein in export of β-1,2-glucan, *J. Bacteriol., 171:* 1609–1615.

Chen, C. M., Misra, T. K., Silver, S., and Rosen, B. P. (1986). Nucleotide sequence of the structural genes for an anion pump, *J. Biol. Chem., 261:* 15030–15038.

Dameron, C. T., Reese, R. N., Mehra, R. K., Kortan, A. R., Carrol, P. J., Steigerwald, M. L., Brus, L. E., and Winge, D. R. (1989). Biosynthesis of cadmium sulfide quantum semiconductor crystallites, *Nature, 338:* 596–597.

Davis, R. H. (1986). Compartmental and regulatory mechanisms in the arginine pathway of *Neurospora crassa* and *Saccharomyces cerevisiae, Microbiol. Rev., 50:* 280–313.

Delhaize, E., Jackson, P. J., Lujan, L. D., and Robinson, N. J. (1989). Poly(γ-glutamycysteiny)glycine synthesis in *Datura innoxia* and binding with cadmium, *Plant Physiol., 89:* 700–706.

Gekeler, W., Grill, E., Winnacker, E.-L., and Zenk, M. H. (1989). Survey of the plant kingdom for the ability to bind heavy metals through phytochelatins, *Z. Naturforsch., Teil C, 44:* 361–369.

Grill, E., Winnacker, E.-L., and Zenk, M. H. (1985). Phytochelatins: The principal heavy-metal complexing peptides of higher plants, *Science, 230:* 674–676.

Grill, E., Gekeler, W., Winnacker, E.-L., and Zenk, M. H. (1986). Homo-phytochelatins are heavy metal–binding peptides of homo-glutathione containing Fabales, *FEBS Lett., 207:* 47–50.

Grill, E., Winnacker, E.-L., and, Zenk, M. H. (1987). Phytochelatins, a class of heavy-metal–binding peptides from plants, are functionally analogous to metallothioneins, *Proc. Natl. Acad. Sci. USA, 84:* 439–443.

Grill, E., Löffler, S., Winnacker, E.-L., and, Zenk, M. H. (1989). Phytochelatins, the heavy-metal–binding peptides of plants, are synthesized from glutathione by a specific γ-glutamylcysteine dipeptidyl transpeptidase (phytochelatin synthase), *Proc. Natl. Acad. Sci. USA, 86:* 6838–6842.

Hamer, D. H. (1986). Metallothionein, *Ann. Rev. Biochem., 55:* 913–939.

Hayashi, Y., Nakagawa, C. W., Mutoh, N., Isobe, M., and Goto, T. (1991). Two pathways in the biosynthesis of cadystins $(\gamma EC)_nG$ in the cell-free system of the fission yeast, *Biochem. Cell Biol., 69:* 115–121.

Hess, J., Wels, W., Vogel, M., and Goebel, W. (1986). Nucleotide sequence of a plasmid-encoded hemolysin determinant and its comparison with a corresponding chromosomal hemolysin sequence, *FEMS Microbiol. Lett., 34:* 1–11.

Jackson, P. J., Unkefer, C. J., Doolen, J. A., Watt, K., and Robinson, N. J. (1987). Poly(γ-glutamylcysteinyl) glycine: Its role in cadmium resistance in plant cells, *Proc. Natl. Acad. Sci. USA, 84:* 6619–6623.

Juang, H.-R., McCue, K. F., and Ow, D. W. (1989). Two purine biosynthetic enzymes that are required for cadmium tolerance in *Schizosaccharomyces pombe* utilize cysteine sulfinate *in vitro, Arch. Biochem., 304:* 392–401.

Juranka, P. F., Zastawny, R. L., and Ling, V. (1989). P-glycoprotein: Multidrug resistance and a superfamily of membrane associated transport proteins, *FASEB J., 3:* 2583–2592.

Kamijo, K., Taketani, S., Yokota, S., Osumi, T., and Hashimoto, T. (1990). The 70 kd peroxisomal membrane protein is a member of the Mdr (P-glycopotein) related ATP binding protein superfamily, *J. Biol. Chem., 265:* 4534–4539.

Klionski, D. J., Herman, P. K., and Emr, S. D. (1990). The fungal vacuole: Composition, function and biogenesis, *Microbiol. Rev., 54:* 266–292.

Kondo, N., Isobe, M., Imai, K., Goto, T., Murasugi, A., and Hayashi, Y. (1983). Structure of cadystin, the unit-peptide of cadmium binding peptides induced in a fission yeast, *Schizosaccharomyces pombe, Tetrahedron Lett., 24:* 925–928.

Kondo, N., Imai, K., Isobe, M., Gotto, T., Murasugi, A., Wada-Nakagawa, C., and Hayashi, Y. (1984). Cadystin A and B, major unit peptides comprising cadmium binding peptides induced in a fission yeast—Separation, revision of structures and synthesis, *Tetrahedron Lett., 25:* 3869–3872.

Laikind, P. K., Seegmiller, L. E., and Gruber, H. E. (1986). Detection of 5′-phosphribosyl-4-(*N*-succinylcarboxamide)-5-aminoimidazole in urine by use of the Bratton-Marshall reaction: Identification of patients deficient in adenylosuccinate lyase activity, *Anal. Biochem., 156:* 81–90.

Loeffler, S., Hochberger, A., Grill, E., Winnaker, E. L., and Zenk, M. H. (1989). Termination of the phytochelatin synthase reaction through sequestration of heavy metals by the reaction product, *FEBS Lett., 258:* 42–46.

Mehra, R. K., Tarbet, E. B., Gray, W. R., and Winge, D. R. (1988). Metal-specific synthesis of two metallothioneins and γ-glutamyl peptides in *Candida glabrata, Proc. Natl. Acad. Sci. USA, 85:* 8815–8819.

Murasugi, A., Wada, C., and Hayashi, Y. (1981). Cadmium binding peptides induced in the fission yeast *Schizosaccharomyces pombe, J. Biochem., 90:* 1561–1564.

Murasugi, A., Wada, C., and Hayashi, Y. (1983). Occurrence of acid labile sulfide in cadmium binding peptide 1 from fission yeast, *J. Biochem., 93:* 661–664.

Mutoh, N., and Hayashi, Y. (1988). Isolation of mutants of *Schizosaccharomyces pombe* unable to synthesize cadystin, small cadmium-binding peptides, *Biochem. Biophys. Res. Commun., 151:* 32–39.

Ortiz, D. F., Kreppel, L., Speiser, D. M., Scheel, G., McDonald, G., and Ow, D. W. (1992). Heavy metal tolerance in the fission yeast requires an ATP-binding cassette-type vacuolar membrane transporter, *EMBO J., 11:* 3491–3499.

Porter, D. J. T., Rudie, N. G., and Bright, H. J. (1983). Nitro analogs of substrates for adenylosuccinate synthetase and adenylosuccinate lyase, *Arch. Biochem. Biophys., 225:* 157–163.

Rauser, W. E. (1990). Phytochelatins, *Ann. Rev. Biochem., 56:* 61–86.

Reese, R. N., and Winge, D. R. (1988). Sulfide stabilization of the cadmium-γ-glutamyl peptide complex of *Schizosaccharomyces pombe, J. Biol. Chem., 263:* 12832–12835.

Reese, R. N., Mehra, R. K., Tarbet, E. B., and Winge, D. R. (1988). Studies on the γ-glutamyl Cu-binding peptide from *Schizosaccharomyces pombe, J. Biol. Chem., 263:* 4186–4192.

Reese R. N., White, C. A., and Winge, D. R. (1992). Cadmium-sulfide crystallites in Cd-$(\gamma EC)_n$G peptide complexes from tomato, *Plant Physiol., 98:* 225–229.

Riordan, J. R., Rommens, J. M., Kerem, B. S., Alon, N., Rozmahel, R., Grzelczak, D., Zielenski, J., Lok, S., Plavsic, N. Chou, J. L., Drumm, M. L., Iannuzzi, M. C., Collins, F. S., and Tsui, L. C. (1990). Identification of the cystic fibrosis gene: Cloning and characterization of complementary DNA, *Science, 245:* 1066–1071.

Robinow, C. F., and Hyams, J. S. (1989). General cytology of fission yeasts, *Molecular Biology of the Fission Yeast* (A. Nassim, P. Young, and B. F. Johnson, eds.), Academic Press, Toronto, pp. 273–331.

Robinson, N. J. (1990). Metal-binding polypeptides in plants, *Heavy Metal Tolerance in Plants: Evolutionary Aspects* (A. J. Shaw, ed.), CRC Press, Boca Raton, FL, pp. 195–214.

Speiser, D. M., Abrahamson, S. L., Banuelos, G., and Ow, D. W. (1992a). *Brassica juncea* produces a phytochelatin-cadmium-sulfide complex, *Plant Physiol., 99:* 817–821.

Speiser, D. M., Ortiz, D. F., Kreppel, L., Scheel, G., McDonald, G., and Ow, D. W. (1992b). Purine biosynthetic genes are required for cadmium tolerance in *Schizosaccharomyces pombe, Mol. Cell. Biol., 12:* 5301–5310.

Stayton, M. M., Rudolph, F. B., and Fromm, H. J. (1983). Regulation, genetics, and properties of adenylosuccinate synthetase: A review, *Curr. Topics Cell. Reg., 22:* 103–141.

Steffens, J. C. (1990). The heavy metal–binding peptides of plants, *Ann. Rev. Plant Physiol. Plant Mol. Biol., 41:* 553–575.

Tomsett, A. B., and Thurman, D. A. (1988). Molecular biology of metal tolerance of plants, *Plant Cell and Environ., 11:* 383–394.

Trowsdale, J., Hanson, I., Mockridge, I., Beck, S., Townsend, A., and Kelly, A. (1990). Sequences encoded in the class II region of the MHC related to the ABC superfamily of transporters, *Nature, 348:* 741–744.

Verklejj, J. A. C., Koevoets, P., Van't Riet, J., Bank, R., Nijdam, Y., and Ernst, H. O. (1990). Poly(γ-glutamylcysteinyl)glycines or phytochelatins and their role in cadmium tolerance of *Silene vulgaris, Plant Cell Environm., 13:* 913–921.

Vögeli-Lange, R., and Wagner, G. J. (1990). Subcellular localization of cadmium and cadmium binding peptides in tobacco leaves, *Plant Physiol., 92:* 1086–1093.

Walker, J. E., Saraste, M., Runswick, J., and Gay, N. J. (1982). Distantly related sequences in the α-subunits and β-subunits of ATP synthase, myosin, kinases and other ATP requiring enzymes and a common nucleotide binding fold, *EMBO J., 1:* 945–951.

13

Superoxide Dismutases in *Saccharomyces cerevisiae*

Francesca Galiazzo, Maria Teresa Carrì, Maria Rosa Ciriolo, and Giuseppe Rotilio
University of Rome "Tor Vergata," Rome, Italy

I. MECHANISMS OF OXYGEN TOXICITY AND ANTIOXIDANT DEFENSE

A. Chemistry of O_2^- and Its Cellular Production in *S. cerevisiae*

Microorganisms were present on earth in the absence of oxygen at the very early stages of the evolution of life. The amount of molecular oxygen in the atmosphere began to increase as the result of the evolution of cyanobacteria, which, using light energy, produced molecular oxygen from water and carbon dioxide. Then molecular oxygen was used by respiring organisms as the terminal electron acceptor in their energy-producing electron transfer chains, a process that involves generation of partially reduced, potentially toxic oxygen by-products (Fridovich, 1983).

Molecular oxygen, or dioxygen, is paramagnetic because of the presence of two unpaired electrons in its ground state, which can be described as a triplet ($^3O_2$). Since most molecules are diamagnetic (singlet), their reactions with oxygen require a change of spin to occur. Such reactions are forbidden and very improbable (spin restriction). Activation of dioxygen can then proceed either via conversion to singlet oxygen ($^1O_2$) or through reduction by organic radicals. In the former case, spin restriction is removed since dioxygen becomes diamagnetic, but this requires much energy (usually light) to bring dioxygen to an excited state with paired electrons. In the latter case, the reactant itself is paramagnetic. This is the most usual way of dioxygen activation in biology and is one of the reasons why univalent reduction of dioxygen is favored. The reactivity of oxygen increases upon acceptance of one, two, or three electrons to form, respectively, superoxide radical (O_2^-), hydrogen peroxide (H_2O_2), and hydroxyl radical ($^{\cdot}OH$). The step-by-step four-electron reduction of molecular oxygen to water is:

$$O_2 \xrightarrow{e^-} O_2^- \xrightarrow{e^-, 2H^+} H_2O_2 \xrightarrow{e^-, H^+} OH^{\cdot} \xrightarrow{e^-, H^+} H_2O$$

Oxygen toxicity is a complex phenomenon, and several hypotheses have been made in order to explain it. A most up-to-date view is that oxygen toxicity is linked to O_2^- production (Fridovich, 1986). The reactivity of O_2^- is highly dependent upon the nature of the solvent: in aqueous media O_2^- is considerably less reactive, its predominant reaction being nonenzymatic dismutation to H_2O_2 and O_2. At pH 7.4 the rate constant for nonenzymatic dismutation is as high as $1 \times 10^5\ M^{-1}s^{-1}$. However, there is substantial evidence that the steady-state level of O_2^- correlates with the manifestations of cellular oxidant stress. Such a contradiction is currently explained by the ability of O_2^- to react with H_2O_2 in the metal-driven Haber-Weiss reaction:

$$O_2^- + H_2O_2 \longrightarrow OH^{\cdot} + OH^- + O_2$$

Since O_2^- is able to reduce both Fe^{3+} and Cu^{2+}, and since its dismutation produces H_2O_2, it is likely that when the intracellular concentration of O_2^- increases, the concentrations of H_2O_2 and $^{\cdot}OH$ will also rise.

$^{\cdot}OH$ can be considered the most important agent producing cellular oxidant damage. The reactivity of $^{\cdot}OH$ is due to its very high standard redox potential (+2.3 V). Within the cell it will react with any biomolecule at diffusion-limited rates. However, because of its reactivity, the average diffusion distance of an $^{\cdot}OH$ radical is only a few nanometers (Fridovich, 1989). On the other hand, DNA and cell membranes are polyanionic structures to which metal cations would adhere. It is therefore likely that effective $^{\cdot}OH$ is generated in sites that are adjacent to these critical targets (Korbashi et al., 1986).

Living cells contain multiple potential sources of active oxygen species: (1) small molecules known to autoxidize at appreciable rates and to produce O_2^-, including hydroquinones (Misra and Fridovich, 1972), flavins (Ballou et al., 1969), catecholamines (Cohen and Heikkila, 1974), thiols (Misra, 1974; Baccanari, 1978), and reduced hemoproteins (Auclair et al., 1978); (2) cytosolic enzymes that catalyze biological oxidations in which oxygen is utilized as a substrate (Malmstrom, 1982); (3) membrane-bound enzymes included in electron transport chains of organelles (such as mitochondria, chloroplasts, and microsomes) that release O_2^- (Chance et al., 1979; Cadenas, 1989); (4) plasma membranes of granulocytes and macrophages that have been shown to contain oxidases that produce O_2^- during the respiratory burst following activation of these specialized cells (Johnson and Ward, 1985).

Intact mitochondria appear to be the major source of O_2^- in *Saccharomyces cerevisiae.* Depending upon the redox state of the respiratory chain components (Boveris and Chance, 1973), mitochondrial O_2^- production accounts for 1–4% of the total oxygen consumption of the cell under normoxic conditions. O_2^- generation by mitochondria of higher eukaryotes can occur at the NADH-Q segment (complex I) or at the QH2:cytochrome *c* segment (complex III) of the respiratory chain (Turrens and Boveris, 1980; Boveris, 1984). In general, the latter is the source of more than 80% of the mitochondrial O_2^- production, although the relative contribution may vary among different cell types (Boveris, 1984).

Electrophilic quinone compounds, either natural cell constituents (such as ubiquinone) or exogenous sources such as menadione and plumbagin, are easily reduced to semiquinones, which in turn reduce O_2 to O_2^- (Thor et al., 1982). Methyl viologen (paraquat), a dipyridyl compound, is also reduced to a semiquinoid form that is a very efficient superoxide generating agent (Rotilio et al., 1985). Intra- and extracellular nonspecific reductases that can act on paraquat have been identified in *Saccharomyces cerevisiae* (Lesuisse et al., 1990).

Sources of H_2O_2 include spontaneous and enzyme-catalyzed dismutation of O_2^-, as well as several oxidases such as D-amino acid oxidases. About 20% of the oxygen taken up by mitochondria diffuses to the suspending medium as H_2O_2.

The ratio of the rates of O_2^- and H_2O_2 generation in yeast submitochondrial particles was found to be quite close to the theoretical value of 2.0. Therefore O_2^- may be considered as the major intermediate in the H_2O_2 production in yeast mitochondria (Boveris, 1978).

B. Role of Superoxide Dismutase

Evidence that oxygen radical species are continuously generated during aerobic life implies that aerobes evolved some strategies to survive in the presence of oxygen.

The superoxide dismutase metalloenzymes protect against the potentially deleterious effects of O_2^- by catalyzing the dismutation of O_2^- at very high rates, thereby maintaining a steady-state concentration of O_2^- lower than that expected for nonenzymatic dismutation, especially at relatively low O_2^- levels, where the bimolecular nonenzymatic reaction is less efficient. A further advantage of enzymatic dismutation is that it does not give rise to other potentially deleterious redox reactions as in the case of the reaction catalyzed by low molecular weight metal complexes.

Three types of superoxide dismutases have been isolated (Bannister et al., 1987). One type contains Cu(II) and Zn(II) at the active site, with copper as the catalytic center. The other types contain either Mn(III) or Fe(III) as the catalytic metal. During the catalytic cycle the metal is reduced by the first O_2^- molecule producing O_2 and then oxidized by a second O_2^- generating H_2O_2.

Cu,Zn superoxide dismutase is usually found in the cytosol of eukaryotic cells. It was first isolated as a green copper protein from bovine erythrocytes (erythrocuprein) and liver (hepatocuprein) by Mann and Keilin in 1939. In 1969 McCord and Fridovich identified cuprein as a superoxide dismutase, being a competitive inhibitor of the reduction of cytochrome *c* by the superoxide-producing xanthine xanthine-oxidase reaction. The Mn superoxide dismutases are characteristically found in bacteria but are also present in eukaryotic mitochondria. Extensive amino acid sequence homology indicates that the bacterial and the mitochondrial enzymes are indeed closely related, and this is most currently explained in terms of the endosymbiotic origin of mitochondria. The manganese enzyme is dimeric in most bacteria and tetrameric in mitochondria; however the tetrameric form has also been found in thermophilic bacteria (Chikata et al., 1975; Sato and Harris, 1977). Fe superoxide dismutases are evolutionarily related to the manganese enzymes in several aspects, including molecular weight, subunit composition, 3-D structure, and ability to bind metals.

An extracellular superoxide dismutase has been isolated from humans (Tibell et al., 1987) and other mammals (Karlsson and Marklund, 1988). It is a secretory tetrameric glycoprotein containing Cu and Zn (Marklund, 1982). Comparison of sequence similarities indicates that the exo- and endocellular Cu,Zn superoxide dismutase isoenzymes arise from a common ancestral enzyme before the evolution of fungi and plants. Recently a similar enzyme has been partially purified from *Neurospora crassa* and *Saccharomyces cerevisiae* (Munkres, 1990). Furthermore, a gene encoding for an extracellular Cu,Zn superoxide dismutase has been identified in *Schistosoma mansoni* (Simurda et al., 1988).

II. MOLECULAR PROPERTIES OF ISOLATED SUPEROXIDE DISMUTASES FROM *S. cerevisiae*

Both a cytosolic Cu,Zn- and a mitochondrial MnSOD have been isolated from *S. cerevisiae.* The two isoenzymes will be discussed in two separate sections.

A. Cu,ZnSOD

1. Purification and Basic Properties

The enzyme was isolated in homogeneous form by Goscin and Fridovich (1972). It was found to display the typical properties of all members of its class, as present in the prototype protein, the bovine erythrocyte enzyme (BSOD; McCord and Fridovich, 1969): approximately two atoms of copper and two atoms of zinc per enzyme molecule; pale blue-green color to a visible absorption band centered at 670 nm (A=231); a UV spectrum with a maximum at 258 nm (A = 11,300), reflecting the absence of tryptophan, the paucity of tyrosine, and the abundance of phenylalanine, as in the majority of Cu,ZnSODs; two identical subunits of approximately 16 kDa not covalently bridged.

The isoelectric point is 4.6, more acidic than that of the bovine enzyme (Marmocchi et al., 1983). The EPR spectrum shows the typical rhombicity and relatively narrow (approximately 13 mT) parallel hyperfine splitting, reflecting the tetrahedral distortion of the copper coordination in these enzymes (Rotilio et al., 1972). However, significantly less copper is usually revealed by EPR with respect to the value detected by atomic absorption spectroscopy (Goscin and Fridovich, 1972), while the two values coincide in nearly all Cu,ZnSODs tested so far. This indicates a tendency of the metal ion to undergo autoreduction in this enzyme, a property likely to be related to its relatively poor stability (see below).

2. Primary Structure

Cu,ZnSODs are highly conserved enzymes. In fact, the amino acid sequence of the yeast enzyme (YSOD) (Steinman, 1980) displayed 55% similarity with respect to the bovine enzyme (BSOD) and nearly 80% with respect to the human enzyme (HSOD) (Fig. 1). In particular, the YSOD monomer consists of 153 residues, one more (after position 37) than HSOD, while BSOD has a further two residues deletion after position 23. These three sequences are good representatives of some points worth remarking in this respect. Cu,ZnYSOD has a free amino terminus, as do all proteins in this class, except those from reptiles, birds, and mammals. YSOD contains only two Cys residues, forming the disulfide bridge conserved in all members of the class, while BSOD contains a free Cys at position 6 and HSOD two Cys residues (6 and 111). This characteristic lack of at least one free thiol group is common to Cu,ZnSODs from plants, fungi, and bacteria.

3. 3-D Structure

The three-dimensional structure of the enzyme has been determined by x-ray crystallography to 2.5 Å resolution (Frigerio et al., 1989; Djinovic et al., 1992). The overall features are quite the same as for the other three

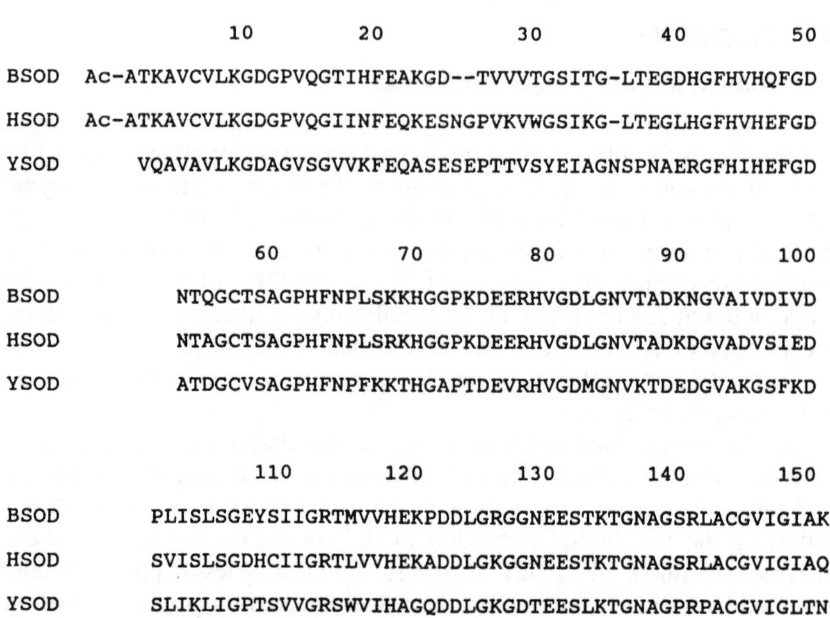

Figure 1 Sequence alignments of bovine, human, and yeast Cu,Zn superoxide dismutases. Residues are numbered according to the bovine sequence.

available 3-D structures: bovine (Tainer et al., 1982), spinach (Kitagawa et al., 1991), and human (Parge et al., 1992): a rigid scaffold made of a flattened antiparallel 8-stranded beta barrel, plus three external loops. The loop regions form the metal binding site and the active site channel and are stabilized by the Cys55-Cys144 disulfide bridge. A superimposition of BSOD and YSOD chains is shown in Fig. 2: the largest deviations can be observed in the regions of aminoacid insertions.

4. Properties of the Metal Binding Sites

The stereochemistry of the Cu and Zn sites is substantially the same as in other structures resolved by x-ray analysis. The intermetal distance is 6.1 Å, similar to that of the spinach enzyme (Kitagawa et al., 1991) but 0.2 Å shorter than that reported for BSOD (Tainer et al., 1982). The zinc coordination geometry (His61,69,78 and Asp81) is distorted tetrahedral, whereas the copper is in distorted square-planar coordination (His44,46,61 and 118, almost exactly in plane, with a solvent peak, compatible with a fifth ligand, at the vertex of the square pyramidal coordination shell, whose base is centered on the copper ion). Thus His61 bridges Cu(II) and Zn(II) also in YSOD, which had been indicated as not having an intermetal histidine

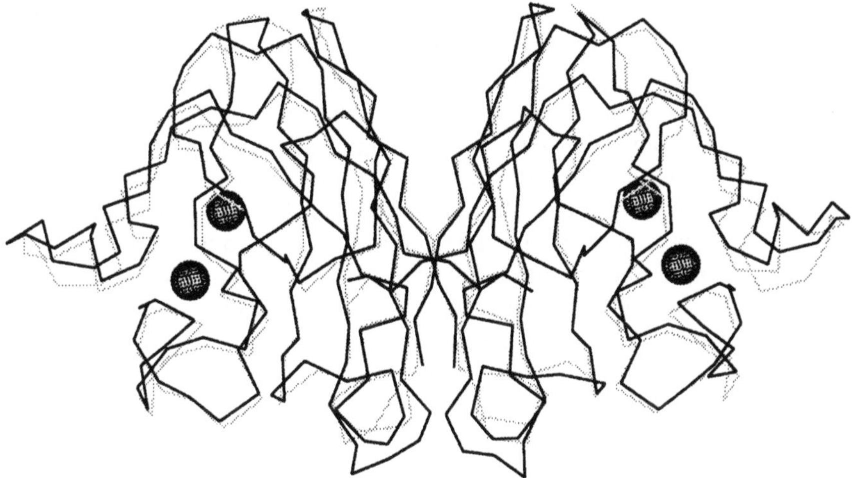

Figure 2 Superimposition of the alpha-carbon trace of *Saccharomyces cerevisiae* (full line) and bovine (dotted line) Cu,Zn superoxide dismutases. Copper atoms (top) and zinc atoms (bottom) are represented by dotted spheres.

bridge (Bauer et al., 1980). Both Zn sites appear to be identical in the crystal structure of the dimer, at variance with previous observations of Johansen's group (Bauer et al., 1980; Dunbar et al., 1984) on the enzyme in solution. In these studies, the zinc had been replaced by ^{111}Cd(II) or Co(II) and examined by perturbed angular correlation of gamma ray spectroscopy or by CD and MCD spectra, respectively. Different coordination geometries were suggested for the zinc site on each subunit, in particular a tetrahedral, high-affinity one, and a pentacoordinate geometry with low affinity. These results seem to be related to the fact that they have been obtained in reconstitution experiments from apoSOD. It is, in fact, established that even in BSOD occupancy of a site during reconstitution of the apoenzyme (Rigo et al., 1977a,b, 1978) or selective metal depletion of the holoenzyme (Cocco et al., 1981) influence the site on the other subunit via conformational effects. Such conformational communication between the two remote (34 Å) active sites is emphasized in freeze-dried samples (Viglino et al., 1981), which is the case for the samples used in the experiments of Johansen's group.

The reduced Cu(I) enzyme, which has not yet been crystallized from any source, has been studied by ^{1}H NMR (Cass et al., 1978). While the resolution of those spectra was too low to give more detailed information than a substantial homology to BSOD, halide ion binding could be analyzed properly enough to establish that they were able to perturb the Cu(I) site in the order $Cl^- = Br^- > I^- > F^-$. In BSOD (Rigo et al., 1977c), I^- has apparently no access to the active site. This difference may be related to the significant differences observed between the x-ray structures of the two enzymes in the loops defining the entrance to the active site channel (Djinovic et al., 1992).

Mention is deserved in this section of a genetically engineered mutant YSOD in which His78 (a zinc ligand) was replaced by Cys (Lu et al., 1992). When copper was added to the apoprotein, giving rise to the Cu_2Cu_2 derivative, the sample became intensely blue, with high extinction absorption bands in the 500–600 nm region and an EPR spectrum characterized by very narrow parallel hyperfine splitting. These properties resemble those typical of blue (Type 1) copper proteins (e.g., azurin, laccase, etc.) and are produced by coordination of copper to sulfur ligands in a distorted tetrahedral geometry, like that of the zinc-binding site in Cu,ZnSOD.

5. *Enzyme Activity*

The catalytic mechanism of bovine Cu,ZnSOD (Fielden et al., 1974) involves alternate reduction and oxidation of the enzyme-bound copper ion by two molecules of superoxide to give dioxygen and hydrogen peroxide. Consequently, exogenous liganas that have access to the copper ion display an inhibitory effect on the enzyme activity to an extent that is a function of their respective dissociation constants. In this respect, CN^- is the most effective inhibitor ($K = 10^5$ M at pH 7.4; Rotilio et al., 1972), and this is considered as a distinctive feature that differentiates Cu,ZnSOD activity from the other members of the family. However, it should be noticed that the two half-reactions of the minimal scheme for the mechanism of action outlined above have identical rates, which are the same as the catalytic rate and are close to the diffusion controlled limits ($K = 2 \times 10^9\ M^{-1}\ s^{-1}$ at pH 5–10 and 0.1 M ionic strength). This, and the concerted inhibition of the rate by increasing ionic strength and pH (above pH 10), led to the suggestion (Argese et al., 1987) that the rate-limiting step of the reaction resides outside the copper site, which changes its chemistry upon reduction (McAdam et al., 1977; Desideri et al., 1992), and is to be found in some event involving the protein moiety of the enzyme. In particular, the rate of the reaction has been proposed to correspond to the rate of the encounter of superoxide with the active site region, which is facilitated by electrostatic attraction to positively charged amino acid side chains surrounding the edges of the active site channel. Among these residues, Arg141 is invariant in Cu,ZnSODs, and its neutralization, by either chemical (Malinowski and Fridovich, 1979a) or genetic (Beyer et al., 1987) engineering inhibits the enzyme activity by 90%. It is located 5.7–5.8 Å apart from the copper and is considered more important in the correct positioning of superoxide and of inhibiting anions (Sette et al., 1992) at the copper site than for steering the substrate from the solvent to the copper. On the other hand, the long-range interaction of the substrate with the protein surface, which generates its facilitated diffusion into the active site, is controlled by charged residues, which have been less strictly conserved during evolution. They consist of pairs of oppositely charged side chains positioned at the edge of the active site funnel, in order to reduce the nonproductive association of the anionic substrate, traveling toward the Arg141-copper sink, with positively charged residues. In BSOD (Getzoff et al., 1983), two such pairs, namely Lys134-Glu131 and Lys120-Glu119, which are located near the entrance of the active site approximately 12 Å apart from the copper, apparently play a determinant role in the steering process. In YSOD the latter pair is nonconservatively substituted by Ala119-Gly120. Surprisingly, the catalytic constant of YSOD is the same as in BSOD, and the pattern of the pH and ionic strength dependence of the enzyme activity is unchanged as well (O'Neill et al., 1988). This finding is explained by the results of a study on six natural variants of Cu,ZnSOD with various degrees of sequence conservation and different total electric charge on the protein (Desideri et al., 1992). Calculations of electric fields surrounding the protein surface have shown that positive values of the electrostatic potential are restricted to the areas in the proximity of the active sites, irrespective of the conservation of the protein net charge and of individual charged residues. Coordinated mutations have maintained a constant distribution of electric fields around the protein surface throughout the evolution of this class of enzymes, so as to give identical efficiency of electrostatic recognition without strict conservation of point charges. On the other hand, measurements of the inhibitory effect of various halides, which act as competitive inhibitors for superoxide (Rigo et al., 1977c), showed that iodide was effective with YSOD but not with BSOD. This result is in line with NMR data (Cass et al., 1978) and with the observed difference of the active site channel entrance in the two enzymes (Djinovic et al., 1992), which may discriminate anionic molecules larger than superoxide.

6. *Stability of the Protein*

Cu,ZnSODs are very stable proteins because of their rigid beta scaffold, the presence of an invariant disulfide bridge between two cysteines far apart from each other (88 residues) in the linear sequence, and the tight intersubunit contact. BSOD is especially stable, and it remains active in 8M urea or 4% SDS. On the other hand, a reversible inactivation by 8M urea was observed with YSOD, which was accompanied by changes of the EPR spectrum of the enzyme-bound copper, indicative of axial distortion of the metal coordination geometry (Barra et al., 1979). Furthermore, while

BSOD exchanges subunits (e.g., with a SOD from another species or a chemically modified form of the same protein) only in the presence of 8M urea (Marmocchi et al., 1978; Malinowski and Fridovich, 1979b), YSOD is able to do so under normal conditions (Arnold and Lepock, 1982). This reflects a greater tendency for the YSOD subunits to separate.

Additional information on these aspects has been provided by studies on heat stability of SODs. Concerning the thermal stability of enzymes, one has to distinguish between the rate of heat inactivation and the intrinsic stability of the protein folded form. The former process is followed by monitoring the enzyme activity at room temperature after timed exposure to high temperature, thus essentially reflecting irreversible thermal denaturation. The latter property is reflected by different values of the melting temperature (Tm) as recorded by differential scanning calorimetry (DSC) experiments. It has been shown by studies on wild-type and mutant BSOD and HSOD (McRee et al., 1990; Hallewell et al., 1991) that replacement of cysteine residues with unreactive side chains increases the stability of the enzyme activity of these SODs to irreversible heat inactivation. This effect is likely to be due to inhibition of the formation of aggregates, cross-linked by intermolecular disulfide bridges, during the refolding process. However, YSOD, which lacks free cysteine residues, although being more thermostable than wild-type HSOD (two cysteines), is not as stable as the HSOD mutant with no free cysteines (R. Hallewell, personal communication). Evidently, other structural features contribute to the overall sensitivity to heat inactivation. On the other hand, the stability of the folded state is less dependent on the presence of free cysteines than on the establishment of inner cavities in the protein core, which result from sterically noncompensating amino acid substitutions. This is shown by the lower Tm of the BSOD with no free cysteine with respect to the wild-type protein, in spite of its higher rate of thermal inactivation (McRee et al., 1990). YSOD melts at a significantly lower temperature than BSOD, i.e., 82°C versus 96°C at pH 5.5 (Roe et al., 1988). It is likely that the absence of the two bulky sulphur atoms of the residues Cys6 and Met115, which face each other across the beta barrel in BSOD, is not compensated, in term of close packing of the protein interior, by the Ala and Val residues, which occupy the same positions in YSOD (Djinovic et al., 1992). The inner cavity left can be one of the structural bases of the lower stability of the folded conformation of YSOD with respect to that of BSOD. Furthermore, it should be recalled that Pro13 of BSOD, which is located in the loop that connects the two first beta strands of the beta barrel, plays a key role in the assembly of beta strands during the folding. This residue is substituted by Gly in YSOD: similar substitutions have been shown to lower the heat stability in other proteins.

B. MnSOD

1. *Purification and Basic Properties*

The enzyme was found to be localized in the mitochondrial matrix of yeast and was purified in a crystallizable form by Ravindranath and Fridovich (1975). The enzyme activity was CN-insensitive, at variance with that of the Cu,Zn enzyme. The molecule was shown to have a mass of 96 kDa and to be composed of four subunits of equal size (24 kDa) and composition. It contained one atom of Mn per subunit, which was suggested to be Mn(III) on the basis of its visible absorption spectrum, consisting of a broad band with a maximum at 460 nm, giving the resting enzyme a characteristic pink color.

2. *Primary Structure*

As soon as primary structures of non-CuSODs became available, a striking similarity was noticed among Fe- and MnSODs of either bacterial or eukaryotic origin, while no homology could be detected with respect to the Cu,Zn enzymes (for a recent summary of data see Beyer et al., 1991). Approximately 28 residues are invariant out of the maximum number of 216, and many positions show conservative replacements. This high degree of conservation of primary structure, which results in highly similar conformations (see below), suggests a common evolutionary origin, and for MnSOD it is strong evidence for the endosymbiotic origin of mitochondria. For yeast (207 residues), both the amino acid (Ditlow et al., 1984) and the nucleotide (Marres et al., 1985) sequences have been determined.

3. *3-D Structure*

The close similarity of primary structures suggests that Mn- and Fe-superoxide dismutases have similar 3-D structures unrelated to those of Cu,ZnSODs. This expectation is fully confirmed by the available high-resolution data, which regard the MnSODs from *T. thermophilus* (1.8 Å resolution; Ludwig et al., 1991), *B. stearothermophilus* (2.4 Å; Parker and Blake, 1988), human mitochondria (2.2 Å; Borgstahl et al., 1992), and the FeSODs from *E. coli* (3.1 Å; Carlioz et al., 1988) and *P. ovalis* (2.1 Å; Stoddard et al., 1990). No fine structure has been reported for the yeast Mn enzyme, although it was the first of its class to be crystallized (Beem et al., 1976). However, the recently resolved 3-D structure of another tetrameric MnSOD from human mitochondria is undoubtedly a reasonably good model for the yeast analogous protein. The subunit fold of MnSODs can be divided into two distinct domains. The N-terminal domain is made up of an N-terminal loop of 10 residues, packed against two long antiparallel helices forming a hairpin-shaped structure. The C-terminal domain is a mixed alpha/beta structure, with the central layer formed by three antiparallel beta strands surrounded by five short helical segments. A dimer interface has the same ligand pattern: it is common to all Fe/MnSODs and forms the metal-binding site. The other interface is unique to the mitochondrial enzyme and is formed by the association of the helical hairpin N-terminal domains into closely packed, intersubunit 4-helix bundles, which are very different from the loose contacts involving loop regions in the same intersubunit interface of the tetrameric bacterial MnSOD.

4. *Properties of the Metal-Binding Site*

The active site metal joins the two domains. The Mn ion—no significant conformational changes are observed when the metal is reduced by dithionite (Stallings et al., 1986)—is coordinated by two amino acid residues from each domain. the N-terminal domain contributes His26 from the first helix and His81 (74 in man) from the second helix. The C-terminal domain has Asp168 (159 in man) and His172 (163 in man). It is interesting to note that Zn in Cu,ZnSODs has the same ligand pattern. In fact, when copper was inserted into the active site of *B. stearothermophilus* SOD in the place of Mn, it gave an identical EPR spectrum to that of Cn,ZnSOD, with copper in the place of Zn (Bannister et al., 1985). A water molecule binds to the metal, giving the active site an overall five-coordinated trigonal bipyramidal geometry.

The active site of Mn/FeSODs is surrounded by hydrophobic residues, including three Tyr, three Trp, and two Phe within 10 Å of the metal center, with extensive aromatic stacking, probably in order to achieve a better stabilization of the Mn(III) state.

Reversible dissociation of the metal has been obtained for MnSODs, but conditions are more critical with respect to Cu,ZnSODs. Usually a chaotropic agent is needed plus a chelating agent. Furthermore, in spite of the identity of binding sites, most MnSODs are active only when reconstituted with Mn, although they bind Fe as well. Correspondingly, FeSODs are activity-specific for iron. Subtle structural differences may be responsible for this behavior. In this respect, it will be important to better characterize the few SODs that give active enzymes with either metal bound to the same protein (Meier et al., 1982).

5. *Enzyme Activity*

The mechanism of action of MnSOD is substantially similar to that of Cu,Zn enzymes: alternate redox cycle of the active metal with equal rates (McAdam et al., 1977) and electrostatic facilitation of enzyme-substrate encounter (Chan et al., 1990). Although studies regarding the latter aspect are less advanced as compared to the Cu,Zn enzyme, it is established that numerous cationic residues that surround the entrance to the active site pocket are conserved in Fe/MnSODs. Among them are Lys28 and Arg189. Interestingly, the latter one is absent in yeast MnSOD, but it is conservatively substituted for by a lysine, rendering this enzyme insensitive to arginine-specific reagents, such as phenylglyoxal, and very sensitive to lysine-modifying compounds.

On the other hand, important features characterize MnSOD. The enzyme is insensitive to cyanide, like FeSOD, but (at variance with it) it is not inactivated by hydrogen peroxide or azide. Both Mn- and FeSOD activities decrease above neutrality, while that of Cu,ZnSOD is constant until pH values go as high as 10. The catalytic constant of Fe/Mn enzymes is approximately one order of magnitude less than that of the copper enzyme. Fe/MnSODs are saturated at superoxide concentrations lower than Cu,ZnSOD, and the Mn enzyme shows a rather complex kinetics, indicating the presence of a slower side reaction with the enzyme (McAdam et al., 1977).

6. *Stability of the Protein*

Fe/MnSODs are less stable enzymes than the Cu,Zn family, probably because of the much less compact folding, containing large amounts of alpha-helix. MnSODs do not display any EPR spectrum when properly purified, but a signal typical of aquo Mn(II) appears after storage or repeated freezing and thawing. They are better stored in the presence of glycerol. Contrary to Cu,ZnSOD, their quaternary assembly is sensitive to chaotropic agents or removal of the active site metal. It should, however, be pointed out that most data refer to the dimeric bacterial type of enzyme, while tetrameric forms may have more stable structures, as shown by the tighter intersubunit contacts found in the human enzyme.

III. REGULATION OF SOD GENES IN *S. cerevisiae*

A. Gene Structure and Localization

1. *SOD1*

Cloning of the gene coding for Cu,Zn superoxide dismutase in *S. cerevisiae* was accomplished by Bermingham-McDonogh et al. (1988) by screening a lambda-gt11 library with a long, unique oligonucleotide. The probe had been designed deducing the sequence from the known amino acid sequence (Johansen et al., 1979) and minimizing recognition of false positives by using a computer-generated yeast codon preference table. The 47-mer oligonucleotide proved to match the real sequence except for 4 mismatches and allowed successful isolation of two clones containing the whole coding region plus 5′ and 3′ flanking sequences.

Bilinski et al. (1985) had previously described a yeast strain lacking Cu,ZnSOD (DSCD1-4a: ura3, his4, arg4, sod1-1), which has a very low level of Cu,ZnSOD activity and does not grow in 100% dioxygen. Bermingham-McDonogh et al. (1988) have been able to correct the characteristic dioxygen sensitivity of this strain by transformation with the Cu,ZnSOD coding sequence inserted in a yeast-*E. coli* shuttle expression vector. Furthermore, they have been able to detect immunoreactive Cu,ZnSOD in the above-mentioned transformant as well as in the parental wild-type strain, but not in the mutant DSCD1-4a.

Analysis of the DNA sequence in the noncoding regions of the gene has yielded several interesting features: (1) a transcription stop-site (TAGATTTATG) positioned 84 bases downstream from the TAA stop codon; (2) a potential transcription start-site (TATATAA) located about 100 bases upstream from the ATG start codon; this does not seem to be the real transcription start site, which has been more recently located at −44 bp (Gralla et al., 1991); (3) a 13-base sequence homology with the MnSOD gene (*SOD2,* see below) at −303 bp. This 13-mer is similarly positioned in *SOD2,* and there is only one mismatch between the two genes, raising the interesting possibility that it may be important in some regulatory mechanism common to both superoxide dismutases.

A 25 bp binding site for the ACE1 (activator of copperthionein expression) protein has been localized in the *SOD1* promoter, about 200 bp upstream from the transcription start-site, showing a 12-out-of-15 base pairs identity with a strong ACE1 binding site in the promoter of the gene *CUP1,* coding for copperthionein. Induction by copper of Cu,ZnSOD (see below) is strictly dependent on the presence of a functional ACE1 protein (Carrì et al., 1991) and a functional ACE1 binding site, as demonstrated by mutagenesis of several important G residues (Gralla et al., 1991).

By sequence homology, *SOD1* promoter seems also to possess two copies of the Ap1-recognition element (ARE), located at −292 and −273 base pairs, a sequence that has been involved in the response of mammalian cells to several stresses including oxidative stress and metal ion poisoning (Schnell et al., 1992).

SOD1 was reported to be present in a single copy in the *S. cerevisiae* genome, as indicated by Southern blot analysis (Bermingham-McDonogh et al., 1988). More recent work has allowed its localization on the right arm of chromosome X, positioned at 11–14 cM proximal to *cdc11* and equidistant to the *cycl-rad7-SUP4-cdc8* cluster (Chang et al., 1991).

2. *SOD2*

Isolation of the gene coding for Mn superoxide dismutase in *S. cerevisiae* has been accomplished seredipitously by Marres et al. (1985) during a study on the regulation of the expression of nuclear genes coding for imported mitochondrial proteins. A clone coding for a 25 kDa polypeptide present in a purified QH_2:cytochrome *c* oxidoreductase preparation, erroneously identified as the Rieske iron-sulfur protein (Van Loon et al., 1983) was successively found out to encode for MnSOD. The early incorrect identification, deriving from the isolation procedure of the iron-sulfur protein, raises the interesting possibility of a functional association of MnSOD with QH_2:cytochrome *c* oxidoreductase, a source of superoxide radicals in mitochondria deriving from the autoxidation of one of the species of bound ubisemiquinones involved in the catalysis of this enzyme.

Expression of the selected genomic clone increased the level of MnSOD in yeast (Marres et al., 1985) and directed the production of a 27 kDa protein in *E. coli,* reacting with antibodies against MnSOD.

Nucleotide sequence analysis of this clone allowed determination of the coding region plus 5′ and 3′ flanking regions of 558 and 785 bp, respectively, and confirmed that MnSOD is synthesized as a larger precursor (Autor, 1982) with a molecular mass of 26,123 Da. The predicted amino acid sequence corresponded with that reported for the purified enzyme (Ditlow et al., 1984), except for an amino-terminal extension of 27 residues and a C-terminal extension of 4 amino acids.

Processing of the 27 N-terminal residues is in agreement with previous observations of a precursor 2 kDa larger than the mature protein. However, C-terminal processing is unlikely, and sequence homology with MnSOD of other species suggests that the four residues (including conserved Ala211 and Lys213) are indeed present in the mature protein.

Transcription of *SOD2* starts about 50 bp upstream from the ATG start codon (Marres et al., 1985) and stops at about 500 bp beyond the TGA stop.

Analysis of the nucleotide sequence in the 5′ flanking region revealed that the *SOD2* promoter contains the usual elements of transcription initiation, capping and translation, plus two additional cis elements potentially relevant for MnSOD regulation of expression: (1) a 13-nucleotide stretch homologous to the *SOD1* promoter (see above); (2) two DNA stretches similar to the CYC1-UAS (iso-1-cytochrome *c* upstream activating sequences) positioned at −341 and −251 bp relative to the ATG start codon. UAS1 and UAS2 are similarly positioned in *CYC1,* a fact that may be significant, since *SOD2* is expressed in parallel with *CYC1* in a mutant that exhibits constitutivity of CYC1 under anaerobic conditions (Lowry and Zitomer, 1984). Similar UASs are found in other heme-containing proteins, regulated by oxygen and by carbon source. Regulation of MnSOD gene expression will be described below.

SOD2 maps to the right arm of chromosome VIII in *S. cerevisiae,* 4 cM distal from *pet1* (Van Loon et al., 1986), and it is immediately proximal to the ERG1 locus (Turi et al., 1991).

B. Regulation of *SOD1* and *SOD2* Gene Expression

Regulation of the expression of SOD genes in *S. cerevisiae* is more a modulation than a dramatic change in the enzymatic activity, with a maximum variation of about 10-fold observed either at the transcriptional or at the posttranscriptional level for both Cu,Zn- and MnSOD. This does not mean that there are not environmental stimuli that affect gene expression, and some of them act with a well-characterized mechanism. The most noticeable change has been observed for MnSOD activity in response to oxygen; therefore this enzyme has been considered as an adaptive enzyme, in contrast to Cu,ZnSOD, which has been regarded as a constitutive enzyme.

The respiratory chain of mitochondria is known to be a source of O_2^- (Boveris, 1978; Boveris and Cadenas, 1982), a fact that could explain the different regulation of the mitochondria-localized enzyme (MnSOD) versus the cytoplasmic variant (Cu,ZnSOD). The molecular mechanism by which oxidative stress rapidly increases the level of MnSOD in mitochondria, however, is poorly understood (the MnSOD gene has a nuclear localization).

The relative enzymatic activity of Cu,ZnSOD is about 80–95% of the total SOD activity of the cell, the remaining 5–20% being accounted for by MnSOD (Lee and Hassan, 1985; Greco et al., 1990; Chang et al., 1991). In this respect it is interesting to evaluate the concentration of the two enzymes in relation to their cellular localization. It is known that Cu,ZnSOD is a cytosolic enzyme, and MnSOD is localized in mitochondria. The mitochondrial volume is a function of the physiological state of the cell, and it represents about 10–12% of the total volume in fully respiring cells (Stevens, 1981). We can deduce that the level of enzymatic activity of the two SODs is almost the same in their own cellular compartments, or Cu,ZnSOD activity is only slightly more than MnSOD if we exclude the volume of other organelles. In this view, growth conditions appear to be relevant, since mitochondria are dynamic structures changing with carbon sources and phase of growth, which therefore are factors affecting SOD expression. Several results obtained on the regulation of the two SODs indicate that they respond in a similar fashion to such stimuli as oxygen and glucose, in agreement with their role as cellular antioxidative defenses. However, SODs also behave independently in response to other factors such as copper concentration in the medium, suggesting a different, additional physiological role of Cu,ZnSOD in the cell.

1. *Regulation of SOD Gene Expression by Oxygen and Heme*

S. cerevisiae can live either in aerobic or in anaerobic conditions. The transition from one physiological state to the other means that several genes are subject (although to a different extent) to regulation by oxygen. Superoxide dismutases, which play an important role in protecting against oxidative damage, are also regulated by oxygen. On the other hand, a large number of genes sensitive to oxygen are regulated by heme, which is functionally closely related to oxygen metabolism and requires oxygen for its biosynthesis. It is therefore likely that superoxide dismutases could be regulated by heme as catalase is.

In an early study, Gregory et al. (1974) reported the effect of oxygen on the level of superoxide dismutases in *S. cerevisiae*. They demonstrated that hyperbaric oxygen produced an equal increase of both SOD activities, with respect to anaerobiosis, suggesting that an equal amont of O_2^- was produced in the cytosol and in mitochondria and that both SODs represent the mechanism of cell defense against oxygen toxicity.

An increase of Cu,Zn- and MnSOD by hyperoxia was also reported in yeast cells grown on ethanol (Westerbeek-Marres et al., 1988). In the absence of oxygen a ninefold decrease of Cu,ZnSOD activity was observed in *S. cerevisiae* cells grown to the late exponential phase, but part of the enzyme was present as an inactive proenzyme requiring copper for activation (Galiazzo et al., 1991). Complete reactivation of the proenzyme by copper indicates that a decreased cellular availability of copper is responsible, in part, for the effect produced by anaerobiosis. A lower copper concentration was indeed detected in anaerobic cells with respect to those grown under air.

The effect of oxygen on the transcription of *SOD1* (coding for Cu,ZnSOD) and *SOD2* (MnSOD) has been studied in yeast cells harboring either *SOD1* or *SOD2* promoters fused in frame with *lacZ* coding sequence (Galiazzo and Labbe-Bois, 1993). *SOD1* and *SOD2* transcription of cells grown to the early exponential phase decreased by factors of 4 and 8, respectively, under anaerobic conditions. This transcriptional regulation matches well the decrease in MnSOD immunoreactive protein (12–15%) reported by Autor (1982) in anaerobic cells. It is worth mentioning that during the transition from the anaerobic to the aerobic condition, maximum induction of *SOD2* and *CYC1* (coding for iso-1-cytochrome *c*) transcription occurs concomitantly (Galiazzo and Labbe-Bois, 1993).

SOD regulation by heme has been investigated by using a mutant deprived of the gene coding for 5-aminolevulinic acid synthase (*Δhem1*). This strain contains only one-third of the MnSOD activity and one-fourth of the immunoreactive protein detected in the wild type (Autor, 1982). However, under anaerobic conditions, MnSOD immunoreactive protein was not active, suggesting that oxygen effect is not mediated by heme.

Northern blot analysis of RNA extracted from *Δhem1* and *β*-galactosidase activity of *SOD1-lacZ* and *SOD2-LacZ* fusions in *Δhem15* (lacking ferrochelatase) have indicated that a decrease in gene transcription is evident for both *SOD1* and *SOD2* (3- and 5-fold, respectively) (Galiazzo and Labbe-Bois, 1993) in the absence of oxygen. The decrease of the β-gal activity for the *SOD2* gene fusion in the absence of oxygen or heme correlates with the values obtained in the same conditions for MnSOD immunoprotein by Autor (1982).

Gene activation by heme is, in most cases, mediated by the trans-acting heme activation protein complexes HAP1 and HAP2,3,4. *SOD2* gene regulation by heme also occurs via HAP1 (Zitomer and Lowry, 1992), and, using a *SOD2-lacZ* fusion, it has been recently demonstrated that *SOD2* transcription is regulated by both complexes, and the binding sites for HAP2,3,4 have been identified (Pinkham, 1991).

SOD2 gene transcription is also affected by the *ROX1* gene product, which is a heme-activated repressor of heme-regulated genes. It has been reported that a *rox1-a1* mutant strain (having a semidominant mutant allele of *ROX1*) shows constitutive expression of *SOD2* (Lowry and Zitomer, 1984).

2. *Effect of Oxidative Stress on SOD Genes*

Experimentally, it is possible to generate oxidative stress *in vivo* by addition of hydrogen peroxide or such other redox cycling compounds as paraquat. The first study conducted on the effect of paraquat on the biosynthesis of superoxide dismutases in *S. cerevisiae* showed that this drug caused induction of Cu,Zn- and MnSOD activities to about the same extent (70% and 100%, respectively), possibly as a consequence of an increased intracellular flux of O_2^- (Lee and Hassan, 1985). These results correlate with those obtained under the same conditions (2.5 mM paraquat in glucose-cultivated cells) on *SOD1* and *SOD2* transcription evaluated using *SOD1*- and *SOD2-lacZ* fusions (50% and 100% increase, respectively) (Galiazzo and Labbe-Bois, 1993). These results indicate that the effect of paraquat is not as relevant in *S. cerevisiae* as it is in *E. coli,* where a 40-fold increase of activity was observed for MnSOD (Hassan and Fridovich, 1977).

Results obtained recently by Schnell et al. (1992) suggest that *PAR1* (*YAP1/SNQ3*) gene product is involved in the regulation of the response to oxidative stress in *S. cerevisiae.* This gene confers resistance to iron chelators in multicopy transformants. Cu,ZnSOD activity of *par1* mutants was reported to be 60% of the wild type and increased by 50% in *PAR1* multicopy transformants. However, *PAR1* is not considered as solely responsible for Cu,ZnSOD induction by oxidative stress, since in its absence Cu,ZnSOD activity is still inducible by oxidative stress agents.

3. *Regulation of SOD Gene Expression by Metals*

The presence of copper at the catalytic site of Cu,ZnSOD suggests that copper availability can limit enzymatic activity, thus exerting a regulation at the posttranslational level. This is indeed the case in anaerobic *S. cerevisiae,* which contains a proenzyme requiring copper for activation (Galiazzo et al., 1991). Furthermore, copper addition to the growth medium acts on *SOD1* transcription. The early observation of an increase of Cu,ZnSOD activity in both aerobic and anaerobic conditions in copper-loaded cells had suggested that copper could also act at the genetic level (Galiazzo et al., 1988). Moreover, when grown in copper-depleted media (50 nM Cu), yeast cells reduce to 50% their Cu,ZnSOD activity, while showing comparable amounts of mRNA and immunoreactive protein, both in the presence and in the absence of oxygen (Greco et al., 1990).

New insight on the mechanism of transcriptional regulation of Cu, ZnSOD has been provided by studies on the costitutive protein ACE1 which, upon binding of Cu(I), activates the promoter of *CUP1,* the gene coding for copperthionein (Thiele, 1988; Furst et al., 1988). Mutants lacking ACE1 are unable to increase *SOD1* transciption in response to copper (Carrì et al., 1991). An ACE1 binding site has been localized in *SOD1* promoter, and its ability to bind ACE1 *in vitro* has been demonstrated (Gralla et al., 1991). Furthermore, Ag(I), which is the only metal capable of unspecifically inducing copperthionein, induces *SOD1* transcription as well (Carrì et al., 1991). All these data indicate that *CUP1* and *SOD1* are coregulated at the transcriptional level by ACE1 and copper. The significance of this coregulation in the physiology of *S. cerevisiae* is not clear. A role of Cu,ZnSOD in metal storage can be envisaged. This hypothesis is supported by the purification of a Ag,Cu,Zn enzyme from yeast supplemented with $AgNo_3$ (unpublished results from this laboratory). Alternatively, copper-thionein could play a role in the protection from oxidative damage. It is worth mentioning that induction of *CUP1* has been observed under oxidative growth conditions and that *CUP1* can suppress the requirement for Cu,ZnSOD in some conditions (Tamai et al., 1992).

Copper does not affect MnSOD transcription; however, when yeast is grown in copper-free media an increase in MnSOD activity seems to compensate for the decrease of Cu,ZnSOD (Greco et al., 1990).

The effect of manganese on yeast SOD activity was also investigated. Previous observations (Archibald and Fridovich, 1981) had indicated that in *Lactobacillus plantarum* manganese can compensate for the absence of superoxide dismutase. In *Saccharomyces cerevisiae* manganese induces only a slight increase of MnSOD activity, essentially due to dialyzable Mn complexes (Galiazzo et al., 1989). Moreover, in *SOD1*-deficient yeast mutants, which are not able to grow in 100% oxygen, Mn accumulation rescued the O_2-sensitive phenotype, and the O_2-scavenging activity detected in these mutants under these conditions was demonstrated to be due to free or loosely bound Mn ions (Chang and Kosman, 1989).

4. *Effect of Nutrients and Phase of Growth on SOD Gene Expression*

Yeast can modify its metabolism in relation to carbon sources utilized for growth. These modifications imply changes in the expression of several genes, essentially related to respiratory activity or fermentation. In particular, a set of genes encoding mitochondrial and respiratory proteins is subject to catabolite repression, i.e., to a decrease in their expression when cells are grown at elevated glucose concentration. This kind of regulation has been shown for MnSOD gene expression.

A tenfold decrease of MnSOD immunoreactive protein has been reported for yeast cells in the early logarithmic phase of growth in 10% glucose, with respect to the growth in 2% ethanol. Under the same conditions, a slight enhancement (40%) was observed for Cu,ZnSOD immunoreactive protein. The effect of glucose on MnSOD was also demonstrated at the transcriptional level in parallel with the cytochrome bc1 complex subunit core 2 (Westerbeek-Marres et al., 1988). However, we have observed that growth on a fermentable carbon source (9% glucose) caused a two- to threefold decrease of both Cu,ZnSOD activity and immunoprotein (Galiazzo et al., 1991).

Catabolite repression of both SODs operates at the transcriptional level; expression of *SOD1*- and *SOD2-lacZ* fusions is about threefold and fivefold lower, respectively, when comparing cells grown on ethanol plus glycerol with those grown on glucose (2% or 10%) (Galiazzo and Labbe-Bois, 1993).

An increase in the activity (50–60%) of both *SOD-lacZ* fusions has been observed during the exponential growth phase on glucose, and a smaller increase (25–40%) has been observed on ethanol. The effect of glucose seems not to be due to derepression, since it is also present in *rho*$^-$ cells (Galiazzo and Labbe-Bois, 1993). The results obtained with the above-mentioned gene fusions showed that, in stationary phase cells grown on glucose, the activity of *SOD1-lacZ* was very low, and that *SOD2-lacZ* did not change with respect to the exponential phase. In the stationary phase, nutrient starvation produces stress conditions, which can induce gene transcription, as for the *CTT1* gene (catalase T) and *CYC7* (iso-2-cytochrome *c*) (Belazzi et al., 1991; Pillar and Bradshaw, 1991), which are also induced by heat shock (Wieser et al., 1991; Pillar and Bradshaw, 1991). This is not the case for both SOD genes, which are not induced in the stationary phase and are insensible to a shift of temperature from 23 to 39°C, suggesting that they are not stress proteins (Galiazzo and Labbe-Bois, 1993). Conflicting results have been reported for the change in the amount of both SODs immunoreactive protein and in the activity level during the transition from the exponential to the stationary growth phase. Westerbeek-Marres et al. (1988) reported that in this latter phase, cells grown on ethanol contained aproximately 50% MnSOD and a double amount of Cu,ZnSOD immunoreactive protein. On the other hand, Chang et al. (1991) reported that the activity of Cu,ZnSOD and MnSOD increase six- and fourfold, respectively, in this phase with respect to the exponential growth phase during cell growth on glucose.

C. Effect of Mutations in *SOD1* and *SOD2*

Direct evidence for a central role of superoxide dismutases in cellular defense against oxygen toxicity has come, in recent years, from studies of *S. cerevisiae* strains mutated in one or both genes coding for superoxide dismutases.

In yeast, *sod1*$^-$ and *sod2*$^-$ strains show the same typical sensitivity to hyperoxia and to redox cycling drugs as paraquat. However, the *sod1*$^-$ phenotype differs from *sod2*$^-$ as far as deficiencies in growth, carbon utilization, and auxotrophies are concerned, possibly in relation to different metabolic functions carried out by the two enzymes.

A *S. cerevisiae* mutant strain lacking Cu,ZnSOD activity (*sod1*$^-$) was described by Bilinski et al. in 1985. This strain carries an allele (*scd1*) that is a recessive chromosomal mutation and is unable to grow in 100% O_2 in rich medium (see below). In a later study, it has been observed that in a *sod1*$^-$ strain the size and the amount of Cu,ZnSOD mRNA are essentially the same as in the wild type. However, three SOD1 immunoreactive polypeptides were detected in extracts from the mutant (one corresponding to the wild type), none of which exhibited superoxide dismutase activity. Furthermore, copper was apparently missing in all SOD1 immunoreactive proteins. Determination of the nucleotide sequence of the mutant allele demonstrated that the failure to bind copper was not due to alterations in the coding sequence, the only changes being observable in the 5′ noncoding region of the gene. These data suggested that mutant polypeptides—due to false starts of transcription—are produced that cannot fold correctly and/or do not incorporate copper in the active site (Chang et al., 1991).

A *sod2*$^-$ mutant was constructed by gene disruption (Van Loon et al., 1986) and demonstrated to lack any protein cross-reacting with antibodies against MnSOD.

More recently, a double mutant (*sod1⁻ sod2⁻*) was constructed by Liu et al. (1992), together with suppressors that have bypassed the SOD defect (see below).

It is worth mentioning that the typical oxygen sensitivity of *sod⁻* strains can be complemented by transformation with *either* SOD gene. For instance, *sod1⁻* phenotype can be reversed to wild type by transformation of the mutant *S. cerevisiae* strain with a plasmid supporting expression of MnSOD from *B. stearothermophilus* in the cytosol (Bowler et al., 1990), suggesting that Cu,ZnSOD can be functionally replaced by MnSOD in this cellular compartment in spite of the evolutionary unrelatedness of the two enzymes.

Mutants in *SOD1* in *S. cerevisiae* show auxotrophy for cysteine or methionine and lysine when grown in synthetic medium under air, but they do not show the same requirements when grown under N_2 (Bilinski et al., 1985; Chang et al., 1991). The Met/Cys auxotrophy has been investigated in full detail by Chang and Kosman (1990), who suggested that the block in converting SO_4^{2-} to S^{2-} in a *sod1⁻* strain resulted in the persistence of the cytotoxic compound SO_3^{2-}. In their view, such mutant strains would require methionine not because of any lack in the synthetic machinery but because Met turns off the assimilation of SO_4^{2-} and this would prevent accumulation of toxic SO_3^{2-}.

sod2⁻ yeast is unaffected by methionine depletion and only slightly inhibited by lack of lysine (Liu et al., 1992).

Ability to grow on different media and in different aeration conditions have been reported for both *sod1⁻* and *sod2⁻* mutants. Doubling time of *sod1⁻* cells (lacking Cu,ZnSOD) is increased when they are grown on 10% glucose, a condition that causes strong repression of MnSOD. On the other hand, the rate of growth of this mutant is not affected when the carbon source is glycerol, which induces MnSOD, possibly compensating for the deleterious effects of the lack of Cu,ZnSOD (Gralla and Valentine, 1991).

sod2⁻ strains exhibit normal growth on glucose at 80% O_2 (Westerbeek-Marres et al., 1988) and are still able to grow—although at a strongly reduced rate—in pure oxygen on such media (Van Loon et al., 1986). However, *sod2⁻* strains fail to grow on ethanol and acetate under air but can duplicate when supplied with glycerol, lactate, or pyruvate (Van Loon et al., 1986; Westerbeek-Marres et al., 1988). Why three- but not two-carbon units can serve as carbon sources for *sod2⁻* strains under all conditions is not clear. A possible explanation is that one of the enzymes necessary for their utilization is sensitive to superoxide-related species.

Null mutants of superoxide dismutase (containing deletions in both SOD1 and SOD2 genes; Liu et al., 1992) are also defective in sporulation, a process known to require oxygen consumption. Sporulation of the single mutants *sod1⁻* and *sod2⁻* decreased 100- and more than 10,000-fold, respectively, suggesting that protection from oxidative damage is essential for some metabolic enzymes in *S. cerevisiae* as it is in *E. coli* (Carlioz and Touati, 1986).

Liu et al. (1992) have also isolated two groups of *S. cerevisiae* mutants (*bsd1* and *bsd2*) that are able to bypass the *sod1⁻ sod2⁻* defective phenotype, suppressing several of the biological defects associated with the inactivation of both superoxide dismutases. The *bsd* mutations are recessive, and suppression seems to arise from mutation of components of the electron transfer system, probably acting upstream of cytochrome *c*, resulting in depression of the O_2^- generating activity of mitochondria and therefore bypassing the need for SOD.

REFERENCES

Archibald, F.S., and Fridovich, I. (1981). Manganese and defenses against oxygen toxicity in *Lactobacillus plantarum, J. Bacteriol., 145*:442.

Argese, E., Viglino, P., Rotilio, G., Scarpa, M., and Rigo, A. (1987). Electrostatic control of the rate-determining step of the copper, zinc superoxide dismutase catalytic reaction, *Biochemistry, 26:* 3224.

Arnold, L.D., and Lepock, J.R. (1982). Reversibility of the thermal denaturation of yeast superoxide dismutase, *FEBS Lett., 146:* 302.

Auclair, C., De Prost, D., and Hakim, J. (1978). Superoxide anion production by liver microsomes from phenobarbital treated rats, *Biochem. Pharmacol., 27:* 335.

Autor, A.P. (1982). Biosynthesis of mitochondrial manganese superoxide dismutase in *Saccharomyces cerevisiae.* Precursor form of mitochondrial manganese superoxide dismutase made in the cytoplasm, *J. Biol. Chem., 257:* 2713.

Baccanari, D.P. (1978). Coupled oxidation of NADPH with thiols at neutral pH, *Arch. Biochem. Biophys., 191:* 351.

Ballou, D., Palmer, G., and Massey, V. (1969). Direct demonstration of superoxide anion production during the oxidation of reduced flavins and of its catalytic decomposition by erythrocuprein, *Biochem. Biophys. Res. Commun., 36:* 898.

Bannister, J.V., Desideri, A., and Rotilio, G. (1985). Replacement of Mn(III) with Cu(II) in *Bacillus stearothermophilus* superoxide dismutase, *FEBS Lett., 188:* 91.

Bannister, J.V., Bannister, W.H., and Rotilio, G. (1987). Aspects of the structure, function and applications of superoxide dismutase, *CRC Crit. Rev. Biochem., 22:* 11–180.

Barra, D., Bossa, F., Marmocchi, F., Martini, F., Rigo, A., and Rotilio, G. (1979). Differential effects of urea on yeast and bovine copper, zinc superoxide dismutases, in relation to the extent of analogy of primary structures, *Biochem. Biophys. Res. Commun., 86:* 1199.

Bauer, R., Demeter, I., Hasemann, V., and Johansen, J.T. (1980). Structural properties of the zinc site in Cu,Zn-superoxide dismutase; perturbed angular correlation of gamma-ray spectroscopy of the Cu,^{111}Cd-superoxide dismutase derivative, *Biochem. Biophys. Res. Commun., 94:* 1296.

Beem, K.M., Richardson, J.S., and Richardson, C.D. (1976). Manganese superoxide dismutases from *Escherichia coli* and from yeast mitochondria: Preliminary X-ray crystallographic studies, *J. Mol. Biol., 105:* 327.

Belazzi, T., Wagner, A., Wieser, R., Shanz, M., Adam, G., Hartig, A., and Ruis, H. (1991). Negative regulation of transcription of the *Saccharomyces cerevisiae* catalase T (CTT1) gene by cAMP is mediated by a positive control element, *EMBO J., 10:*585.

Bermingham-McDonogh, O., Gralla, E.B., and Valentine, J.S. (1988). The copper, zinc-superoxide dismutase gene of *Saccharomyces cerevisiae:* Cloning, sequencing, and biological activity, *Proc. Natl. Acad. Sci. USA, 85:* 4789.

Beyer, W.F., Fridovich, I., Mullenbach, G.T., and Hallewell, R. (1987). Examination of the role of Arginine-143 in the human copper and zinc superoxide dismutase by site-specific mutagenesis, *J. Biol. Chem., 262:* 11182.

Beyer, W., Imlay, J., and Fridovich, I. (1991). Superoxide dismutases, *Progr. Nucl. Acid Res. Mol. Biol., 40:* 221.

Bilinski, T., Krawiec, Z., Liczmanski, A., and Litwinska, J. (1985). Is hydroxyl radical generated by the Fenton reaction in vivo? *Biochem. Biophys. Res. Commun., 130:* 533.

Borgsthal, S.E.O., Parge, H.E., Hickey, M.J., Beyer, W.F., Hallewell, R.A., and Tainer, J.A. (1992). The structure of human mitochondrial manganese superoxide dismutase reveals a novel tetrameric interface of two 4-helix bundles, *Cell, 71:* 107.

Boveris, A. (1978). Production of superoxide anion and hydrogen peroxide in yeast mitochondria, *Biochemistry and Genetics of Yeasts* (M. Bacila, B.L. Horecker, and A.O.M. Stoppani, eds.), Academic Press, New York, pp. 65–80.

Boveris, A. (1984). Determination of the production of superoxide radical and hydrogen peroxide in mitochondria, *Methods Enzymol., 105:* 429.

Boveris, A., and Cadenas, E. (1982). Production of superoxide radicals and hydrogen peroxide in mitochondria, *Superoxide Dismutases* (L.W. Oberley, ed.), vol. II, CRC Press, Boca Raton, FL, pp. 15–30.

Boveris, A., and Chance, B. (1973). The mitochondrial generation of hydrogen peroxide. General properties and effect of hyperbaric oxygen, *Biochem. J., 134:* 707.

Bowler, C., van Kaer, L., van Camp, W., van Montagu, M., Inzé, D., and Dhaese, P. (1990). Characterization of the *Bacillus stearothermophilus* manganese superoxide dismutase gene and its ability to complement copper/zinc superoxide dismutase deficiency in *Saccharomyces cerevisiae, J. Bacteriol., 172:* 1539.

Cadenas, E. (1989). Biochemistry of oxygen toxicity, *Ann. Rev. Biochem., 58:* 79.

Carlioz, A., and Touati, D. (1986). Isolation of superoxide dismutase mutants in *Escherichia coli:* Is superoxide dismutase necessary for aerobic life? *EMBO J., 5:* 623.

Carlioz, A., Ludwig, M.L., Stallings, W.C., Fee, J.A., Steinman, H.M., and

Touati, D. (1988). Iron superoxide dismutase nucleotide sequence of the gene from *Escherichia coli* K12 and correlation with crystal structures, *J. Biol. Chem., 263:* 1555.

Carrì, M.T., Galiazzo, F., Ciriolo, M.R., and Rotilio, G. (1991). Evidence for co-regulation of Cu,Zn superoxide dismutase and metallothionein gene expression in yeast through transcriptional control by copper *via* the ACE1 factor, *FEBS Lett., 278:* 263.

Cass, A.E.G., O'Hill, H.A., Hasemann, V., and Johansen, J.T. (1978). ^{1}H Nuclear magnetic resonance spectroscopy of yeast copper-zinc superoxide dismutase. Structural homology with the bovine enzime, *Carlsberg Res. Commun., 43:* 439.

Chan, V.W.F., Bjerrum, M.J., and Borders, C.L. (1990). Evidence that chemical modification of a positively charged residue at position 189 causes the loss of catalytic activity of iron-containing and manganese-containing superoxide dismutases, *Arch. Biochem. Biophys., 279:* 195.

Chance, B., Sies, H., and Boveris, A. (1979). Hydroperoxide metabolism in mammalian organs, *Physiol. Rev., 59:* 527.

Chang, E.C., and Kosman, D.J. (1989). Intracellular Mn(II)-associated superoxide scavenging activity protects Cu,Zn superoxide dismutase-deficient *Saccharomyces cerevisiae* against dioxygen stress, *J. Biol. Chem., 264:* 12172.

Chang, E.C., and Kosman, D.J. (1990). O_2-dependent methionine auxotrophy in Cu,Zn superoxide dismutase-deficient mutants of *Saccharomyces cerevisiae, J. Bacteriol., 172:* 1840.

Chang, E.C., Crawford, B.F., Hong, Z., Bilinski, T., and Kosman, D.J. (1991). Genetic and biochemical characterization of Cu,Zn superoxide dismutase mutants in *Saccharomyces cerevisiae, J. Biol. Chem., 266:* 4417.

Chikata, Y., Kusunose, E., and Ichihara, K. (1975). Purification of superoxide dismutases from *Mycobacterium phlei, Osaka City Med. J., 21:* 127.

Cocco, D., Calabrese, L., Rigo, A., Marmocchi, F., and Rotilio, G. (1981). Preparation of selectively metal-free and metal-substituted derivatives by reaction of Cu-Zn superoxide dismutase with diethyldithiocarbamate, *Biochem. J., 199:* 675.

Cohen, G., and Heikkila, R.E. (1974). The generation of hydrogen peroxide, superoxide radical, and hydroxyl radical by 6-hydroxydopamine, dialuric acid, and related cytotoxic agents, *J. Biol. Chem., 249:* 2447.

Desideri, A., Falconi, M., Polticelli, F., Bolognesi, M., Djinovic, K., and Rotilio, G. (1992). Evolutionary conservativeness of electric field in the Cu,Zn superoxide dismutase active site, *J. Mol. Biol., 223:* 337.

Ditlow, C., Johansen, J.T., Martin, B.M., and Svendsen, I.B. (1984). The complete amino acid sequence of manganese-superoxide dismutase from *Saccharomyces cerevisiae. Carlsberg Res. Commun., 47:* 81.

Djinovic, K., Gatti, G., Coda, A., Antolini, G., Pelosi, G., Desideri, A., Falconi, M., Marmocchi, F., Rotilio, G., and Bolognesi, M. (1992). Crystal structure of yeast Cu,Zn superoxide dismutase, *J. Mol. Biol., 225:* 791.

Dunbar, J.C., Holmquist, B., and Johansen, J.T. (1984). Asymmetric active site structures in yeast dicopper dizink superoxide dismutase. 1. Reconstitution of apo-superoxide dismutase, *Biochemistry, 23:* 4324.

Fielden, E.M., Roberts, P.B., Bray, R.C., Lowe, D.J., Mautner, G.M., Rotilio, G., and Calabrese, L. (1974). The mechanism of action of superoxide dismutase from pulse radiolysis and electron paramagnetic resonance, *Biochem. J., 139:* 49.

Fridovich, I. (1983). Superoxide radical: An endogenous toxicant, *Ann. Rev. Pharmacol. Toxicol., 23:* 239.

Fridovich, I. (1986). Biological effects of the superoxide radical, *Arch. Biochem. Biophys., 247:* 1.

Fridovich, I. (1989). Superoxide dismutases. An adaptation to a paramagnetic gas, *J. Biol. Chem., 264:* 7761.

Frigerio, F., Falconi, M., Gatti, G., Bolognesi, M., Desideri, A., Marmocchi, F., and Rotilio, G. (1989). Crystallographic characterization and the three-dimensional model of yeast Cu,Zn superoxide dismutase, *Biochem. Biophys. Res. Commun., 160:* 677.

Furst, P., Hu, S., Hackett, R., and Hamer, D. (1988). Copper activates metallothionein gene transcription by altering the conformation of a specific DNA binding protein, *Cell, 55:* 705.

Galiazzo, F., and Labbe-Bois, R. (1993). Regulation of Cu,Zn- and Mn-superoxide dismutase transcription in *Saccharomyces cerevisiae, FEBS Lett., 315:* 197.

Galiazzo, F., Schiesser, A., and Rotilio, G. (1988). Oxygen-independent induction of enzyme activities related to oxygen metabolism in yeast by copper, *Biochim. Biophys. Acta, 965:* 46.

Galiazzo, F., Pedersen, J., Civitareale, P., Schiesser, A., and Rotilio, G. (1989). Manganese accumulation in yeast cells. Electron-spin-resonance characterization and superoxide dismutase activity, *Biol. Metals., 2:* 6.

Galiazzo, F., Ciriolo, M.R., Carrì, M.T., Civitareale, P., Marcocci, L., Marmocchi, F., and Rotilio, G. (1991). Activation and induction by copper of Cu/Zn superoxide dismutase in *Saccharomyces cerevisiae.* Presence of an inactive proenzyme in anaerobic yeast, *Eur. J. Biochem., 196:* 545.

Getzoff, E.D., Tainer, J.A., Weiner, P.K., Kollman, P.A., Richardson, J., and Richardson, D.C. (1983). Electrostatic recognition between superoxide and copper, zinc superoxide dismutase, *Nature, 306:* 287.

Goscin, S.A., and Fridovich, I. (1972). The purification and properties of superoxide dismutase from *Saccharomyces cerevisiae, Biochim. Biophys. Acta, 289:* 276.

Gralla, E.B., and Valentine, J.S. (1991). Null mutants of *Saccharomyces cerevisiae* Cu,Zn superoxide dismutase: Characterization and spontaneous mutation rates, *J. Bacteriol., 173:* 5918.

Gralla, E.B., Thiele, D.J., Silar, P., and Valentine, J.S. (1991). ACE1, a copper-dependent transcription factor, activates expression of the yeast copper,zinc superoxide dismutase gene, *Proc. Natl. Acad. Sci. USA, 88:* 8558.

Greco, M.A., Hrab, D.I., Magner, W., and Kosman, D.J. (1990). Cu,Zn superoxide dismutase and copper deprivation and toxicity in *Saccharomyces cerevisiae, J. Bacteriol., 172:* 317.

Gregory, E.M., Goscin, S.A., and Fridovich, I. (1974). Superoxide dismutase and oxygen toxicity in a eukaryote, *J. Bacteriol., 177:* 456.

Hallewell, R.A., Imlay, K.C., Lee, P., Fong, N.M., Gallegos, C., Getzoff, E., Tainer, J.A., Cabelli, D.E., Tekamp-Olson, P, Mullenbach, G.T., and Cousens, L.S. (1991). Thermostabilization of recombinant human and bovine superoxide dismutases by replacement of free cysteines, *Biochem, Biophys. Res. Commun., 181:* 474.

Hassan, H.M., and Fridovich, I. (1977). Regulation of the synthesis of superoxide dismutase in *Escherichia coli.* Induction by methyl viologen, *J. Biol. Chem., 252:* 7667.

Johansen, J.T., Overballe-Petersen, C., Martin, B., Hasemann, V., and Svendsen, I. (1979). The complete amino acid sequence of copper,zinc superoxide dismutase from *Saccharomyces cerevisiae, Carlsberg Res. Commun., 44:*201.

Johnson, K.J., and Ward, P.A. (1985). Inflammation and active oxygen species, *Superoxide Dismutases* (L.W. Oberley, ed.), vol. 3, CRC Press, Boca Raton, FL, pp. 129–142.

Karlsson, K., and Marklund, S.L. (1988). Extracellular superoxide dismutase in the vascular system of mammals, *Biochem. J., 255:* 223.

Kitagawa, Y., Tanaka, N., Hata, Y., Kusonoki, M., Lee, G., Katsube, Y., Asada, K., Aibara, S., and Morita, Y. (1991). Three-dimensional structure of Cu,Zn superoxide dismutase from spinach at 2.0 Å resolution, *J. Biochem., 109:* 447.

Korbashi, P., Katzhendler, J., and Chevion, M. (1986). Iron mediates paraquat toxicity in *Escherichia coli, J. Biol. Chem., 261:* 12472.

Lee, F.J., and Hassan, H.M. (1985). Biosynthesis of superoxide dismutase in *Saccharomyces cerevisiae:* Effect of paraquat and copper, *J. Free Radicals Biol. Med., 1:* 319.

Lesuisse, E., Crichton, R.R., and Labbe, P. (1990). Iron-reductases in yeast *Saccharomyces cerevisiae, Biochim. Biophys. Acta, 1038:* 253.

Liu, X.F., Elashvili, I., Gralla, E.B., Valentine, J.S., Lapinsakas, P., and Culotta, V.C. (1992). Yeast lacking superoxide dismutase, *J. Biol. Chem., 267:* 18298.

Lowry, C.V., and Zitomer, R.S. (1984). Oxygen regulation of anaerobic and aerobic genes mediated by a common factor in yeast, *Proc. Natl. Acad. Sci. USA, 81:* 6129.

Lu, Y., Gralla, E.B., Roe, J.A., and Valentine, J.S. (1992). Redesign of a type 2 into a type 1 copper protein: Construction and characterization of yeast copper-zinc superoxide dismutase mutants, *J. Amer. Chem. Soc., 114:* 3560.

Ludwig, M.L., Metzger, A.L., Pattridge, K.A., and Stallings, W.C. (1991) Manganese superoxide dismutase from *Thermus thermophilus:* A structural model refined at 1.8 Å resolution, *J. Mol. Biol., 219:* 335.

McAdam, M.E., Fielden, E.M., Lavelle, F., Calabrese, L., Cocco, D., and Rotilio, G. (1977). The involvement of the bridging imidazolate in the catalytic mechanism of action of bovine superoxide dismutase, *Biochmem. J., 167:* 271.

McCord, J., and Fridovich, I. (1969). Superoxide dismutase. An enzymatic function for erythrocuprein (hemocuprein), *J. Biol. Chem., 244:* 6049.

McRee, D.E., Redford, S.M., Getzoff, E.D., Lepock, J.R., Hallewell, R.A., and Tainer, J.A. (1990). Changes in the crytallographic structure and thermostability of a Cu,Zn superoxide dismutase mutant resulting from the removal of a buried cysteine, *J. Biol. Chem., 265:* 14234.

Malinowski, D.P., and Fridovich, I. (1979a). Chemical modification of arginine at the active site of the bovine erythrocyte superoxide dismutase, *Biochemistry, 18:* 5909.

Malinowski, D.P., and Fridovich, I. (1979b). Subunit association and side-chain reactivities of bovine erythrocyte superoxide dismutase in denaturing solvents, *Biochemistry, 18:* 5055.

Malmstrom, B.G. (1982). Enzymology of oxygen, *Ann. Rev. Biochem., 51:* 21.

Mann, T., and Keilin, D. (1939). Haemocuprein and hepatocuprein, copper-protein compounds of blood and liver in mammals, *Proc. R. Soc. B., 126: 303.*

Marklund, S.L. (1982). Human copper-containing superoxide dismutase of high molecular weight, *Proc. Natl. Acad. Sci., 79:* 7634.

Marmocchi, F., Venardi, G., Bossa, F., Rigo, A., and Rotilio, G. (1978). Dissociation of Cu-Zn superoxide dismutase into monomers by urea, *FEBS Lett., 94:* 109.

Marmocchi, F., Argese, E., Rigo, A., Mavelli, I., Rossi, L., and Rotilio, G. (1983). A comparative study of bovine, porcine and yeast superoxide dismutases, *Mol. Cell. Biol., 51:* 161.

Marres, C.A.M., van Loon, A.P.G.M., Oudshoorn, P., van Steeg, H., Grivell, L.A., and Slater, E.C. (1985). Nucleotide sequence analysis of the nuclear gene coding for manganese superoxide dismutase of yeast mitochondria, a gene previously assumed to code for the Rieske iron-sulfur protein, *Eur. J. Biochem., 147:* 153.

Meier, B., Barra, D., Bossa, F., Calabrese, L., and Rotilio, G. (1982). Synthesis of either Fe- or Mn-superoxide dismutase with an apparently identical protein moiety by an anaerobic bacterium dependent on the metal supplied, *J. Biol. Chem., 257:* 13977.

Misra, H.P. (1974). Generation of superoxide free radical during the autoxidation of thiols, *J. Biol. Chem., 249:* 2151.

Misra, H.P., and Fridovich, I. (1972). The univalent reduction of oxygen by reduced flavins and quinones, *J. Biol. Chem., 247:* 188.

Munkres, K.D. (1990). Purification of exocellular superoxide dismutases, *Methods Enzymol., 186:* 249.

O'Neill, P., Davies, S., Fielden, M., Calabrese, L., Capo, C., Marmocchi, F., Natoli, G., and Rotilio, G. (1988). The effects of pH and various salts upon the activity of a series of superoxide dismutases, *Biochem. J., 251:* 41.

Parge, H.E., Hallewell, R.A., and Tainer, J.A. (1992). Atomic structures of wild-type and thermostable mutant recombinant human Cu,Zn SOD, *Proc. Natl. Acad. Sci. USA, 89:* 6109.

Parker, M.W., and Blake, C.C.F. (1988). Crystal structure of manganese superoxide dismutase from *Bacillus stearothermophilus* at 2.4 Å resolution, *J. Mol. Biol., 199:* 649.

Pillar, T.M., and Bradshaw, R.E. (1991). Heat shock and stationary phase induce transcription of the *Saccharomyces cerevisiae* iso-2-cytochrome c gene, *Curr. Genet., 20:* 185.

Pinkham, J.L. (1991). Transcriptional regulation of the manganous superoxide dismutase gene of *S. cerevisiae,* Albany Conference, Sept. 12–15, "Molecular and Cellular Responses to Oxygen," p. 41.

Ravindranath, S.D., and Fridovich, I. (1975). Isolation and characterization of a manganese-containing superoxide dismutase from yeast, *J. Biol. Chem., 250:* 6107.

Rigo, A., Viglino, P., Calabrese, L., Cocco, D., and Rotilio, G. (1977a). The binding of copper ions to copper-free bovine superoxide dismutase. Copper distribution in protein samples recombined with less than stoichiometric copper/protein ratios, *Biochem. J., 161:* 27.

Rigo, A., Terenzi, M., Viglino, P., Calabrese, L., Cocco, D., and Rotilio, G. (1977b). The binding of copper ions to copper-free bovine superoxide dismutase. Properties of the protein recombined with increasing amounts of copper ions, *Biochem. J. 161:* 31.

Rigo, A., Stevanato, R., and Viglino, P. (1977c). Competitive inhibition of Cu,Zn superoxide dismutase by monovalent anions, *Biochem. Biophys. Res. Commun., 79:* 776.

Rigo, A., Viglino, P., Brunori, M., Cocco, D., Calabrese, L., and Rotilio, G. (1978). The binding of copper ions to copper-free bovine superoxide dismutase, *Biochem. J., 169:* 277.

Roe, J.A., Butler, A., Scholler, D.M., Valentine, J.S., Marky, L., and Breslauer, K.J. (1988). Differential scanning calorimetry of Cu,Zn superoxide dismutase, the apoprotein, and its zinc-substituted derivatives, *Biochemistry, 27:* 950.

Rotilio, G., Morpurgo, L., Giovagnoli, G., Calabrese, L., and Mondovì B. (1972). Studies on the metal sites of copper proteins. Symmetry of copper in bovine superoxide dismutase and its functional significance, *Biochemistry, 11:* 2187.

Rotilio, G., Mavelli, I., Rossi, L., and Ciriolo, M.R. (1985). Biochemical mechanism of oxidative damage by redox-cycling drugs, *Environ. Health Perspect., 64:* 258.

Sato, S., and Harris, J.I. (1977). Superoxide dismutase from *Thermus aquaticus.* Isolation and characterization of manganese and apo enzymes, *Eur. J. Biochem., 73:* 373.

Schnell, N., Krems, B., and Entian, K.-D. (1992). The PAR1 (YAP1/SNQ3) gene of *Saccharomyces cerevisiae,* a *c-jun* homologue, is involved in oxygen metabolism, *Curr. Genet., 21:*269.

Sette, M., Paci, M., Desideri, A., and Rotilio, G. (1992). Formate as an NMR probe of anion binding to Cu,Zn and Cu,Co bovine erythrocyte superoxide dismutases, *Biochemistry, 31:* 12410.

Simurda, M.C., Keulen, H.V., Rekosh, D.M., and Lo Verde, P.T. (1988). *Schistosoma mansoni:* Identification and analysis of an mRNA and a gene encoding superoxide dismutase (Cu/Zn), *Exp. Parasitol., 67:* 73.

Stallings, W.C., Pattridge, K.A., and Ludwig, M.L. (1986). The active center of *T. thermophilus* Mn superoxide dismutase and of *E. coli* Fe superoxide dismutase: Current status of the crystallography, *Superoxide and Superoxide Dismutase in Chemistry, Biology and Medicine* (G. Rotilio, ed.), Elsevier, p. 195.

Steinman, H.M. (1980). The amino acid sequence of copper-zinc superoxide dismutase from baker's yeast, *J. Biol. Chem., 225:* 6758.

Stevens, B.J. (1981). Mitochondrial structure, *The Molecular Biology of the Yeast Saccharamoyces: Life Cycle and Inheritance* (J.N. Strathern et al., eds.), Cold Spring Harbor Laboratory, Cold Spring Harbor, New York, p. 471.

Stoddard, B.L., Howell, P.L., Ringe, D., and Petsko, G. (1990). The 2.1 Å resolution structure of iron superoxide dismutase from *Pseudomonas ovalis, Biochemistry, 29:* 8885.

Tainer, J.A., Getzoff, E.D., Beem, K.M., Richardson, J.S., and Richardson, D.C. (1982). Determination and analysis of the 2 Å structure of copper,zinc superoxide dismutase, *J. Mol. Biol., 160:* 181.

Tamai, T.K., Gralla, E.B., Ellerby, L., Valentine, J.S., and Thiele, D.J. (1992). The role of metallothionein in oxidative stress, Sixteenth International Conference on Yeast Genetics and Molecular Biology, Vienna, Austria, August 15–21.

Thiele, D.J. (1988). ACE1 regulates expression of the *Saccharomyces cerevisiae* metallothionein gene, *Mol. Cell. Biol., 8:* 2745.

Thor, H., Smith, M.T., Hartzell, P., Bellomo, G., Jewell, S.A., and Orrenius, S. (1982). The metabolism of menadione by isolated hepatocytes. A study of the implications of oxidative stress in intact cells, *J. Biol. Chem., 257:* 12419.

Tibell, L., Hjalmarsson, K., Edlund, T., Skogman, G., Engstrom, A., and Marklund, S.L. (1987). Expression of human extracellular superoxide dismutase in Chinese hamster ovary cells and characterization of the product, *Proc. Natl. Acad. Sci. USA, 84:* 6634.

Turi, T.G., Kalb, V.F., and Loper, J.C. (1991). Cytochrome P450 lanosterol 14alfa-demethylase (ERG1) and manganese superoxide dismutase (SOD1) are adjacent genes in *Saccharomyces cerevisiae, Yeast, 7:* 627.

Turrens, J.F., and Boveris, A. (1980). Generation of superoxide anion by the NADH dehydrogenase of bovine heart mitochondria, *Biochem. J., 191:* 421.

Van Loon, A.P.G.M., Maarse, A.C., Riezman, H., and Grivell, L.A. (1983). Isolation, characterization and regulation of expression of the nuclear genes for the Core II and Rieske iron-sulfur proteins of the yeast ubiquinol-cytochrome c reductase, *Gene, 26:* 261.

Van Loon, A.P.G.M., Pesold-Hurt, B., and Schatz, G. (1986). A yeast mutant lacking mitochondrial manganese-superoxide dismutase is hypersensitive to oxygen, *Proc. Natl. Acad. Sci. USA, 83:* 3820.

Viglino, P., Rigo, A., Argese, E., Calabrese, L., Cocco, D., and Rotilio, G. (1981). ^{19}F relaxation as a probe of the oxidation state of Cu,Zn superoxide dismutase. Studies on the enzyme steady-state turnover, *Biochem. Biophys. Res. Comm., 100:* 125.

Westerbeek-Marres, C.A.M., Moore, M.M., and Autor, A.P. (1988). Regulation of manganese superoxide dismutase in *Saccharomyces cerevisiae:* The role of respiratory chain activity, *Eur. J. Biochem., 174:* 611.

Wieser, R., Adam, G., Wagner, A., Schuller, C., Marchler, G., Ruis, H., Krawieck, Z., and Bilinski, T. (1991). Heat shock factor-independent heat control of transcription of the CTT1 gene encoding of the cytosolic catalase T of *Saccharomyces cerevisiae, J. Biol. Chem., 266:* 12406.

Zitomer, R.S., and Lowry, C.V. (1992). Regulation of gene expression by oxygen in *Saccharomyces cerevisiae, Micobiol. Rev., 56:* 1.

14

Evolution and Biological Roles of Fungal Superoxide Dismutases

Donald O. Natvig, William H. Dvorachek, Jr., and Kenneth Sylvester
University of New Mexico, Albuquerque, New Mexico

I. INTRODUCTION

A. Scope

Although superoxide dismutases (SODs) are among the best studied proteins, many questions remain regarding the distribution and importance of different SOD types. Among fungi, the SODs of *Saccharomyces cerevisiae* and *Neurospora crassa* have been most studied. We have sought here to review contributions that fungal studies have made to understanding SOD distribution, function, and evolution. In addition, we have attempted to identify important unresolved questions, particularly in those areas where fungi other than *S. cerevisiae* and *N. crassa* may serve important roles as experimental organisms.

Although some overlap with a recent excellent review (Gralla and Kosman, 1992) and with a contribution to this volume (Falliazzo et al.) is inevitable, we have attempted to focus on topics related to our own recent studies with *N. crassa*. We have not attempted a review of the somewhat complex issue of SOD gene regulation, which has been reviewed recently by Gralla and Kosman (1992).

B. Superoxide Toxicity and Superoxide Dismutases

Superoxide dismutases are metalloenzymes that catalyze the reaction

$$O_2^- + O_2^- + 2H^+ \rightarrow H_2O_2 + O_2$$

This activity was first reported as an *in vitro* activity exhibited by bovine erythrocuprein (now called CuZn superoxide dismutase) (McCord and Fridovich, 1969). Its discovery led to the hypothesis that superoxide (O_2^-) is a primary agent of oxygen toxicity in aerobic cells, with the further suggestion that SOD is a first-line defense against this toxicity (McCord et al., 1971). This hypothesis has been controversial, because the importance of O_2^- as an intracellular toxin has been questioned (Fee, 1982; Sawyer and Valentine, 1981). Doubts as to whether superoxide dismutation is the *in vivo* function of this group of proteins persisted for more than a decade. In the past decade, however, acceptance has grown as evidence for this role has accumulated.

The intracellular production of O_2^- during normal aerobic metabolism is now well documented. Superoxide can be generated either enzymatically or by autoxidation of small molecules (Fridovich, 1984). Intracellular production of O_2^- can also be increased artificially by a large number of redox-active compounds (Hassan and Fridovich, 1978, 1979a). These include synthetic compounds, most notably paraquat, and others that are naturally occurring. Paraquat and other compounds have played critical roles in experimental studies, because they provide a means to control intracellular levels of O_2^-.

There is a growing list of specialized biochemical systems that suggest direct or indirect influences of O_2^- on microbial environments. These include the well-studied oxidative burst of neutrophils (e.g., Babior, 1988), the hypersensitive response of certain plant tissues to fungal, bacterial, or viral infection (e.g., Doke and Ohashi, 1988), and the release of O_2^- by plant roots into the environment of root colonizing bacteria and fungi (Katsuwon et al., 1993).

Substantial evidence points to the toxicity of O_2^-. Superoxide has been implicated, directly or indirectly, in damage to all major classes of macromolecules (Fridovich, 1986a; Brawn and Fridovich, 1981; Fridovich, 1981; Farr et al., 1986). The precise mode of toxicity of O_2^- has remained elusive, however, in part because the toxic effects can be indirect. It has been proposed that much of the damage mediated by O_2^-, particularly damage to DNA, is actually caused by the hydroxyl radical (OH·), generated in the presence of metal catalysts, O_2^-, and H_2O_2 by the Haber-Weiss reaction (Haber and Weiss, 1934):

$$O_2^- + H_2O_2 \rightarrow O_2 + OH^- + OH\cdot$$

Whatever the primary agent of toxicity, it is clear that superoxide and its by-products are harmful and that organisms growing in aerobic environments are under selective pressure to develop mechanisms to protect themselves from superoxide toxicity. Superoxide dismutases constitute an important first-line defense.

C. The Importance of Fungal Studies

Studies employing *Saccharomyces cerevisiae* and *Neurospora crassa* are making important contributions toward understanding the biological roles of SODs. This is particularly true of studies that have employed mutant alleles of one or more SOD type. As is discussed in greater detail below, not only has the characterization of these mutants helped establish the importance of SODs for growth and survival, but also specific phenotypes of these mutants have helped to identify likely important targets of superoxide-mediated toxicity. As tissue-simple eukaryotes, fungi are ideal organisms to explore such basic consequences of O_2^- toxicity as damage to DNA and membrane components, disruption of metabolic processes, and induction of defensive genes such as those that encode SODs.

Fungal studies are also proving important in the study of SOD evolution in terms of gene and protein structures, distribution of different SOD types in the biological world, and intracellular and extracellular targeting of SODs.

Finally, fungal studies will likely prove important in the study of ecological aspects of O_2^- toxicity. There is an ever-increasing number of reports of O_2^- production as a defense against pathogenic organisms, for example. Superoxide dismutases therefore become important as agents of defense against O_2^- (for both host and parasite) and as *in vitro* tools for assessing the importance of O_2^- in specific processes.

II. DISTRIBUTION OF SUPEROXIDE DISMUTASES IN FUNGI AND OTHER ORGANISMS

A. SOD Families

It is now quite clear that from an evolutionary point of view individual SOD genes exist as members of one of two SOD gene families. One family is composed of enzymes that require both Cu and Zn (CuZnSOD).

CuZnSODs are homodimeric proteins with subunit weights of 16 kDa. The other family contains enzymes that require either Fe or Mn (FeSODs and MnSODs). FeSODs and MnSODs are clearly related to each other but are clearly unrelated to CuZnSODs (Steinman, 1978). FeSOD and MnSOD amino-acid sequences exhibit clear similarities (Smith and Doolittle, 1992); Fe and Mn polypeptides in *E. coli* can interact to form a functional dimer (Hassan and Fridovich, 1977; Dougherty et al., 1978); and in a few known bacterial enzymes, Fe and Mn atoms are interchangeable (Martin et al., 1986; Pennington and Gregory, 1986). FeSODs and MnSODs are homodimers or homotetramers with subunit weights of 22 kDa.

B. Structure and Catalysis

Although CuZnSODs and Mn/FeSODs are clearly members of very different protein families, the general mechanism of catalysis is the same. All SODs employ alternate reduction and oxidation of transition metals in two half-cycle reactions that combine to produce O_2 and H_2O_2 from two molecules of O_2^-. This can be summarized as follows (E-M = enzyme-metal complex):

(1) $O_2^- + \text{E-M}^n \rightarrow \text{E-M}^{n-1} + O_2$
(2) $O_2^- + 2H^+ + \text{E-M}^{n-1} \rightarrow \text{E-M}^n + H_2O_2$

The relevant metals (and valence states) are Cu(II) ↔ Cu(I), Fe(III) ↔ Fe(II), and Mn(III) ↔ Mn(II).

As the previous discussion suggests, copper has the primary catalytic role of the two metals associated with the CuZnSOD polypeptide. The zinc atom is generally assigned a structural function, although a subsidiary role for zinc in catalysis has been proposed (Tainer et al., 1983).

The rate of catalysis by CuZnSODs depends not only on events at the active site but also on electrostatic interactions that attract and guide O_2^- to the active site. The result is a catalytic rate that approaches diffusion limits (Getzoff et al., 1983). The electrostatic guidance of O_2^- to the active-site copper depends on a spatial distribution of charges near the active-site channel. The conservation of this charge distribution among diverse CuZnSODs is reflected in charge-compensating amino-acid substitutions, and not simply in specific conserved residues (Desideri et al., 1992). An analogous guidance system for O_2^- has been proposed for MnSODs (Borgstahl et al., 1992).

C. Subcellular Localization and SOD Specialization

Evolution has taken many liberties with members of both SOD families. This is especially true with regard to subcellular targeting. A complex and ever-changing picture has emerged with respect to the targeting of SODs to organelles and across cell membranes.

CuZnSODs, once believed to be restricted to the cytosols of eukaryotes (where they appear to reside in *S. cerevisiae* and *N. crassa*), are now known to occur in a variety of cellular compartments across the biological world. These compartments include chloroplasts (Kanematsu and Asada, 1990) and the peroxisomes of mammals (Keller et al., 1991). Although CuZn SODs have been reported rarely from prokaryotes, they do occur. When they occur in prokaryotes, they appear to be targeted for the periplasmic space (Steinman, 1987). A modified CuZnSOD is synthesized and exported from certain mammalian cells. The gene for this extracellular SOD (EC-SOD) in humans has been cloned and sequenced (Hjalmarsson et al., 1987). A gene that may be homologous has also been characterized from *Schistosoma mansoni* (Simurda et al., 1988).

It is interesting that the major intracellular CuZnSOD is peroxisomal in mammals but cytosolic in fungi. The cytosolic location of this protein in *S. cerevisiae* is suggested by both experiment and inference based on the absence of a peroxisome-targeting sequence (Keller et al., 1991). The human protein has a putative peroxisome-targeting tripeptide (shown below in bold) nine residues from the C terminus. The *S. cerevisiae* and *N. crassa* CuZnSODs lack similar peroxisome-targeting consensus sequences in this otherwise highly conserved portion of the protein. The homologous regions for the CuZnSODs from the three organisms are (see Chary et al., 1990)

Ala-Gly-**Ser-Arg-Leu**-Ala-Cys	(human, residues 140–146)
Ala-Gly-Pro-Arg-Pro-Ala-Cys	(*N. crassa* and *S. cerevisiae*)

Mitochondria typically or always possess a matrix MnSOD. The amino-acid sequences of these proteins show clear homology with the MnSODs of bacteria (Fridovich, 1986b). Functional mitochondrial MnSODs exist as homotetramers, in contrast with typical bacterial MnSODs, which function as homodimers. Tetrameric bacterial MnSODs have been reported, however (Barkley and Gregory, 1990).

Reports of nonmitochondrial, eukaryotic MnSODs include a peroxisomal MnSOD from carnation (Droillard and Paulin, 1990) and cytosolic MnSOD in the fungus *Dactylium dendroides* (Shatzman and Kosman, 1979).

Although more commonly reported from bacteria, FeSODs have been reported from plant chloroplasts (e.g., van Camp et al., 1990), from plant peroxisomes and mitochondria (Droillard and Paulin, 1990), and from *Tetrahymena pyriformis* (Barra et al., 1990).

This diversity of distribution among SODs clearly represents a great deal of evolutionary change. This is particularly true for amino-acid sequences in regions that result in subcellular targeting. What is not clear is the extent to which the present complex distribution of SOD types reflects natural selection for SODs with different biological properties versus evolutionary happenstance. It is not sufficient to hypothesize that the current distribution of SODs simply reflects evolutionary history in the context of endosymbiotic origins of organelles. All characterized eukaryotic SOD genes are nuclear, regardless of the final location of the enzyme within the cell. Moreover, although historically eukaryotic CuZnSOD has been considered to be cytosolic, and MnSOD mitochondrial, it is apparent that evolution is capable of altering the subcellular location of either SOD type.

Given the ability of evolution to alter the subcellular targeting of SODs, it is possible to hypothesize that members of the two SOD families have been conserved across diverse groups of organisms because CuZn and MnSODs have different specialized roles yet to be identified.

D. How Many Different Fungal SODs?

The number of different SOD types to be found among fungi remains an open question. Genes for two different SODs have now been characterized for both *S. cerevisiae* and *N. crassa* (Bermingham-McDonogh et al., 1988; Marres et al., 1985; Chary et al., 1990; Dvorachek and Natvig, unpublished). Each organism possesses a gene encoding cytosolic CuZnSOD and another encoding mitochondrial MnSOD (for *N. crassa* the mitochondrial location is inferred from the nucleotide sequence). For a number of reasons it is possible to hypothesize the existence of additional genes in either these or other fungi.

Henry et al. (1980) reported the isolation of a CuZnSOD from the mitochondrial intermembrane space of *N. crassa*. The existence of a mitochondrial CuZnSOD has also been supported for *N. crassa* by Munkres (1992). Our own genetic and biochemical studies of the *N. crassa sod-1* gene, which encodes CuZnSOD, and mutants carrying null alleles of this gene, have failed to corroborate the suggested presence of a mitochondrion-specific CuZnSOD. If a separate gene exists for such a SOD, it (1) is expressed at a very low level and (2) is not closely related to *sod-1* (Chary et al., 1990; Chary et al., submitted). We have not, however, ruled out the possibility of a distantly related, poorly expressed gene (see below).

A CuZnSOD associated with mitochondria has also been reported for *S. cerevisiae* (Weisiger and Fridovich, 1973). Again, there is as yet no genetic evidence for a separate gene encoding this enzyme.

Exocellular (extracellular) SOD activities have been reported for both *S. cerevisiae* and *N. crassa* (Munkres, 1990). It has not yet been established that these proteins are members of one of the two known SOD families. The identification of extracellular SOD activity associated with fungi is particularly interesting in the context of the possible ancient origin of the extracellular SODs of schistosomes and mammals (discussed below).

Although to date only two *N. crassa* SOD genes have been confirmed by our cloning and sequencing studies (Chary et al., 1990; Dvorachek and Natvig, unpublished), several observations from our own studies have reinforced the need to explore the possible existence of additional SOD genes in this organism. During the characterization of *sod-1,* blot hybridization analysis revealed two sequences in the *N. crassa* genome that hybridized weakly with the synthetic oligonucleotides employed to identify *sod-1* (Chary et al., 1990). Likewise, our screen of the cosmid library of Vollmer and Yanofksy (1986) that resulted in the cloning of a gene for *N. crassa* MnSOD, designated *sod-2,* revealed several cosmids that hybridized with the synthetic probes employed to identify *sod-2,* and which mapped to linkage groups other than that carrying *sod-2* (i.e., linkage group VR). These results are in contrast, however, with the fact that in genomic blot hybridizations with restriction endonuclease digested DNA, our *sod-2* clone hybridizes to only one genomic sequence.

Among other fungi, substantial work is needed. For example, Natvig (1982) reported finding MnSOD, but not CuZnSOD, activity in protein extracts from certain flagellate fungi, and CuZnSOD, but not MnSOD, was found in others. This apparent contrast in the principal SOD type employed by these fungal groups is intriguing, regardless of whether the failure to find a particular SOD in a given organism represents an absence of, or instead, a low specific activity for that SOD.

III. COMPARATIVE ANALYSIS OF SOD GENE AND PROTEIN STRUCTURES

Nucleotide and/or amino-acid sequences for Mn and CuZn SODs have been published for a wide diversity of organisms (see Smith and Doolittle, 1992). Among fungi, complete nucleotide sequences have been published for CuZnSOD genes from *S. cerevisiae* and *N. crassa* and for the MnSOD gene from *S. cerevisiae* (Bermingham-McDonogh et al., 1988; Chary et al., 1990; Marres et al., 1985). We have recently cloned and sequenced the *N. crassa* MnSOD gene. The complete nucleotide sequence of the latter gene will be reported at a later date. However, we have included general features about this gene and its deduced polypeptide in the following discussion.

In mycelial extracts from wild type *N. crassa,* the majority of SOD activity (up to 90%) is from the CuZn enzyme. In possessing no third-position A bases, the nucleotide sequence of the *sod-1* gene in fact represents the most extreme example of a type of codon bias exhibited by highly expressed genes in this organism (Chary et al., 1990).

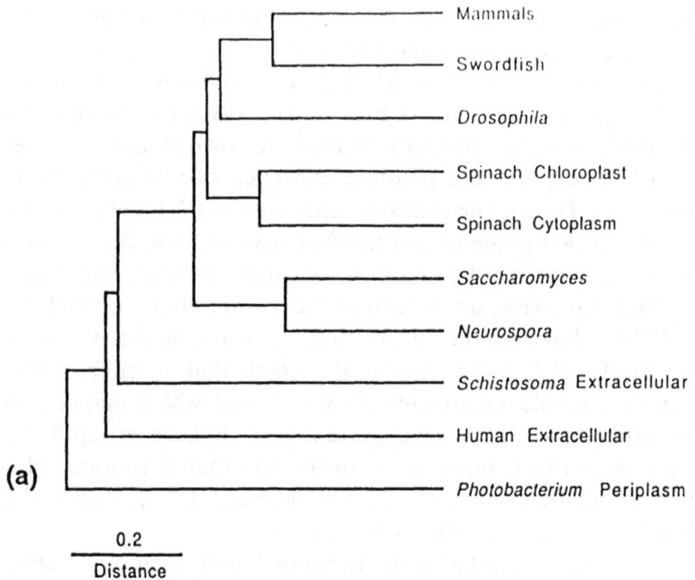
Mammals
Swordfish
Drosophila
Spinach Chloroplast
Spinach Cytoplasm
Saccharomyces
Neurospora
Schistosoma Extracellular
Human Extracellular
Photobacterium Periplasm
(a)
0.2
Distance

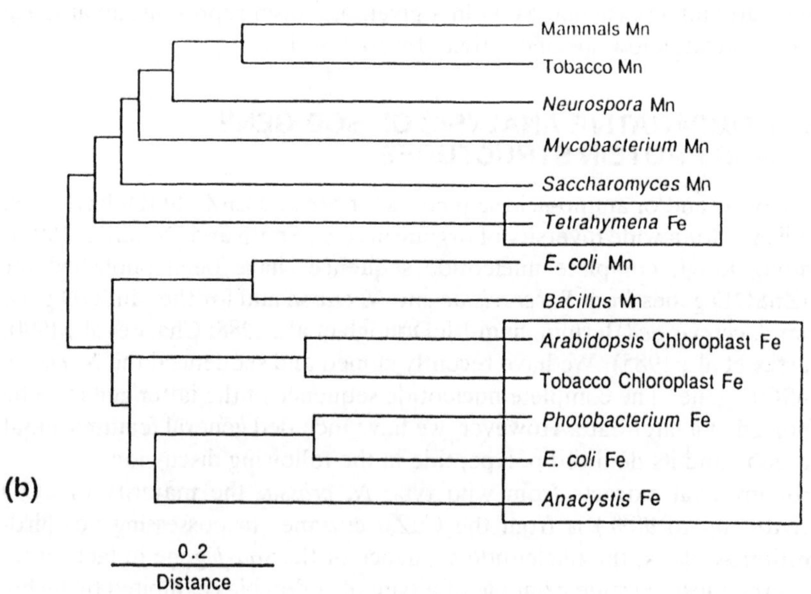
Mammals Mn
Tobacco Mn
Neurospora Mn
Mycobacterium Mn
Saccharomyces Mn
Tetrahymena Fe
E. coli Mn
Bacillus Mn
Arabidopsis Chloroplast Fe
Tobacco Chloroplast Fe
Photobacterium Fe
E. coli Fe
Anacystis Fe
(b)
0.2
Distance

The *N. crassa* CuZnSOD gene possesses three introns, whereas the equivalent *S. cerevisiae* gene contains no introns. Although the presence of introns is characteristic of animal CuZnSOD genes, none of the *N. crassa* introns corresponds in position to an intron present in a previously published CuZnSOD gene nucleotide sequence (Chary et al., 1990).

Although the *N. crassa sod-1* gene is more like that of complex eukaryotes in gross structure, the encoded amino-acid sequence is more similar to that of *S. cerevisiae* than it is to the sequence of any other known CuZnSOD (Fig. 1a). This relationship is reflected in all tree-building methods, including distance and parsimony methods, employed by others (Smith and Doolittle, 1992) and ourselves. Recent sequence comparisons with other genes and proteins have suggested a large evolutionary divergence between *N. crassa* and *S. cerevisiae,* in some cases suggesting a divergence as great as occurs among eukaryotes known to be distantly related (Smith, 1989; Williams et al., 1992). The relatively close similarity between the sequences of *N. crassa* and *S. cerevisiae* CuZnSODs therefore stands in marked contrast to results obtained with other sequences.

The *N. crassa* and *S. cerevisiae* MnSOD deduced amino-acid sequences are quite divergent and, in tree-building analyses, present a picture very different from that observed for CuZn SODs (Fig. 1b). The *N. crassa* sequence groups closer to the MnSODs of animals and plants than to that of *S. cerevisiae.* This is true for both distance and parsimony trees, although the specific affinities vary somewhat with the type of analysis. The problem of establishing the level of similarity between *N. crassa* and *S. cerevisiae* MnSODs is complicated by the possibility that *N. crassa* may contain multiple MnSOD genes. It is necessary to consider the possibility that *N. crassa* possesses a member of the MnSOD gene family that is more closely related to the yeast gene than is *sod-2,* although we consider this to be unlikely.

With respect to introns, the MnSOD genes of *N. crassa* and *S. cerevisiae* present a picture similar to that observed for the CuZnSOD genes. The *N. crassa* gene possesses two introns, whereas the *S. cerevisiae* has neither. As is the case with the *N. crassa* CuZnSOD gene, neither of the MnSOD

Figure 1 Trees presenting possible relationships among selected CuZn and MnSODs. The trees were constructed using the Kitsch program (Phylip version 3.3 from J. Felsenstein). Input data were distance values based on overall sequence identity. Amino acid sequence alignments (available from the authors on request) were those of Smith and Doolittle (1992) with minor modifications. (a) cuZnSOD evolution; (b) MnFeSOD evolution.

introns corresponds in position to an intron present in animal MnSOD genes (Dvorachek and Natvig, unpublished).

The trees presented in Figs. 1a,b illustrate unanswered questions regarding SOD evolution that additional fungal studies will help to address. One unresolved issue is reflected in the positions of the human and schistosome extracellular CuZnSODs. The placement of these sequences in tree-building analyses suggests an ancient origin for each, allowing the possibility that related extracellular SODs will be found among fungi. An alternative to the hypothesis of ancient origin is that these genes diverged recently and independently from genes for cytosolic CuZnSOD and were driven rapidly in this divergence by natural selection. Relevant to this, we consider it an open question whether the human and schistosome extracellular SODs arose separately. Although the distance analysis presented in Fig. 1a suggests separate origins, parsimony analysis places these two extracellular SODs on a branch together, separate from other CuZnSODs (Smith and Doolittle, 1992; Natvig and Dvorachek, unpublished).

The most ancient split in the MnSOD tree appears to divide two groups of MnSODs, suggesting that FeSODs may have arisen after the advent of MnSODs. This is contrary to the general view—based on the distribution of SOD types and arguments concerning the availability of reduced metal ions on early earth—that the ancestral SOD for this family was an FeSOD or cambialistic SOD (Gralla and Kosman, 1992; Martin et al., 1986). This distance tree must be considered tentative with respect to deep branches. Other tree-building methods have produced trees in which the most ancient split separates FeSODs from MnSODs (Smith and Doolittle, 1992).

The *Tetrahymena pyriformis* protein is one of only a few characterized eukaryotic FeSODs and may well have been derived relatively recently from an MnSOD (Barra et al., 1990). Additional eukaryotic sequences from the Mn/Fe family, particularly those from fungi, should help to establish better the correct placement of the FeSODs from *T. pyriformis* and other eukaryotes.

Future studies with fungi may also play important roles in defining the evolutionary history of the CuZnSOD family, particularly with respect to extracellular SODs. It is also likely that future studies of fungal SODs will expand the known distribution of SODs with respect to SOD type and subcellular targeting.

IV. STUDIES WITH SOD MUTANTS

A. History

More than a century ago Pasteur recognized that certain microorganisms are killed in the presence of oxygen (see Pasteur, 1876, pp. 302–303). The mechanisms of oxygen toxicity remained elusive, however, and the study of oxygen toxicity remained at least somewhat esoteric, until the past two decades. The field received new life with the discovery of bovine superoxide dismutase in the late 1960s (McCord and Fridovich, 1969). This discovery was important on two fronts. First, studies of the enzyme led quite quickly to a new testable hypothesis for oxygen toxicity, based on superoxide (McCord et al., 1971). Second, purified bovine CuZnSOD provided a tool in the form of a stable enzyme that greatly facilitated the *in vitro* study of O_2^-. The renaissance in the study of oxygen toxicity, directly traceable to the discovery of superoxide dismutase, has now expanded into one of the most important subdisciplines of biological science, with implications for fields from forest ecology to human health (e.g., Rosen et al., 1993).

The superoxide-based theory of oxygen toxicity (McCord et al., 1971) led to two testable hypotheses concerning the role and importance of superoxide dismutase. The first hypothesis was that superoxide dismutase is required for aerobic survival of all living organisms. The hypothesis was based on the distribution of SOD activity among aerobic, aerotolerant, and anaerobic bacteria. Initial studies indicated that SOD activity was present in aerobes but absent in obligate anaerobes (McCord et al., 1971). This hypothesis was later found not to be correct in the strict sense. Carlioz and Touati (1986) at the University of Paris demonstrated that *E. coli* mutants completely null for SOD activity can grow aerobically under laboratory conditions. There is a caveat, however, in that such mutants exhibit conditional oxygen lethality and do not grow aerobically on minimal medium, a phenomenon that has been termed aerobic auxotrophy.

The second hypothesis was nonoverlapping with respect to the first. It stated that SOD is not of significant importance in protecting organisms from the toxic effects of oxygen. This argument was based on the apparent lack of reactivity of O_2^- with biological molecules (Sawyer and Valentine, 1981), and it led to speculation that dismutation of O_2^- was not the biological function of these proteins (Fee, 1982). This hypothesis was also incorrect. The evidence is now overwhelming that SODs are important in protecting against oxygen toxicity. Most significant is the fact that SOD-deficient organisms are sensitive to agents of superoxide stress, as was first shown for *E. coli* (Carlioz and Touati, 1986). A second important line of evidence has come from gene substitution experiments. Several years ago we demonstrated that expression of human CuZnSOD could reverse the active oxygen-sensitive phenotype of *E. coli* cells that lack Mn and Fe SODs (Natvig et al., 1987). A reciprocal experiment demonstrated that MnSOD from *Bacillus stearothermophilus* can replace CuZnSOD activity in *S. cerevisiae* (Bowler et al., 1990).

If both of these hypotheses are incorrect in their purest forms, then it must be concluded that the truth is somewhere in between. Clearly, SODs are important in protecting cells from oxygen toxicity. What is becoming quite clear, however, is that SODs are not "magic-bullet" enzymes that serve as the only protection against cellular oxidative stress (see, for example, Farr and Kogoma, 1991, for a discussion of auxiliary defenses). Particularly in eukaryotes, many questions remain regarding the relative importance of different SOD types.

B. Mutant Studies

Combined genetic and biochemical studies employing SOD null mutants are proving extremely valuable toward the goals of establishing the biological roles of different SODs and identifying important targets of superoxide-mediated stress. Mutants have now been reported for *E. coli* (Carlioz and Touati, 1986), *S. cerevisiae* (Bilinski et al., 1985; van Loon et al., 1986), and *Drosophila melanogaster* (Phillips et al., 1989). We have recently constructed and characterized a null mutant for *N. crassa* CuZnSOD encoded by the *sod-1* gene (Chary *et al.,* submitted). The studies of these mutants

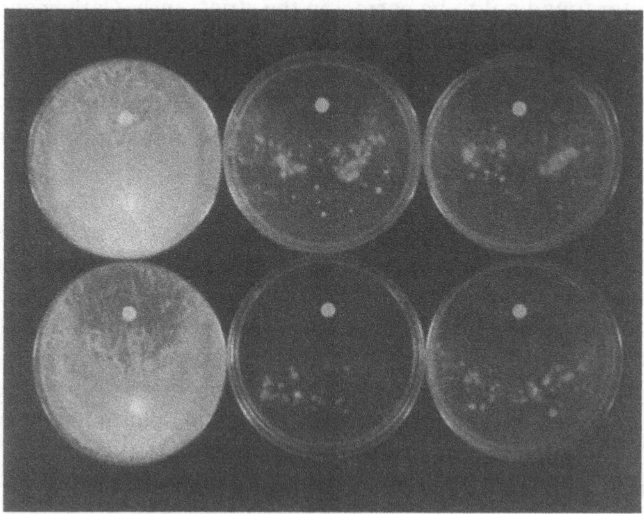

Figure 2 Sensitivity to paraquat of wild type (left) N. crassa and mutants (center and right) null for the CuZnSOD gene (sod-1). The strains were inoculated onto solid medium opposite a filter-paper disk that had been saturated with either 100 mM (top row) or 500 mM (bottom row) paraquat. Plates were incubated for 4 days at room temperature.

have been performed in diverse laboratories and are not directly comparable with respect to all phenotypic characteristics examined. Nevertheless, substantial comparison can be made.

The disruption of even a single SOD gene has pleiotropic effects (Table 1). This is not surprising given that very diverse cellular molecules have been reported as direct or indirect targets of damage from O_2^- (Fridovich, 1986a). It is of interest, therefore, to identify similarities and differences among the known SOD mutants. To date, a sensitivity to paraquat and other superoxide-generating agents stands out as a common characteristic in published studies. It is also the most dramatic characteristic of the *N. crassa sod-1* mutant (Fig. 2) (Chary et al., submitted). This universal sensitivity of SOD mutants to superoxide-generating compounds gives strong support to the important role of SODs in preventing superoxide-mediated damage.

The value of mutant studies is further evident in that characterization of SOD mutants has revealed the importance of previously overlooked tar-

Table 1 SOD Mutant Phenotypes

Organism[a]	SOD type and gene	Growth on minimal medium	Increased spontaneous mutation rate	Increased sensitivity to ionizing radiation	Slow or abnormal growth	Increased sensitivity to paraquat and other redox compounds
E. coli[b]	Mn and Fe (*sodA sodB*)	No	Yes	No	Yes	Yes
S. cerevisiae	CuZn (*sod1*)[c]	No	Yes	Unk	Yes	Yes
S. cerevisiae	Mn(*sod2*)[c]	Yes	Unk	Unk	Yes[d]	Yes
N. crassa	CuZn (*sod-1*)	Yes	Yes	No	Yes	Yes
D. melanogaster	CuZn (*Sod*)	NA	Unk	Yes	Yes	Yes

[a]References not cited here are presented in the text.
[b]Information on the phenotypes of *E. coli sodA* and *sodB* single mutants is presented in Carlioz and Touati (1986) and Farr et al. (1986).
[c]Here we accept the gene designations recommended by Gralla and Kosman (1992).
[d]Growth is poor or absent on two-carbon, non-fermentable substrates but is nearly wild type on six- and three-carbon substrates (van Loon *et al.*, 1986; Westerbeek *et al.*, 1988).
NA = not applicable. Unk = unknown.

gets of superoxide-mediated damage. Most notable are the aerobic auxotrophies exhibited by the *E. coli sodA sodB* double mutant (Carlioz and Touati, 1986) and the *S. cerevisiae sod1* (CuZnSOD) mutant (Bilinski et al., 1985). Although this appears at first examination to be a shared phenotype—neither mutant will grow aerobically on defined media lacking supplemental amino acids—the specific cause of the phenotype appears to be different for each mutant. This is evident first in that different amino acids are required by the mutants to reverse oxygen sensitivity and second in that biochemical investigations have pointed to different mechanisms in the two organisms.

The aerobic auxotrophy of the *E. coli* mutant arises from a requirement for branched-chain amino acids (Carlioz and Touati, 1986). This appears to result entirely or largely from the sensitivity to O_2^- of the iron-sulfur center of α,β-dihydroxyisovalerate dehydratase (Kuo et al., 1987).

The *S. cerevisiae sod1* mutant has a requirement under aerobic conditions for either cysteine or methionine and for lysine. The toxicity of oxygen in the absence of methionine or cysteine appears not to be a synthetic requirement for these amino acids. Rather, oxygen toxicity in their absence has been attributed to the production of toxic levels of the sulfur trioxy radical (Chang and Kosman, 1990; Gralla and Kosman, 1992), which is avoided by the suppression of sulfate assimilation by methionine and cysteine.

Growing evidence points to a link between DNA damage and superoxide-mediated toxicity (see Farr and Kogama, 1991). As a result, SOD mutants have been examined for increased spontaneous mutation rate and sensitivity to various mutagens. The *E. coli* (Farr et al., 1986) *sodA sodB* mutant, *S. cerevisiae sod1* mutant (Gralla and Valentine, 1991), and the *N. crassa sod-1* (Chary et al., submitted) all exhibit increased spontaneous mutation rates. These increased rates range from 4- to 40-fold as measured by forward or reverse mutation. Direct comparisons of the levels of increase are not practical, because each organism has been examined under a different experimental system.

Interestingly, a consistent pattern does not exist with respect to sensitivity of SOD mutants to ionizing radiation. This is true even though production of O_2^- is believed to accompany exposure to ionizing radiation (Totter, 1980). The *E. coli sodA sodB* and the *N. crassa sod-1* mutants are as resistant or more resistant to killing by ionizing radiation than are their respective wild types (Scott et al., 1989; Chary et al., submitted). In contrast, in *D. melanogaster* there appears to be a positive correlation between the specific activity of CuZnSOD and resistance to ionizing radiation (Peng et al., 1986). This important difference between *D. melanogaster* and the two microorganisms has not been explained. It could signal the complexity of intracellular interactions between O_2^- and other forms of active oxygen, particularly OH· (see Scott et al., 1989). Alternatively, it is possible that the *E. coli* and *N. crassa* mutants are induced for DNA repair responses that protect against the damage caused by ionizing radiation.

C. A Natural Selection Based Hypothesis for the Role of SODs

That SODs are of fundamental importance can be deduced from their nearly ubiquitous presence among living organisms, supported by the diversity of types among the two major SOD families. It is therefore interesting that attempts to understand the role of SODs have led to a complex picture. This is true, first, in the sense that the proposed targets of superoxide damage are varied, with the possibility that the importance of any one target is perhaps not constant throughout the biological world. It is also true in the sense that the SOD mutations investigated to date exhibit pleiotropic effects, several of which are not constant across organisms. It is further intriguing that a lethal SOD mutation has not yet been identified among microorganisms.

These observations have led us to consider the possibility that the role of SODs in most organisms is to increase fitness (in the Darwinian sense). This hypothesis is consistent with the fact that SOD mutants of *E. coli, S. cerevisiae,* and *N. crassa* grow more slowly than wild type, but nonetheless survive, under standard laboratory conditions. It is also consistent with the likelihood that O_2^- has manifold targets in the biological world, with the importance of any one specific target varying among organisms depending on molecular design. For example, the iron-sulfur cluster of *E. coli* dihydroxy acid dehydratase is sensitive to O_2^-, resulting in aerobic auxotrophy, whereas the iron-sulfur cluster of the spinach enzyme apparently is not sensitive (Kuo et al., 1987; Flint and Emptage, 1988).

This hypothesis suggests that viability of the average cell is not dependent on the presence of a specific SOD. Instead, a given SOD may increase the fitness of a given organism by either decreasing the possibility of a lethal event (mutation, for example) or by increasing growth rate.

V. ECOLOGICAL CONSIDERATIONS

A fascinating and still under-investigated field is developing that recognizes the potential importance of environmental O_2^- and other forms of active oxygen that impact the growth of microorganisms in nature. Growing evidence suggests that the production of active-oxygen species can be employed as a defensive mechanism. As a result, both the defending organism and the organism being defended against benefit from mechanisms of defense against active oxygen. The production of O_2^- by mammalian leukocytes has been well documented (e.g., Babior, 1988). Analogous mechanisms have been reported for plant tissues, presumably as a means of deterring bacterial and fungal pathogens (Thompson et al., 1987).

There are two general mechanisms of O_2^- warfare that can be imagined. One is the direct production of O_2^- that is then released into the combat zone. The second is the production of compounds by one organism that will result in the production of O_2^- in the cells of a second organism. Paraquat is a human-made representative of the latter mechanism. There is a growing list of secondary plant compounds that may serve as naturally occurring examples. Plumbagin may be one such compound (Hassan and Fridovich, 1979a; Fieser and Dunn, 1936).

The generation of O_2^- by plant cells in response to germinating spores of the pathogenic fungus *Phytophthora infestans* has been reported by two groups (Chai and Doke, 1987; Ivanova et al., 1991). Chai and Doke observed two bursts of O_2^- production in certain potato leaf tissues after exposure to germinating zoospores of *P. infestans*. The first burst was observed in tissues exposed to zoospores or zoospore germination fluids and appeared to be a general response that was also exhibited by the leaves of several nonhost plant species. The second burst was specific for tissues undergoing a hypersensitive response upon penetration of fungal hyphae.

These observations suggest that production of O_2^- may be part of the plant defense system against pathogens. It also implies that survival of pathogen cells and host cells may be dependent on SOD to tolerate the defense response. The possibility of O_2^- in the extracellular environment raises again the question of extracellular fungal SODs, in part because it is generally assumed that O_2^- cannot cross the cell membrane (Hassan and Fridovich, 1979b). Relevant to this is a recent report that intracellular SOD does not increase the survival of bacteria in contact with bean roots releasing superoxide (Katsuwon et al., 1993).

VI. FUNGAL SOD GENE CLONING STRATEGIES

We are strong advocates of molecular-genetic approaches to the study of SODs. We hope the value of such studies is evident from the previous discussions. Accordingly, we hope the following discussion will be of value to other laboratories with interest in SOD gene cloning.

Although identification of the first fungal SOD mutant, that of *S. cerevisiae* by Bilinski et al. (1985), employed a screen for paraquat sensitivity, successful gene cloning efforts have employed synthetic oligonucleotides instead of genetic screens (Bermingham-McDonogh et al., 1988; Chary et al., 1990; Dvorachek and Natvig, unpublished). The general method (Chary et al., 1990) that we have employed to clone genes for *N. crassa* CuZnSOD and MnSOD should in theory work for any organism.

Our cloning strategy has been to screen genomic libraries with long (59–60 bases), labeled oligonucleotides, designed to be complementary to highly conserved regions of the gene. The oligonucleotide preparations are not degenerate, so a degree of mismatch is expected, particularly at codon third positions. For each gene, we have employed two such oligonucleotides, one complementary to a region near the 5′ end of the gene and another complementary to a 3′ region. Genomic libraries, or individual clones, are then probed once with each oligonucleotide. We have employed standard Southern-blot hybridization conditions (Maniatis et al., 1982) to probe successfully cosmid and plasmid libraries. The stringency of the hybridization is lowered by hybridizing at 53–55° C. The identification of clones that hybridize with both probes eliminates the possibility of a false-positive signal.

We have to date constructed four such oligonucleotides, two each for the *N. crassa sod-1* and *sod-2* genes. Each of the four has worked very well. In constructing these probes, we have attempted to incorporate *N. crassa* codon preferences. In general, this should be unnecessary. Because *N. crassa* has two different classes of codon bias, we in fact estimated quite wrongly in the case of *sod-1,* but this was of no consequence (Chary et al., 1990). It is also interesting that one of the *sod-2* probes worked even though the corresponding gene region was interrupted in the middle by a sizable intron.

For the *N. crassa sod-1* gene we constructed probes with reference to the *N. crassa* CuZnSOD amino-acid sequence (Lerch and Schenk, 1985). For the *N. crassa sod-2* gene we based probes on regions of MnSOD that are highly conserved between *S. cerevisiae* and humans. The two approaches worked equally well. The construction of our *sod-1* probes has been published (Chary et al., 1990).

Readers interested in SOD gene cloning should also refer to the recent paper by Smith and Doolittle (1992). These authors employed a PCR based method to clone portions of MnSOD cDNAs from several animal species. The strategy involved construction of degenerate PCR primers for conserved gene regions.

ACKNOWLEDGMENTS

We would like to thank Dr. Anne Anderson for providing a copy of a paper in press. Our current research is supported by grant MCB-9022177 from the National Science Foundation.

REFERENCES

Babior, B. M. (1988). The respiratory burst, *Oxygen Radicals in Biology and Medicine* (M. G. Simic, K. A. Taylor, J. F. Ward, and C. von Sonntag, eds.), Plenum Press, New York, pp. 815–821.

Barkley, K. B., and Gregory, E. M. (1990). Tetrameric manganese superoxide dismutases from anaerobic *Actinomyces, Arch. Biochem. Biophys., 280:* 192–200.

Barra, D., Schinina, M. E., Bossa, F., Puget, K., Durosay, P., Guissani, A., and Michelson, A. M. (1990). A tetrameric iron superoxide dismutase from the eucaryote *Tetrahymena pyriformis, J. Biol. Chem., 265:* 17680–17687.

Bermingham-McDonogh, O., Gralla, E. B., and Valentine, J. S. (1988). The copper, zinc-superoxide dismutase gene of *Saccharomyces cerevisiae:* Cloning, sequencing, and biological activity, *Proc. Natl. Acad. Sci. USA, 85:* 4789–4793.

Bilinski, T., Krawiec, Z., Liczmanski, A., and Litwinska, J. (1985). Is hydroxyl radical generated by the Fenton reaction in vivo? *Biochem. Biophys. Res. Comm.,* 130: 533–539.

Borgstahl, G. E. O., Parge, H. E., Hickey, M. J., Beyer, W. F., Jr., Hallewell, R. A., and Tainer, J. A. (1992). The structure of human mitochondrial manganese superoxide dismutase reveals a novel tetrameric interface of two 4-helix bundles, *Cell, 71:* 107–118.

Bowler, C., Van Kaer, L., Van Camp, W., Van Montagu, M., Inze, D., and Dhaese, P. (1990). Characterization of the *Bacillus stearothermophilus* manganese superoxide dismutase gene and its ability to complement copper/zinc superoxide dismutase deficiency in *Saccharomyces cerevisiae, J. Bacteriol., 172:* 1539–1546.

Brawn, K., and Fridovich, I. (1981). DNA stand-scission by enzymatically generated oxygen radicals, *Arch. Biochem. Biophys., 206:* 414–419.

Carlioz, A., and Touati, D. (1986). Isolation of superoxide dismutase mutants in *Escherichia coli:* Is superoxide dismutase necessary for aerobic Life? *EMBO. J., 5:* 623–630.

Chai, H. B., and Doke, N. (1987). Superoxide anion generation: A response of potato leaves to infection with *Phytophthora infestans, Phytopathology, 77:* 645–649.

Chang, E. C., and Kosman, D. J. (1990). O_2-dependent methionine auxotrophy in Cu, Zn superoxide dismutase-deficient mutants of *Saccharomyces cerevisiae, J. Bacteriol., 172:* 1840–1845.

Chary, P., Hallewell, R. A., and Natvig, D. O. (1990). Structure, exon pattern and chromosome mapping of the gene for cytosolic copper-zinc superoxide dismutase (*sod-1*) from *Neurospora crassa, J. Biol. Chem., 265:* 18961–18967.

Chary, P., Dillon, D., Schroeder, A. L., and Natvig, D. O. Superoxide dismutase null mutants of *Neurospora crassa:* Oxidative stress sensitivity, spontaneous mutation rate and response to mutagens, submitted for publication.

Desideri, A., Falconi, M., Polticelli, F., Bolognesi, M., Djinovic, K., and Rotilio, G. (1992). Evolutionary conservativeness of electric field in the Cu,Zn superoxide dismutase active site. Evidence for co-ordinated mutation of charged amino acid residues. *J. Mol. Biol. 223:* 337–342.

Doke, N., and Ohashi, Y. (1988). Involvement of an O_2^- generating system in the induction of necrotic lesions on tobacco leaves infected with tobacco mosaic virus, *Physiol. Mol. Plant Pathol., 32:* 163–175.

Dougherty, H. W., Sadowski, S. J., and Baker, E. E. (1978). A new iron-containing superoxide dismutase from *Escherichia coli, J. Biol. Chem., 253:* 5220–5223.

Droillard M.-J., and Paulin, A. (1990). Isozymes of superoxide dismutase in mitochondria and peroxisomes isolated from petals of carnation (*Dianthus caryophyllus*) during senescence, *Plant Physiol., 94:* 1187–1192.

Farr, S. B., D'Ari, R., and Touati, D. (1986). Oxygen dependent mutagenesis in *Escherichia coli* lacking superoxide dismutase, *Proc. Natl. Acad. Sci. USA, 83:* 8268–8272.

Farr, S. B., and Kogoma, T. (1991). Oxidative stress responses in *Escherichia coli* and *Salmonella typhimurium, Microbiol. Rev., 55:* 561–585.

Fee, J. A. (1982). Is superoxide important in oxygen poisoning? *Trends Biochem. Sci., 7:* 84–86.

Fieser, L. F., and Dunn, J. T. (1936). Synthesis of plumbagin, *J. Am. Chem. Soc., 58:* 572–575.

Flint, D. H., and Emptage, M. H. (1988). Dihydroxy acid dehydratase from spinach contains a [2Fe-2S] cluster, *J. Biol. Chem., 263:* 3558–3564.

Fridovich, I. (1981). Superoxide radical and superoxide dismutases, *Oxygen and Living Processes* (D. L. Gilbert, ed.), Springer-Verlag, Berlin and New York, pp. 250–272.

Fridovich, I. (1984). Overview: Biological sources of O_2^-, *Meth. Enzymol., 105:* 59–61.

Fridovich, I. (1986a). Biological effects of the superoxide radical, *Arch. Biochem. Biophys., 247:* 1–11.

Fridovich, I. (1986b). Superoxide dismutases, *Adv. Enzymol. Relat. Areas Mol. Biol., 58:* 61–97.

Getzoff, E. D., Tainer, J. A., Weiner, P. K., Kollman, P. A., Richardson, J. S., and Richardson, D. C. (1983). Electrostatic recognition between superoxide and copper, zinc superoxide dismutase, *Nature, 306:* 287–290.

Gralla, E. B., and Kosman, D. J. (1992). Molecular genetics of superoxide dismutases in yeasts and related fungi, *Adv. Genetics, 30:* 251–319.

Gralla, E. B., and Valentine, J. S. (1991). Null mutants of *Saccharomyces cerevisiae* Cu, Zn superoxide dismutase: Characterization and spontaneous mutation rates, *J. Bacteriol. 173:* 5918–5920.

Haber, F., and Weiss, J. (1934). The catalytic decomposition of hydrogen peroxide by iron salts, *Proc. R. Soc. London* (A), *147:* 332–351.

Hassan, H. M., and I. Fridovich. (1977). Enzymatic defenses against the toxicity of oxygen and of streptonigrin in *Escherichia coli, J. Bacteriol., 129:* 1574–1583.

Hassan, H. M., and Fridovich I. (1978). Superoxide radical and the oxygen enhancement of the toxicity of paraquat in *E. coli, J. Biol. Chem., 253:* 8143–8147.

Hassan, H.M., and Fridovich, I. (1979a). Intracellular production of superoxide radicals and H_2O_2 by redox active compounds, *Arch. Biochem. Biophys., 196:* 385–395.

Hassan, H. M., and Fridovich, I. (1979b). Paraquat and *Escherichia coli:* Mechanism of production of extracellular superoxide, *J. Biol. Chem., 254:* 10846–10856.

Henry, L. E. A., Cammack, R., Schwitzguebel, J. P., Palmer, J. M., and Hall, D. O. (1980). Intracellular localization, isolation and characterization of two distinct varieties of superoxide dismutase from *Neurospora crassa, Biochem. J., 187:* 321–328.

Hjalmarsson, K., Marklund, S. L., Engstrom, A., and Edlund, T. (1987). Isolation and sequence of complementary DNA encoding human extracellular superoxide dismutase, *Proc. Natl. Acad. Sci. USA, 84:* 6340–6344.

Ivanova, D. G., Gughova, N. V., Merzlyak, M. N., and Rassadina, G. V. (1991). Effect of *Phytophthora infestans* infection on superoxide dismutase dependent cytochrome c-reducing activities of leaves as related to resistance of potato plants to late blight, *Plant Science, 78:* 151–156.

Kanematsu, S., and Asada, K. (1990). Characteristic amino acid sequences of chloroplast and cytosol isozymes of CuZn-superoxide dismutase in spinach, rice and horsetail, *Plant Cell Physiol., 31:* 99–112.

Katsuwon, J., Zdor, R., and Anderson, A. J. (1993). Superoxide dismutase activity in root-colonizing pseudomonads, *Can. J. Microbiol.,* in press.

Keller, G.-A., Warner, T. G, Steimer, K. S, and Hallewell, R. A. (1991). Cu,Zn superoxide dismutase is a peroxisomal enzyme in human fibroblasts and hepatoma cells, *Proc. Natl. Acad. Sci. USA, 88:* 7381–7385.

Kuo, C. F., Mashino, T., and Fridovich, I. (1987). α,β-dihydroxyisovalerate dehydratase: A superoxide sensitive enzyme, *J. Biol. Chem., 262:* 4724–4727.

Lerch, K., and Schenk, E. (1985). Primary structure of copper-zinc superoxide dismutase from *Neurospora crassa, J. Biol. Chem., 260:* 9559–9566.

McCord, J. M., and Fridovich, I. (1969). Superoxide dismutase an enzymic function for erythrocuprein (hemocuprein), *J. Biol. Chem., 244:* 6044–6055.

McCord, J. M., Keele, B. B., Jr., and Fridovich, I. (1971). An enzyme-based theory of obligate anaerobiosis: The physiological function of superoxide dismutase, *Proc. Natl. Acad. Sci. USA., 68:* 1024–1027.

Maniatis, T., Fritsch, E. F., and Sambrook, J. (1982). *Molecular Cloning,* Cold Spring Harbor Laboratory, New York, 545 pp.

Marres, C. A. M., Van Loon, A. P. G. M, Oudshoorn, P., Van Steeg, H., Grivell, L. A., and Slater, E. C. (1985). Nucleotide sequence analysis of the nuclear gene coding for manganese superoxide dismutase of yeast mitochondria, a gene previously assumed to code for the Rieske iron-sulphur protein, *Eur. J. Biochem., 147:* 153–161.

Martin, M. E., Byers, B. R., Olson, M. O. J., Salin, M. L., Arceneaux, J. E. L., and Tolbert, C. (1986). A *Streptococcus mutans* superoxide dismutase that is active with either manganese or iron as a cofactor, *J. Biol. Chem., 261:* 9361–9367.

Munkres, K. D. (1990). Purification of exocellular superoxide dismutases, *Meth. Enzymol., 186:* 249–260.

Munkres, K. D. (1992). Selection and analysis of superoxide dismutase mutants of *Neurospora, Free Radicals Bio. Med., 13:* 305–318.

Natvig, D. O. (1982). Comparative biochemistry of oxygen toxicity in lactic acid forming aquatic fungi, *Arch. Microbiol., 132:* 107–114.

Natvig, D. O., Imlay, K., Touati, D., and Hallewell, R. (1987). Human copper-zinc superoxide dismutase complements superoxide dismutase-deficient *Escherichia coli* mutants, *J. Biol. Chem., 262:* 14697–14701.

Pasteur, L. (1876). *Studies on Fermentation: The diseases of Beer, Their Causes, and the Means of Preventing Them* (F. Faulkner and D. C. Robb, trs.), American Library Service, New York, 418 pp.

Peng, T. X., Moya, A., and Ayala, F. J. (1986). Irradiation-resistance conferred by superoxide dismutase: Possible adaptive role of a natural polymorphism in *Drosophila melanogaster, Proc. Natl. Acad. Sci. USA, 83:* 684–687.

Pennington, C. D., and Gregory, E. M. (1986). Isolation and reconstitution of iron- and manganese-containing superoxide dismutases from *Bacteroides thetaiotaomicron, J. Bacteriol., 166:* 528–532.

Phillips, J. P., Campbell, S. D., Michaud, D., Charbonneau, M., and Hilliker, A. J. (1989). Null mutation of copper/zinc superoxide dismutase in *Drosophila* confers hypersensitivity to paraquat and reduced longevity, *Proc. Natl. Acad. Sci. USA, 86:* 2761–2765.

Rosen, D. R., Siddique, T., Patterson, D., Figlewicz, D. A., Sapp, P., Hentati, A., Donaldson, D., Goto, J., O'Regan, J. P., Deng, H.-X., Rahmani, Z., Krizus, A., McKenna-Yasek, D., Cayabyab, A., Gaston, S. M., Berger, R., Tanzi, R. E., Halperin, J. J., Herzfeldt, B., Van den Berg, R., Hung, W.-Y., Bird, T., Deng, G., Mulder, D. W., Smith, C., Laing, N. G., Soriano, E., Pericak-Vance, M. A., Haines, J., Rouleau, G. A., Gusella, J. S., Horvitz, H. R., and Brown, R. H., Jr. (1993). Mutations on Cu/Zn superoxide dismutase gene are associated with familial amyotrophic lateral sclerosis, *Nature, 362:*59–62.

Sawyer, D. T., and Valentine, J. S. (1981). How super is superoxide? *Acc. Chem. Res., 14:* 393–400.

Scott, M. D., Meshnick, S. R., and Eaton, J. W. (1989). Superoxide dismutase amplifies organismal sensitivity to ionizing radiation, *J. Biol. Chem., 264:* 2498–2501.

Shatzman, A. R., and Kosman, D. J. (1979). Biosynthesis and cellular distribution of the two superoxide dismutases of *Dactylium dendroides, J. Bacteriol., 137:* 313–320.

Simurda, M. C., van Keulen, H., Rekosh, D. M., and LoVerde, P. T. (1988). *Schistosoma mansoni:* Identification and analysis of an mRNA and a gene encoding superoxide dismutase (Cu/Zn), *Exp. Parasit., 67:* 73–84.

Smith, M. W., and Doolittle, R. F. (1992). A comparison of evolutionary rates of the two major kinds of superoxide dismutase, *J. Mol. Evol. 34:* 175–184.

Smith, T. L. (1989). Disparate evolution of yeasts and filamentous fungi indicated by phylogenetic analysis of glyceraldehyde-3-phosphate dehydrogenase genes, *Proc. Natl. Acad. Sci. USA, 86:* 7063–7066.

Steinman, H. M. (1978). The amino acid sequence of mangano superoxide dismutase from *Escherichia coli* B, *J. Biol. Chem., 253:* 8708–8720.

Steinman, H. M. (1987). Bacteriocuprein superoxide dismutase of *Photobacterium leiognathi:* Isolation and sequence of the gene and evidence for a precursor form, *J. Biol. Chem., 262:* 1882–1887.

Tainer, J. A., Getzoff, E. D., Richardson, J. S., and Richardson, D. C. (1983). Structure and mechanism of copper, zinc superoxide dismutase, *Nature, 306:* 284–287.

Thompson, J. E., Legge, R. L., and Barber, R. F. (1987). The role of free radicals in senescence and wounding, *New Phytol., 105:* 317–344.

Totter, J. R. (1980). Spontaneous cancer and its possible relationship to oxygen metabolism, *Proc. Natl. Acad. Sci. USA, 77:* 1763–1767.

Van Camp, W., Bowler, C., Villarroel, R., Tsang, E. W. T., van Montagu, M. V., and Inze, D. (1990). Characterization of iron superoxide dismutase cDNAs from plants obtained by genetic complementation of *Escherichia coli, Proc. Natl. Acad. Sci. USA, 87:* 9903–9907.

Van Loon, A. P. G. M., Pesold-Hurt, B., and Schatz, G. (1986). A yeast mutant lacking mitochondrial manganese-superoxide dismutase is hypersensitive to oxygen, *Proc. Natl. Acad. Sci. USA, 83:*3820–3824.

Vollmer, S. J., and Yanofsky, C. (1986). Efficient cloning of genes of *Neurospora crassa, Proc. Natl. Acad. Sci. USA, 83:* 4869–4873.

Weisiger, R. A., and Fridovich, I. (1973). Mitochondrial superoxide dismutase site of synthesis and intramitochondrial localization, *J. Biol. Chem., 248:* 4793–4796.

Westerbeek-Marres, C. A., Moore, M. M., and Autor, A. (1988). Regulation of manganese superoxide dismutase in *Saccharomyces cerevisiae.* The role of respiratory chain activity. *Eur. J. Biochem., 174:* 611–620.

Williams, L. J., Barnett, G. R., Ristow, J. L., Pitkin, J., Perriere, M., and Davis, R. H. (1992). Ornithine decarboxylase gene of *Neurospora crassa:* Isolation, sequence, and polyamine-mediated regulation of its mRNA, *Mol. Cell. Biol., 12:* 347–359.

15

Ferrochelatase in *Saccharomyces cerevisiae*

Rosine Labbe-Bois and Jean-Michel Camadro
Institut Jacques Monod, University of Paris VII, Paris, France

I. INTRODUCTION

Ferrochelatase (EC 4. 99. 1. 1, protoheme ferrolyase) is the enzyme that catalyzes the insertion of ferrous iron into protoporphyrin IX to form protoheme (Fig. 1; see Nomenclature at the end of this chapter). This is the last of the eight enzymic steps in the heme biosynthetic pathway (Fig. 2; see Dailey, 1990 and Jordan, 1991 for reviews). The first committed step of the pathway is the synthesis of 5-aminolevulinic acid (ALA). Most organisms, except plants and some bacteria, do this by condensing glycine with succinyl-CoA. Eight molecules of ALA are then converted to one molecule of protoporphyrin IX in a series of six reactions, two of which require molecular oxygen. The synthesis of protoheme can be written as

$$8 \text{ glycine} + 8 \text{ succinylCoA} + m\,O_2 + 1\,Fe^{2+} \rightarrow$$
$$1 \text{ protoheme} + (8 \text{ CoA}, 4\,NH_4^+, 14\,CO_2, n\,H_2O)$$

Protoheme and its derivatives heme *c* and heme *a* serve as prosthetic groups of hemoproteins and enzymes that are involved in such important cell functions as respiration (cytochromes of the respiratory chains), oxygen transport (hemoglobin, myoglobin), detoxification of oxygen by-products (catalases, peroxidases), xenobiotics and sterol metabolism (cytochromes

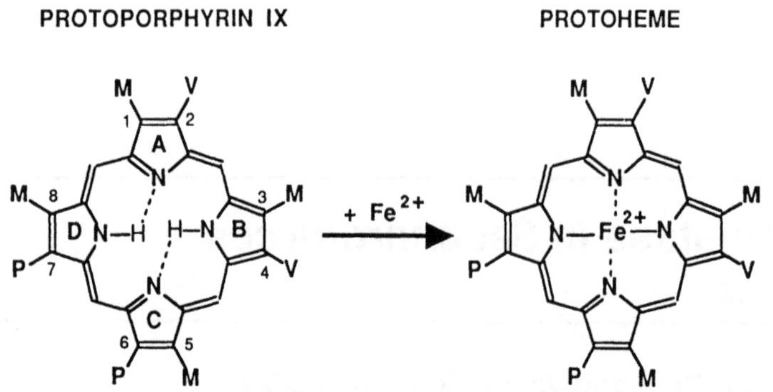

PROTOPORPHYRIN IX PROTOHEME

$+ Fe^{2+}$

Uroporphyrin III	: 1, 3, 5, 8 -Ac;	2,4,6,7 -P	
Coproporphyrin III	: 1, 3, 5, 8 -M;	2,4,6,7 -P	
Protoporphyrin IX	: 1, 3, 5, 8 -M;	6,7 -P;	2,4 -V
Deuteroporphyrin IX	: 1, 3, 5, 8 -M;	6,7 -P;	2,4 -H
Mesoporphyrin IX	: 1, 3, 5, 8 -M;	6,7 -P;	2,4 -Et

Heme c : 1, 3, 5, 8 -M; 6,7 -P; 2,4 -CH2-CH2-S-Cys(protein)
Heme a : 1, 3, 5 -M; 8 -formyl; 6,7 -P; 4 -V; 2 -hydroxyethyl farnesyl

Ac, acetyl; Et, ethyl; M, methyl; P, propionyl; V, vinyl

Figure 1 Reaction catalyzed by ferrochelatase and structures of some porphyrins and heme c and a.

P450), NO synthesis (NO synthase), tryptophane degradation (tryptophane pyrrolase), etc. The total amount of hemoproteins made by a cell depends on its type and physiological status. It can range from a few percent to over half the total cell protein. As a consequence, much of the total cell iron may be in the form of iron-porphyrin chelates, and the synthesis of these chelates may account for a large proportion of the iron used by the cell.

The yeast *Saccharomyces cerevisiae,* a "small" eukaryotic, unicellular organism, is well suited to studies on the relationships between iron metabolism and protoheme synthesis. It is stable in both haploid and diploid states and is easy to handle both biochemically and genetically. Most importantly, it is a facultative anaerobe that can grow in the absence of oxygen using the energy provided by the fermentation of sugars, such as glucose or galactose. The fact that a functional mitochondrial respiratory chain is not essential for life makes it possible to isolate viable mutants that are deficient in protoheme synthesis. Therefore, mutants with totally or partially impaired ferrochelatase function can be obtained and used to study structure-function relationships of ferrochelatase *in vivo* and *in vitro.* These heme-deficient mutants have also proved invaluable for analyzing the regu-

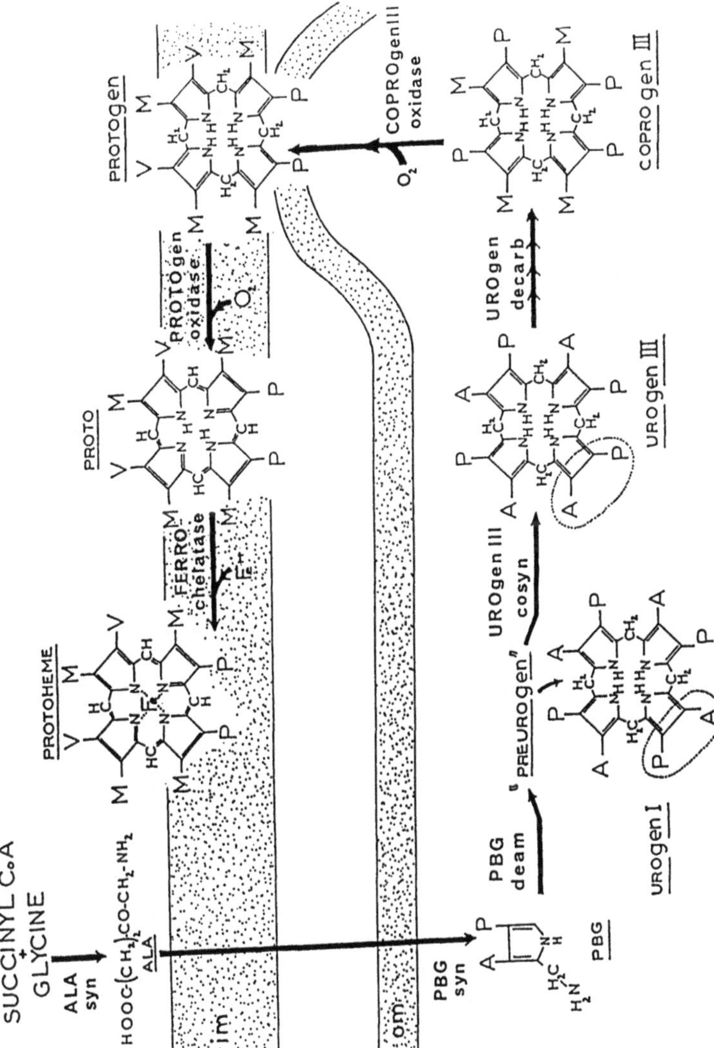

Figure 2 Heme biosynthetic pathway in *Saccharomyces cerevisiae*. im, om: inner and outer mitochondrial membranes. ALA: 5-aminolevulinic acid; PBG: porphobilinogen; urogen, coprogen, protogen: uro-, copro-, proto-porphyrinogen; proto: protoporphyrin IX. Syn: synthase; deam: deaminase; cosyn: cosynthase; decarb: decarboxylase.

latory role played by heme in the transcription of many (apohemo) proteins (Zitomer and Lowry, 1992; Pinkham and Keng, this volume).

This review begins with an analysis of ferrochelatase function *in vivo* in *S. cerevisiae*. The next section deals with the kinetic properties of membrane-bound ferrochelatase, followed by the characteristics of the purified enzyme. The last section examines the molecular genetics of *S. cerevisiae* ferrochelatase with a discussion focusing on its mechanism of action.

II. PHYSIOLOGICAL FUNCTION OF YEAST FERROCHELATASE

A. *In Vivo* Activity

Ferrochelatase provides protoheme for the various apocytochromes and apohemoproteins that require it as a prosthetic group for their biological functions. These hemoproteins are located in different cellular compartments (see Labbe-Bois and Labbe, 1990 for references). In addition to the cytochromes a + a3, b, c, and c1 associated with the respiratory chain bound to the inner mitochondrial membrane, yeast mitochondria contain cytochrome c peroxidase and cytochrome b2 that lie in the intermembrane space. Cytochromes b5 and P450 are associated with the membranes of the endoplasmic reticulum. Catalase A is present in the peroxisomes, while catalase T is cytosolic. There is also a flavohemoglobin in the cytosol whose function is still unclear (Oshino et al., 1973) and whose gene has been recently cloned and sequenced (Zhu and Riggs, 1992). This flavohemoglobin is probably the pigment P420 described earlier.

The protoheme in cytochromes c and c1 is further covalently attached to apoproteins via thioether bonds between the vinyl substituents of protoheme and cysteine residues of the proteins (heme *c*, Fig. 1). These reactions are catalyzed by cytochrome c- and cytochrome c1-heme lyases and take place during the import processes of the apoproteins into the mitochondria (Dumont et al., 1987; Zollner et al., 1992 and references therein). Heme *a*, present in cytochromes a + a3, differs from protoheme by having a formyl group in place of a methyl at position 8 and by the addition of a hydroxyfarnesyl group to the vinyl group at position 2 of the protoporphyrin ring (Fig. 1). The enzymic steps specific to heme *a* formation remain unclear. Mevalonic acid is known to be a precursor of the farnesyl moiety (Weinstein et al., 1986 and references therein), and very recently a gene has been identified in *E. coli* (homologous to the yeast *COX10* gene) encoding a protoheme IX farnesyltransferase (Saiki et al., 1992 and references therein). Protoheme is thus probably the precursor for heme *a* formation, rather than protoporphyrin. The liver ferrochelatase cannot incorporate iron into porphyrin *a in vitro* (Porra and Jones, 1963b). The last heme-like compound found in yeast is siroheme; this modified-uroporphyrin iron chelate is associated with sulfite reductase involved in sulfate assimilation and methionine biosynthesis. It is highly unlikely that ferrochelatase participates in its formation, since uroporphyrin is not a substrate for the enzyme and a lack of ferrochelatase does not lead to methionine auxotrophy.

Spectral analysis of whole cells (and subcellular fractions) indicates that other metalloporphyrins are present under some circumstances. Anaerobically grown cells contain Zn-protoporphyrin, especially when in stationary phase of growth (Chaix and Labbe, 1965; Labbe-Bois and Volland, 1977). Resting cells, incubated under some conditions, accumulate considerable amounts of Zn-protoporphyrin, which is found bound to the mitochondrial membranes (Gilardi et al., 1971). The formation of Zn-protoporphyrin is catalyzed by ferrochelatase (see Sec. III). In contrast, the Zn-uroporphyrin and Zn-coproporphyrin that accumulate in mutants lacking uroporphyrinogen decarboxylase and coproporphyrinogen oxidase activities, respectively, are formed nonenzymically (Rytka et al., 1984; Bilinski et al., 1981). Protoporphyrin does not accumulate in the cells and is not excreted into the culture medium under normal growth conditions of wild-type cells. The small amount present in stationary cultures represents less than 1% of the heme content. This indicates that the metabolite flow through the last two steps of the pathway is tightly controlled, without any leak of the protoporphyrin intermediate.

The activity of ferrochelatase *in vivo* is estimated by measuring the amount of heme and Zn-protoporphyrin present in the cells. This is done spectrophotometrically after converting all protoheme + heme *c* + heme *a* to their pyridine hemochromogens, or by estimating each of the various cytochromes individually. These measurements are not very precise and only provide semiquantitative estimations of the total heme. Most importantly, they can be distorted by the facts that (1) the presence of a spectrally detectable hemoprotein depends on the availability of both the apoprotein and the heme moiety, and (2) the syntheses of most apoproteins are regulated at the transcriptional level by oxygen, heme and glucose, either positively or negatively (Zitomer and Lowry, 1992). Therefore these spectral determinations probably also reflect the demand for heme made by the various apoproteins synthesized at different rates depending on the physiological conditions of the cells.

The total amount of heme-proteins, which are mainly accounted for by the cytochromes of the respiratory chain, is greatest in derepressed cells, that is, cells grown with vigorous aeration and a nonfermentable carbon source (such as ethanol, glycerol, or lactate) (Fig. 3d; Table 1). Glucose represses mitochondria development and the synthesis of mitochondrial

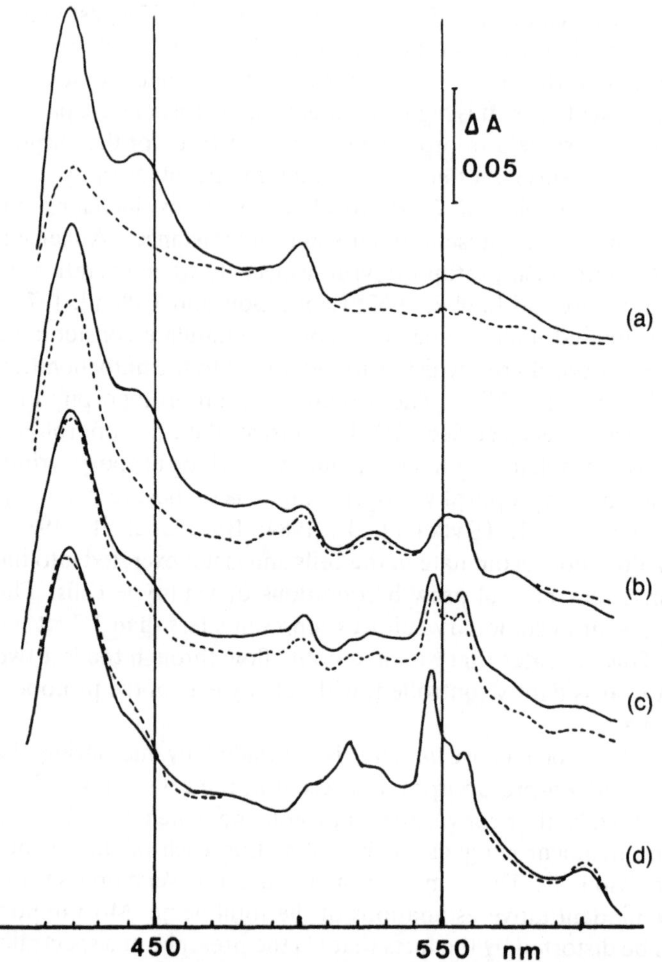

Figure 3 Low temperature (liquid nitrogen) spectrophotometric analysis of hemoproteins made by *S. cerevisiae* cells grown under different conditions. Anaerobiosis with 5% glucose (a), semiaerobiosis (low aeration) with 5% (b) or 1% glucose (c), and aerobiosis (vigorous aeration) with 1% glucose or 2% glycerol (d). The cells (40 mg dry weight) are filtered at the end of their exponential growth from aliquots of the culture (---) eventually gassed for 3 min with CO (———). Reduction is achieved by endogenous substrates.

Table 1 ***In vivo* and *in vitro* Ferrochelatase Activity in *S. cerevisiae* Grown Under Different Conditions**

Growth conditions	Total heme content	Ferrochelatase activity[a]	
		In vitro	*In vivo*
Ethanol	150–300	15–20	0.12–0.2
Glucose	60–120	10–15	0.1–0.2
Anaerobiosis	25–50	10–15	0.04–0.08
Heme-deficiency	0	10–15	0
Resting conditions	(Zn-proto)	10–15	0.03–0.04

[a]Expressed as nmol heme (or Zn-proto in resting conditions) per h per mg of membrane protein (*in vitro*) or per mg of total cell protein (*in vivo*).

cytochromes and other heme-proteins, while production of cytochrome P450 and hemoglobin (P420 pigment) is increased (Fig. 3b, c; Table 1). The only spectrally detectable heme-proteins in cells grown anaerobically are the cytochromes b5 and P450, and hemoglobin (P420 pigment) (Fig. 3a); the amount of heme falls to 10–20% of the maximal values observed in derepressed cells (Table 1). Within the limits of error of the methods used for these measurements, there is no evidence for the overproduction of "unassigned heme" (i.e., heme not bound to specific apoproteins), when apoprotein synthesis is repressed by glucose, turned off by anaerobiosis, or absent in mutants affected in their structural genes. This indicates a tight control of heme synthesis and coordination with apoprotein synthesis.

The rate of heme synthesis *in vivo,* corresponding to the *in vivo* ferrochelatase activity, can be calculated from the total heme cell content, taking into account the known cell-generation time and assuming no heme catabolism in growing cells. The results shown in Table 1 are compared to the ferrochelatase activities measured *in vitro* on mitochondrial membranes. The main conclusions are that (1) *in vitro* ferrochelatase activity does not change significantly in response to the growth conditions, that is, to total heme production, and that (2) ferrochelatase functions *in vivo* at only a small percentage of its maximal velocity measured *in vitro.* What, then, does limit the functioning of ferrochelatase *in vivo:* protoporphyrin production, iron supply, or the demand made by the different apoproteins?

B. Protoporphyrin IX Production: The Heme Biosynthetic Pathway

The heme biosynthetic pathway in *S. cerevisiae* has the same eight enzymic steps as in mammalian cells, with the same subcellular compartmentation, except that the coproporphyrinogen oxidase is in the cytosol, whereas it is in the mitochondrial intermembrane space in mammalian cells (Fig. 2). There are many steps along this pathway that may control the rate and the amount of protoporphyrin, and possibly of heme, made. Since this has been reviewed recently (Labbe-Bois and Labbe, 1990), only the main features of the pathway and new data relevant to this review will be presented.

ALA synthase (encoded by the *HEM1* gene), the first committed enzyme located in the mitochondrial matrix, requires an adequate supply of its substrates, glycine (K_M = 3 mM) and succinyl-CoA (K_M = 2 μM). Succinyl-CoA is provided by the Krebs cycle, and a disturbance of this cycle might lead to a decrease in ALA production. A severe shortage of succinyl-CoA supply due to a defect in alpha-ketoglutarate dehydrogenase has been shown to lead to the absence of cytochromes (Labbe-Bois et al., 1983). There is no evidence that ALA synthase is feedback inhibited by heme, the end product of the pathway (Volland and Felix, 1984). Neither does heme seem to affect *HEM1* gene expression, which is controlled by a complex array of activation and repression leading to a constitutive expression (Keng and Guarente, 1987).

The synthesis of heme in cells growing under anaerobic conditions is puzzling, since oxygen is required as electron acceptor for the two oxidases of the pathway. The steady-state levels of most of the different enzymic activities of the heme pathway measured in cell-free extracts of cells growing under different metabolic conditions (glucose-derepression or glucose-repression, anaerobiosis) do not change much as a function of total heme made. The exception is coproporphyrinogen oxidase, which is dramatically increased in anaerobic cells (Labbe-Bois and Labbe, 1990). A similar increase also occurs in cells made totally heme-deficient by a mutational block in any enzyme of the pathway (Urban-Grimal and Labbe-Bois, 1981). This negative regulation of coproporphyrinogen oxidase by oxygen and heme operates at a transcriptional level. Deletion of regulatory sequence(s) in the promoter leads to a loss of induction of coproporphyrinogen oxidase activity by anaerobiosis and causes a considerable drop of heme formation in anaerobic cells (Zagorec and Labbe-Bois, 1986; Zagorec et al., 1988). This means that coproporphyrinogen oxidase is rate limiting for heme production when its substrate oxygen becomes limiting, and the cell responds to oxygen limitation by increasing the amount of the enzyme. The physiological significance of the regulation by heme in aerobiosis is less clear. Current work on the mechanism(s) of this regulation (Verdière et al., 1991; Keng, 1992) suggests that it might provide for regulatory linkage of heme synthesis to apohemoprotein synthesis via the common heme-dependent transcriptional activator product of the gene *CYP1* (*HAP1*).

Oxygen is also required by the penultimate enzyme, protoporphyrinogen oxidase, which oxidizes protoporphyrinogen IX to yield protoporphyrin IX (Camadro et al., manuscript in preparation). This enzyme activity, like ferrochelatase, is in excess over the other enzymes of the pathway and functions *in vivo* at well below its maximal velocity measured *in vitro*. This may explain why the enzyme is not induced under oxygen limitation, in contrast to coproporphyrinogen oxidase. The formation of heme in cells grown in the absence of oxygen can thus be explained by (1) the high coproporphyrinogen and protoporphyrinogen oxidases activities and their high affinity for oxygen (Camadro et al., 1986; Camadro et al., manuscript in preparation) and (2) the fact that it is practically impossible to avoid traces of contaminating air in the culture due to its diffusion through the tubings and the glass walls.

Although the first six enzymes of the pathway operate at about their maximal velocity *in vivo*, the overexpression of only one of them does not increase the amount of protoporphyrin or heme made. This was shown by transforming yeast cells with multicopy plasmids carrying the structural genes for ALA synthase (Arrese et al., 1983), uroporphyrinogen decarboxylase, or coproporphyrinogen oxidase (Labbe-Bois, unpublished work). Cells overproducing ferrochelatase are also undistinguishable from the wild-type cells when assayed for their content of total hemes and cytochromes (Labbe-Bois, 1990). Additional copies of the *HEM2* and *HEM3* genes result in increased activity of PBG synthase and PBG deaminase, respectively; but the effect on the heme content was not documented (Myers et al., 1987; Keng et al., 1992). These results indicate that heme production cannot be increased by increasing the flow of metabolite through a single rate-limiting step. It would be interesting to know if the simultaneous overexpression of groups or all the enzymes of the pathway would lead to increased protoporphyrin and heme formation. This approach will be feasible with the cloning of the two *HEM* genes that have not yet been isolated, those coding for uroporphyrinogen III synthase and protoporphyrinogen oxidase. It is worth noting here that supplementing the culture medium with ALA, protoporphyrin IX, or protoheme does not increase the cell hemoprotein content in standard strains.

In contrast, the heme production is subnormal in mutant cells with partially impaired ALA synthase (Urban-Grimal and Volland, unpublished results), uroporphyrinogen decarboxylase (Rytka et al., 1984; Chelstowska et al., 1992), coproporphyrinogen oxidase (Bilinski et al., 1981), or ferrochelatase (Abbas and Labbe-Bois, 1993) function. The heme deficiency is correlated with the severity of the enzyme defect. These mutants grown on glucose make less heme and accumulate less porphyrins than do cells growing on ethanol. However, the sum of heme + porphyrins is the same as the amount of heme made by the wild-type cells grown on glucose or ethanol, suggesting that glucose diminishes the production of tetrapyrroles. This can occur at the level of succinyl-CoA formation through repression of the Krebs cycle enzymes by glucose, and/or at the level of the enzymes of the heme pathway (the steady-state levels of some enzymic activities of the pathway are about twofold lower in extracts of cells grown with glucose than extracts of cells grown with ethanol). Surprisingly, almost normal amounts of heme (90%) are made, but uroporphyrin accumulates, in mutants that contain only 10% of the normal uroporphyrinogen decarboxylase level due to a specific transcriptional defect (Zoladek, Cheltowska, Labbe-Bois, and Rytka, manuscript in preparation). This indicates that the flow of protoporphyrin is almost normal, despite a 90% reduction in uroporphyrinogen decarboxylase activity.

Hence it is clear that a normal rate of protoporphyrin synthesis is required for ferrochelatase function. But we still do not know if an increase in this rate would result in more heme and hemoprotein formation.

C. Relationship to Iron Status

There is little information on the correlation between the iron status of yeast cells and their heme production. Light (1972) presented data on cells of *Torulopsis utilis* (an obligate aerobe yeast species) growing in chemostat culture in a chemically defined medium. Decreasing the concentration of iron in the flowing medium led initially to a 10-fold decrease in cellular nonheme iron, while the total heme content was unchanged. Then the contents of nonheme and heme iron, which were very similar, decreased in parallel: for 0.3 μM iron, total heme dropped 4-fold and total nonheme iron dropped 40-fold. Iron limitation led to the disappearance of the electron paramagnetic resonance signals from iron-sulfur proteins in *C. utilis* submitochondrial particles (Ohnishi et al., 1971). In rat muscle mitochondria, the concentrations of both cytochromes and Fe-S clusters associated with the respiratory chain were decreased by a severe dietary iron deficiency (Maguire et al., 1982). Wright and Honek (1989) showed that cytochrome P450 was no longer present in *S. cerevisiae* cells growing with 100 μM 2,2′-dipyridyl, a concentration that caused less than 50% growth inhibition. We have observed that total heme iron was approximately 10% of total cellular iron in *S. cerevisiae* cells during their exponential phase of growth on glucose synthetic (1 μM iron) or complete (10 μM iron) medium. Adding the nonpermeant iron chelator bathophenantroline sulfonate or ferrozine to the culture resulted in a considerable decrease in total iron (8–10-fold) after 6 h, while the total heme was only slightly (20%) lower. Interestingly, porphyrin was not accumulated or excreted under severe iron deficiency that reduced heme production 3–5-fold, suggesting that protoporphyrin production was also limited. This might occur at the coproporphyrinogen oxidase, an iron-containing enzyme, although it is not known whether iron is involved in the catalytic activity (Camadro et al., 1986) or at the ALA synthase, although it is not known whether its synthesis in yeast is regulated by iron, as is the erythroid (but not the liver) mammalian ALA synthase (Cox et al., 1991; Dandekar et al., 1991). The third possibility is that the supply of the substrates for ALA synthase, especially succinylCoA, is limited via the functioning of the iron-containing enzymes of the Krebs cycle. On the other hand, increasing the concentration of iron in the culture medium (up to 200 μM) did not increase the cellular heme level, although doubling cellular nonheme iron content, either in standard wild-type strains or in a ferrochelatase mutant dramatically impaired in its affinity for iron (Abbas and Labbe-Bois, 1993). Taken together, these results indicate a tight control of intracellular iron flow that is apparently independent of ferrochelatase functioning. No coupling of iron uptake and heme formation was shown in *R. sphaeroides,* either (Moody and Dailey, 1985).

The nature and the location of the iron pool used by ferrochelatase are still mysterious. Since ferrochelatase is associated with the inner mitochondrial membrane, with its active site on the matrix side of that membrane, we favor the hypothesis of an iron pool located in the mitochondrial matrix. This is consistent with the fact that iron deficiency reduces contents of both heme and mitochondrial Fe-S clusters. This would make good biochemical sense by linking protoporphyrin production, heme synthesis, and iron uptake and intracellular routing, mediated possibly by citric acid (see Lesuisse and Labbe, this volume), as discussed by Williams (1982). Such a "free" iron pool (nonheme, non-Fe-S; 1 nmol/mg protein) located in the mitochondrial inner compartment (inner membrane + matrix), half in the reduced form, and available for heme synthesis, was found in isolated rat liver mitochondria (Tangeras et al., 1980; Tangeras, 1985 and 1986). The rate of heme synthesis from this endogenous iron pool was low, reflecting its slow mobilization by chelators; but, as noted by Tangeras (1985), "it exceeds by a factor of 5 the rate of heme synthesis necessary to support the turnover of hemoproteins in the hepatocytes."

If this intramitochondrial iron pool is the proximate iron donor for ferrochelatase, then one may ask how iron is transported into the mitochondria and where it is reduced if it is taken up from an oxidized storage form or low molecular mass pool. There is some evidence that iron can be reduced by the mitochondrial respiratory chain in yeast and mammals, but the side of the inner mitochondrial membrane on which this occurs remains controversial (Lesuisse and Labbe, this volume; Jones and Jones, 1969; Barnes et al.,

1972; Koller and Romslo, 1977; Funk et al., 1986). There is also some evidence for an energy-dependent transport of ferrous iron across the inner membrane (liver), but its mechanism is still unknown (Flatmark and Romslo, 1975; Koller and Romslo, 1977). Indeed, it is highly likely that mitochondria possess a specific transporter for such a hazardous charged ion as iron (ferric or ferrous), which can be very harmful in the presence of O_2^- and H_2O_2, which are produced mainly by the mitochondria. But the demonstration of its existence will have to wait for biochemical or genetic identification, as with the cloning of the yeast *COT1* gene encoding a putative mitochondrial cobalt transporter (Conklin et al., 1992).

The nature of the intracellular donor(s) of iron to the mitochondria is still unclear. High rates of iron-porphyrin formation are measured in suspensions of isolated liver or yeast mitochondria in the presence of various exogenous ferric iron complexes, including ferritin and low molecular weight complexes (Funk et al., 1986 and references therein; Weaver and Pollack, 1990). Interesting though these facts are, they do not answer the question of the physiological *in vivo* iron donor. Actually, this problem is difficult to assess with disruptive methods. The precise nature of the iron donor(s) to mitochondria, as well as the iron species of the intramitochondrial pool that serve(s) as substrate(s) for ferrochelatase may remain an enigma for some time yet.

D. Fate of Heme and Demand Made by the Apohemoproteins

The heme made in the mitochondria is used in the processing and assembly of holohemoproteins located in different cellular compartments. Heme must also be available for its role in transcriptional control. There must therefore be transport mechanism(s) and carrier protein(s) for the intracellular routing of heme. For instance, the efflux of heme (and other metalloporphyrins) made intramitochondrially by isolated mitochondria is enhanced in the presence of heme-binding proteins. But these problems are very poorly understood at present (see Muller-Eberhard and Vincent, 1985 and Smith, 1990 for reviews). It is not known whether ferroheme is released from ferrochelatase on the matrix side of or in the inner mitochondrial membrane, or if it stays in the reduced form during its intracellular traffic. It is not known whether heme is transported outside of the mitochondria by passive diffusion through the membranes (via contact sites?) or by a specific transport system. Recently, a complex transporter system has been characterized genetically in *Rhodobacter capsulatus* (and *Bradyrhizobium japonicum*) that may be involved in the export of heme from the cytosol to the periplasmic space (Beckman et al., 1992 and references therein). But no mitochondrial heme transporter has been documented as yet in eukaryotes.

Since heme is a lipophilic molecule that is potentially harmful as a Fenton reagent and dimerizes and aggregates readily in an aqueous environment, it is not unreasonable to assume that it must be associated with carrier protein(s) during its transit to the apoproteins into which it is finally incorporated. But no such protein has yet been positively identified, although there is indirect evidence for their involvement. Pulse-labeling studies with radioactive ALA have shown that the newly synthesized heme first appears in hepatocytes "in transit" in a small cytosolic pool before its incorporation into membrane hemoproteins (references in Muller-Eberhard and Vincent, 1985 and Smith, 1990). Exogenous heme mixes with that pool of precursor heme for hemoprotein formation, and it has been suggested that this "free" heme pool is identical to the previously postulated "regulatory heme" pool. No such studies are available for *S. cerevisiae,* where pulse-labeling studies are made difficult, if not impossible, by the presence of a large intracellular pool of endogenous ALA (Labbe-Bois and Labbe, 1990), the long time (about 2 h) needed for protoplast formation and recovery of the cellular fractions, and the long time taken for heme formation (1–2 h) when *hem1* mutants are supplemented by exogenous ALA (Schmalix et al., 1986).

The problem of the metabolic disposition of heme in yeast could be addressed indirectly by asking the question whether there is competition between the different apohemoproteins for heme. But no systematic study has been carried out, and the published information is fragmentary and often contradictory. We have observed that leaky *hem* mutants can grow in the absence of unsaturated fatty acids and ergosterol, indicating that sufficient heme is made to provide the cells with the functional microsomal cytochromes b5 and P450 required for fatty acid desaturation and ergosterol biosynthesis. But no mitochondrial cytochromes were spectrally detectable, making these mutants unable to grow on glycerol. Meussdoerffer and Fiechter (1986) analyzed the content of mitochondrial and P450 cytochromes under heme limitation and concluded that competition for heme between cytochromes is rather unlikely. Weber et al. (1992) reached the opposite conclusion and showed an inverse relationship between the cytochromes c and P450 contents in cells growing in a chemostat under various dilution rates of the glucose medium; but their results could also be explained by an effect of glucose on the transcription of the apocytochromes c and P450. Anaerobically grown cells that make less heme, but enough for cytochromes b5 and P450 and for hemoglobin (P420) formation, contain apocytochrome c peroxidase, but not holoprotein (Djavadi-Ohaniance et al., 1978). This is puzzling and suggests a different disposal of heme in anaerobic and aerobic cells.

Unbalanced heme use could also be produced by overexpressing genes encoding different apohemoproteins placed under control of inducible yeast promoters. This has been done for the overexpression of heterologous hemoglobin, cytochrome b5, and cytochrome P450 (Wagenbach et al., 1991; Coghlan et al., 1992; Vergères and Waskell, 1992; Weber et al., 1992; Peyronneau et al., 1992 and references therein). Although the conditions used in these studies were very different, the production of holohemoproteins induced in stationary cells appeared to be limited by the availability of endogenous heme. But the amount of the yeast cytochromes made under these conditions was not documented, which precludes any conclusion. Nevertheless, careful examination of the hemoproteins made in cells overexpressing a heterologous hemoprotein might give information on the fate of heme and any competition for its further use. Of course, all these studies are valid only if there is no catabolism of heme, which appears to be the case (Labbe-Bois and Labbe, 1990).

III. MEMBRANE-BOUND YEAST FERROCHELATASE

The functioning of ferrochelatase raises many problems due to (1) the topology of the enzyme within the inner mitochondrial membrane and the influence of the (phospho)lipid environment on the enzyme activity, and (2) the physiological supply and the physicochemical properties of the substrates, protoporphyrin IX and ferrous iron, in terms of hydrophobicity, charge, stability, and competition between different metals.

A. The Membrane Context and Enzyme Topology

Early work by Nishida and Labbe (1959) showed that ferrochelatase activity is associated with mitochondrial membranes. The enzyme in rat liver cells was then found to be associated with the inner mitochondrial membrane (McKay et al., 1969), with the active site on the matrix side of the membrane (Jones and Jones, 1969; Harbin and Dailey, 1985; Deybach et al., 1985). Harbin and Dailey (1985), using a membrane-impermeable lysine reagent to label covalently bovine liver ferrochelatase, concluded that ferrochelatase is a transmembrane protein.

Ferrochelatase is also associated with the inner membrane of the mitochondria in yeast cells (Jomary and Labbe, unpublished results), and the topology of the enzyme is assumed to be the same as in mammalian cells. However, both the protein sequence data (Labbe-Bois, 1990; Sec. V) and alkaline extraction experiments (Volland and Urban-Grimal, 1988) strongly indicate that ferrochelatase is a peripheral rather than an integral membrane-bound protein.

Labbe et al. (1968) showed that the membrane-bound yeast ferrochelatase is completely inactivated by extraction of the lipids from the membranes. However, ferrochelatase activity could be recovered by adding back the phospholipid extract to the assay medium. Such phospholipid dependence of ferrochelatase activity has been reported for ferrochelatases of other origins (Sawada et al., 1969; Yoneyama et al., 1969; Takeshita et al., 1970).

B. Ferrochelatase Assay

The published procedures for assaying ferrochelatase *in vitro* all involve incubating the enzyme fraction (membrane suspensions) with known concentrations of protoporphyrin IX, or alternate porphyrin and ferrous iron, or alternate metal, and quantification of either the metallo-porphyrin produced or the disappearance of the tetrapyrrole substrate with time. Protoporphyrin IX and ferrous iron are not easy to use for ferrochelatase assays because

(1) Protoporphyrin IX tends to aggregate rapidly in aqueous solutions, forming dimers and micellelike polymers that are not substrates for ferrochelatase (Margalit et al., 1983). In order to keep protoporphyrin in the monomeric state necessary for ferrochelatase assay, protoporphyrin IX must be pseudosolubilized in detergent or liposome micelles. This "solubilization" consists of a reversible binding of protoporphyrin to the detergent micelles (Rotenberg and Margalit, 1987). The protoporphyrin stock solutions are not very stable, due to the equilibrium between the two phenomena, aggregation and "solubilization." In our experience, the best detergent is Tween 80, which has a low critical micellar concentration (Camadro and Labbe, 1982). Because of these problems many authors use more hydrophilic dicarboxylic porphyrins such as meso- or deuteroporphyrin that aggregate less than protoporphyrin.

(2) Ferrous iron is readily oxidized to ferric iron by molecular oxygen. As a result, stable divalent cations such as cobalt, zinc, or nickel are widely used as substrates. The use of ferrous iron requires anaerobic conditions during the assay. This is usually achieved by flushing nitrogen into the reaction mixture in specialized vessels ("Thunberg tubes"; Porra et al., 1967; Labbe et al., 1968) or by consuming dissolved oxygen by enzymatic systems, such as the endogenous respiration of the mitochondrial membranes with succinate, the addition of whole yeast cells with glucose (Jones and Jones, 1969; Camadro and Labbe, 1981), or the addition of glucose + glucose oxidase + catalase (Camadro and Labbe, 1988).

The methods used to assay ferrochelatase activity can be divided into three main groups:

(1) Radiochemical assays (Nishida and Labbe, 1959; Bottomley, 1968), in which a radioactive isotope of iron or cobalt is incorporated into protoporphyrin. The synthesized metalloporphyrin is extracted in the presence of carrier hemin and the radioactivity measured.

(2) Indirect spectrophotometric assays (Porra and Jones, 1963a), in which a stable chromophore is formed at the end of the enzymatic reaction by coordination of pyridine on chelated iron, and quantified by the difference in absorbance of its reduced versus oxidized spectrum. This assay is still popular and useful, but the substrate concentrations required are in the millimolar range and the limit of detection of heme formed is in the micromolar range.

This type of end-point assay requires appropriate controls to ensure that the activity is measured under initial velocity conditions. The pyridine-hemochromogen method also requires substantial quantities of enzyme to be reliable.

(3) Direct spectrophotometric assays (Jones and Jones, 1969), in which the disappearance of the tetrapyrrolic substrate is measured by the changes in absorbance at an appropriate wavelength pair. The substrate concentrations required are in the micromolar range, and the limit of detection of heme formed, deduced from porphyrin disappearance, is in the 20–50 nanomolar range. The sensitivity of this assay is greatly improved by spectrofluorometric monitoring of the porphyrin or metalloporphyrin (Zn-protoporphyrin) concentration changes during the enzyme reaction, and picomoles of products can be readily detected (Camadro et al., 1984; Camadro and Labbe, 1988).

Clearly, the results will depend greatly on the assay conditions and the sensitivity of the assay procedures, rather than on the ferrochelatase species under study. It is not surprising that large differences, up to three orders of magnitude, have been reported for the K_M values during the two last decades. This is illustrated by the change in the K_M for protoporphyrin IX of yeast ferrochelatase measured, from 66 μM (pyridine hemochromogen assay, Labbe et al., 1968), to 5.9 μM (direct spectrophotometric assay, Camadro and Labbe, 1982), and most recently to 0.03–0.05 μM (direct spectrofluorometric assay, Abbas and Labbe-Bois, 1993).

C. Kinetic Properties: Iron and Zinc as Competing Substrates

The central question for early studies of yeast ferrochelatase was the synthesis of zinc-protoporphyrin by yeast cells and the possibility of different enzymes catalyzing the synthesis of protoheme and zinc-protoporphyrin. This question governed our choice to use only protoporphyrin IX, ferrous iron, and zinc as substrates for the ferrochelatase reaction.

The early results obtained on yeast ferrochelatase (Chaix and Labbe, 1965; Labbe et al., 1968) clearly established the enzymatic nature of both iron chelation and zinc chelation into protoporphyrin IX. It was also shown that high concentrations of zinc inhibited iron chelation. A more sensitive direct spectrophotometric assay later showed that iron and zinc do indeed act as reciprocal competitive substrates (Camadro and Labbe, 1982). Ferrous iron ($K_{M\ app}$ = 2.9 μM) is a strong competitive inhibitor of zinc chelation (K_I = 0.05 μM), whereas zinc ($K_{M\ app}$ = 2.8 μM) is a weaker competitor of iron chelation (K_I = 1.5 μM). Ferric iron is also a competitive inhibitor of both iron- and zinc-chelatase activities; but the high apparent K_I values (around 100 μM) make the physiological significance of such an inhibition doubtful. It appears, then, that the balance between iron- and zinc-chelatase activities depends on the concentrations of the metal substrates and the redox state of the iron. We thus propose that, under normal conditions of growth, the zinc-chelatase activity is inhibited because of a constant supply of ferrous iron to mitochondria; but the ferrous iron pool may be reduced or even depleted in "resting" cells releasing the inhibition of zinc-chelatase activity. A single enzyme in this process was shown to be responsible when a yeast strain carrying a single nuclear recessive mutation abolishing ferrochelatase activity (Urban-Grimal and Labbe-Bois, 1981) could not catalyze zinc-protoporphyrin or protoheme synthesis either *in vivo* or *in vitro*.

During the course of this study (Camadro and Labbe, 1982) it was found that yeast mitochondrial membranes contained "free" iron (0.5–0.6 nmol/mg protein) and zinc (0.7–1.7 nmol/mg protein) that were very slowly accessible to the ferrochelatase reaction performed in the absence of added metals. The amount of this "free" iron is identical to that described in rat liver mitochondria (Tangeras et al., 1980; Tangeras, 1985). Both endogenous iron and zinc are chelatable by EDTA, which completely inhibits iron- and zinc-chelatase activities. Large amounts of endogenous iron and zinc that are substrates for ferrochelatase are also present in human liver mitochondrial membranes (Camadro et al., 1984). The presence of these endogenous metals, often saturating, adds a further layer of complexity to the kinetic analysis of ferrochelatase, especially when very sensitive spectrofluorimetric methods are used. This means that the apparent K_M values have to be recalculated to take into account the concentrations and inhibitory effects of these endogenous metals. When this is done, the K_M values for iron and zinc are 1.6 and 0.15 μM for the yeast enzyme and 0.35 and 0.08 μM for the human enzyme. Only heme is made under normal conditions in yeast and humans, despite the higher affinity of ferrochelatase for zinc, probably because the zinc-chelatase activity is strongly inhibited by ferrous iron.

The presence of endogenous iron and zinc in crude mitochondrial membranes raises the questions of their origin and of their physiological significance. During disruption of the cells, membrane proteins and phospholipids may trap these metals that are normally located in different intracellular compartments such as the mitochondrial matrix, the cytosol, or the vacuole. Therefore we need to ask whether the iron and zinc found associated with the mitochondrial membranes are truly the physiological donors for ferrochelatase or whether their presence is merely artifactual. We do not know the answer at present.

D. Origin of Protoporphyrin IX

The interactions between the membrane lipid phase and the hydrophobic substrate of ferrochelatase, protoporphyrin IX, may be another important point for regulating ferrochelatase activity. This point has not been fully studied, mainly because of the obligatory presence of detergents required to "solubilize" protoporphyrin IX. This led us to investigate the ferrochelatase reactivity toward protoporphyrin IX generated enzymatically within the mitochondrial membrane.

Protoporphyrin is synthesized *in vivo* by the membrane-bound enzyme protoporphyrinogen oxidase. This enzyme is an intrinsic protein of the inner mitochondrial membrane in mammalian cells (Deybach et al., 1985; Ferreira et al., 1988) and in yeast cells (Camadro et al., manuscript in preparation). Although the orientation of its active site is not yet clear (in the membrane bilayer or on the side of the membrane facing the intermembrane space), an agreement seems to exist for locating the newly synthesized protoporphyrin within the membrane. Hence we have studied the kinetic properties of ferrochelatase with protoporphyrin IX generated *in situ* from coproporphyrinogen III by the activities of coproporphyrinogen oxidase and protoporphyrinogen oxidase.

Ferrochelatase activity was measured in a two-step assay, with fluorimetric detection of protoporphyrin IX synthesis and its subsequent utilization by ferrochelatase, as shown in Fig. 4. Variable amounts of protoporphyrin IX were synthesized within the membrane fractions when coproporphyrinogen oxidase and coproporphyrinogen III were added while ferrochelatase activity was inhibited by excess chelating agent (EDTA, 1 mM). The reaction was stopped by adding succinate to consume oxygen via the high respiratory activity of the mitochondrial membranes: anaerobic conditions block the activities of coproporphyrinogen oxidase

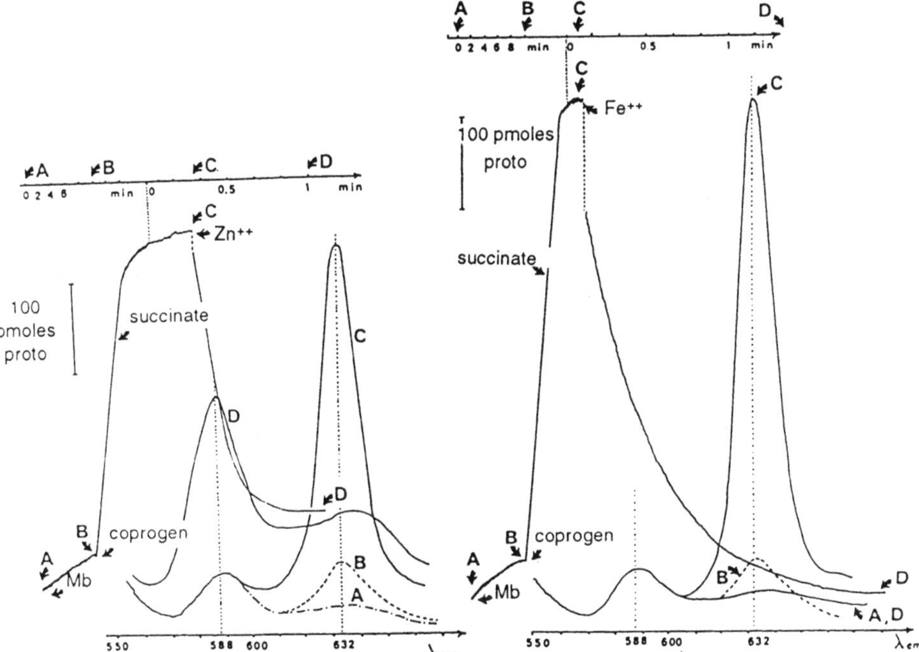

Figure 4 Ferrochelatase activity measured with protoporphyrin IX generated in the mitochondrial membranes. The incubation medium (3 mL) contained 0.1 M potassium phosphate buffer pH 7.6, 5 mM DTT, 1 mM EDTA, 7 units/mL coproporphyrinogen oxidase, 5 mg mitochondrial membrane protein carrying 60 units of protoporphyrinogen oxidase and ferrochelatase. After 10 min preincubation (A), protoporphyrin IX synthesis was initiated (B) by addition of coproporphyrinogen III (coprogen) (5 μM final concentration). Coproporphyrinogen and protoporphyrinogen oxidase activities were stopped by enzymic consumption of dissolved oxygen after addition of 0.1 M succinate (C). Different amounts of protoporphyrin were synthesized by varying the time interval between the initiation of protoporphyrin synthesis and the addition of succinate (compare protoporphyrin fluorescence intensity between left and right panel). Ferrochelatase activity was then initiated by addition of 2 mM zinc (left panel) or ferrous iron (right panel). Protoporphyrin fluorescence changes were recorded at λ exc = 410 nm and λ em = 632 nm (left part of each panel). Fluorescence emission spectra (λ exc = 410 nm) (right part of each panel) were recorded at the different periods of the reaction A, B, C, and D. Emission maxima are 632 nm for protoporphyrin and 588 nm for zinc-protoporphyrin; protoheme does not fluoresce.

and protoporphyrinogen oxidase. The protoporphyrin IX that accumulated within the membrane had absorption and fluorescence spectra of a neutral (with respect to the pyrrolic nitrogen) protoporphyrin monomer dissolved in a nonpolar solvent (Fig. 4). This suggests that the protoporphyrin hydrophobic core is embedded in the hydrophobic domain of the membrane lipid bilayer with the two deprotonated propionic side chains lying near the polar heads of the bilayer, as proposed for protoporphyrin binding to liposomes (Brault et al., 1986; Ricchelli et al., 1991). Ferrochelatase activity was then measured by adding excess metal ion, zinc or ferrous iron, and recording either the appearance of zinc-protoporphyrin fluorescence or the loss of protoporphyrin IX fluorescence (Fig. 4).

This experimental approach allowed us to measure a $K_M = 0.075\ \mu M$ for the generated protoporphyrin IX, very close to that measured with exogenous protoporphyrin (0.03–0.05 μM). The chelator-metal equilibrium displacement technique used did not permit us to measure the K_M for zinc or iron.

The main conclusion of these studies is that ferrochelatase has the same affinity toward both enzymatically generated protoporphyrin and exogenously added protoporphyrin, provided that physiological concentrations of protoporphyrin IX (nanomolar to micromolar range) are used. This probably reflects the fact that the porphyrin accessible to ferrochelatase is sequestered by the most nonpolar part of the membrane in both cases, either after synthesis or after binding to the membrane. It is thus very likely that the active site of ferrochelatase is at least partly facing the membrane rather than the aqueous region surrounding the membrane.

IV. PURIFIED YEAST FERROCHELATASE

A. Purification

Riethmüller and Tuppy (1964) first described a 75-fold purification of the enzyme from yeast mitochondrial membranes after Triton X100 extraction, adsorption onto silica gel, and DEAE-Sephadex chromatography. Yeast ferrochelatase was later purified to homogeneity (Camadro and Labbe, 1988) by taking advantage of the unusual binding properties of ferrochelatase on the immobilized dye Cibacron Blue GF-3A (Blue-Sepharose). This system was first used by Taketani and Tokunaga (1981) to purify rat liver ferrochelatase. Ferrochelatase binds to Blue-Sepharose in the presence of a nonionic detergent but is eluted only in the presence of an ionic detergent. Several modifications to the original protocol were necessary to make this step efficient for the purification of yeast ferrochelatase. The critical change was to use a nonionic detergent (Tween 80) to solubilize the enzyme from the membranes: anionic or cationic detergents preclude the binding of yeast ferrochelatase to Blue-Sepharose. The reason(s) for such difference of reactivity toward Blue-Separose is not clear. Cibacron Blue GF-3A has no effect on yeast ferrochelatase activity measured *in vitro,* so it is likely that it does not interact with the active site of the enzyme. The adsorption of the enzyme to the dye must be a complex phenomenon involving both pseudoaffinity and hydrophobic interactions. Ionic detergents may, for example, cosolubilize some compounds (proteins or lipids) with higher affinities for the dye than ferrochelatase, preventing its binding.

Yeast ferrochelatase was purified to homogeneity (1800-fold) from mitochondrial membranes with an overall yield of 43%. It appears to be a hydrophobic protein that requires detergent for its extraction from the membranes and throughout the whole purification procedure.

B. Molecular Properties and Biogenesis

The purified enzyme has a pI of 6.3 and is a 40-kDa polypeptide after SDS-polyacrylamide gel electrophoresis. This relative molecular mass is close to that reported for other eukaryotic ferrochelatases (Mr = 40,000–42,000) (Taketani and Tokunaga, 1981, 1982; Dailey et al., 1986a) but different from that of the enzyme purified from the bacteria *A. itersonii* (Mr = 50,000) and *R. spheroides* (Mr = 115,000) (Dailey, 1986). However, bacterial ferrochelatase gene sequences (Sec. V) indicate that the deduced relative molecular mass of the protein is also in the range of Mr = 36,000 to 40,000.

The estimation of ferrochelatase molecular weight by gel permeation chromatography seems to depend on the concentration of sodium cholate in the column buffer. Yeast ferrochelatase, like the rat and bovine liver enzymes, elutes with a high molecular weight (>240,000) at low cholate concentration, suggesting an oligomeric form of the native enzyme or aggregates. But at higher cholate concentration, the bovine enzyme is eluted with a Mr = 40,000, indicating that the enzyme is active as a monomer that is prevented from aggregating by high cholate concentration (Dailey and Fleming, 1983; Bloomer et al., 1987). This is consistent with the fact that one molecule of protoporphyrin or of the competitive inhibitor *N*-methylprotoporphyrin binds per 40,000 Da (Dailey, 1985), and that reaction of one sulfhydryl group per enzyme molecule inactivates bovine ferrochelatase (Dailey, 1984). The functional size of ferrochelatase *in situ* has been addressed by radiation inactivation analysis. Early work with dried acetone-extracted chicken erythrocytes indicated a target size of 320 kDa (Tanaka et al., 1976). Recently, a mass of 82 ± 13 kDa was found for the functional unit required for enzymic activity in intact and sodium cholate–solubilized bovine liver mitochondria, suggesting that two interacting 40-kDa subunits (homo- or heterodimer) are required for membrane-bound ferrochelatase function (Straka et al., 1991a). Obviously, the problem of the native form of ferrochelatase is still open.

Purified yeast ferrochelatase has an absolute requirement for fatty acids to be active (Camadro and Labbe, 1988). All the fatty acids tested, myristic, palmitic, and oleic acids, are almost equally efficient in activating the enzyme; however, some nonenzymatic protoheme and zinc-protoporphyrin synthesis were observed with oleic acid in the presence of dithiothreitol, as previously reported by Taketani and Tokunaga (1984). Yeast ferrochelatase is always assayed in the presence of the neutral detergent Tween 80, an ester of oleic acid, that does not support the nonenzymatic incorporation of metals into protoporphyrin and does not activate the purified ferrochelatase, suggesting that the free carboxylic acid groups of fatty acids are necessary for activation. Taketani and Tokunaga (1981) first showed that purified rat liver ferrochelatase is lipid-dependent: although the enzyme contains fatty acids, its activity was enhanced (15-fold) by exogenous fatty acids but not by neutral lipids. In contrast, the lipid-free purified bovine ferrochelatase was activated very little (2-fold) by fatty acids (Taketani and Tokunaga, 1982). The role of fatty acids in the activity of purified ferrochelatase remains unclear. They may help in the delivery of substrates to the active site of the enzyme, or induce the proper conformation of the enzyme by mimicking its membrane environment. It is all the more puzzling that, paradoxically, two yeast mutant ferrochelatases are inhibited by palmitic acid while they are still associated with the membranes (Abbas and Labbe-Bois, 1993).

Polyclonal antibodies were raised against purified yeast ferrochelatase. They did not inhibit the enzyme activity. They were used to show that the protein is synthesized as a higher molecular weight precursor (Mr = 44,000) (1) from *in vitro* translation of total yeast mRNAs and (2) *in vivo,* where it is rapidly processed (with a half-life of about 30 s) to the mature form (Mr = 40,000) (Camadro and Labbe, 1988). Interestingly, yeast ferrochelatase has been produced in insect cells infected with a baculovirus expression vector carrying the yeast ferrochelatase gene (Eldridge and Dailey, 1992). The protein was active, associated with the total cellular membrane fraction, and had a molecular weight of 44,000. This indicates that there was a defect of proteolytic processing, which could be due to differences in the mitochondrial processing systems of insect cells and yeast or to saturation of the import-processing machinery by the large quantities of overexpressed heterologous ferrochelatase.

C. Catalytic and Kinetic Properties

All the kinetic studies on purified yeast ferrochelatase were performed using a sensitive direct spectrofluorimetric assay. Most of the chemicals used appeared to be contaminated with metal ions, especially zinc at concentrations high enough to saturate the enzyme during zinc-chelatase measurements and to inhibit competitively iron incorporation into protoporphyrin. Accurate determinations of the kinetic constants of ferrochelatase for metals were possible only when the incubation media had been freed of endogenous metals by specific chelation and extraction (Camadro and Labbe, 1988).

The maximum velocities of purified yeast enzyme are 35,000 nmol heme/h/mg enzyme and 27,000 nmol zinc-protoporphyrin/h/mg enzyme at 30°C. The turnover numbers of the enzyme are 0.4 s^{-1} for zinc-protoporphyrin and 0.3 s^{-1} for heme synthesis. These are of the same order of magnitude as those for the rat, mouse, bovine, and chicken enzymes.

Yeast ferrochelatase activity is inhibited by thiol-reacting reagents. The hydrophobic inhibitor *p*-chloromercuribenzoate is 1000-fold more efficient than is the polar iodoacetamide, and inhibition is prevented by excess dithiothreitol. Ferrochelatases from various sources are also inhibited by sulfhydryl reagents. Modification of a single sulfhydryl group is sufficient to inactivate the bovine (Dailey, 1984) and *R. sphaeroides* (Dailey et al., 1986b) enzymes, and ferrous iron but not the porphyrin substrate protects the enzymes from inactivation. This led to the proposal that at least one, and perhaps two, cystein sulfhydryl groups are implicated in the binding of the metal ion at the active site.

The Lineweaver-Burk plots of the reciprocal of the initial velocities ($1/V$) versus the reciprocal of substrate concentrations ($1/S$) give families of converging linear plots indicating a sequential mechanism of ferrochelatase reaction, with binding of the substrates prior to product release. The convergence of the lines on the x-axis indicates that the Michaelis constants for the substrates (K_M) are identical to their dissociation constants (K_D). The Michaelis constant of the purified yeast enzyme for protoporphyrin is 0.09 μM with both iron and zinc. This value correlates well with the dissociation constant K_D = 0.15 μM determined by fluorescence anisotropy measurements for the mouse enzyme (Dailey et al., 1989). The affinity of purified yeast ferrochelatase for protoporphyrin is of the same order of magnitude as that of the membrane-bound enzyme with either endogenous or exogenous protoporphyrin (0.07–0.03 μM). This suggests that the microenvironment of the active site in the detergent micelles and the distribution of protoporphyrin within these micelles mimic those prevailing in the mitochondrial membranes. The K_M values for ferrous iron and zinc are 0.16 and 0.25 μM, respectively, very similar to the values measured for the membrane-bound enzyme.

The mechanism of ferrochelatase reaction, ordered versus random, is still controversial. Under our experimental conditions, both iron and zinc are simultaneously incorporated into protoporphyrin by purified yeast ferrochelatase yielding alternative substrate inhibition. The fact that the plots of $1/V$ against $1/S$ for protoporphyrin are linear suggests a random bi-bi reaction mechanism (see Fromm, 1975 for detailed discussion): the binding of either substrate is random, and the binding of the first substrate does not influence the binding of the second. But this criterion is not, in itself, sufficient to establish unambiguously the nature of the reaction mechanism. Kinetic studies of enzyme inhibition by the products of the reaction or by known inhibitors are critical to establish a reaction mechanism. On the basis of kinetic studies with competitive inhibitors, mainly manganese and *N*-methylprotoporphyrin, Dailey and Fleming (1983) proposed that the bovine enzyme catalyzes an ordered reaction, where the metal ion binds to the active site prior to protoporphyrin and protoheme is released prior to protons. This model implies that the binding of the first substrate, the metal ion, is necessary for the binding of the second substrate, protoporphyrin. However, Dailey et al. (1989) were able to measure the dissociation constants of various porphyrins for purified mouse ferrochelatase in the absence of metals. The binding of the second substrate in an ordered mechanism may occur if this substrate is able to form an abortive complex with the enzyme; it must then be an inhibitor of the reaction, with a dissociation constant equal to the inhibitory constant. No such inhibition of ferrochelatase by protoporphyrin has been reported, indicating that such an abortive complex is not formed.

Recent analysis of yeast mutant ferrochelatases revealed that single amino-acid substitutions can increase the K_M for both metal and porphyrin substrates (Abbas and Labbe-Bois, 1993). This suggests that the two binding sites are not independent of each other.

V. MOLECULAR GENETICS OF FERROCHELATASE

A. Amino-Acid Sequences of Yeast Ferrochelatase and Other Ferrochelatases

Today, thanks to molecular genetics, the amino-acid sequence of ferrochelatase, deduced from the nucleotide sequence, is known for seven different species. The ferrochelatase gene from *S. cerevisiae, HEM15,* was the first to be isolated by functional complementation of a *hem15* mutant (Labbe-Bois, 1990; Gokhman and Zamir, 1990). The cloned genomic DNA fragment contains a single large open reading frame of 1179 nucleotides. The closest "TATA"-like sequence is at position −242 relative to the initiating ATG (A: +1). The initiation of transcription was analyzed by mapping the 5′ termini of the *HEM15* mRNA by the primer extension method. A major start was found at position −67 and two minor ones in a 10-nucleotide region centered at −110 (Chelstowska and Labbe-Bois, unpublished). The region downstream of the two tandem stop codons contains a yeast consensus sequence involved in transcription termination and polyadenylation. The organization of the *HEM15* gene is indeed classical for a yeast gene, and it is present as a single copy in the yeast genome. It is not linked to any other known *HEM* gene. Preliminary mapping experiments show that a *HEM15* probe hybridizes at a position corresponding to chromosomes VII and XV, which were not separated under the conditions used for pulsed-field gel electrophoresis (Labbe-Bois, unpublished).

The long open reading frame encodes a protein of 393 amino acids with a calculated molecular weight of 44,545, consistent with the size of the precursor form of yeast ferrochelatase (Mr = 44,000). The deduced amino-acid sequence from Asn-32 to Lys-52 is identical with the 20-residue amino-terminal sequence determined for the purified mature protein (Camadro and Labbe, 1988) (Fig. 5). This indicates that the amino-acid sequence from Met-1 to Gln-31 represents the presequence of the precursor form of the enzyme. This 31-residue presequence has all features characteristic of proteins destined for the mitochondrial matrix or inner membrane. It also contains the three-amino-acid motif, Arg at −10, Phe at −8, and Thr at −5 with respect to the cleavage site. This motif is found in leader peptides that are processed in two proteolytic steps (references in Isaya et al., 1992). Hence, yeast ferrochelatase might undergo two cleavages. The first would be between Ser-23 and Phe-24, at a cleavage site XRX/XS typical of the mitochondrial processing peptidase, leading to an intermediate form with an octapeptide extension. The second cleavage by an intermediate peptidase would yield the mature protein. It is somewhat striking that the amino-acid sequence of the rat mitochondrial intermediate peptidase is similar to the amino-acid sequence of an open reading frame on yeast chromosome III (Isaya et al., 1992). It would be interesting to know if the yeast protein is necessary for the correct processing of ferrochelatase.

Yeast mature ferrochelatase is a 362 amino-acid protein with a calculated molecular weight of 40,900, similar to the experimental value of 40,000. It contains 25.4% charged (acidic + basic) residues, 32.8% polar residues, and 44.7% nonpolar hydrophobic residues. The hydropathy profile shows that the protein is rather hydrophilic, with only two stretches of significant hydrophobicity centered near positions 235 and 305. There appears to be no membrane-spanning segment, in agreement with the peripheral character of ferrochelatase.

As is often the case, the amino-acid sequence of yeast ferrochelatase itself is not very talkative. However, comparing it with homologous sequences may give more information. This is now possible, as ferrochelatase genes or cDNAs have been isolated and sequenced from two mammals, the

```
    1
E
B                                                                                M
Y                                                          MLSRTIRTQ GSFLRRSQLT
H                          MRSLGANMA AALRAAGVLL RDPLASSSWR VCQPWRWKSG AAAAAVTTET
M                          MLSASANMA AALRAAGALL REPLVHGSSR ACQPWRCQSG AAVAA-TTEK
A   MQATALSSGF NPLTKRKDHR FPRSCSQRNS LSLIQCDIKE RSFGESMTIT NRGLSFKTNV FEQARSVTGD

    71
E                   MRQTKT GILLANLGTP DAPTPEAVKR YLKQFLSDRR VVDTS--RLL WWPLLRGVIF
B   STAAPNETTQ PTVRSGQKRV GVLLVNLGTP DTADAPGVRV YLKEFLSDAR VIEDQ--GLV WKVVLNGIIL
S                        M GLLVMAYGTP YK--EEDIER YYTHIRRGRK PEPEM----- LQDL------
Y   ITRSFSVTFN MQNAQKRSPT GIVLMNMGGP SK--VEETYD FLYQLFADND LIPIS---AK YQTTIAKYIA
H   AQHAQGAKPQ VQPQKRKPKT GILMLNMGGP ET--LGDVHD FLLRLFLDQD LMTLP----- IQNKLAPFIA
M   VHHAKTTKPQ AQPERRKPKT GILMLNMGGP ET--LGEVQD FLQRLFLDRD LMTLP----- IQNKLAPFIA
A   CSYDETSAKA RSHVVAEDKI GVLLLNLGGP ET--LNDVQP FLYNLFADPD IIRLPRPFQF LQGTIAKFIS
                          *....+ *.*            .+    +                    .  +
                               I

    141
E   PLRSPRVAKL YASVWME--G GSPLMVYSRQ QQQALAQRLP EMP-VAL--- ---GMSYGSP SLESAVDELL
B   RQRPRSKALD YQKIWNNEKN ESPLKTITRS QSAKLAAALS DRDHVVVDW- ---AMRYGNP SIKSGIDALI
S   KDR------- YEAI----GG ISPLAQITEQ QAHNLEQHLN EIQDEITF-K AYIGLKHIEP FIEDAVAEMH
Y   KFRTPKIEKQ YREI----GG GSPIRKWSEY QATEVCKILD KTCPETAPHK PYVAFRYAKP LTAETYKQML
H   KRRTPKIQEQ YRRI----GG GSPIKIWTSK QGEGMVKLLD ELSPNTAPHK YYIGFRYVHP LTEEAIEEME
M   KRRTPKIQE- -RRI----GG GSPIKMWTSK QGEGMVKLLD ELSPATAPHK YYIGFRYVHP LTEEAIEEME
A   VVRAPKSKEG YAAI----GG GSPLRKITDE QADAIKMSLQ AKN-IAAN-- VYVGMRYWYP FTEEAVQQIK
      *         +  +     + **.   .   *   .   *                .  +  *     .   .
                                  II

    211
E   AEHVDHIVVL PLYPQFSCST VGAVWDELAR ILARKRSIPG ISF--IRDYA DNHDYINALA NSV---RASF
B   -GGMRPHLAV PLYPQYSAST SATVCDEVFR VLARLRAQPT LRV--TPPYY EDEAYIEALA VSI---ETHL
S   KDGITEAVSI VLAPHFSTFS VQSYNKRAKE EAEKLGGL-T ITS--VESWY DEPKFVTYWV DRVKETYASM
Y   KDGVKKAVAF SQYPHFSYST TGSSINELWR QIKALDSERS ISWSVIDRWP TNEGLIKAFS ENITKKLQEF
H   RDGLERAIAF TQYPQYSCST TGSSLNAIYR YYNQVGRKPT MKWSTIDRWP THHLLIQCFA DHILKELDHF
M   RDGLERAIAF TQYPQYSCST TGSSLNAIYR YYNEVGQKPT MKWSTIDRWP THPLLIQCFA DHILKELNHF
A   KDKITRLVVL PLYPQYSIST TGSSIRVLQD LFRKDPYLAG VPVAIIKSWY QRRGYVNSMA DLIEKELQTF
       .   .     +*  .*  ++   .    .            . .       .       .       .
                  III

    281
E   AKHG-EPDLL LLSYHGIPQR YA-DEGDDYP QRCRTTTREL ASALGM--AP EKVMMTFQSR FG-REFWIMP
B   ATLPFKPELI VASFHGMPKS YV-DKGDPYQ EHCIATTEAL RAARRL--DA SKLLLTFQSR FG-NDEWLQP
S   PEDERENAML IVSAHSLPEK IK-EFGDPYP DQLHESAKLI AEGAGV---- SEYAVGWQSE GNTPDPWLGP
Y   PQPVRDKVVL LFSAHSLPMD VV-NTGDAYP AEVAATVYNI MQKLKF---K NPYRLVWQSQ VG-PKPWLGA
H   PLEKRSEVVI LFSAHSLPMS VV-NRGDPYP QEVSATVQKV MERLEY---C NPYRLVWQSK VG-PMPWLGP
M   PEEKRSEVVI LFSAHSLPMS VV-NRGDPYP QEVGATVHKV MEKLGY---P NPYRLVWQSK VG-PVPWLGP
A   SDPK--EVMI FFSAHGVPVS YVENAGDPYQ KQMEECIDLI MEELKARGVL NDHKLAYQSR VG-PVQWLKP
    .       ..   * *..*      . ** *            .                . .**   +    ** +
                    IV                                                    V

    351
E   YTDETLKMLG EKGVGHIQVM CP-GFAADCL ETLEEIAEQN RE-VFLGAGG KKYEYIPALN ATPEHIEMMA
B   YTDKTMERLA KEGVRRIAVV TP-GFAADCL ETLEEIAQEN AE-IFKHNGG ETFSAIPCLN DSEPGMDVIR
S   DVQDLTRDLF EQKGYQAFVY VPVGFVADHL EVLYDNDYEC KV-VTDDIG- ASYYRPEMPN AKPEFIDALA
Y   QTAEIAEFLG PKVDG-LMFI -PIAFTSDHI ETLHEIDLGV IG---ESEYK DKFKRCESLN GNQTFIEGMA
H   QTDESIKGLC ERGRKNILLV -PIAFTSDHI ETLYELDIEY SQVLAKECGV ENIRRAESLN GNPLFSKALA
M   QTDEAIKGLC ERGRKNILLV -PIAFTSDHI ETLYELDIEY SQVLAQKCGA ENIRRAESLN GNPLFSKALA
A   YTDEVLVDLG KSGVKSLLAV -PVSFVSEHI ETLEEIDMEY RE-LALESGV ENWGRVPALG LTPSFITDLA
     +      *        .     *..* .+ . *+* +           .    +          ++         .+
                                   VI

    421
E   NLVAAYR
B   TLVLRELQGW I
S   TVVLKKLGR
Y   DLVKSHLQSN QLYSNQLPLD FALGKSNDPV KD-LSLVFGN HEST
H   DLVHSHIQSN ELCSKQLTLS CPLCVNPVCR ET---KSFFT SQQL
M   DLVHSHIQSN KLCSTQLSLN CPLCVNPVCR KT---KSFFT SQQL
A   DAVIESLPSA EAMSNPNAVV DSEDSESSDA FSYIVKMFFG SILAFVLLLS PKMFHAFRNL
     .*         + * ++.+   .+             +  *+  +
```

mouse (Taketani et al., 1990; Brenner and Frasier, 1991) and the human (Nakahashi et al., 1990) from the bacteria *Escherichia coli* (Miyamoto et al., 1991; Nakahigashi et al., 1991), *Bradyrhizobium japonicum* (Frustaci and O'Brian, 1992), and *Bacillus subtilis* (Hansson and Hederstedt, 1992), and from the plant *Arabidopsis* (Smith and Labbe-Bois, unpublished). Figure 5 shows the alignment of these seven ferrochelatase sequences. The eukaryotic enzymes are very similar (46% identity between yeast and human), as are the gram-negative bacterial enzymes (43% identity between *E. coli* and *B. japonicum*). But the yeast protein has only 25% identity with the *E. coli* one. The lower similarity was found for the *B. subtilis* protein, especially at its *N*-terminus, which aligns poorly. All mature ferrochelatases are about the same size, except that the eukaryotic ones have a 30- to 50-amino-acid extension at the C-terminus. All are apparently devoid of transmembrane segments, which raises the question of the nature of their strong interactions with the membranes: ionic and/or penetration into the hydrophobic core of the phospholipid bilayer?

The alignment of the seven ferrochelatase sequences reveals an overall 8% identity and 24% similarity, taking into account the conservative replacements (with respect to the properties of the side chains) extending over the entire length of the proteins. No significant similarities with other proteins in data banks have been reported. The 27 amino acids that are common to all sequences probably play essential roles in the structure or function of ferrochelatase, although evolutionary invariance does not necessarily imply functional invariance, as demonstrated for cytochrome c (Hampsey et al., 1988). The majority of the invariant positions are clus-

---

Figure 5 Alignment of ferrochelatase amino-acid sequences. Ferrochelatases are from (E) *Escherichia coli* (Miyamoto et al., 1991), (B) *Bradyrhizobium japonicum* (Frustacci and O'Brian, 1992), (S) *Bacillus subtilis* (Hansson and Hederstedt, 1992), (Y) *Saccharomyces cerevisiae* (Labbe-Bois, 1990), (H) human (Nakahashi et al., 1990), (M) mouse (Taketani et al., 1990; Brenner and Frasier, 1991), and (A) *Arabidopsis* (Smith and Labbe-Bois, unpublished work). Asterisks (*) mark identical residues in all proteins. Crosses (+) mark residues identical in all but one protein. Points (.) indicate conservative substitutions. Underlined amino acids mark positions where substitutions have been identified in mutant ferrochelatases from *E. coli* (Miyamoto et al., 1991), humans (Lamoril et al., 1991; Brenner et al., 1992), and *S. cerevisiae* (Abbas and Labbe-Bois, 1993, and unpublished results). Vertical arrows indicate the cleavage sites of the precursor forms of yeast, human, and mouse enzymes. Vertical lines along the human sequence mark the exon boundaries (Taketani et al., 1992). The six blocks I to VI, which contain almost all invariant residues, are underlined. Numbering starts with the first methionine of *Arabidopsis* ferrochelatase.

tered within six blocks (I to VI in Fig. 5), suggesting that some of these blocks might form part of the active site in the interior of the protein, while the regions connecting them, in which most of the insertions/deletions occur, might form the surface of the protein. The exon/intron organization of the human ferrochelatase gene is known (Taketani et al., 1992): there is no correspondence between exons and these blocks (Fig. 5). But we do not know whether these blocks correspond to units of protein structure-function; and even if this were the case, exons do not always correlate with such units (Traut, 1988). The sequences were also analyzed for secondary structure predictions. These predictions are rarely more than 60% correct in predicting the conformation at a given position for a given protein; but if the predicted secondary structures are conserved in a set of aligned homologous sequences, then the accuracy of the consensus conformation is greatly improved (Niermann and Kirschner, 1991). There are two predicted alpha-helices common to all proteins, one before block IV and another between blocks IV and V. They might act as core structural elements. The six blocks appear to have consensus conformations. Block I is mainly a highly hydrophobic beta-strand, for which it is tempting to speculate a role in the overall structure or perhaps in the protein's association with the membrane. Block II begins as a random coil and finishes as a 10-residue alpha-helix. The second half of block III is random + turn(s). Block IV has the consensus conformation helix-strand-turn, while block V is almost entirely random + turns. Block VI begins in a 6–8-residue hydrophobic beta-strand, then turns to finish with a 9–10-residue amphiphilic alpha-helix with one side highly negatively charged. This helix might play a role in the electrostatic steering of the charged ferrous iron to the active site.

The 27 invariant amino acids include 3 Gly and 6 Pro, which most probably are crucial for the proper folding of the backbone structure. Pro is often found at bends between ordered structures, and Gly is often found within densely packed regions at sites that cannot accommodate side chains. Aliphatic (4 Leu, 1 Val) and aromatic (1 Trp, 1 Phe) amino-acid side chains are probably required for the hydrophobic lining of the active site necessary for the proper alignment of the porphyrin. This point will be discussed further in the section on the active site of the enzyme, together with the nature of the residues that might be implicated in the binding of metal (Sec. C).

B. Analysis of Mutant Ferrochelatases

The availability of the ferrochelatase nucleotide sequences made it possible to study the molecular basis of the ferrochelatase defects described in mammals and microorganisms, thus helping investigate the structural and functional requirements of the enzyme. The mutations were identified by sequencing the mutant ferrochelatase alleles amplified *in vitro* from cDNA or genomic DNA. It did not come as a great surprise that most of the mutations caused amino-acid changes at invariant or highly conserved residues, or in evolutionarily conserved regions.

Ferrochelatase activity is decreased in the human and bovine hereditary disease (dominant or recessive, with variable penetrance) erythropoietic protoporphyria, in which excessive protoporphyrin accumulates in various tissues, causing cutaneous photosensitivity and, more rarely, hepatic complications. Two different mutations, one in each allele, were identified in a patient who had lost 94% of ferrochelatase activity: they caused Gly-55 to Cys and Met-267 to Ile substitutions (positions 76 and 299 in the numeration of Fig. 5) (Lamoril et al., 1991). The cDNA from each of the patient's parents carried one of the two mutations, leading to a 50% loss of ferrochelatase activity. The apparent K_M values for both substrates of the residual ferrochelatase were unchanged, but the activity was more thermosensitive, usually an indication of an unstable protein (Deybach et al., 1986). This is too little information to allow speculation on the structural or functional significance of these mutations. The same is true for a mutation leading to the replacement of Phe-417 with Ser (position 458 in Fig. 5) identified in a cell line from a protoporphyric patient (Brenner et al., 1992). When this mutant cDNA is expressed in *E. coli,* the ferrochelatase activity is only 4–5% of that measured with the wild-type cDNA, although the amino-acid change occurs in the C-terminus extension that is absent in the *E. coli* enzyme. Other defective ferrochelatases, made in normal amounts, have been described in human and bovine protoporphyria, but the nature of the mutations is not yet documented (Bloomer et al., 1987; Blom et al., 1990; Straka et al., 1991b). On the other hand, a number of mutations causing amino-acid substitutions (underlined in Fig. 5) have been reported for *E. coli* ferrochelatase; they apparently lead to a partial loss of activity, but the nature of the enzyme dysfunction has not been analyzed (Miyamoto et al., 1991; Nakahigashi et al., 1991).

Two totally heme-deficient mutants of *S. cerevisiae* make neither ferrochelatase mRNA nor immunodetectable protein, probably owing to a nonsense mutation and a small deletion (Camadro and Labbe, 1988; Labbe-Bois, 1990, unpublished data). But four mutants were isolated in a search for mutants partially defective in ferrochelatase. These accumulate protoporphyrin, are photosensitive, fluoresce under UV light, but maintain sufficient heme synthesis to grow on nonfermentable carbon sources such as ethanol or glycerol (Kurlandzka and Rytka, 1985). A single nucleotide change, causing an amino acid substitution, was found in each mutant: Gly-47 to Ser, Ser-102 to Phe, Ser-169 to Phe, and Ser-174 to Pro (positions 98, 162, 229, and 234 in

Fig. 5). The effects of the two latter mutations, located near each other, on ferrochelatase function were investigated in some detail, both *in vivo* and *in vitro* (Abbas and Labbe-Bois, 1993). Normal amounts of normal-sized ferrochelatase are made by the two mutant cells. Replacement of Ser-169 with Phe causes a 10-fold increase in V_{max} and a 45- and 35-fold increase in the K_M for protoporphyrin and metal, respectively. Replacement of Ser-174 by Pro produces the same effects, but to a lesser degree. We estimated that the mutant enzymes functioned *in vivo* 30 and 4.5 times less efficiently than the wild-type enzyme, which correlated remarkably well with the relative increase in their K_M for metal, of 35 and 4. This suggests that iron is limiting for heme production in the mutant cells and that no compensatory mechanism increases the concentration of iron in the mitochondrial membrane or matrix.

The inverse relationship between the affinity of these yeast mutant enzymes for protoporphyrin and the rate of heme formation is similar to the situation described in the analysis of porphyrin specificity of ferrochelatase. Hydrophobic substituents on pyrroles A and B lower both the V_{max} and the K_M for porphyrin of the mammalian enzyme (Honeybourne et al., 1979; Dailey and Fleming, 1983). This suggests that the release of heme is the rate-limiting step of the overall reaction. It also indicates that the region of the enzyme affected by the two mutations, which is rich in conservative aliphatic and aromatic residues, might contribute to the hydrophobic interaction(s) and van der Waals contact(s) of the active site with the porphyrin vinyl group(s). This region (block III) is predicted to be a random coil; it is also rich in hydroxylated residues that might be part of the metal-binding domain. The mutations have probably affected the structure or the proper geometry of the binding sites of both substrates.

C. Active Site and Mechanism of Action

Models have been proposed for the active site and the mechanism of action of ferrochelatase, on the basis of spectroscopic and kinetic studies, chemical modifications of specific amino-acid residues, from the analysis of the strong inhibition by *N*-alkylporphyrins, and taking into consideration the general features of the nonenzymic metal ion incorporation into porphyrins (Lavallee, 1988; Dailey et al., 1989; Dailey, 1990). Since all eukaryotic and bacterial ferrochelatases studied so far present very similar catalytic properties, it seems reasonable to assume that the major aspects of their reaction mechanisms are identical.

The information concerning the structure of the active site of ferrochelatase came from spectroscopic and inhibition studies. By examining the fluorescence properties of the protein tryptophan residue(s) and of a fluorescent probe bound to cysteine residue(s), Dailey (1985) proposed that the active site is a hydrophobic pocket relatively poorly accessible to external medium. We have proposed that it is, at least partly, in contact with the membrane, thus facilitating the direct transfer of protoporphyrin from the membrane, where it is synthesized, into the active site of the enzyme (Sec. III.D). In fact, the active site probably contains two domains, one hydrophobic to accommodate the nonpolar porphyrin, and another hydrophilic for the binding of the polar charged metal.

The porphyrin binding site appears to be highly selective for the position and the nature of the side-chain substituents on the porphyrin macrocycle (see Fig. 1). The position of the propionate side chains at positions 6 and 7 is critical for the porphyrin to serve as a substrate. Displacing the 6-propionate to position 5 lowers both the V_{max} and the K_M, while its absence abolishes the activity; moving propionate from position 7 to position 8 completely abrogates activity (Honeybourne et al., 1979). Esterification of the two propionates also leads to a loss of activity (Porra and Jones, 1963b). The function of these propionyl carboxylic groups is not clear. They might assist in the proper orientation of porphyrin at the active site, by charge-pair interactions and/or hydrogen bonding with protein residues. Arginine residues might be implicated in these interactions, because (1) their chemical modification causes the rapid loss of ferrochelatase activity, and (2) the enzyme is protected from inactivation when porphyrin is present during the reaction with arginyl reagents (Dailey and Fleming, 1986; Dailey et al., 1986b). One Arg appears to be conserved in the alignment of Fig. 5 (position 143). However, this might be just a matter of chance, if one considers the poor alignment and the many insertions/deletions present in that region of the protein. An alternate role of the propionyl carboxylic groups might be to assist metal binding. The rate of the nonenzymic porphyrin metallation increases with the number of pendant carboxylate groups, "perhaps because these groups could themselves coordinate iron and create a greater local concentration of iron" (Lavallee, 1988).

The nature of the porphyrin substituents at positions 2 and 4 on pyrroles A and B is important for producing a porphyrin either substrate or competitive inhibitor of ferrochelatase, regardless of its binding ability (Porra and Jones, 1963b; Honeybourne et al., 1979; Taketani and Tokunaga, 1982; Dailey and Fleming, 1983; Dailey et al., 1989). Porphyrins with bulky or charged substituents at positions 2 and 4 are very poor substrates and competitive inhibitors. The 2- and 4-substituents also affect the inhibitory activity of the *N*-alkylporphyrins (De Matteis et al., 1985; Mccluskey et al., 1989). These porphyrins have an alkyl substituent to one of the pyrrolenic nitrogen atoms. They are produced during the oxidative metabolism of some drugs by cytochrome P450 and have been shown to be potent competitive inhibitors (K_I = 7–10 nM) (with respect to porphyrin) of ferrochelatase *in vivo* and *in vitro*. The properties of these compounds and their inhibitory activities have been examined in great detail. One of their most interesting features is the large distortion of the porphyrin macrocycle, with the *N*-alkylated pyrrole ring tilted out of the porphyrin plane, "that exposes nitrogen lone pair electrons to an incoming metal ion" (Lavallee, 1988). They are thought to mimic the transition-state of the enzymic reaction. This hypothesis has recently received strong support with the finding that an antibody elicited to the distorted *N*-methylmesoporphyrin (transition-state analog) catalyzed efficient metal ion chelation by the planar mesoporphyrin (Cochran and Schultz, 1990). Interestingly, the antibody did not catalyze the metallation of protoporphyrin or deuteroporphyrin, thus emphasizing a role for the 2- and 4-substituents in the optimal binding of the porphyrin into the ferrochelatase active site. In their models, Lavallee (1988) and Dailey (1990) proposed that metallation is mediated via distortion of the A or B pyrrole ring, and that aromatic residues present in the porphyrin binding site, on the side of the pocket opposite to the iron atom, might "aid in ring bending and/or stabilization of the transition-state puckered porphyrin." Two hydrophobic aromatic residues are invariant in the ferrochelatase sequences: Trp-347 and Phe-375 (Fig. 5). Trp-347, adjacent to an invariant Leu, is particularly attractive, because it is located in a predicted coil region (block V) which might be flexible enough to promote or stabilize pyrrole ring tilting after porphyrin binding.

Much less is known concerning the metal binding site. Inhibition studies by sulfhydryl reagents implicate cysteine residue(s) in the binding of metal. But chemical modifications of amino-acid side chains can cause steric hindrance and/or unexpected changes in protein structure, whose effects are difficult to distinguish from those actually due to modification of a ligand residue. The fact that no cysteine is conserved in the seven known ferrochelatase sequences (Fig. 5) makes it very unlikely that cysteine(s) plays a role in metal binding. Unfortunately, there are no spectroscopic data to help in defining the coordination geometry of ferrous iron in ferrochelatase. Lower oxidation states, high spin states (for iron), and tetrahedral rather than octahedral geometry favor rapid ligand dissociation, then rapid metalloporphyrin formation (Lavallee, 1988). Therefore the enzyme probably binds the metal with residues that provide a four-coordinate geometry and that can dissociate readily. A number of such residues exist among the invariant amino acids of ferrochelatase. They are His-295, Asp-307, Tyr-309, and Glu-381 (in the numeration of Fig. 5). The first three are clustered in block IV, His at a predicted helix-terminus and Asp and Tyr on the following coil. It is worth remembering here that amino acids important for the activity are often found in loops connecting ordered structures.

Alternatively, the participation of one (or more) of the conserved Ser cannot be ruled out if this Ser is in an environment stabilizing its deprotonated form. These potential metal ligands would constitute an anionic domain well suited for binding the metal cation.

Detailed analysis of the activity of mutant enzymes, in which the invariant amino acids discussed above have been replaced by site-directed mutagenesis, should help clarify the actual role played by these residues. A more refined model could then be constructed for the structure of the active site of ferrochelatase, until it can be confronted with the crystal structure of the enzyme. Only then will we understand the way in which ferrochelatase binds its substrates, distorts the porphyrin ring, promotes ligand exchange and iron insertion into porphyrin, and releases heme.

NOMENCLATURE

Protoheme, heme, heme *b:* iron-protoporphyrin IX chelate. In principle, heme or protoheme is used when iron is in the reduced state and hemin is used when iron is in the oxidized state. However, if not specified in the text (ferri-, ferroheme), we have used the term heme or protoheme loosely whatever the oxidation state of iron. Total heme refers to heme *a* + heme *b* + heme *c*.

ACKNOWLEDGMENTS

The work from this laboratory was funded by the Centre National de la Recherche Scientifique and the Université Paris 7. We thank O. Parkes for his help in the preparation of the manuscript.

REFERENCES

Abbas, A., and Labbe-Bois, R. (1993). Structure-function studies of yeast ferrochelatase. Identification and functional analysis of amino acid substitutions that increase V_{max} and K_M for both substrates, *J. Biol. Chem., 268:* 8541–8546.

Arrese, M., Carvajal, E., Robison, S., Sambunaris, A., Panek, A., and Mattoon, J. (1983). Cloning of the 5-aminolevulinic acid synthase structural gene in yeast, *Curr. Genet., 7:* 175–183.

Barnes, R., Connelly, J. L., and Jones, O. T. G. (1972). The utilization of iron and its complexes by mammalian mitochondria, *Biochem. J., 128:* 1043–1055.

Beckman, D. L., Trawick, D. R., and Kranz, R. G. (1992). Bacterial cytochromes c biogenesis, *Genes Develop., 6:* 268–283.

Bilinski, T., Litwinska, J., Lukaskiewicz, J., Rytka, J., Simon, M., and Labbe-Bois, R. (1981). Characterization of two mutant strains of *Saccharomyces cerevisiae* deficient in coproporphyrinogen oxidase activity, *J. Gen. Microbiol., 122:* 79–87.

Blom, C., Klasen, E. C., and Van Steveninck, J. (1990). Different characteristics of ferrochelatase in cultured fibroblasts of erythropoietic protoporphyria patients and normal controls, *Biochim. Biophys. Acta, 1039:* 339–342.

Bloomer, J. R., Hill, H. D., Morton, K. O., Anderson-Burnham, L. A., and Straka, J. G. (1987). The enzyme defect in bovine protoporphyria. Studies with purified ferrochelatase, *J. Biol. Chem., 262:* 667–671.

Bottomley, S. S. (1968). Characterization and measurement of heme synthetase in normal human bone marrow, *Blood, 31:* 314–322.

Brault, D., Vever-Bizet, C., and Le Doan, T. (1986). Spectrofluorimetric study of porphyrin incorporation into membrane models, evidence for pH effects, *Biochim. Biophys. Acta, 857:* 238–250.

Brenner, D. A., and Frasier, F. (1991). Cloning of murine ferrochelatase, *Proc. Natl. Acad. Sci. USA, 88:* 849–853.

Brenner, D. A., Didier, J. M., Frasier, F., Christensen, S. R., Evans, G. A., and Dailey, H. A. (1992). A molecular defect in human protoporphyria, *Am. J. Hum. Genet., 50:* 1203–1210.

Camadro, J. M., and Labbe, P., (1981). A simple ferrochelatase assay, *Biochimie, 63:* 463–465.

Camadro, J. M., and Labbe, P. (1982). Kinetic studies of ferrochelatase in yeast. Zinc or iron as competing substrates, *Biochim. Biophys. Acta, 707:* 280–288.

Camadro, J. M., and Labbe, P. (1988). Purification and properties of ferrochelatase from the yeast *Saccharomyces cerevisiae.* Evidence for a precursor form of the protein, *J. Biol. Chem., 263:* 11675–11682.

Camadro, J. M., Abraham, N. G., and Levere, R. D. (1984). Kinetic studies on human liver ferrochelatase. Role of endogenous metals, *J. Biol. Chem., 259:* 5678–5682.

Camadro, J. M., Chambon, H., Jolles, J., and Labbe, P. (1986). Purification and properties of coproporphyrinogen oxidase from the yeast *Saccharomyces cerevisiae, Eur. J. Biochem., 156:* 579–587.

Chaix, P., and Labbe, P. (1965). A propos de l'interprétation du spectre d'absorption de cellules de levures récoltées à la fin ou après la phase exponentielle de leur croissance anaérobie, *Colloque International sur les Mécanismes de Régulation des Activités Cellulaires chez les Microorganismes,* CNRS, Paris, pp. 481–489.

Chelstowska, A., Zoladek, T., Garey, J., Kushner, J., Rytka, J., and Labbe-Bois, R. (1992). Identification of amino acid changes affecting yeast uroporphyrinogen decarboxylase activity by sequence analysis of *hem12* mutant alleles, *Biochem. J., 288:* 753–757.

Cochran, A. G., and Shultz, P. G. (1990). Antibody-catalyzed porphyrin metallation, *Science, 249:* 781–783.

Coghlan, D., Jones, G., Denton, K. A., Wilson, M. T., Chan, B., Harris, R., Woodrow, J. R., and Ogden, J. E. (1992). Structural and functional characterisation of recombinant human haemoglobin A expressed in *Saccharomyces cerevisiae, Eur. J. Biochem., 207:* 931–936.

Conklin, D. S., McMaster, J. A., Culbertson, M. R., and Kung, C. (1992). *COT1,* a gene involved in cobalt accumulation in *Saccharomyces cerevisiae, Mol. Cell. Biol., 12:* 3678–3688.

Cox, T. C., Bawden, M. J., Martin, A., and May, B. K. (1991). Human erythroid 5-aminolevulinate synthase: Promoter analysis and identification of an iron-responsive element in the mRNA, *EMBO J., 10:* 1891–1902.

Dailey, H. A. (1984). Effect of sulfhydryl group modification on the activity of bovine ferrochelatase, *J. Biol. Chem., 259:* 2711–2715.

Dailey, H. A. (1985). Spectroscopic examination of the active site of bovine ferrochelatase, *Biochemistry, 24:* 1287–1291.

Dailey, H. A. (1986). Purification and characterization of bacterial ferrochelatase, *Methods Enzymol., 123:* 408–415.

Dailey, H. A., ed. (1990). *Biosynthesis of Heme and Chlorophylls,* McGraw-Hill, New York.

Dailey, H. A., and Fleming, J. E. (1983). Bovine ferrochelatase. Kinetic analysis of inhibition by N-methylprotoporphyrin, manganese, and heme, *J. Biol. Chem., 258:* 11453–11459.

Dailey, H. A., and Fleming, J. E. (1986). The role of arginyl residues in porphyrin binding to ferrochelatase, *J. Biol. Chem., 261:* 7902–7905.

Dailey, H. A., Fleming, J. E., and Harbin, B. M. (1986a). Purification and characterization of mammalian and chicken ferrochelatase, *Methods Enzymol., 123:* 401–408.

Dailey, H. A., Fleming, J. E., and Harbin, B. M. (1986b). Ferrochelatase from *Rhodopseudomonas sphaeroides:* Substrate specificity and role of sulfhydryl and arginyl residues, *J. Bacteriol., 165:* 1–5.

Dailey, H. A., Jones, C. S., and Karr, S. W. (1989). Interaction of free porphyrins and metalloporphyrins with mouse ferrochelatase. A model for the active site of ferrochelatase, *Biochim. Biophys. Acta, 999:* 7–11.

Dandekar, T., Stripecke, R., Gray, N. K., Goossen, B., Constable, A., Johansson, H. E., and Hentze, M. W. (1991). Identification of a novel iron-responsive element in murine and human erythroid 5-aminolevulinic acid synthase mRNA, *EMBO J., 10:* 1903–1909.

De Matteis, F., Gibbs, A. H., and Harvey, C. (1985). Studies on the inhibition of ferrochelatase by *N*-alkylated dicarboxylic porphyrins. Steric factors involved and evidence that the inhibition is reversible, *Biochem. J., 226:* 537–544.

Deybach, J. C., Da Silva, V., Grandchamp, B., and Nordmann, Y. (1985). The mitochondrial location of protoporphyrinogen oxidase, *Eur. J. Biochem., 149:* 431–435.

Deybach, J. C., Da Silva, V., Pasquier, Y., and Nordmann, Y. (1986). Ferrochelatase in human erythropoietic protoporphyria: The first case of a homozygous form of the enzyme deficiency, *Porphyrins and Porphyria* (Y. Nordmann, ed.), Colloque INSERM, John Libbey Eurotext, *134:* 163–173.

Djavadi-Ohaniance, L., Rudin, Y., and Schatz, G. (1978). Identification of enzymically inactive apocytochrome c peroxidase in anaerobically grown *Saccharomyces cerevisiae, J. Biol. Chem., 253:* 4402–4407.

Dumont, M. E., Ernst, J. F., Hempsey, D. M., and Sherman, F. (1987). Identification and sequence of the gene encoding cytochrome c heme lyase in the yeast *Saccharomyces cerevisiae, EMBO J., 6:* 235–241.

Eldridge, M. G., and Dailey, H. A. (1992). Yeast ferrochelatase: Expression in a baculovirus system and purification of the expression protein, *Prot. Sci., 1:* 271–277.

Ferreira, G. C., Andrew, T. L., Karr, S. W., and Dailey, H. A. (1988). Organization of the terminal two enzymes of the heme biosynthetic pathway. Orientation of protoporphyrinogen oxidase and evidence for a membrane complex, *J. Biol. Chem., 263:* 3835–3839.

Flatmark, T., and Romslo, I. (1975). Energy-dependent accumulation of iron by isolated rat liver mitochondria. Requirement of reducing equivalents and evidence for a unidirectional flux of Fe (II) across the inner membrane, *J. Biol. Chem., 250:* 6433–6438.

Fromm, H. (1975). Initial rate enzyme kinetics, *Molecular Biology, Biochemistry and Biophysics, 22,* Springer-Verlag, Berlin.

Frustaci, J. M., and O'Brian, M. R. (1992). Characterization of a *Bradyrhizobium japonicum* ferrochelatase mutant and isolation of the *hemH* gene, *J. Bacteriol., 174:* 4223–4229.

Funk, F., Lecrenier, C., Lesuisse, E., Crichton, R. R., and Schneider, W. (1986). A comparative study on iron sources for mitochondrial haem synthesis including ferritin and models of transit pool species, *Eur. J. Biochem., 157:* 303–309.

Gilardi, A., Djavani-Ohaniance, L., Labbe, P., and Chaix, P. (1971). Effet de l'accumulation de Zn-protoporphyrine par la cellule de levure sur la synthèse et le fonctionnement de son système respiratoire, *Biochim. Biophys. Acta, 234:* 446–457.

Gokhman, I., and Zamir, A. (1990). The nucleotide sequence of the ferrochelatase and tRNAval gene region from *Saccharomyces cerevisiae, Nucleic Acids Res., 18:* 6130.

Hampsey, D. M., Das, G., and Sherman, F. (1988). Yeast iso-1-cytochrome *c:* Genetic analysis of structural requirements, *FEBS Lett., 231:* 275–283.

Hansson, M., and Hederstedt, L. (1992). Cloning and characterization of the *Bacillus subtilis hemEHY* gene cluster, which encodes protoheme IX biosynthetic enzymes, *J. Bacteriol., 174:* 8081–8093.

Harbin, B. M., and Dailey, H. A. (1985). Orientation of ferrochelatase in bovine liver mitochondria, *Biochemistry, 24:* 366–370.

Honeybourne, C. L., Jackson, J. T., and Jones, O. T. G. (1979). The interaction of mitochondrial ferrochelatase with a range of porphyrin substrates, *FEBS Lett., 98:* 207–210.

Isaya, G., Kalousek, F., and Rosenberg, L. E. (1992). Sequence analysis of rat mitochondrial intermediate peptidase: Similarity to zinc metallopeptidases and to a putative yeast homologue, *Proc. Natl. Acad. Sci. USA, 89:* 8317–8321.

Jones, M. S., and Jones, O. T. G. (1969). The structural organization of haem synthesis in rat liver mitochondria, *Biochem. J., 113:* 507–514.

Jordan, P. M., ed. (1991). *Biosynthesis of Tetrapyrroles.* New Comprehensive Biochemistry, vol. 19 (A. Neuberger and L. L. M. van Deenen, general eds.), Elsevier.

Keng, T. (1992). *HAP1* and *ROX1* form a regulatory pathway in the repression of *HEM13* transcription in *Saccharomyces cerevisiae, Mol. Cell. Biol., 12:* 2616–2623.

Keng, T., and Guarente, L. (1987). Constitutive expression of the yeast *HEM1* gene is actually a composite of activation and repression, *Proc. Natl. Acad. Sci. USA, 84:* 9113–9117.

Keng, T., Richard, C., and Larocque, R. (1992). Structure and regulation of yeast *HEM3,* the gene for porphobilinogen deaminase, *Mol. Gen. Genet., 234:* 233–243.

Koller, M. E., and Romslo, I. (1977). Studies on the ferrochelatase activity of mitochondria and submitochondrial particles with special reference to the regulatory function of the mitochondrial inner membrane, *Biochim. Biophys. Acta, 461:* 283–296.

Kurlandzka, A., and Rytka, J. (1985). Mutants of *Saccharomyces cerevisiae* partially defective in the last steps of the haem biosynthetic pathway: Isolation and genetical characterization, *J. Gen. Microbiol., 131:* 2909–2918.

Labbe, P., Volland, C., and Chaix, P. (1968). Etude de l'activité ferrochélatase des mitochondries de levure, *Biochim. Biophys. Acta, 159:* 527–539.

Labbe-Bois, R. (1990). The ferrochelatase from *Saccharomyces cerevisiae.* Sequence, disruption, and expression of its structural gene *HEM15, J. Biol. Chem., 265:* 7278–7283.

Labbe-Bois, R., and Labbe, P. (1990). Tetrapyrrole and heme biosynthesis in the yeast *Saccharomyces cerevisiae, Biosynthesis of Heme and Chlorophylls* (H. A. Dailey, ed.), McGraw-Hill, New York, pp. 235–285.

Labbe-Bois, R., and Volland, C. (1977). Changes in the activities of the protoheme-synthesizing system during the growth of yeast under different conditions, *Arch. Biochem. Biophys., 179:* 565–577.

Labbe-Bois, R., Urban-Grimal, D., Volland, C., Camadro, J. M., and Dehoux, P. (1983). About the regulation of protoheme synthesis in the yeast *Saccharomyces cerevisiae, Mitochondria 1983,* Walter de Gruyter, Berlin, pp. 523–534.

Lamoril, J., Boulechfar, S., de Verneuil, H., Grandchamp, B., Nordmann Y., and Deybach, J. C. (1991). Human erythropoietic protoporphyria: Two point mutations in the ferrochelatase gene, *Biochem. Biophys. Res. Comm., 181:* 594–599.

Lavallee, D. K. (1988). Porphyrin metalation reactions in biochemistry, *Mechanistic Principles of Enzyme Activity* (J. F. Liebman and A. Greenberg, eds), VCH, pp. 279–314.

Light, P. A. (1972). Influence of environment on mitochondrial function in yeast, *J. Appl. Chem. Biotechnol., 22:* 509–526.

Mccluskey, S. A., Whitney, R. A., and Marks, G. S. (1989). Evidence for the stereoselective inhibition of chick embryo hepatic ferrochelatase by *N*-alkylated porphyrins, *Mol. Pharmacol., 36:* 608–614.

McKay, R., Druyan, R., Getz, G. S., and Rabinowitz, M. (1969). Intramitochondrial localization of 5-aminolevulinate synthetase and ferrochelatase in rat liver, *Biochem. J., 114:* 455–461.

Maguire, J. J., Davies, K. J. A., Dallman, P. R., and Packer, L. (1982). Effect of dietary iron deficiency on iron-sulfur proteins and bioenergetic functions of skeletal muscle mitochondria, *Biochim. Biophys. Acta, 679:* 210–220.

Margalit, R., Shaklai, N., and Cohen, S. (1983). Fluorimetric studies on the dimerization equilibrium of protoporphyrin IX and its haemato derivative, *Biochem. J., 209:* 547–552.

Meussdoerffer, F., and Fiechter, A. (1986). Effect of mitochondrial cytochromes and haem content on cytochrome P450 in *Saccharomyces cerevisiae, J. Gen. Microbiol., 132:* 2187–2193.

Miyamoto, K., Nakahigashi, K., Nishimura, K., and Inokuchi, H. (1991). Isolation and characterization of visible light-sensitive mutants of *Escherichia coli K12, J. Mol. Biol., 219:* 393–398.

Moody, M. D., and Dailey, H. A. (1985). Iron transport and its relation to heme biosynthesis in *Rhodopseudomonas sphaeroides, J. Bacteriol., 161:* 1074–1079.

Muller-Eberhard, U., and Vincent, S. H. (1985). Concepts of heme distribution within hepatocytes, *Biochem. Pharmacol., 34:* 719–725.

Myers, A. M., Crivellone, M. D., Koerner, T. J., and Tzagoloff, A. (1987). Characterization of the yeast *HEM2* gene and transcriptional regulation of *COX5* and *COR1* by heme, *J. Biol. Chem., 262:* 16822–16829.

Nakahashi, Y., Taketani, S., Okuda, M., Inoue, K., and Tokunaga, R. (1990). Molecular cloning and sequence analysis of cDNA encoding human ferrochelatase, *Biochem. Biophys. Res. Commun., 173:* 748–755.

Nakahigashi, K., Nishimura, K. Miyamoto, K., and Inokuchi, H. (1991). Photosensitivity of a protoporphyrin-accumulating, light-sensitive mutant (*visA*) of *Escherichia coli* K-12, *Proc. Natl. Acad. Sci. USA, 88:* 10520–10524.

Niermann, T., and Kirschner, K. (1991). Use of homologous sequences to improve protein secondary structure prediction, *Methods Enzymol., 202:* 45–59.

Nishida, G., and Labbe, R. F. (1959). Heme biosynthesis: On the incorporation of iron into protoporphyrin, *Biochim. Biophys. Acta, 31:* 519–524.

Ohnishi, T., Asakura, T., Yonetani, T., and Chance, B. (1971). Electron paramagnetic resonance studies at temperatures below 77°K on iron-sulfur proteins of yeast and bovine heart submitochondrial particles, *J. Biol. Chem., 246:* 5960–5964.

Oshino, R., Oshino, N., Chance, B., and Hagihara, B. (1973). Studies on yeast hemoglobin. The properties of yeast hemoglobin and its physiological function in the cell, *Eur. J. Biochem. 35:* 23–33.

Peyronneau, M. A., Renaud, J. P., Truan, G., Urban, P., Pompon, D., and Mansuy, D. (1992). Optimization of yeast-expressed human liver cytochrome P450 3A4 catalytic activities by coexpressing NADPH-cytochrome P450 reductase and cytochrome b5, *Eur. J. Biochem., 207:* 109–116.

Porra, R. J., and Jones, O. T. G. (1963a). Studies on ferrochelatase. 1. Assay and properties of ferrochelatase from pig liver mitochondrial extract, *Biochem. J., 87:* 181–185.

Porra, R. J., and Jones, O. T. G. (1963b). Studies on ferrochelatase. 2. An investigation of the role of ferrochelatase in the biosynthesis of various haem prosthetic groups, *Biochem. J., 87:* 186–192.

Porra, R. J., Vitols, K. S., Labbe, R. F., and Newton, N. A. (1967). Studies on ferrochelatase. The effects of thiols and other factors on the determination of activity, *Biochem. J., 104:* 321–327.

Ricchelli, F., Jori, G., Gobbo, S., and Tronchin, M. (1991). Liposomes as models to study the distribution of porphyrins in cell membranes, *Biochim. Biophys. Acta, 1065:* 42–48.

Riethmüller, G., and Tuppy, H. (1964). Häem synthetase (ferrochelatase) in *Saccharomyces cerevisiae* nach aeroben und anaeroben wachstum, *Biochem. Zeitschrift, 340:* 413–420.

Rotenberg, M., and Margalit, R. (1987). Porphyrin-membrane interactions: Binding or partition? *Biochim. Biophys. Acta, 905:* 173–180.

Rytka, J., Bilinski, T., and Labbe-Bois, R. (1984). Modified uroporphyrinogen decarboxylase activity in a yeast mutant which mimics porphyria cutanea tarda, *Biochem. J., 218:* 405–413.

Saiki, K., Mogi, T., and Anraku, Y. (1992). Heme O biosynthesis in *Escherichia coli:* The *cyoE* gene in the cytochrome *bo* operon encodes a protoheme IX farnesyltransferase, *Biochem. Biophys. Res. Commun., 189:* 1491–1497.

Sawada, H., Takeshita, M., Sugita, Y., and Yoneyama, Y. (1969). Effect of lipids on protoheme ferro-lyase, *Biochim. Biophys. Acta, 178:* 145–155.

Schmalix, W., Oechsner U., Magdolen, V., and Bandlow, W. (1986). Kinetics of the intracellular availability of heme after supplementing a heme-deficient yeast mutant with 5-aminolevulinate, *Biol. Chem. Hoppe-Seyler, 367:* 379–385.

Smith, A. (1990). Transport of tetrapyrroles: Mechanisms and biological and regulatory consequences, *Biosynthesis of Heme and Chlorophylls* (H. A. Dailey, ed.), McGraw-Hill, New York, pp. 435–490.

Straka, J. G., Bloomer, J. R., and Kempner, E. S. (1991a). The functional size of ferrochelatase determined in situ by radiation inactivation, *J. Biol. Chem., 266:* 24637–24641.

Straka, J. G., Hill, H. D., Krikava, J. M., Kools, A. M., and Bloomer, J. R. (1991b). Immunochemical studies of ferrochelatase protein: Characterization of the normal and mutant protein in bovine and human protoporphyria, *Am. J. Hum. Genet., 48:* 72–78.

Takeshita, M., Sugita, Y., and Yoneyama, Y. (1970). Relation between electrophoretic charge of phospholipids and the activating effect on protoheme ferro-lyase, *Biochim. Biophys. Acta, 202:* 544–546.

Taketani, S., and Tokunaga, R. (1981). Rat liver ferrochelatase. Purification, properties, and stimulation by fatty acids, *J. Biol. Chem., 256:* 12748–12753.

Taketani, S., and Tokunaga, R. (1982). Purification and substrate specificity of bovine liver ferrochelatase, *Eur. J. Biochem., 127:* 443–447.

Taketani, S., and Tokunaga, R. (1984). Non-enzymatic heme formation in the presence of fatty acids and thiol reductants, *Biochim. Biophys. Acta, 798:* 226–230.

Taketani, S., Nakahashi, Y., Osumi, T., and Tokunaga, R. (1990). Molecular cloning, sequencing, and expression of mouse ferrochelatase, *J. Biol. Chem., 265:* 19377–19380.

Taketani, S., Inazawa, J., Nakahashi, Y., Abe, T., and Tokunaga, R. (1992). Structure of the human ferrochelatase gene. Exon/intron gene organization and location of the gene to chromosome 18, *Eur. J. Biochem., 205:* 217–222.

Tanaka, S., Nagahama, S., Takeshita, M., and Yoneyama, Y. (1976). Molecular weight estimation of protoheme ferro-lyase by radiation inactivation, *J. Biochem., 80:* 1067–1071.

Tangeras, A. (1985). Mitochondrial iron not bound in heme and iron-sulfur centers and its availability for heme synthesis in vitro, *Biochim. Biophys. Acta, 843:* 199–207.

Tangeras, A. (1986). Effect of decreased ferrochelatase activity on iron and porphyrin content in mitochondria of mice with porphyria induced by griseofulvin, *Biochim. Biophys. Acta, 882:* 77–84.

Tangeras, A., Flatmark, T., Backstrom, D., and Ehrenberg, A. (1980). Mitochondrial iron not bound in heme and iron-sulfur centers. Estimation, compartmentation and redox state, *Biochim. Biophys. Acta, 589:* 162–175.

Traut, T. W. (1988). Do exons code for structural or functional units in proteins? *Proc. Natl. Acad. Sci. USA, 85:* 2944–2948.

Urban-Grimal, D., and Labbe-Bois, R. (1981). Genetic and biochemical characterization of mutants of *Saccharomyces cerevisiae* blocked in six different steps of heme biosynthesis, *Mol. Gen. Genet., 183:* 85–92.

Verdière, J., Gaisne, M., and Labbe-Bois, R. (1991). *CYP1* (*HAP1*) is a determinant effector of alternative expression of heme-dependent transcription in yeast, *Mol. Gen. Genet., 228:* 300–306.

Vergères, G., and Waskell, L. (1992). Expression of cytochrome b5 in yeast and characterization of mutants of the membrane-anchoring domain, *J. Biol. Chem., 267:* 12583–12591.

Volland, C., and Felix, F. (1984). Isolation and properties of 5-aminolevulinate synthase from the yeast *Saccharomyces cerevisiae, Eur. J. Biochem., 142:* 551–557.

Volland, C., and Urban-Grimal, D. (1988). The presequence of yeast 5-aminolevulinate synthase is not required for targeting to mitochondria, *J. Biol. Chem., 263:* 8294–8299.

Wagenbach, M., O'Rourke, K., Vitez, L., Wieczorek, A., Hoffman, S., Durfee, S., Tedesco, J., and Stetler, G. (1991). Synthesis of wild type and mutant human hemoglobins in *Saccharomyces cerevisiae, Biotechnology, 9:* 57–61.

Weaver, J., and Pollack, S. (1990). Two types of receptors for iron in mitochondria, *Biochem. J., 271:* 463–466.

Weber, J. M., Ponti, C. G., Käppeli, O., and Reiser, J. (1992). Factors affecting homologous overexpression of the *Saccharomyces cerevisiae* lanosterol 14alpha-demethylase gene, *Yeast, 8:* 519–533.

Weinstein, J. D., Branchaud, R., Beale, S. I., Bement, W. J., and Sinclair, P. R. (1986). Biosynthesis of the farnesyl moiety of heme a from exogenous mevalonic acid by cultured chick liver cells, *Arch. Biochem. Biophys., 245:* 44–50.

Williams, R. J. P. (1982). Free manganese (II) and iron (II) cations can act as intracellular cell controls, *FEBS Lett., 140:* 3–10.

Wright, G. D., and Honek, J. F. (1989). Effects of iron binding agents on *Saccharomyces cerevisiae* growth and cytochrome P450 content, *Can. J. Microbiol., 35:* 945–950.

Yoneyama, Y., Sawada, H., Takeshita, M., and Sugita, Y. (1969). The role of lipids in heme synthesis, *Lipids, 4:* 321–326.

Zagorec, M., and Labbe-Bois, R. (1986). Negative control of yeast coproporphyrinogen oxidase synthesis by heme and oxygen, *J. Biol. Chem., 261:* 2506–2509.

Zagorec, M., Buhler, J. M., Treich, I., Keng, T., Guarente, L., and Labbe-Bois, R. (1988). Isolation, sequence and regulation by oxygen of the yeast *HEM13* gene coding for coproporphyrinogen oxidase, *J. Biol. Chem., 263:* 9718–9724.

Zhu, H., and Riggs, A. F. (1992). Yeast flavohemoglobin is an ancient protein related to globins and a reductase family, *Proc. Natl. Acad. Sci. USA, 89:* 5015–5019.

Zitomer, R. S., and Lowry, C. V. (1992). Regulation of gene expression by oxygen in *Saccharomyces cerevisiae, Microbiol. Rev., 56:* 1–11.

Zollner, A., Rodel, G., and Haid, A. (1992). Molecular cloning and characterization of the *Saccharomyces cerevisiae CYT2* gene encoding cytochrome-c1-heme lyase, *Eur. J. Biochem., 207:* 1093–1100.

16

Heme-Mediated Gene Regulation in *Saccharomyces cerevisiae*

Jennifer L. Pinkham
University of Massachusetts, Amherst, Massachusetts

Teresa Keng
McGill University, Montreal, Quebec, Canada

I. INTRODUCTION

Prized by humankind for its ability to ferment glucose, the yeast *Saccharomyces cerevisiae* generates ATP preferentially from substrate level phosphorylation. When no fermentable carbon source is available, the organism converts its ATP production rapidly to oxidative phosphorylation within the mitochondrial compartment. This conversion requires the transcription of nuclear genes specifying cytochromes, the constituents of the electron transfer complex, the enzymes of the tricarboxylic acid (TCA) cycle, and a rapid increase in mitochondrial gene expression. Since the biosynthesis of heme requires oxygen, and the cytochromes required for the transport of electrons from NADH to molecular oxygen are hemoproteins, heme plays a pivotal role in coordinating the availability of oxygen with respiratory function. Heme mediates the effects of oxygen and carbon source on gene expression. Genes regulated by oxygen tension are almost always similarly regulated by heme. Recent review articles have discussed heme biosynthesis (Labbe-Bois and Labbe, 1990), oxygen regulation of gene expression (Zitomer and Lowry, 1992), and the coordinate regulation of nuclear and mitochondrial genes (Forsburg and Guarente, 1989b). In this chapter, we will focus on heme-mediated regulation of transcription and update this rapidly moving field. We will also describe heme-mediated, posttranscriptional regulation of gene expression in *Saccharomyces cerevisiae.*

II. HEMOPROTEINS IN YEAST

Hemoproteins are found in nearly all subcellular compartments in yeast and are encoded by both nuclear and mitochondrial genes. Expression of many of the hemoproteins is regulated by heme. The mitochondrial hemoproteins include cytochromes associated with the electron transport complex, which are located on the inner mitochondrial membrane. Cytochrome oxidase contains cytochromes *c* and *a*. Two isologs of cytochrome *c* are encoded by unlinked, nuclear genes *CYC1* and *CYC7* (Montgomery et al., 1980). The complete biosynthetic pathway for heme *a* in yeast has not been worked out. However, heme *a* has been shown to be associated with the cytochrome oxidase subunit I homolog from the photosynthetic bacterium *Rhodobacter sphaeroides* (Shapleigh et al., 1992). Since the bacterial subunit I shares homologies with the eukaryotic cytochrome oxidase subunit I specified by the mitochondrial gene *COX1* (*oxi3*) (Raitio et al., 1987), heme *a* is likely to be associated with the yeast subunit I polypeptide. Coenzyme QH_2-cytochrome *c* reductase has two associated cytochromes, cytochrome c_1 (*CYT1*) (Sadler et al., 1984; Schneider and Guarente, 1991) and cytochrome *b*, the product of the mitochondrial *COB* gene (Nobrega and Tzagoloff, 1980). Within the mitochondrial intermembrane space are found cytochrome *c* peroxidase (*CCP1*), which decomposes hydrogen peroxide produced in the matrix (Kaput et al., 1982; Kaput et al., 1989), and cytochrome b_2 (*CYB2*), also known as L-lactate dehydrogenase, which catalyzes the transfer of electrons from L-(+)-lactate to cytochrome *c* (Capellière-Blandin, 1982; Guiard, 1985). The expression of nuclear-encoded cytochromes is regulated by heme levels.

Hemoproteins found outside of the mitochondria are catalases and enzymes of biosynthetic pathways. *S. cerevisiae* contains two catalases, catalase A (*CTA1*), which is peroxisomal, and catalase T (*CTT1*), which is cytoplasmic (Hartig and Ruis, 1986; Cohen et al., 1988). These enzymes scavenge hydrogen peroxide generated outside of the mitochondria. The microsomal fraction contains cytochrome b_5, which is required for oxidative desaturation of fatty acids (Tamura et al., 1976), and cytochrome P450 (*ERG11*), the enzyme that demethylates lanosterol and catalyzes the first step in ergosterol synthesis (Turi et al., 1991). Thus heme auxotrophs may be grown fermentatively in media supplemented with Tween 80 (oleic acid) and ergosterol. Methionine biosynthesis is linked to heme because sulfite reductase contains siroheme formed from uroporphyrinogen III (Yoshimoto and Sato, 1970). Therefore mutants blocked in the first three steps of heme biosynthesis are methionine auxotrophs.

In addition to the hemoproteins with enzymatic functions, another heme-binding protein is a transcription activator and presumably is located in the nucleus. The gene specifying this protein is known as *CYP1* (Clavilier et al., 1969) or *HAP1* for heme activator protein (Guarente et al., 1984). The activity of HAP1 protein will be discussed in Sec. VI.

The expression of hemoproteins must be differentially regulated because they participate in such a wide variety of processes. The biosynthetic pathways for methionine, unsaturated fatty acids, and ergosterol must be constitutive regardless of oxygen tension or carbon source in the growth medium. In contrast, the cytochromes and hemoproteins with respiratory functions are needed at low levels during fermentative growth and at high levels for respiratory growth conditions. Heme-mediated regulation at almost every level of gene expression has been observed. The effects of heme on translation, protein stability, and enzyme complex assembly will be described in Sec. IV. The mechanisms of heme-mediated transcription activation and repression will be described in Secs. III and VI.

III. HEME-REGULATED TRANSCRIPTION

Heme is an indicator of carbon source and oxygen availability in yeast cells; strains deficient in heme biosynthesis grown without heme supplementation exhibit the same inability to express respiratory genes as wild-type strains grown anaerobically. To study the effects of heme on expression, one typically uses a *hem1* mutant strain. This strain is defective in δ-aminolevulinate (ALA) synthase, the first enzyme in the heme biosynthetic pathway, and requires the presence of ALA in the media for growth. Under these conditions, ALA can be taken up and converted to heme. Alternatively, *hem1* mutants can grow in media supplemented with Tween 80, ergosterol, and methionine, which represent the products of pathways that require heme or heme biosynthetic intermediates. Under these conditions the cells are heme deficient. Heme is necessary for the transcription of genes encoding respiratory enzymes (Guarente and Mason, 1983). Heme is also required for the repression of anaerobically expressed genes (Lowry and Lieber, 1986). The consequence for the cell is that oxygen and heme act through positive regulators to induce the transcription of genes necessary for respiration and through repressors to prevent the expression of genes that function only during anaerobiosis. Table 1 is a compilation of genes known to be regulated by heme or heme-regulatory proteins in *S. cerevisiae*.

Table 1 Genes Regulated by Heme and Oxygen

Gene	Enzyme/protein	Heme regulatory protein	Reference[a]
Activated by heme			
CYC1	Iso-1-cytochrome *c*	HAP1, HAP2/3/4	(1–5)
COX4	Cytochrome oxidase, subunit IV	HAP2/3/4	(6, 7)
COX5a	Cytochrome oxidase, subunit V_a, isolog	HAP2/3/4	(8)
COX6	Cytochrome oxidase, subunit VI	HAP2/3/4	(9, 10)
CYB2	Cytochrome b_2	HAP1	(11)
CYT1	Cytochrome c_1	HAP1, HAP2/3/4	(12)
COR1	QH_2-cytochrome c oxidoreductase, subunit I	?	(13)
COR2	QH_2-cytochrome c oxidoreductase, subunit II	HAP1, HAP2/3/4	(14)
QCR8[b]	QH_2-cytochrome c oxidoreductase, subunit VIII	HAP2/3/4	(15)
CTA1	Catalase A	?	(16)
CTT1	Catalase T	HAP1	(16, 17)
SOD2	Manganous superoxide dismutase	HAP1, HAP2/3/4	(18, 19)
HMG1	3-Hydroxy-3-methylglutaryl CoA[c] reductase, isozyme	HAP1	(20)
TIF51A	eIF-5A, isolog	?	(18, 21)
ROX1	Heme-induced transcription repressor	HAP1	(22, 23)
Repressed by heme			
COX5b	Cytochrome oxidase, subunit V_b, isolog	ROX1	(24)
HMG2	3-Hydroxy-3-methylglutaryl CoA reductase, isozyme	?	(20)
ERG11	Cytochrome P450 lanosterol 14 α-demethylase	ROX1, HAP1	(25, 26)
HEM13	Coproporphyringen oxidase	ROX1, HAP1	(23, 26, 27)
ANB1	eIF-5A, isolog	ROX1, HAP1	(28, 29)

Neither heme-activated nor heme-repressed but controlled by heme regulatory proteins			
CYC7	Iso-2-cytochrome *c*	HAP1, ROX1	(30, 31)
HEM1	δ-Aminolevulinate synthase	HAP2/3/4	(6)
HEM3	Porphobilinogen deaminase	HAP2/3/4	(32)
ACO1	Aconitase	HAP2/3/4	(33)
CIT1	Mitochondrial citrate synthase	HAP2/3/4	(34)
KGD1	α-Ketoglutarate dehydrogenase	HAP2/3/4	(35)
KGD2	Dihydrolipoyl transuccinylase	HAP2/3/4	(36)
LPD1	Lipoamide dehydrogenase	HAP2/3/4	(37)

[a]References: (1) Guarente and Mason (1983); (2) Guarente et al. (1984); (3) Pfeifer et al. (1987); (4) Olesen et al. (1987); (5) Forsburg and Guarente (1989); (6) Keng and Guarente (1987); (7) Schneider (1989); (8) Trueblood et al. (1988); (9) Trawick et al. (1989); (10) Trawick et al. (1992); (11) Lodi and Guiard (1991); (12) Schneider and Guarente (1991); (13) Meyers et al. (1987); (14) Dorsman and Grivell (1990); (15) de Winde and Grivell (1992); (16) Hörtner et al. (1982); (17) Winkler et al. (1988); (18) Lowry and Zitomer (1984); (19) J. Pinkham, unpublished observations; (20) Thorsness et al. (1989); (21) Schnier et al. (1991); (22) Lowry and Zitomer (1988); (23) Keng (1992); (24) Hodge et al (1989); (25) Turi and Loper (1992); (26) Verdière et al. (1991); (27) Zagorec et al. (1986); (28) Mehta and Smith (1989); (29) Lowry et al. (1990); (30) Zitomer et al. (1987); (31) Prezant et al. (1987); (32) Keng et al. (1992); (33) Gangloff et al. (1990); (34) Kell et al. (1992); (35) Repetto and Tzagoloff (1989); (36) Repetto and Tzagoloff (1990); (37) Bowman et al. (1992).

[b]The transcription of *QCR8* has not been tested for heme induction, but its product is functionally related to the *COR2*-encoded protein.

[c]Coenzyme A.

A. Heme-Activation of *CYC1*

The *CYC1* gene, encoding iso-1-apocytochrome *c*, is the prototype of a heme-inducible, carbon source–regulated gene in *S. cerevisiae*. *CYC1* transcription has been studied using a fusion of the *CYC1* promoter including the initiating ATG codon to a reporter gene, *lacZ*, encoding β-galactosidase (Guarente, 1983). The DNA sequences required for *CYC1* activation lie in the 300 base pairs (bp) upstream of the mRNA initiation sites; therefore this DNA is called an upstream activation sequence or UAS (Zitomer et al., 1979; Guarente, 1983; Guarente et al., 1984). UASs of yeast are binding sites for transcription activator proteins and are similar to the enhancers of mammalian genes, since they are orientation independent and work at variable distances from the transcription initiation site.

Transcription of *CYC1* is induced 200-fold when cells are shifted from glucose media to growth in media with lactate, a nonfermentable carbon source (Guarente et al., 1984). Both the basal level of *CYC1* transcription in glucose-grown cells and the induced transcription in lactate are heme dependent (Guarente and Mason, 1983; Guarente et al., 1984). *CYC1* transcription is regulated by two UASs: UAS1 mediates the heme induction in glucose-grown cells, and UAS2 mediates the induction in response to growth on a nonfermentable carbon source. UAS1 is the primary source of *CYC1* transcription activation in glucose-grown cells and is derepressed 10-fold during respiratory growth. This 10-fold derepression can be achieved by adding heme or heme analogs to fermenting cells. Thus the direct link between heme and the transcription of respiratory genes was made by the demonstration that heme is necessary for transcription of *CYC1* (Guarente and Mason, 1983). UAS2 is only slightly active in glucose-grown cells, but it mediates a 100-fold induction of *CYC1* transcription under respiratory growth conditions (Guarente and Mason, 1983; Guarente et al., 1984). The transcriptional activator proteins encoded by the *HAP* genes bind to UAS1 and UAS2 (Pfeifer et al., 1987a; Olesen et al., 1987; Forsburg and Guarente, 1989a). Thus UAS1 bound by the HAP1 protein coordinates *CYC1* transcription with heme availability, and UAS2, recognized by the HAP2/3/4 complex, coordinates the transcription of respiratory enzymes. Each site contributes equally to the derepressed transcription of *CYC1*. The organization of the regulatory region of the *CYC1* gene is shown in Fig. 1.

B. Heme Repression of *ANB1*

The best studied example of a heme-repressed gene is *ANB1* (Mehta and Smith, 1989; Lowry et al., 1990). *ANB1* codes for an isolog of the translation initiation factor eIF-5A, which is expressed only during anaerobic growth. The gene maps to the region 5′ of the *CYC1* gene, and *CYC1* and

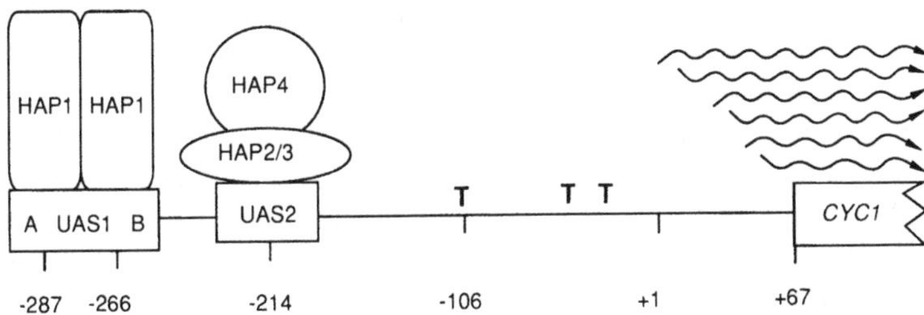

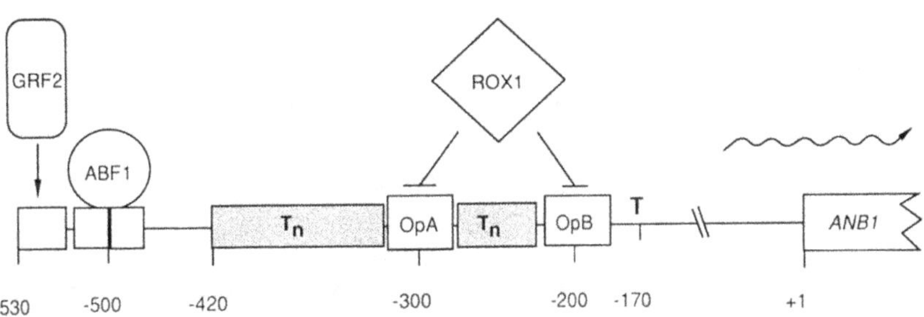

Figure 1 **The structure of the transcription regulatory regions of the heme-activated *CYC1* gene and the heme-repressed *ANB1* gene from *S. cerevisiae.* Panel A displays the regulatory region of the *CYC1* gene. The bold T's represent the three functional TATA elements that are responsible for the six distinct clusters of mRNA initiation sites shown by the wavy arrows (Hahn et al., 1985). The HAP proteins are shown bound to their respective recognition sites. UAS1 region A and region B are shown with a HAP1 protein bound at each site. The distances are given in base pairs from the start of mRNA initiation with +1 at first nucleotide of the first transcript. Panel A is a compilation of results reported in Guarente et al. (1984), Lalonde et al. (1986), Pfeifer et al. (1987a), Olesen et al. (1987), and Forsburg and Guarente (1988, 1989a). Panel B depicts the *ANB1* regulatory region, and the distances are given in base pairs with the A of the translation initiation codon at +1. The bold T shows a presumptive TATA element, as the precise mRNA initiation site for *ANB1* has not been reported. The filled boxes labeled T_n signify regions of poly(dT-dA). The operator sites are represented by OpA and OpB, and each contains two consensus sequences. The ABF1 (GF1) factor has been shown to bind in the *ANB1* regulatory region. Deletion of the GRF2 site reduces *ANB1* transcription. Panel B is compiled from data reported in Dorsman et al. (1988), Mehta and Smith (1989), Lowry et al. (1990), and Zitomer and Lowry (1992).**

ANB1 are divergently and independently transcribed (Lowry and Zitomer, 1984). The *ANB1* promoter is comprised of two elements: the UAS mediates constitutive transcription via a region of poly(dA-dT) sequence, and the upstream repressor sites (URSs), also referred to as operator sites, are believed to be the binding sites for a repressor protein ROX1. The general transcription factor ABF1 has also been shown to bind a site in the *ANB1*UAS (Dorsman et al., 1988; Diffley and Stillman, 1988). Poly(dA-dT) sequences in other yeast promoters provide constitutive, bidirectional transcription (Struhl, 1985). The ROX1 binding sites lie 3′ from the poly(dA-dT), and ROX1 appears to be a very efficient repressor because no *ANB1* transcript can be detected during aerobic growth (Lowry and Zitomer, 1984; Mehta and Smith, 1989). The mechanism of ROX1-mediated repression is currently under investigation. The structure of the regulatory region of the *ANB1* gene is shown in Fig. 1.

C. Heme-Activated and Heme-Repressed Isologs

In *S. cerevisiae,* heme availability determines the expression of protein isologs, which are expressed from unlinked, nuclear genes. One of these paired genes is induced by heme, while the other is repressed. Predictably, most of the proteins specified by these genes are involved in respiration or in a biosynthetic pathway that supplies respiratory cofactors. For example, *CYC1* and *CYC7* encode two isozymes of apocytochrome *c*. *CYC1* is highly inducible by the HAP proteins in response to heme, oxygen, and nonfermentable carbon sources and not transcribed in the absence of heme (Guarente and Mason, 1983; Guarente et al., 1984). In contrast, *CYC7* is transcribed at a low constitutive level (Laz et al., 1984). When the UAS region of *CYC7* was dissected, however, both the heme-dependent activator, HAP1, and the heme-dependent repressor, ROX1, were found to modulate *CYC7* transcription (Prezant et al., 1987; Zitomer et al., 1987). The apoprotein specified by *CYC7* is stable in the absence of heme and therefore provides active cytochrome *c* at the transition from fermentative growth to respiratory growth. However, the *CYC1* apoprotein is unstable in the absence of heme and is synthesized only when heme and oxygen are available (Matner and Sherman, 1982; Laz et al., 1984).

The subunit V of cytochrome oxidase is expressed apparently constitutively and is encoded by two genes, *COX5a* and *COX5b* (Cumsky et al., 1985, 1987). *COX5a* is activated by the HAP2/3/4 complex (Trueblood et al., 1988), and *COX5b* is repressed during aerobic growth by ROX1 (Hodge et al., 1990). Cytochrome oxidase containing the V_b subunit shows an increased turnover number and a faster rate of heme *a* oxidation than does the isoenzyme with the V_a subunit (Waterland et al., 1991). The physiological significance of these different kinetic properties may be related to which cytochrome *c* isozyme is incorporated into cytochrome oxidase and is currently under investigation.

HMG1 and *HMG2* code for 3-hydroxy-3-methylglutaryl coenzyme A (HMG-CoA) reductase, which catalyzes the rate-limiting step in sterol biosynthesis (Basson et al., 1986). The transcription of *HMG1* is induced by heme via HAP1. *HMG2* is repressed by heme during aerobic growth and transcribed during anaerobic growth, making ergosterol synthesis constitutive (Thorsness et al., 1989). Although it is not known whether the activities of these two isozymes differ, it is possible that *HMG1* and *HMG2* gene products are not localized to the same place in the cell. Ubiquinone and the alkyl side chain of heme *a*, two cofactors in the electron transport complex, are derived from a pathway that branches off from sterol biosynthesis (Keyhani and Keyhani, 1978; Olson and Rudney, 1983). The requirement for mevalonate, the product of HMG-CoA reductase, as a precursor for the synthesis of these two cofactors is greater in respiring cells, and it may be supplied more efficiently by the *HMG1*-encoded isozyme.

The *S. cerevisiae* genes specifying the translation initiation factor eIF-5A are *TIF51A* and *ANB1* (Mehta et al., 1990; Schnier et al., 1991). *TIF51A* is expressed only during aerobic growth, while *ANB1* is repressed during aerobic growth by ROX1 and is the only isolog expressed during anaerobic growth (Lowry and Zitomer, 1984; Lowry and Lieber, 1986; Schnier et al., 1991; Mehta and Smith, 1989). However, when the *TIF51A* gene is deleted, *ANB1* transcription, which is normally completely repressed by ROX1 during heme-sufficient, aerobic growth, becomes detectable (Schnier et al., 1991). This suggests that *TIF51A* is involved in repressing *ANB1*. The mechanism for this repression could be a reduced efficiency of translation initiation of either *ROX1* or *HAP1* mRNA (see Sec. VI) when the *ANB1*-encoded eIF-5A is expressed. It is tempting to speculate that the eIF-5A isologs are involved in posttranscriptional regulation of specific genes in response to heme and oxygen. However, the oxygen-dependent, translational regulation of *PET494,* a nuclear gene encoding a translational activator for cytochrome oxidase subunit III in the mitochondria, is independent of heme (Marykwas and Fox, 1989). Thus any functional difference between the two eIF-5A isologs has yet to be documented.

IV. OTHER EFFECTS: HEME-REGULATED TRANSLATION, TRANSPORT, PROTEIN STABILITY, AND ASSEMBLY

In addition to regulation on the transcriptional level, heme and oxygen are postulated to be involved in translational regulation of gene expression.

Translation of the *CTT1* mRNA, in particular, is regulated by heme levels (Hamilton et al., 1982). This was demonstrated using mRNA-dependent *in vitro* translation systems prepared from *hem1* mutants grown in the presence or absence of the heme precursor δ-aminolevulinate. While translation of most proteins occurs with equal efficiency in both extracts, translation of *CTT1* mRNA occurs with a reduced efficiency in the extract prepared from cells grown in the absence of δ-aminolevulinate. This decreased expression of catalase T is due to the absence of heme in the extract, as addition of heme to the translation reaction resulted in increased expression of catalase T protein.

Heme-regulated translation also occurs in reticulocyte lysates. Protein synthesis in reticulocytes is regulated by heme levels in that heme depletion leads to the activation of a protein kinase called the hemin-regulated inhibitor that phosphorylates eIF2α at Ser-51 and inhibits translation initiation (reviewed by London et al., 1987; Pathak et al., 1988). Phosphorylation of eIF2α has also been observed to affect gene expression in yeast. In response to amino acid starvation, yeast cells respond by turning on transcription of a large number of genes that encode amino acid biosynthetic enzymes (reviewed by Hinnebusch, 1988). This response is dependent upon the *GCN4* gene product, a transcriptional activator. Expression of *GCN4* is regulatetd on the translational level. Within the leader RNA region of *GCN4* are four short open reading frames that restrict translation of the *GCN4* coding region under conditions when amino acids are abundant. Upon amino acid starvation, ribosomes are able to translate the *GCN4* coding region. This activation of *GCN4* expression is dependent upon many factors including the GCN2 protein (Roussou et al., 1988). *GCN2* encodes a protein kinase that can phosphorylate eIF2α at Ser-51 in response to amino acid starvation (Dever et al., 1992). Thus the translational control of *GCN4* expression closely parallels that of protein synthesis in reticulocyte lysate systems. Moreover, the sequence of the catalytic domain of GCN2 more closely resembles that of kinases that phosphorylate eIF2α than other eukaryotic protein kinases (Chen et al., 1991; Roussou et al., 1988). Whether GCN2 kinase activity is stimulated by heme depletion remains to be determined.

In addition to regulating expression of *CYC1* on the transcriptional level, heme also plays an important role in transport of apocytochrome *c* into the mitochondrion as well as in determining the stability of apo-1-cytochrome *c*. Mutations in the *CYC3* locus result in a complete absence of functional iso-1-cytochrome *c* and iso-2-cytochrome *c* (Matner and Sherman, 1982). Wild-type levels of cytochrome *c* mRNAs were found in *cyc3* mutant strains, indicating that *CYC3* is not involved in regulating transcription of *CYC1* or *CYC7* (Laz et al., 1984). However, although apo-iso-2-cytochrome *c* protein could be detected in *cyc3* mutant strains, apo-iso-1-cytochrome *c* appeared to be unstable in these strains and could not be detected (Matner and Sherman, 1982). *CYC3* was subsequently shown to encode the enzyme cytochrome *c* heme lyase, which catalyzes the attachment of heme to apocytochrome *c* (Dumont et al., 1987). An analogous enzyme is required for import of cytochrome *c* into the mitochondrion of *Neurospora crassa* (Drygas et al., 1989). In both organisms, this enzyme is required for proper localization of cytochrome *c* to the mitochondrial intermembrane space. In the absence of heme attachment, apocytochrome *c* is only partially inserted through the outer membrane where it is susceptible to exogenous proteases (Nicholson et al., 1988). The translocation of apocytochrome *c* across the outer membrane of the mitochondrion is coupled to its covalent attachment to heme. This process requires the reduction of heme in the presence of NADH and can be inhibited by the heme analog deuterohemin (Nicholson et al., 1987; Nicholson and Neupert, 1989). Thus not only is heme required for the transcription of *CYC1* and *CYC7,* it is also needed for the mitochondrial localization and stability of cytochrome *c*.

Heme is also necessary for the accumulation and assembly of the yeast cytochrome oxidase subunits in *S. cerevisiae* (Saltzgaber-Müller and Schatz, 1978). Cytochrome oxidase is composed of at least ten subunits (LaMarche et al., 1992; Taanman and Capaldi, 1992). Subunits I, II, and III are mitochondrially encoded, while the remaining subunits are nuclear-encoded. In the absence of heme, isolated mitochondria contain nearly normal levels of subunits II and III but only barely detectable amounts of subunit I. Thus heme is required for the accumulation of the mitochondrially encoded subunit I, but not of subunits II and III. Heme deficiency also leads to reduced levels of subunit VI, and subunits IV, V, and VII cannot be detected. The reduction in these nuclear-encoded subunits is due, in part, to the reduction in heme-dependent transcription of the *COX4, COX5a,* and *COX6* genes (Trueblood et al., 1988; Schneider, 1989; Hodge et al., 1989; Trawick et al., 1989). The structure of cytochrome oxidase also appears to be altered in heme-deficient cells because antibodies directed against subunit VI coimmunoprecipitate subunits I, II, and III from mitochondria prepared from heme-sufficient cells, whereas from heme-deficient cells subunit VI alone is immunoprecipitated. These results suggest that heme is required not only for the synthesis of various subunits of cytochrome oxidase but also for the association of these subunits in the assembly of the complex.

V. The Heme Biosynthetic Pathway

If heme were to serve as the intracellular indicator of oxygen tension and carbon source, how do the heme biosynthetic enzymes themselves respond to

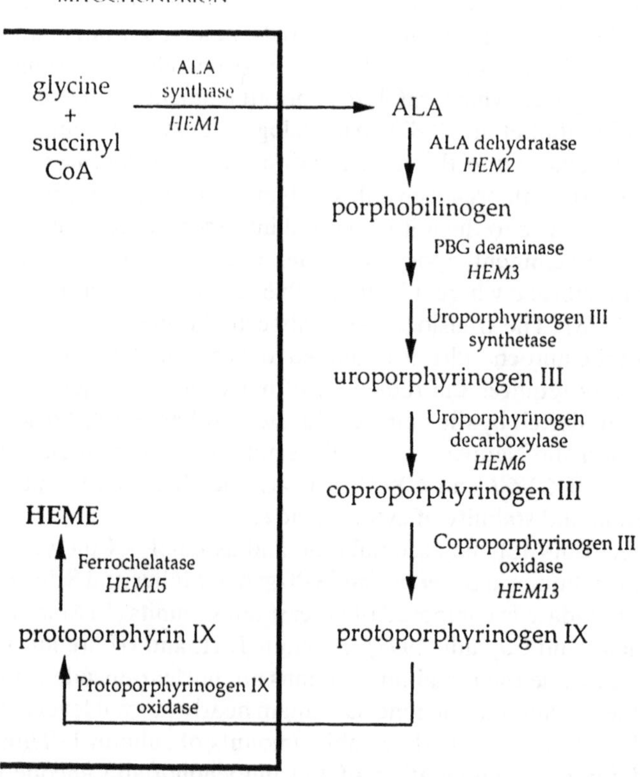

Figure 2 The heme biosynthetic pathway in *Saccharomyces cerevisiae*. ALA synthase is located within the mitochondrial matrix. Protoporphyrinogen IX oxidase and ferrochelatase are located within the mitochondrial inner membrane. ALA, δ-aminolevulinic acid; PBG, porphobilinogen.

such changes in the environment? The heme biosynthetic pathway consists of eight enzymatic steps (reviewed by Labbe-Bois and Labbe, 1990) (Fig. 2). These enzymes are partitioned between the cytoplasmic and mitochondrial compartments. In spite of their different intracellular locations, the enzymes examined thus far are encoded by nuclear genes. The pathway is identical to that found in other eukaryotic cells except for coproporphyrinogen oxidase, the sixth enzyme in the pathway, which is cytoplasmic in yeast but mitochondrial in other eukaryotic systems (Camadro et al., 1986).

Mutants defective in six of the eight enzymatic steps in heme biosynthesis have been isolated (Gollub et al., 1977; Grimal and Labbe-Bois, 1980; Urban-Grimal and Labbe-Bois, 1981; Kurlandzka and Rytka, 1985;

Keng et al., 1992). These mutants have varying phenotypes, ranging from accumulation of porphyrin intermediates and inability to grow on a nonfermentable carbon source to complete heme auxotrophy. The isolation of these mutants allowed the cloning of genes that encode heme biosynthetic enzymes and the analysis of how the synthesis and activity of heme biosynthetic enzymes are regulated. Although the enzymatic activity of some of the enzymes appears to be altered by changes in oxygen, in heme levels, or by carbon source, transcription studies of most of the genes encoding heme biosynthetic enzymes reveal no dramatic effects of carbon source. However, transcription of one gene (*HEM13*) is regulated by oxygen levels, and the transcription of three genes (*HEM1, HEM3, HEM13*) involves factors that also regulate expression of heme-regulated genes.

A. δ-Aminolevulinate Synthase

δ-Aminolevulinate (ALA) synthase is the first enzyme in the heme biosynthetic pathway and is localized within the mitochondrial matrix. It catalyzes the formation of ALA from glycine and succinyl-CoA. The enzyme is encoded by the *HEM1* gene (Arrese et al., 1983; Bard and Ingolia, 1984; Keng et al., 1986; Urban-Grimal et al., 1986). Surprisingly, the presence of *HEM1* on a multicopy plasmid in yeast cells led to an increase in ALA synthase activity and intracellular levels of ALA but did not lead to increased levels of heme, porphyrins, or cytochromes (Arrese et al., 1983). This suggests that ALA synthase is not the rate-limiting step in heme biosynthesis in yeast.

Studies on regulation of ALA synthase activity present a confusing picture of how levels of this enzyme respond to changes in carbon source or oxygen levels. In growth on glucose, ALA synthase activity appears to increase at the end of the exponential phase but decreases to low levels when cells are fully derepressed (Jayaraman et al., 1971; Labbe-Bois and Volland, 1977). A transient increase in enzyme activity is also observed when cells are shifted from growth in glucose to growth in a derepressing medium containing a nonfermentable carbon source and low glucose (Mahler and Lin, 1974). This increase is prevented by the addition of cycloheximide. In addition, ALA synthase activity in yeast, unlike that of higher eukaryotic systems, is not feedback inhibited by heme, the end product of the pathway. Normal levels of ALA synthase activity were found in mutants that are unable to make heme (Urban-Grimal and Labbe-Bois, 1981). Thus there appears to be no clear relationship between the level of ALA synthase activity and heme or oxygen tension and carbon source.

Transcription of *HEM1* has been analyzed using *HEM1-lacZ* fusions and is nearly constitutive (Keng and Guarente, 1987). Expression is derepressed at most twofold in cells grown in the presence of lactate compared to glucose. In addition, the presence or absence of heme or its biosynthetic intermediates does not affect the level of *HEM1* expression. Transcription of *HEM1* is activated by the HAP2/3/4 complex and is dependent upon the sequence TCATTGGT, the consensus binding sequence for the HAP2/3/4 complex. Paradoxically, expression of other HAP2/3/4-dependent genes is heme dependent and dramatically induced during respiration, while expression of *HEM1* is not affected by heme levels or carbon sources. Further examination of the regulatory sequences of *HEM1* identified a negative regulatory site that represses transcription during respiratory growth. We believe that constitutive expression of *HEM1* is a result of interactions between different regulatory systems.

B. δ-Aminolevulinate Dehydratase

ALA dehydratase has been postulated to be the rate-limiting step in the heme biosynthetic pathway. Levulinic acid is a competitive inhibitor of this enzyme, and addition of this antimetabolite to cells results in an increase in intracellular ALA levels and a simultaneous decrease in mitochondrial cytochromes (Malamud et al., 1979). In addition, transfer of cells from a repressing glucose medium to a derepressing medium or from anaerobic to aerobic conditions resulted in an increase in ALA dehydratase activity. The increase observed was between 2.5- and 4-fold, depending upon the particular carbon source used (Labbe, 1971; Jayaraman et al., 1971; Labbe et al., 1972; Poulson, 1976; Labbe-Bois and Volland, 1977; Mahler and Lin, 1978). This increase in ALA dehydratase activity requires both RNA and protein synthesis but is independent of mitochondrial function (Labbe-Bois and Volland, 1977; Mahler and Lin, 1978).

The *HEM2* gene, encoding ALA dehydratase, has been cloned, and transcription of *HEM2* has been examined (Myers et al., 1987; T. Keng, unpublished observations). Expression of *HEM2* is not affected by carbon source or intracellular heme levels. It is also not subject to regulation by any of the *HAP* gene products.

C. Porphobilinogen Deaminase

Yeast porphobilinogen (PBG) deaminase has not been purified, and it remains poorly characterized. This enzyme catalyzes the formation of preuroporphyrinogen, a linear hydroxymethylbilane, which then cyclizes nonenzymatically to form uroporphyrinogen I. The enzyme activity appears to be slightly derepressed in cells grown aerobically (Labbe, 1971; Labbe-Bois and Volland, 1977). PBG deaminase activity is absent in some mutants that are unable to synthesize ALA or PBG, suggesting that PBG may be involved in the induction of the enzyme (Gollub et al., 1977; Urban-Grimal and Labbe-Bois, 1981). *HEM3,* encoding PBG deaminase, has recently been cloned and characterized (Keng et al., 1992). Transcription of *HEM3* is not affected by carbon source or by intracellular heme levels or levels of heme biosynthetic intermediates. The absence of PBG deaminase activity in mutants unable to make ALA or PBG is probably due to the inability of these mutant strains to synthesize dipyrromethane, the cofactor required for PBG deaminase activity. Dipyrromethane is derived from PBG and is essential for the activity of the bacterial enzyme (Jordan et al., 1988). *HEM3* expression requires the HAP2/3/4 transcription complex, and a binding site for the HAP2/3/4 complex, TTATTGGT, is found within the regulatory sequences of *HEM3* (Keng et al., 1992). Thus the constitutive expression of *HEM3* resembles that of *HEM1* and is unlike the heme- and carbon source-regulated expression of other HAP2/3/4-regulated promoters.

D. Uroporphyrinogen III Synthase

This enzymatic activity remains poorly characterized in yeast, but it is believed to convert uroporphyrinogen I to uroporphyrinogen III. In the one report of its activity, the levels of this enzymatic activity appeared to be in excess over that of PBG deaminase (Jordan and Berry, 1980). There is no indication that it plays a rate-limiting role in heme biosynthesis. No mutants defective in uroporphyrinogen III synthase activity have been isolated to date.

E. Uroporphyrinogen Decarboxylase

Uroporphyrinogen decarboxylase converts uroporphyrinogen to coproporphyrinogen by the sequential removal of the four carboxylic groups of the carboxymethyl side chains of uroporphyrinogen. Neither uroporphyrinogen nor any of the intermediates in the decarboxylation reaction accumulate to any significant extent in wild-type cells under a variety of growth conditions (Labbe-Bois and Labbe, 1990). This suggests that the enzyme is probably not a rate-limiting step in the pathway. The enzyme is encoded by the *HEM12* (*HEM6*) gene (Garey et al., 1992; DiFlumeri et al., 1993). This gene has been isolated, and its expression has been characterized. It is induced at most twofold by a nonfermentable carbon source and is not regulated by heme levels. Expression was also not affected in a *hap2* mutant (DiFlumeri et al., 1993).

F. Coproporphyrinogen III Oxidase

Coproporphyrinogen III oxidase converts coproporphyrinogen III to protoporphyrinogen IX in the presence of oxygen. This enzyme is cytosolic in yeast but is mitochondrial in other eukaryotic systems (Camadro et al., 1986). Early studies have indicated that the enzymatic activity was increased 10-fold in cells grown under anaerobic conditions (Miyake and Sugimura, 1968). In addition, enzymatic activity was increased in strains that are defective in heme biosynthesis, regardless of the precise block in the pathway (Labbe-Bois et al., 1980; Urban-Grimal and Labbe-Bois, 1981; Rytka et al., 1984). These observations suggest that synthesis of coproporphyrinogen oxidase is repressed by heme and oxygen. In mutants that are unable to induce *HEM13* expression in response to growth in anaerobic conditions, the levels of heme are significantly reduced (Zagorec et al., 1988). This indicates that under anaerobic conditions, coproporphyrinogen oxidase is the rate-limiting step in the heme biosynthetic pathway. Mutants defective in coproporphyrinogen oxidase activity have been isolated, and the defects in these strains map to the *HEM13* locus. Transcription of *HEM13* is repressed by oxygen via heme (Zagorec and Labbe-Bois, 1986; Zagorec et al., 1988). Addition of heme to aerobically grown cells can bring about repression, while aerobic growth of cells under heme-deficient conditions induces *HEM13* expression. Repression of *HEM13* transcription is mediated by the ROX1 repressor, while induction of *HEM13* transcription in the absence of heme requires the participation of the HAP1 activator (Verdière et al., 1991; Keng, 1992).

G. Protoporphyrinogen IX Oxidase

Protoporphyrinogen III oxidase converts protoporphyrinogen IX to protoporphyrin IX in an oxygen-dependent reaction. This enzyme activity is only twofold increased in cells grown in the presence of a nonfermentable carbon source (Poulson and Polglase, 1974). It is tempting to speculate that heme is involved in repression of the enzymatic activity or enzyme synthesis because mutants defective in any step of the heme biosynthetic pathway have elevated levels of protoporphyrinogen IX oxidase activity (Camadro et al., 1982). A *hem14* mutant was isolated that is defective in protoporphyrinogen IX oxidase activity (Urban-Grimal and Labbe-Bois, 1981; Camadro et al., 1982). This mutant lacks cytochromes and accumulated protoporphyrin. Surprisingly, normal levels of protoporphyrinogen oxidase mRNA and enzyme were found in the mutant. However, the mutant has significantly lower levels of free iron in the mitochondria than wild-type cells. Thus it appears that the primary defect in this mutant is in iron metabolism. How iron metabolism is able to affect protoporphyrinogen IX oxidase activity is unclear at present.

H. Ferrochelatase

Ferrochelatase attaches the ferrous iron to protoporphyrin IX, the final step in heme biosynthesis. The steady-state level of ferrochelatase activity was observed to be twofold to fourfold higher in cells grown in ethanol compared to that in cells grown in glucose or under anaerobic conditions (Labbe, 1971; Labbe et al., 1972; Poulson, 1976). This increase appears to be partly due to an increase in enzyme synthesis and partly due to an increase in enzymatic activity (Labbe-Bois, 1990). Enzyme activity is dependent upon the presence of lipids, and addition of a crude extract of lipids prepared from mitochondria of cells grown under inducing conditions to ferrochelatase isolated from repressed cells resulted in a stimulation of ferrochelatase activity (Camadro and Labbe, 1988). Thus the enzymatic activity may be influenced by the lipid composition of the membrane in which the enzyme resides. Ferrochelatase does not appear to be the rate-limiting step in the pathway, as its over-expression in yeast strains did not lead to any distinguishing phenotypes related to intracellular pools of heme biosynthetic intermediates, hemes, or cytochromes (Labbe-Bois, 1990).

HEM15, encoding ferrochelatase, has been cloned, and its expression has been investigated (Labbe-Bois, 1990). *HEM15* mRNA levels are unaffected by intracellular heme levels or by mutations in the *HAP1* or *HAP2* genes (T. Keng, unpublished observations). However, transcription of *HEM15* is increased 1.5- to 2-fold in ethanol, consistent with the observed increase in enzymatic activity (Labbe-Bois, 1990). Preliminary results suggest that *HEM15* expression may be induced in the presence of low iron (D. Eide and T. Keng, unpublished observation).

VI. HEME-DEPENDENT TRANSCRIPTION REGULATORS

A. HAP1 (CYP1) Protein

Mutations in *HAP1* (*CYP1*) were first isolated by Clavilier et al. (1969) in a selection for mutants that have elevated expression of *CYC7* (*CYP3*). These mutant alleles, represented by the *CYP1-18* (*HAP1-18*) allele, are semidominant over the wild type (Clavilier et al., 1976). Other mutant alleles, which are recessive, have also been isolated (Verdière and Petrochillo, 1979). These mutations resulted in decreased expression of *CYC7*. Analysis of strains containing both classes of *cyp1* (*hap1*) mutations revealed that expression of *CYC1* is also decreased in these strains. Other mutations in *HAP1* (*CYP1*) were isolated by Guarente and his coworkers in their studies on expression of *CYC1* (Guarente et al., 1984). Transcription of *CYC1* is induced by heme and by nonfermentable carbon sources and is mediated by two upstream activation sites, UAS1 and UAS2 (Guarente and Mason, 1983; Guarente et al., 1984). In glucose, expression of *CYC1* is mainly due to expression from UAS1. In a *hap1-1* mutant, expression from UAS1 is greatly diminished. *HAP1* and *CYP1* are allelic and will be referred to as *HAP1* (Verdière et al., 1986).

HAP1 has been cloned and sequenced by two different groups (Creusot et al., 1988; Verdière et al., 1988; Pfeifer et al., 1989). Transcription of the *HAP1* gene is not heme regulated. The HAP1 protein is an activator that binds to UAS1 of *CYC1 in vitro* (Pfeifer et al., 1987a). The DNA binding activity is increased 10-fold by the addition of heme to the binding reactions, suggesting that the heme-dependent expression from UAS1 is due to heme-dependent binding of the HAP1 activator at UAS1. Induction of transcription from UAS1 by nonfermentable carbon sources also appears to be mediated by HAP1, since fully derepressed, UAS1-dependent transcription is obtained by adding heme analogs to glucose-grown cells (Guarente et al., 1984).

1. HAP1 Structure

The HAP1 protein is composed of 1483 amino acids (Creusot et al., 1988; Verdière et al., 1988; Pfeifer et al., 1989). Within the amino-terminal 148 amino acids is a cysteine-rich zinc-finger motif that is homologous to the DNA-binding domain of the GAL4 protein and other yeast transcription activators. Zinc and the integrity of the zinc finger are required for HAP1 function, as substitution of Pro-79, with Ser-79, within the zinc-finger region, resulted in a mutant HAP1 that displayed zinc-dependent DNA-binding activity *in vitro* (Kim et al., 1990). Moreover, mutations in Cys-64 or Cys-81, which constitute residues postulated to be required for zinc binding, abolished DNA-binding activity of HAP1 (Pfeifer et al., 1989).

The heme-responsiveness of the DNA-binding activity of HAP1 was investigated through the construction of deletion derivatives of HAP1. A truncated form of HAP1 with the amino-terminal 445 residues can still bind DNA in a heme-dependent manner. However, shorter derivatives with only the amino-terminal 148 or 244 amino acids can bind DNA independently of heme *in vitro* (Pfeifer et al., 1989). In addition, an internal deletion of amino acids 245 to 445 resulted in a constitutive, heme-independent phenotype *in vivo*. Within the region of amino acids 245 and 445 are seven repeats of the sequences Lys/Arg-Cys-Pro-Val/Ile-Asp-His. This region of evenly spaced Cys and His residues is implicated in heme binding in other hemoproteins and is believed to be the heme binding site of HAP1 (Salemme, 1977). Binding to heme may change the conformation of the HAP1 protein and may permit recognition of the DNA. This idea is supported by the recent isolation of mutant alleles of *HAP1* that allow completely heme-independent transcription of *CYC1*. The mutation in one of these alleles is a Gly-235 to Asp-235 substitution between the DNA-binding and proposed heme-binding domains in HAP1 (S. Ushinsky and T. Keng, submitted). This substitution may lock the HAP1 protein into a conformation that allows DNA-binding independent of heme.

The activation domain of HAP1 lies between residues 1308 and 1483 (Pfeifer et al., 1989). Deletion of this region results in a HAP1 protein that is able to bind DNA *in vitro* but unable to activate transcription *in vivo*. This domain is highly acidic, making it similar to activation domains in other proteins such as GAL4. In particular, the region between 1385 and 1483 carries a net charge of −12.

2. *HAP1 DNA-Binding Activity*

The HAP1 binding sites within the regulatory sequences of *CYC1*UAS1 and *CYC7* have been determined (Pfeifer et al., 1987a, 1987b) (Fig. 3). In

*CYC1*UAS1A	►	-293	TTTCA**CC**GATCTTTC**C**G**G**G
*CYC1*UAS1B	►	-269	TGGCC**GGG**G**TTTA**C**GGA**C**G**A**T**G**A**
CYT1	►	-511	**G**GCGG**CCGG**T**A**TTTCC**GG**CGGC
CTT1	◄	-449	GGAAT**GG**AG**AT**A**A**C**GG**AGG**TT**CTA
CYB2	◄	-218	GGGGCAAGGAGATATCGGCAGG
CYC7	◄	-229	GC**TAA**TA**GC**GA**TAA**TA**GC**GAGGGC

Figure 3 The HAP1 binding sites of *CYC1*, *CYT1*, *CTT1*, *CYB2*, and *CYC7*. The sequences shown are the regions protected by HAP1 in DNAse footprinting experiments and are aligned to show maximum homology. The *CYC7* binding site does not show significant homology to the other binding sites. The arrowheads on the left indicate the direction in which the sequences are found in the regulatory regions of each gene. The numbers indicate the position of the first base shown. The bases shown in bold are the residues that when methylated or ethylated prevent HAP1 binding. Methylation interference experiments have not been performed for the HAP1 binding site in *CYB2*. The HAP1 binding site in *CYT1* is an inverted repeat, as indicated by the arrows. The *CYC7* HAP1 binding site is a direct repeat and is also indicated with arrows. This figure is compiled from observations presented in Pfeifer et al. (1987a, 1987b), Winkler et al. (1988), Schneider and Guarente (1991), and Lodi and Guiard (1991).

mutagenesis studies of *CYC1*UAS1, two regions, region A, centered at position −287 upstream of the *CYC1* transcription initiation site, and region B, centered around position −266, were identified to be important for UAS1 activity *in vivo* (Lalonde et al., 1986). Point mutations in either site substantially reduced transcription from UAS1. The sequences surrounding region A and region B constitute imperfect direct repeats and may be duplicated protein binding sites. Analysis of HAP1 binding to UAS1 identified two HAP1-dependent complexes (Pfeifer et al., 1987a). Formation of both complexes is affected by mutations in region B. DNAse footprinting analysis revealed that the smaller complex corresponded to HAP1 bound to region B and that the larger complex corresponded to proteins bound to both regions A and B. A second factor RAF (Region A Factor) was originally postulated to bind to region A in this larger complex, since the protected sequence of region A is smaller than that of region B, and methylation interference footprints over the two regions are different. The postulated RAF was later shown to be HAP1 (Kim et al., 1990). Region A may be a weak HAP1 binding site, and the differences in DNAse footprints and methylation interference footprints in regions A and B may reflect poor HAP1 binding to region A. These observations suggest some cooperativity between the HAP1 proteins bound at the two UAS1 sites, and HAP1 binding to region A may require HAP1 to be bound at region B, since no HAP1-UAS1 region A complexes can be detected in gel retardation assays with wild-type HAP1.

In addition to HAP1, another factor called RC2 interacts with region B of *CYC1*UAS1 (Arcangioli and Lescure, 1985; Pfeifer et al., 1987a). Mutations in region B also affected formation of the RC2-DNA complex (Lalonde et al., 1986). RC2 activity can only be detected in extracts prepared from cells grown aerobically in the presence of heme. Addition of heme to extracts prepared from cells grown anaerobically or in the absence of heme does not restore binding activity, suggesting that either the synthesis or the stability of RC2 is heme dependent (Arcangioli and Lescure, 1985; Pfeifer et al., 1987a). The function of RC2 in expression from CYC1UAS1 *in vivo* remains unclear.

Binding of HAP1 to the *CYC7* site has been investigated and is also simulated approximately 10-fold by heme (Pfeifer et al., 1987b). The HAP1 binding site in *CYC7* is centered on direct repeats of 9 bp that bear no similarity to the HAP1 binding site in *CYC1*UAS1 (Fig. 3). Despite the dissimilarity, the *CYC1* and *CYC7* sites have comparable affinities for the protein and compete for binding to HAP1 (Pfeifer et al., 1987b). How HAP1 is able to recognize two different DNA sequences was probed by analysis of the *HAP1-18* (*CYP1-18*) mutation. In strains containing this mutant allele of *HAP1*, expression of *CYC7* is dramatically increased, while expression of *CYC1* is significantly reduced (Clavilier et al., 1976; Verdière et al., 1986). In extracts prepared from the *HAP1-18* mutant strain, the HAP1-18 protein is unable to bind to *CYC1*UAS1 but retains *CYC7*UAS binding activity (Pfeifer et al., 1987b). The *HAP1-18* mutation is a Ser-63 to Arg-63 substitution in the HAP1 DNA-binding domain (Verdière et al., 1988; Pfeifer et al., 1989). This residue is positioned immediately amino-terminal to the cysteine-rich finger and is postulated to be involved in recognition of *CYC1*UAS1. Because transcription of *CYC7* is increased in the *HAP1-18* mutant, Ser-63 is also postulated to affect the function of activation domain. Furthermore, the increased expression of *CYC7* in the *HAP1-18* mutant strain is not a result of increased affinity or specificity of the mutant HAP1-18 protein for the *CYC7* binding site nor of an increased occupancy of *CYC7*UAS, because UAS1 and other *CYC1*UAS1-like elements in the yeast genome do not compete for HAP1 binding (Kim and Guarente, 1989).

Studies also suggest that HAP1 has different properties when bound to *CYC1*UAS1 and *CYC7* (Guarente et al., 1984; Prezant et al., 1987). In particular, although HAP1 binds equally well to binding sites in both *CYC7*UAS and UAS1, transcription driven by the *CYC7* binding site is much lower than that observed for the binding site in *CYC1*UAS1B. This observation was further investigated through the construction of internally deleted mutants of HAP1 protein and the examination of their effects on transcription from *CYC7* and *CYC1*UAS1 (Kim et al., 1990). Activity of these *HAP1* mutants at UAS1 increases with deletion size, while the activity at *CYC7* decreases. The largest deletion consisted of the amino-terminal DNA-binding domain fused directly to the carboxyl-terminal activation domain. This derivative, called mini-HAP1, bound equally well to both sites *in vitro,* but *in vivo* mini-HAP1 activated high levels of expression from UAS1 but was unable to activate expression of *CYC7.* When the acidic domain of mini-HAP1 was replaced with the acidic activation domain of GAL4, the HAP1-GAL4 fusions could activate expression at both *CYC1*UAS1 and *CYC7.* This indicates that the selectivity of activation of transcription by mini-HAP1 is specified by the acidic activation domain of HAP1. Taken together, these findings suggest that both the DNA-binding domain and the activation domain of HAP1 affect transcription activation and that the conformation of the DNA-binding domain, and hence the ability of HAP1 to activate transcription, differs when HAP1 is bound at *CYC1*UAS1 or *CYC7.*

To analyze further how HAP1 binds to sites in *CYC1*UAS1 and *CYC7,* saturation mutagenesis was performed on the DNA-binding domain of HAP1 (Turcotte and Guarente, 1992). In addition to obtaining mutants that were defective in transcription driven by both regulatory sequences, mutants that were only defective in *CYC7* expression were also isolated.

Surprisingly, several mutant alleles among both classes of mutants produced HAP1 that had wild-type binding activity. The mutations in these "positive control" mutants were localized to residues adjacent to the zinc finger, more than 1200 amino acids away from the acidic activation domain of HAP1. The isolation of positive control *HAP1* mutants that can selectively affect expression from *CYC7* but not *CYC1*UAS1 supports the notion that binding to different DNA sequences affects the ability of HAP1 to activate transcription.

3. Genes Under HAP1 Control

In addition to regulating expression from *CYC7* and *CYC1,* HAP1 is also involved in regulating expression of a number of other genes. They include *CTT1,* encoding catalase T, *CYT1,* encoding cytochrome c_1, and *CYB2,* encoding cytochrome b_2 (Winkler et al., 1988; Schneider and Guarente, 1991; Lodi and Guiard, 1991). Expression of these genes, like that of *CYC1,* is also induced by heme and by nonfermentable carbon sources (Hörtner et al., 1982; Schneider and Guarente, 1991; Lodi and Guiard, 1991). HAP1 binding to the regulatory regions of these genes has also been demonstrated. The HAP1 binding sites of *CTT1, CYT1,* and *CYB2* are similar to each other and to *CYC1*UAS1, and they appear to be completely different from the binding site in *CYC7* (Fig. 3) (Winkler et al., 1988; Schneider and Guarente, 1991; Lodi and Guiard, 1991).

HAP1 binding to these other regulatory sequences gives a DNAse I footprint of approximately 23 bp. The HAP1 binding site in *CYB2* is highly homologous to that of *CTT1;* alignment of these two sequences shows a region in which 14 out of 20 bases are identical (Winkler et al., 1988; Lodi and Guiard, 1991). Methylation interference footprints have not been obtained for HAP1 binding to *CYB2* sequences. However, the analysis of the *CTT1* site indicates that there is some similarity of methylation-interference contacts between *CYC1*UAS1B and *CTT1* binding sites (Fig. 3). The regulatory sequences of *CYT1* contain a single HAP1 binding site, represented by imperfect inverted repeats. HAP1 binds to this site with a higher affinity than to UAS1A or UAS1B (Schneider and Guarente, 1991). Alignment of the *CYT1* and UAS1B sequences shows a region in which 12 out of 15 bases are identical. Although this homology lies within the DNAse I footprints of both sites, the footprints are offset by several nucleotides. In addition, methylation/ethylation interference footprints indicate that there are a few similarities but many differences between HAP1 contacts at the two sites. Thus qualitative differences in HAP1 binding are observed among the different regulatory sequences. These qualitative differences may result in different levels of HAP1-mediated transcription from these different promoters.

HAP1 is also required for the expression of *HEM13,* encoding coproporphyrinogen oxidase, the sixth enzyme in the heme biosynthetic pathway (Verdière et al., 1991; Keng, 1992). Expression of *HEM13* is repressed by oxygen and by heme (Zagorec et al., 1988). This repression requires the ROX1 repressor as well as HAP1; in a *hap1* mutant in the presence of heme, *HEM13* mRNA levels are elevated compared to wild type. This repressing activity of HAP1 is indirect. It has been demonstrated that HAP1 is required for the transcription of *ROX1* in the presence of heme, and ROX1 is required for the repression of *HEM13* expression (Keng, 1992). This finding is supported by the observation that the DNA binding domain of HAP1 is necessary for the repression of *HEM13* expression under aerobic, heme-sufficient conditions; replacement of the DNA-binding domain of HAP1 with that of another transcriptional activator, PPR1, resulted in an elevation of *HEM13* mRNA levels under repressing conditions (Verdière et al., 1991). Furthermore, in a *hap1* mutant strain in which the expression of *ROX1* has been made independent of HAP1 and heme, *HEM13* mRNA levels are restored to the repressed levels. Thus the heme-dependent, DNA-binding activity of HAP1 is required for *ROX1* transcription (Keng, 1992).

Surprisingly, HAP1 is also required for *HEM13* expression in the absence of heme; expression of *HEM13* cannot be fully induced in a *hap1* mutant (Verdière et al., 1991; Keng, 1992). Although we do not understand how HAP1 functions under heme-deficient conditions, and do not know whether HAP1 acts directly to activate *HEM13* expression under heme-deficient conditions, this finding is intriguing and may represent a completely novel activity for HAP1.

B. HAP2/3/4

The transcription activation of *CYC1* that occurs when yeast cells adjust from fermentative growth to respiration is mediated through the proteins that bind to UAS2. This derepression increases the levels of iso-1-cytochrome *c* in respiring yeast cells and is typical of the expression of respiratory genes. A point mutation in UAS2 called UP1 resulted in a 10-fold increase in UAS2-mediated transcription in both glucose and derepressed growth conditions (Guarente et al., 1984). To understand better the mechanism of UAS2 activity, and to begin to identify the proteins that recognize UAS2, a UAS2UP1-*lacZ* fusion gene was used to screen for mutants that were defective in UAS2UP1-mediated transcription. The pleiotropic *hap2-1* mutant displayed a large reduction in *CYC1*UAS2UP1-*lacZ* fusion gene expression on glucose media. Not only was the expression of the UAS2UP1-*lacZ* fusion gene reduced 20-fold in glucose, but the mutant strain was respiratory deficient and grew poorly on lactate (Guarente et al., 1984). These observations suggest that *HAP2* coordinates transcription of *CYC1* with the other genes required for respiration and oxidative phosphorylation. Using the *CYC1*UAS2UP1-*lacZ* fusion, the *hap3-1* and *hap4-2* mutants with phenotypes similar to *hap2-1* were subsequently isolated (Hahn et al., 1988; Forsburg and Guarente, 1989a). The unlinked, wild-type *HAP2, HAP3,* and *HAP4* genes were cloned by complementation of the respiratory deficient phenotype of the respective mutants (Pinkham and Guarente, 1985; Hahn et al., 1988: Forsburg and Guarente, 1989a).

1. The HAP2/3/4 Complex

The proteins encoded by the *HAP2, HAP3,* and *HAP4* genes form a heterotrimeric complex that binds to UAS2UP1 (Olesen et al., 1987; Forsburg and Guarente, 1989a). The *HAP2* gene specifies a 265 amino acid protein with a region of basic residues that is necessary for transcription activation and a region of polyglutamine whose function remains unknown (Pinkham et al., 1987; Olesen and Guarente, 1990). *HAP3* encodes a 144 amino acid protein (Hahn et al., 1988), and *HAP4* codes for the largest subunit of the complex, a 554 amino acid polypeptide. The carboxyl-terminal 130 amino acids of HAP4 contain two regions of remarkably acidic residues. Such acidic domains are typical of eukaryotic transcriptional activators. Thus the DNA-binding domain and transcriptional activation domains of the complex are located in separate polypeptides, and the respective subunits of the HAP2/3/4 complex must also contain specific protein-protein interacting domains.

The HAP2/3/4 binding site was generally defined by deletions and *in vitro* mutagenesis of the *CYC1*UAS region (Guarente et al., 1984). It was more precisely delineated by linker-scanning mutagenesis (Forsburg and Guarente, 1988) and by analysis of methylation interference footprints of HAP2/3/4-UAS2UP1 complex purified by gel retardation assays (Olesen et al., 1987). The heterotrimeric complex binds with high affinity to UAS2UP1, but it forms no detectable complex with the wild-type UAS2 site (Olesen et al., 1987; Forsburg and Guarente, 1989a). Thus the G to A transition that distinguishes UAS2UP1, TGATTGGT, from wild-type UAS2, TGGTTGGT, also defines a contact site for the complex on the DNA (Olesen et al., 1987). The consensus HAP2/3/4 recognition site, TNATTGGT, has been deduced from inspection and deletion analysis of the UAS regions of several *HAP2* and *HAP3* regulated genes (Keng and Guarente, 1987; Olesen et al., 1987).

The functional domains of the HAP2/3/4 complex have been assigned by constructing deletions and fusion proteins with the LexA protein and the GAL4 proteins (Pinkham et al., 1987; Forsburg and Guarente, 1989a;

Olesen and Guarente, 1990). LexA has a well-characterized DNA-binding domain but no transcription activation capability (Brent and Ptashne, 1985), and GAL4 contains a DNA-binding domain and a transcription activation domain that function independently and may be conveniently subcloned (Ma and Ptashne, 1987). Binding of the HAP2/3/4 complex to the DNA requires both the HAP2 and HAP3 subunits, because a HAP2-GAL4 fusion protein, which contains the acidic, transcription activation domain of GAL4 fused to the carboxyl terminus of HAP2, still requires *HAP3* but bypasses the requirement for *HAP4* and activates transcription of HAP2/3/4-dependent genes (Olesen and Guarente, 1990). The asymmetry of the HAP2/3/4 recognition sequence is consistent with the heteromeric HAP2 and HAP3 subunits together forming the DNA-binding domain. A 65 amino-acid essential core of the HAP2 protein has been identified that is required for DNA binding and assembly of the HAP2/3/4 complex (Olesen and Guarente, 1990). The DNA-binding domain of HAP2 is not similar to other characterized DNA-binding domains (Olesen and Guarente, 1990). However, this essential core is 82% conserved in the *HAP2* homolog, *Php2,* from *Schizosaccharomyces pombe* (Olesen et al., 1991). Furthermore, *Php2* may be part of a conserved regulatory circuit, since *Php2* is also necessary for mitochondrial function in *S. pombe.* Deletions of the carboxyl-terminus of HAP4 destroyed transcription activation of the HAP2/3/4 complex (Forsburg and Guarente, 1989a). A HAP4-GAL4 fusion protein, in which the acidic domain of GAL4 replaced the HAP4 activation domain, restored transcription activation function to the complex. The HAP2/3/4 complex does not require DNA for association, since a functional complex with DNA-binding activity can be recovered after column chromatography in a purification procedure (Hahn and Guarente, 1988). Thus the subunit interaction and DNA-binding domains have been characterized for HAP2, and the subunit interaction and transcription activation domains have also been assigned in HAP4.

Transcription from UAS2UP1, like UAS1, is completely dependent on heme (Guarente and Mason, 1983), but the mechanism of the heme-dependent HAP2/3/4 activity is not understood. The transcriptional activity from UAS2UP1, in contrast to UAS1, is not increased by heme or heme analogs added to the growth media (Guarente et al., 1984), and the *in vitro* binding of the heterotrimeric HAP2/3/4 complex to its recognition sequence is not enhanced by exogenously added heme. *In vivo* experiments with a LexA-HAP2 fusion protein showed that the HAP2/3/4 complex bound to the LexA operator sequence provides heme-dependent transcription from the LexA operator, suggesting that heme-dependence is intrinsic to the HAP2/3/4 heterotrimer (Olesen and Guarente, 1990). However, the steady-state abundance of the *HAP2* transcript is not affected by intracellular heme levels, since wild-type mRNA levels are detected from strains mutant in heme biosynthesis (Pinkham and Guarente, 1985). The effect of intracellular heme on *HAP3* and *HAP4* transcript levels is not known, and the stability of the HAP2/3/4 complex or its subunits under heme-deficient growth conditions has not been studied. The apparent heme-independent transcription of some HAP2/3/4-regulated genes, discussed in the following section, may be contributed by other proteins interacting at separate sites in those promoters.

The transcription of the *HAP2, HAP3,* and *HAP4* genes is not autoregulated. Although steady-state levels of the *HAP2* and *HAP4* transcripts are elevated about fivefold in cells grown in lactate media (Pinkham and Guarente, 1985; Forsburg and Guarente, 1989a), *HAP3* transcript accumulation is not affected by carbon source (Hahn et al., 1988). *HAP2* expression may not be regulated simply at the transcriptional level. The *HAP2* transcript from a multicopy plasmid showed only a twofold induction in lactate media (Pinkham and Guarente, 1985), and a bifunctional *HAP2-lacZ* fusion in multicopy displayed a similar derepression ratio (Pinkham and Guarente, unpublished observations). It has been proposed that HAP4 is the limiting subunit of the active HAP2/3/4 complex, and that increased expression of HAP2/3/4-regulated genes under inducing conditions is due to increased levels of HAP4 (Forsburg and Guarente, 1989a). However, gel retardation assays with extracts from wild-type strains overproducing either HAP2, from a *GAL10*UAS fusion, or HAP4, from its native UAS on a multicopy plasmid, show an increase of the HAP2/3/4 complex bound to UAS2UP1 DNA (Olesen et al., 1987; Forsburg and Guarente, 1989a). Thus while *HAP4* transcription is clearly regulated by carbon source (Forsburg and Guarente, 1989a), the biochemical evidence suggests that both HAP2 and HAP4 proteins are responsible for the increase in complex formation and the transcription induction during respiratory growth.

2. *Genes Under HAP2/3/4 Control*

The respiratory deficiency of strains mutant in *hap2, hap3,* or *hap4* is the consequence of reduced expression of nuclear genes encoding cytochromes and subunits of the oxidative phosphorylation enzyme complexes. A list of genes regulated by HAP2/3/4 is shown in Table 2, which includes a large number of genes encoding respiratory proteins and enzymes of the TCA cycle that are imported into the mitochondria. A gene is determined to be under HAP2/3/4 regulation if its expression is reduced in *hap2* or *hap3* mutant strains and if the TNATTGGT consensus sequence, or a close relative, in either orientation is present in the UAS region.

However, HAP2/3/4 regulation in many cases differs from the well-characterized regulation of *CYC1*UAS2UP1, and HAP2/3/4-regulated

Table 2 HAP2/3/4 Regulated Genes

Category I Heme-dependent, carbon source–regulated	Category II Heme-independent, carbon source–regulated	Category III Heme-independent, carbon source–independent
CYC1	*ACO1*	*HEM1*
COX4	*CIT1*	*HEM3*
COX5a	*KGD1*	
COX6	*KGD2*	
CYT1	*LPD1*	
CYB2		
COR2		
QCR8[a]		
SOD2		

[a]See Table 1, note b.

genes can be placed in three categories. Genes in the first category are heme regulated, induced by growth under respiratory conditions, and are genes encoding cytochromes and subunits of the oxidative phosphorylation enzymes. Genes in the second category specify enzymes of the TCA cycle that are induced during respiratory growth, but they are not heme regulated. Genes in the third category are neither heme regulated nor lactate inducible. Genes in this group include *HEM1* and *HEM3,* which encode constitutively expressed enzymes of the heme biosynthetic pathway (Table 2).

The best studied genes in category I are *CYC1* and *COX6.* In a study of linker scanning mutations, the *COX6*UAS region was shown to contain two heme-responsive elements and a separate HAP2/3/4-recognition sequence. The heme-responsive elements of the *COX6*UAS behave like repressor sites, since deletion or disruption of these sites gives high levels of *COX6* transcription in heme-deficient growth conditions (Trawick et al., 1992). The apparent heme-dependent activation of *COX6* results from an alleviation of repression in place under heme-sufficient growth conditions. Derepressed transcription, as well as the basal glucose transcription, of *COX6* requires the HAP2/3/4 site and an adjacent binding site for the general transcription factor ABF1. Furthermore, although *hap2* and *hap3* mutants show a 10-fold reduction in COX6 expression, *in vitro* gel retardation assays failed to show HAP2/3/4 binding to the *COX6*UAS sequence (Trawick et al., 1992). In contrast to the HAP2/3/4-mediated regulation of the *CYC1*UAS2UP1, the HAP2/3/4 complex appears to act in concert with ABF1 at its recognition site in the *COX6* promoter.

The genes in category II specify TCA cycle enzymes. Transcription of these genes is derepressed during growth with nonfermentable (lactate) or semifermentable (galactose) carbon sources. HAP2/3/4-mediated regulation of these genes coordinates the expression of the TCA cycle enzymes with heme synthesis and oxidative phosphorylation. HAP2/3/4-recognition sites are present in at least one copy in the UASs of this class of genes. Transcript accumulation and the expression of β-galactosidase fusion constructs of the category II genes are reduced 5- to 10-fold in *hap2* and *hap3* mutant strains grown under derepressing conditions in galactose, or resuspended in lactate for several hours, but glucose-repressed expression is unaffected. The transcription of *CIT1* and *ACO1* is regulated negatively by glutamate and decreases further in glucose media when glutamate is present (Gangloff et al., 1990; Kell et al., 1992). The repression of these two genes reflects the biosynthesis of glutamate from the TCA cycle intermediate α-ketoglutarate, and although the mechanism is not yet understood, glutamate regulation of *CIT1* is independent of HAP2/3/4. The expression of most category II genes has been studied in only *hap2* and *hap3* mutants, but *LPD1* and *CIT1* transcription has been shown to require *HAP4* (Bowman et al., 1992; M. Rosenkrantz, unpublished observations).

A third category of HAP2/3/4-regulated genes includes *HEM1* and *HEM3,* which code for heme biosynthetic enzymes (Keng and Guarente, 1987; Keng et al., 1992) (Sec. V). These genes are constitutively transcribed at a low level and show a fivefold reduction of expression in *hap2* or *hap3* mutant strains. The UAS region of *HEM1* contains a HAP2/3/4 binding sequence whose activity is modified by a repression site that lies between the HAP2/3/4 site and the TATA element of the *HEM1* promoter. When the *HEM1* HAP2/3/4 site was tested alone, it displayed a 10-fold derepression in lactate and a high basal transcription level under heme-deficient growth conditions compared to the *CYC1*UAS2UP1. Thus HAP2/3/4 activity in the *HEM1* promoter is most likely reduced by the presence of a repressor in lactate. In addition, some other factor(s) probably supplies a basal, heme-independent transcription activity (Keng and Guarente, 1987).

3. *Mammalian HAP2 and HAP3 Homologs*

The CCAAT-box transcription element found in the promoters of some mammalian and viral genes binds two proteins from Hela cell extracts, CP1A and CP1B, as a heterodimer (Chodosh et al., 1988a). The DNA binding characteristics of the CP1 complex and the HAP2/HAP3 complex are nearly identical, and *in vitro* gel retardation assays showed that the subunits of these complexes are functionally interchangeable (Chodosh et al., 1988b); CP1B may be replaced by HAP2, and CP1A by HAP3. Although different physiological signals control the activities of these two complexes, the HAP2/HAP3 DNA-binding and the subunit interaction domains have been conserved in the mammalian CP1A and CP1B proteins.

C. ROX1 Protein

Mutations in *ROX1* (regulation by oxygen) were first isolated by Lowry and Zitomer in a selection for mutants that could no longer repress transcription of the anaerobic gene, *ANB1,* under aerobic conditions (Lowry and Zitomer, 1984; Lowry and Zitomer, 1988). Two classes of mutants were identified. Mutations in one class, represented by the *rox1-a1* allele, were semidominant and pleiotropic, affecting expression of not only *ANB1* but also heme-induced genes such as *CYC1, TIF51A,* and *SOD2,* encoding manganous superoxide dismutase. In *rox1-a1* mutants, expression of *CYC1, TIF51A,* and *SOD2* became oxygen- and heme-independent and was no longer repressed under anaerobic conditions. Mutations in the other class, represented by *rox1-b3,* were recessive and only affected expression of heme-repressed genes such as *ANB1;* expression of the heme-dependent genes such as *CYC1* was not affected. Mutations in both classes were shown to be allelic through genetic analysis. *ROX1* is required for the repression of other oxygen and heme-repressed genes such as *COX5b* and *HEM13* (Hodge et al., 1989; Keng, 1992).

REO1 (regulator of expression of oxidase), a gene involved in repression of *COX5b* expression, was identified by Trueblood and Poyton (1988). Mutants of *reo1* were isolated as strains with increased levels of *COX5b* under aerobic conditions (Trueblood et al., 1988). The *reo1* mutations were recessive and were found to affect repression of *ANB1* expression as well. Initial analysis indicated that the *reo1* mutation was in a different complementation group from *rox1*. However, it has been demonstrated recently that the *reo1* mutation is complemented by the cloned *ROX1* gene, suggesting that *REO1* and *ROX1* are allelic (M. Cumsky, unpublished observations).

The cloned *ROX1* gene allowed the construction of a strain with a null *rox1* allele. A strain with this allele behaved like the *rox1-b3* mutation in that expression of *ANB1* was rendered constitutively high, while expression of *CYC1* and other heme-induced genes was unaffected (Lowry and Zitomer, 1988). This observation indicates that expression of *CYC1* normally does not involve the *ROX1* gene product and seems at odds with the pleiotropic phenotype observed with the *rox1-1a* mutant strain. Expression of *CYC7* was also examined in the *rox1* null strain because its expression had been demonstrated to be regulated by both activation and repression mechanisms (Zitomer et al., 1987). The activation mechanism involves HAP1. The *rox1* null mutant had levels of *CYC7* mRNA that were higher than wild-type under aerobic conditions, while mRNA levels under anaerobic conditions were unaffected. This suggests that ROX1 is involved in repression of *CYC7* expression under aerobic conditions (Zitomer and Lowry, 1988).

1. Regulation of ROX1 Expression

Transcription of *ROX1* is induced by oxygen via heme and requires the HAP1 activator (Keng, 1992). Under aerobic conditions, HAP1 activates the transcription of *ROX1,* and the ROX1 protein mediates the heme-dependent repression of anaerobic genes. Heme is needed only for the transcription of *ROX1* and is not required for the function of its gene product. Fusion of the *ROX1* transcription unit to the upstream regulatory sites of *GAL10* resulted in *ROX1* transcription that is galactose-inducible and heme- and HAP1-independent. In a strain containing such a fusion, induction of *ROX1* transcription by galactose results in repression of *HEM13* expression even when cells are grown in the absence of heme (Keng, 1992).

The transcription of the mutant *rox1-a1* gene was examined to understand better the semidominance and pleiotropy of the *rox1-a1* allele (Lowry and Zitomer, 1988). A strain containing this mutant allele synthesizes a full-length *rox1-a1* transcript that is constitutively expressed. Thus transcription of *ROX1* is affected by this allele of *ROX1* in the same way as expression of other heme-induced genes. This observation suggests that the original *rox1-a1* strain contains two different mutations; one mutation prevents the repressing activity of the *ROX1* gene product, and the other mutation makes the synthesis of the *ROX1* transcript heme-independent. These two mutations must be closely linked, as the original genetic analysis of the *rox1-a1* allele indicated the cosegregation of constitutive expression of *ANB1* with constitutive expression of the heme-induced genes (Lowry and Zitomer, 1988).

Why does the *rox1-a1* mutation greatly increase the expression of heme-induced genes when a *rox1* null mutation has only minimal effects on their expression? One possibility is that the machinery involved in sensing heme levels in the cell is a complex that contains several different components, including both ROX1 and activator proteins such as HAP1. In the presence of heme, the complex is able to activate expression of heme-induced genes such as *TIF51A* and *CYC1.* This activation can take place even in the absence of ROX1. In the absence of heme, the complex will repress transcription of genes such as *ANB1.* This repression is dependent upon ROX1. The *rox1-a1* gene product may be defective in its repressing function resulting in constitutive transcription of the heme-repressed genes. However, the ROX1-a1 protein may still be able to associate with the other proteins in the complex and allow the complex to activate expression of the heme-induced genes even in the absence of heme. The precise mutation in the *roxl-al* allele is unknown at present, but its identification would give clues to its pleiotropic nature.

2. *The ROX1 Binding Site*

ROX1 acts through two negative regulatory or operator sites in the regulatory sequences of *ANB1* (Lowry et al., 1990) (Fig. 1). These operators affect *ANB1* expression to varying degrees. Deletion of the A operator sequence resulted in a 50-fold increase in *ANB1* expression under repressing conditions, while deletion of operator B resulted in only a fivefold increase in *ANB1* expression. Each of these two negative regulatory sites contains two copies of the consensus operator sequence YYYATTGTTCTC (Fig. 4). These operator sequences are located upstream of the TATA element and transcription initiation sites of *ANB1* but downstream of the upstream activation sequences of *ANB1,* which consist of a poly(dA-dT) sequence and potential GRF2 and ABF1 binding sites (Lowry et al., 1990). These sequences are postulated to mediate constitutive transcription activation. Thus regulation of *ANB1* expression by oxygen and heme is mediated via the negative regulatory sites. The importance of the consensus operator sequences in repression was highlighted when an oligonucleotide with the sequence CCCATTGTTCTC was shown to restore ROX1-dependent repression of *ANB1* expression to a regulatory region that contained deletions for both A and B operator sequences. Two copies of the operator consensus conferred a much stronger repression than one copy of the sequence. Curiously, the same double-operator oligonucleotide alone had no effect on transcription from *GAL1*UAS, although DNA fragments from *ANB1,* comprising either the A or the B operator sequence with flanking sequences, were able to confer ROX1-dependent repression upon the same UAS. These observations suggest that repression by the consensus operator sequence requires flanking sequences.

The importance of the operator consensus element in *ANB1* was independently demonstrated by Mehta and Smith (1989). A fusion was constructed that contained regulatory sequences from *ANB1,* including the operator A region linked to the TATA element and initiation sites of *CYC1*. Expression from this construct was monitored with the reporter gene *lacZ* and was found to be repressed under anaerobic conditions. Multiple point mutations were introduced into the *ANB1* sequence, and mutations that resulted in elevated expression of this fusion under aerobic conditions were isolated and further analyzed. Many of these mutations were localized to within the consensus operator element. In particular, mutation of the highly conserved T in the eighth position of the consensus element to G or A resulted in dramatic increases in expression under aerobic conditions.

The consensus operator sequence has also been found in the regulatory sequences of other heme-repressed, ROX1-regulated genes (Fig. 4). In *COX5b,* two copies of the consensus can be identified. One copy of the repeat lies within URS_{5b}, a 44 bp region that functions as a repression site for *COX5b* expression (Hodge et al., 1990). Deletion of this region resulted in a high level of expression of *COX5b* under aerobic conditions. Moreover, this region is able to repress transcription from heterologous activation sequences; when this site is positioned downstream of the the *CYC1*UAS or the *LEU2*UAS, transcription driven by these activation sequences was decreased five- to sevenfold. Interestingly, the repression mediated by URS_{5b} in these heterologous promoter constructs is not alleviated by growth of the cells under heme-deficient conditions. The other copy of the consensus sequence in *COX5b* lies downstream of the TATA

Consensus ROX1 Operator Sequence

Gene		Sequence		Repression
ANB1	-321	gttttTCCATTG**T**TCGT	-305	O_2, Heme, *ROX1*
	-290	tttgcCCTATTGTTCTC	-274	
	-223	cctatTCCATTGTTCTC	-207	
	-202	gtaaaCTCATTGTTGCT	-186	
HEM13	-480	taattTCAATTGTTTAG	-464	O_2, Heme, *ROX1*
	-286	cgcctTTTCTGGTTCTC	-270	
	-234	tcttaTGCTTTGTTCAA	-250 (<-)	
	-191	ctttgCCCATTGTTCTC	-175	
COX5b	-233	gatttTGTATTGTTCGA	-217	O_2, Heme, *ROX1*
	- 68	attggTCTATTGTTTAA	- 84 (<-)	
CYC7	-131	agatcAGAATAGTTCTC	-115	*ROX1*
Consensus		YYYATTGTTCTC		

Figure 4 The consensus ROX1 operator sequences found in the regulatory regions of *ANB1, HEM13, COX5b,* and *CYC7.* Bases found within the consensus operator sequences are shown in upper case letters. Flanking sequences are shown as lower case letters. The numbers indicate the positions at which the sequences are found in each regulatory region, with the A residue of the translation initiation codon at position +1. The arrows indicate the orientation in which the sequences are found in each gene. Factors that regulate expression of each gene are indicated. The T shown in bold in the *ANB1* sequence indicates the residue that when mutated resulted in elevated expression of an *ANB1-CYC1-lacZ* fusion (Mehta and Smith, 1989). The derived consensus sequence is indicated. Sequences are taken from Zitomer et al. (1987), Zagorec et al. (1988), Lowry et al. (1990), and Hodge et al. (1990).

elements and its contribution to regulation of *COX5b* expression has not yet been investigated.

Four repeats of the operator consensus element are found in the regulatory sequences of *HEM13* (Fig. 4). The function of each of these repeats has been systematically examined (T. Keng and C. Richard, unpublished observations). Three of the repeats each repress *HEM13* expression threefold. Deletion of all three repeats resulted in constitutive expression of *HEM13*. The remaining copy of the consensus element between positions −286 and −270 is not functional in heme-dependent repression of *HEM13* expression, as its deletion had no effect on expression. One copy of the consensus operator element has been found in the regulatory sequences of *CYC7*. The functionality of this repeat has not yet been investigated.

The *ROX1* gene has been sequenced, and the amino acid sequence of the ROX1 protein has been deduced (Balasubramanian et al., 1993). The protein consists of 368 amino acids. The amino-terminal one-third of the protein contains a domain that is basic and resembles a region found in the HMG (high mobility group) class of nonhistone chromatin proteins. We speculate that this basic region of the ROX1 protein might be directly involved in DNA binding.

VII. OTHER REGULATORY PROTEINS

The transcriptional regulation by heme is influenced by regulatory proteins for other pathways. The transcription of the glucose-repressed gene, *SUC2,* encoding invertase, is regulated by a kinase specified by the gene *SNF1* (sucrose nonfermenting) (Carlson et al., 1981; Celenza and Carlson, 1986). *SSN6* (suppressor of snf) plays a negative role in *SUC2* expression, as *ssn6* mutants show constitutive expression of *SUC2* (Trumbly, 1986). Mutations in *snf1* are suppressed by *ssn6* mutations, suggesting that *SNF1* and *SSN6* define a regulatory pathway in *SUC2* expression (Carlson et al., 1984). *SNF1* and *SSN6* also play roles in the expression of heme-regulated genes. Mutant alleles of *SSN6* (*CYC8*) were also isolated as mutants with elevated expression of iso-2-cytochrome *c* (Rothstein and Sherman, 1980). In addition, in a *snf1* mutant the HAP2/3/4-mediated derepression of *CYC1* and *COX6* expression in lactate is lost, while in a *ssn6* mutant, the expression of these genes becomes constitutively high (Wright and Poyton, 1990). The *SSN6* gene product acts epistatically to *SNF1*. This enhancement of HAP2/3/4 activity by *SSN6* may be accomplished by a direct interaction with the complex, or indirectly, by affecting an interacting factor. Alternatively, the transcriptional regulation of the *HAP2* and *HAP4* genes may be rendered constitutive in a *ssn6* mutant.

A. SSN6 Protein

Yeast strains mutant for *ssn6* have pleiotropic phenotypes besides the constitutive expression of glucose-repressible genes. They are flocculant and show an inability to maintain stably minichromosome plasmids (Schultz et al., 1990). *MATα ssn6* strains are unable to mate because they fail to repress genes that are normally expressed only in the *MAT***a** strains, and homozygous *ssn6* diploids are unable to sporulate (Shultz et al., 1990). The protein encoded by *SSN6* is nuclear but has no detectable DNA-binding activity (Schultz et al., 1990). SSN6 is a member of a family of proteins defined by a repeated amino acid sequence called the tetratricopeptide repeat (TPR). Other members of the TPR family found in fungi are the *SKI3* gene, which represses the replication of double-stranded RNA viruses in *S. cerevisiae* (Rhee et al., 1989), and four genes with mitotic functions: *CDC16* and *CDC23* from *S. cerevisiae* (Sikorski et al., 1990), *nuc2⁺* from *S. pombe* (Hirano et al., 1988) and *bimA* from *Aspergillus nidulans* (Morris, 1976; Sikorski et al., 1990). It has been proposed that the TPR domain functions in a specific protein-protein interaction (Hirano et al., 1990).

B. TUP1 Protein

The *TUP1* gene specifies a protein involved in glucose repression and heme-mediated repression. *tup1* (thymidine uptake) mutants were first isolated as yeast strains that had acquired the ability to take up dTMP from the media (Wickner, 1974). Mutant alleles of *TUP1* have been isolated from a wide variety of mutant screens, and they confer pleiotropic phenotypes similar to those observed for *ssn6* mutants, including flocculence (Fugita et al., 1990), *MATα*-specific sterility (MacKay, 1983), sporulation defects in homozygous diploids, and defects in minichromosme plasmid maintenance (Thrash-Bingham and Fangman, 1989). Mutant alleles of *TUP1* have also been isolated from strains that over-express *CYC7* (Rothstein and Sherman, 1980), and mutants of *tup1* are defective in the repression of *ANB1* during heme-sufficient growth (Zhang et al., 1991). Thus *TUP1* augments ROX1 repressing activity.

TUP1 specifies a protein with sequence similarity to the family of transducins, a β subunit of G-proteins. However, the TUP1 protein is large, 713 amino acids, and it is not believed to be a component of a typical heterotrimeric G-protein complex (Williams and Trumbly, 1990; Zhang et al., 1991). *TUP1* and *SSN6* gene products form a high molecular weight complex *in vivo* (Williams et al., 1991). The mechanism of *SSN6* and *TUP1* regulation of gene expression is under investigation. Many of the *tup1* and *ssn6* phenotypes affect pathways that are regulated by repression mechanisms. Thus TUP1 and SSN6 may be components of the general repression machinery. Since deletions of *TUP1* or *SSN6* are viable, indicating that these genes do not regulate essential processes in yeast, it is also possible that they are part of a signal transduction pathway that senses the metabolic state of the cell and coordinates carbon source metabolism with the energy-producing pathways.

C. ROX3

A number of *rox* mutants that display increased *CYC7* transcription have been isolated (Rosenblum-Vos et al., 1991). Of these, the *rox3* mutant has been further characterized. Strains mutant for *rox3* also have decreased anaerobic expression of *ANB1*. Steady-state *ROX3* transcript accumulation is greater in anaerobically growing cells, but it is not repressed by heme. The *ROX3* gene is essential for viability, and the protein it encodes is localized to the yeast nucleus. Therefore *ROX3* is believed to specify a general transcription factor (Rosenblum-Vos et al., 1991).

VIII. SUMMARY

Heme-mediated transcription regulation in *S. cerevisiae* is integrated into the regulatory pathways that respond to changes in carbon source and oxygen availability. HAP1 regulates the transcription of genes encoding both mitochondrial and cytoplasmic hemoproteins. The HAP1 protein binds heme, and HAP1 binding at its DNA recognition site requires heme. The DNA-binding domain of HAP1 affects its ability to activate transcription. Thus HAP1 has at least three domains that interact to accomplish heme-mediated transcription activation. The link between oxygen availability and HAP1 activity is through the biosynthesis of heme, which requires oxygen. The HAP/2/3/4 complex regulates the expression of genes whose products are involved in respiration and are imported into the mitochondria. Genes regulated by the HAP2/3/4 complex are induced by nonfermentable carbon sources and by heme. The transcription of the *HAP2* and *HAP4* genes is also induced by nonfermentable carbon sources. Therefore the transcriptional regulation of the *HAP2* and *HAP4* genes may be responsible for the carbon source regulation of HAP2/3/4-regulated genes. The heme-dependence of HAP2/3/4 regulation appears to be intrinsic to the complex, but flanking sequences and general transcription factors can influence both the heme- and the carbon source-dependence of the transcription activation mediated by the HAP2/3/4 complex. The product of the *SSN6* gene, which is a regulator of glucose-repression, affects the activity or synthesis of the HAP2/3/4 complex. Thus the transcription of respiratory genes is integrated with the regulation of glucose repression through the interaction of SSN6 with the HAP2/3/4 complex. Heme-mediated repression of transcription of anaerobic genes is accomplished through the ROX1 protein. *ROX1* gene transcription is activated by the HAP1 protein and therefore requires heme, but ROX1 does not require heme for its repressing activity. The repressing activity of ROX1 is augmented by the glucose-repression regulator TUP1, and ROX1 may be part of a complex as well. The SSN6 and TUP1 proteins associate to form a complex, and either as a complex or separately regulate both the activation by HAP2/3/4 and the repressing activity of ROX1. A better understanding of the interactions among these regulatory proteins is needed to provide the details of the mechanisms of heme-mediated gene regulation in *S. cerevisiae.*

ACKNOWLEDGMENTS

We are grateful to Susan Forsburg, Thomas Mason, and Mark Rosenkrantz for helpful discussions, and to Leonard Guarente in whose laboratory our work in this field was initiated. We thank Michael Cumsky, David Eide, Susan Forsburg, and Mark Rosenkrantz for sharing unpublished information, and many members of the yeast community for insightful discussions over the years. This work was supported by the National Institutes of Health (J. L. P.) and by the Medical Research Council of Canada (T. K.).

REFERENCES

Arcangioli, B., and Lescure, B. (1985). Identification of proteins involved in the regulation of yeast iso-1-cytochrome *c* expression by oxygen, *EMBO J., 4:* 2627–2633.

Arrese, M., Carvajal, E., Robison, S., Sambunaris, A., Panek, A., and Mattoon, J. (1983). Cloning of the δ-aminolevulinic acid synthase structural gene in yeast, *Curr. Genet., 7:* 175–183.

Balasubramanian, B., Lowry, C. V., and Zitomer, R. S. (1993). The Rox1 repressor of the *Saccharomyces cerevisiae* hypoxic genes is a specific DNA-binding protein with a high-mobility-group motif. *Mol. Cell. Biol., 13:* 6071–6078.

Bard, M., and Ingolia, T. D. (1984). Plasmid-mediated complementation of a δ-aminolevulinic-acid-requiring *Saccharomyces cerevisiae, Gene, 28:* 195–199.

Basson, M. E., Thorsness, M., and Rine J. (1986). *Saccharomyces cerevisiae* contains two genes encoding 3-hydroxy-3-methylglutaryl coenzyme A reductase, *Proc. Natl. Acad. Sci. USA, 83:* 5563–5567.

Bowman, S. B., Zaman, Z., Collin, L. P., Brown, A. J. P., and Dawes, I. W. (1992). Positive regulation of the *LPD1* gene of *Saccharomyces cerevisiae* by the HAP2/HAP3/HAP4 activation system, *Mol. Gen. Genet., 231:* 296–303.

Brent, R., and Ptashne, M. (1985). A eukaryotic transcriptional activator bearing the DNA specificity of a prokaryotic repressor, *Cell, 43:* 729–736.

Camadro, J. M., and Labbe, P. (1988). Purification and properties of ferrochelatase from the yeast *Saccharomyces cerevisiae.* Evidence for a precursor form of the protein, *J. Biol. Chem., 263:* 11675–11682.

Camadro, J. M., Urban-Grimal, D., and Labbe, P. (1982). A new assay for protoporphyrinogen oxidase—Evidence for a total deficiency in that activity in a heme-less mutant of *Saccharomyces cerevisiae, Biochem. Biophys. Res. Comm., 106:* 724–730.

Camadro, J. M., Chambon, H., Jolles, J., and Labbe, P. (1986). Purification and properties of coproporphyrinogen oxidase from the yeast *Saccharomyces cerevisiae, Eur. J. Biochem., 156:* 579–587.

Capellière-Blandin, C. (1982). Transient kinetics of the one-electron transfer reaction between reduced flavocytochrome b_2 and oxidised cytochrome *c, Eur. J. Biochem., 128:* 533–542.

Carlson, M., Osmond, B. C., and Botstein, D. (1981). Mutants of yeast defective in sucrose utilization, *Genetics, 98:* 25–40.

Carlson, M., Osmond, B. C., Neigeborn, L., and Botstein, D. (1984). A suppressor of *snf1* mutations causes constitutive high-level invertase synthesis in yeast, *Genetics, 107:* 19–32.

Celenza, J. L., and Carlson, M. (1986). A yeast gene that is essential for release from glucose repression encodes a protein kinase, *Science, 233:* 1175–1180.

Chen, J.-J., Throop, M. S., Gehrke, L., Kuo, I., Pal, J. K., Brodskyu, M., and London, I. M. (1991). Cloning of the cDNA of the heme-regulated eukaryotic initiation factor 2α (eIF-2α) kinase of rabbit reticulocytes: Homology to yeast GCN2 protein kinase and human double-stranded-RNA-dependent eIF-2α kinases, *Proc. Natl. Acad. Sci. USA, 88:* 7729–7733.

Chodosh, L. A., Baldwin, A. S., Carthew, R. W., and Sharp, P. A. (1988a). Human CCAAT binding proteins have heterologous subunits, *Cell, 53:* 11–24.

Chodosh, L. A., Olesen, J. T., Hahn, S., Baldwin, A. S., Guarente, L., and Sharp, P. A. (1988b). A yeast and human CCAAT binding protein have heterologous subunits that are functionally interchangeable, *Cell, 53:* 25–35.

Clavilier, L., Péré, G., and Slonimski, P. P. (1969). Mise en évidence de plusieurs loci indépendants impliqués dans la synthèse de l'iso-2-cytochrome *c* chez la levure, *Molec. Gen. Genet., 104:* 195–218.

Clavilier, L., Péré-Aubert, G., Somlo, M., and Slonimski, P. P. (1976). Réseau d'interactions entre des gènes non liés: Régulation synergique ou antagoniste de la synthèse de l'iso-1-cytochrome *c,* de l'iso-2-cytochrome *c,* et du cytochrome b_2, *Biochimie, 58:* 155–172.

Cohen, G., Rapatz, W., and Ruis, H. (1988). Sequence of the *Saccharomyces cerevisiae CTA1* gene and amino acid sequence of catalase A derived from it, *Eur. J. Biochem., 176:* 159–163.

Creusot, F., Verdière, J., Gaisne, M., and Slonimski, P. P. (1988). CYP1 (HAP1) regulator of oxygen-dependent gene expression in yeast. I. Overall organization of the protein sequence displays several novel structural domains, *J. Mol. Biol., 204:* 263–276.

Cumsky, M. G., Ko, C., Trueblood, C. E., and Poyton, R. O. (1985). Two nonidentical forms of subunit V are functional in yeast cytochrome *c* oxidase, *Proc. Natl. Acad. Sci. USA, 82:* 2235–2239.

Cumsky, M. G., Trueblood, C. E., Ko, C., and Poyton, R. O. (1987). Structural analysis of two genes encoding divergent forms of cytochrome *c* oxidase subunit V, *Mol. Cell., Biol., 7:* 3511–3519.

Dever, T. E., Feng, L., Wek, R. C., Cigan, A. M., Donahue, T. F., and Hinnebusch, A. G. (1992). Phosphorylation of initiation factor 2α by protein kinase GCN2 mediates gene-specific translational control of *GCN4* in yeast, *Cell, 68:* 585–596.

de Winde, J. H., and Grivell, L. A. (1992). Global regulation of mitochondrial biogenesis in *Saccharomyces cerevisiae:* ABF1 and CPF 1 play opposite roles in regulating expression of the *QCR8* gene, which encodes subunit subunit VIII of the mitochondrial ubiquinol-cytochrome *c* oxidoreductase, *Mol. Cell. Biol., 12:* 2872–2883.

Diffley, J. F. X., and Stillman, B. (1988). Purification of a yeast protein that binds to origins of DNA replication and a transcriptional silencer, *Proc. Natl. Acad. Sci. USA, 85:* 2120–2124.

DiFlumeri, C., Larocque, R., and Keng, T. (1993). Molecular analysis of *HEM6 (HEM12)* in *Saccharomyces cerevisiae,* the gene for uroporphyrinogen decarboxylase, *Yeast 9:* 613–623.

Dorsman, J. C., and Grivell, L. A. (1990). Expression of the yeast gene encoding subunit II of yeast QH_2: Cytochrome *c* oxidoreductase is regulated by multiple factors, *Curr. Genet., 17:* 459–464.

Dorsman, J. C., van Heewijk, W. C., and Grivell, L. A. (1988). Identification of two factors which bind to the upstream sequences of a number of nuclear genes coding for mitochondrial proteins and to genetic elements important for cell division in yeast, *Nucleic Acids Res., 16:* 7287–7301.

Drygas, M. E., Lambowitz, A. M., and Nargang, F. E. (1989). Cloning and analysis of the *Neurospora crassa* gene for cytochrome *c* heme lyase, *J. Biol. Chem., 264:* 17897–17906.

Dumont, M. E., Ernst, J. F., Hampsey, D. M., and Sherman, F. (1987). Identification and sequence of the gene encoding cytochrome *c* heme lyase in the yeast *Saccharomyces cerevisiae, EMBO J., 6:* 235–241.

Forsburg, S. L., and Guarente, L. (1988). Mutational analysis of upstream activation sequence 2 of the *CYC1* gene of *Saccharomyces cerevisiae:* A HAP2-HAP3-responsive site, *Mol. Cell. Biol., 8:* 647–654.

Forsburg, S. L., and Guarente, L. (1989a). Identification and characterization of HAP4: A third component of the CCAAT-bound HAP2/HAP3 heteromer, *Genes Dev., 3:* 1166–11778.

Forsburg, S. L., and Guarente, L. (1989b). Communication between mitochondria and the nucleus in regulation of cytochrome genes in the yeast *Saccharomyces cerevisiae, Ann. Rev. Cell Biol., 5:* 153–180.

Fugita, A., Shinichi, M., Kuhara, S., Misumi, Y., and Kobayashi, H. (1990). Cloning of the yeast *SFL2* gene: Its disruption results in pleiotropic phenotypes characteristic for *tup1* mutants, *Gene, 89:* 93–99.

Gangloff, S. P., Marguet, D., and Lauquin, G. J.-M. (1990). Molecular cloning of the yeast mitochondrial aconitase gene (*ACO1*) and evidence of a synergistic regulation of expression by glucose plus glutamate, *Mol. Cell. Biol., 10:* 3551–3561.

Garey, J. R., Labbe-Bois, R., Chelstowska, A., Rytka, J., Harrison, L., Kushner, J., and Labbe, P. (1992). Uroporphyrinogen decarboxylase in *Saccharomyces cerevisiae—HEM13* gene sequence and evidence for two conserved glycines essential for enzymatic activity, *Eur. J. Biochem., 205:* 1011–1016.

Gollub, E. G., Liu, K.-P., Dayan, J., Adlersbérg, M., and Sprinson, D. B. (1977). Yeast mutants deficient in heme biosynthesis and a heme mutant additionally blocked in cyclization of 2,3 oxidosqualene, *J. Biol. Chem., 252:* 2846–2854.

Grimal, D., and Labbe-Bois, R. (1980). An enrichment method for heme-less mutants of *Saccharomyces cerevisiae* based on photodynamic properties of Zn-protoporphyrin, *Mol. Gen. Genet., 178:* 713–716.

Guarente, L. (1983). Yeast promoters and *lacZ* fusions designed to study expression of cloned genes in yeast, *Methods Enzymol., 101:* 181–191.

Guarente, L., and Mason, T. (1983). Heme regulates transcription of the *CYC1* gene of *S. cerevisiae* via an upstream activation site, *Cell, 32:* 1279–1286.

Guarente, L., Lalonde, B., Gifford, P., and Alani, E. (1984). Distinctly regulated tandem upstream activation sites mediate catabolite repression of the *CYC1* gene in *S. cerevisiae, Cell, 36:* 503–511.

Guiard, B. (1985). Structure, expression and regulation of a nuclear gene encoding a mitochondrial protein: The yeast L(+)-lactate cytochrome *c* oxidoreductase (cytochrome b_2), *EMBO J., 4:* 3265–3272.

Hahn, S., and Guarente, L. (1988). Yeast HAP2 and HAP3: Transcriptional activators in a heteromeric complex, *Science, 240:* 317–321.

Hahn, S., Hoar, E. T., and Guarente, L. (1985). Each of three TATA elements specifies a subset of the transcription initiation sites in the *CYC1* promoter of *Saccharomyces cerevisiae, Proc. Natl. Acad. Sci. USA, 82:* 8562–8566.

Hahn, S., Pinkham, J., Wei, R., Miller, R., and Guarente, L. (1988). The *HAP3* regulatory locus of *Saccharomyces cerevisiae* encodes divergent overlapping transcripts, *Mol. Cell. Biol., 8:* 655–663.

Hamilton, B., Hofbauer, R., and Ruis, H. (1982). Translational control of catalase synthesis by hemin in the yeast *Saccharomyces cerevisiae, Proc. Natl. Acad. Sci. USA, 79:* 7609–7613.

Hartig, A., and Ruis, H. (1986). Nucleotide sequence of the *Saccharomyces cerevisiae CTT1* gene and deduced amino-acid sequence of yeast catalase T, *Eur. J. Biochem., 160:* 487–490.

Hinnebusch, A. G. (1988). Mechanisms of gene regulation in the general control of amino acid biosynthesis in *Saccharomyces cerevisiae, Microbiol. Rev., 52:* 248–273.

Hirano, T., Hiraoka, Y., and Yanagida, M. (1988). A temperature-sensitive mutation of the *Schizosaccharomyces pombe gene nuc2*$^+$ that encodes a nuclear scaffold-like protein blocks spindle elongation in mitotic anaphase, *J. Cell. Biol., 106:* 1171–1183.

Hirano, T., Kinoshita, N., Morikawa, K., and Yanagida, M. (1990). Snap helix with knob and hole: Essential repeats in *S. pombe* nuclear protein *nuc2*$^+$, *Cell, 60:* 319–328.

Hodge, M. R., Kim, G., Singh, G., and Cumsky, M. (1989). Inverse regulation of the yeast *COX5* genes by oxygen and heme, *Mol. Cell. Biol., 9:* 1958–1964.

Hodge, M. R., Singh, K., and Cumsky, M. (1990). Upstream activation and repression elements control transcription of the yeast *COX5b* gene, *Mol. Cell. Biol., 10:* 5510–5520.

Hörtner, H., Ammerer, G., Hartter, E., Hamilton, B., Rytka, J., Bilinkski, T., and Ruis, H. (1982). Regulation of synthesis of catalases and iso-1-cytochrome *c* in *Saccharomyces cerevisiae* by glucose, oxygen and heme, *Eur. J. Biochem., 128:* 179–184.

Jayaraman, J., Padmanaban, G., Malathi, K., and Sarma, P. S. (1971). Haem synthesis during mitochondriogenesis in yeast, *Biochem. J., 121:* 531–535.

Jordan, P. M., and Berry, A. (1980). Preuroporphyrinogen, a universal intermediate in the biosynthesis of uroporphyrinogen III, *FEBS Lett., 112:* 86–88.

Jordan, P. M., Warren, M. J., Williams, H. J., Stolowich, N. J., Roessner, C. A., Grant, S. K., and Scott, A. I. (1988). Identification of a cysteine residue as the binding site for the dipyrromethane cofactor at the active site of *Escherichia coli* porphobilinogen deaminase, *FEBS Lett., 235:* 189–193.

Kaput, J., Goltz, S., and Blobel, G. (1982). Nucleotide sequence of the yeast nuclear gene for cytochrome *c* peroxidase precursor: Implications of the presequence for protein transport into mitochondria, *J. Biol. Chem., 257:* 15054–15058.

Kaput, J., Brandriss, M. C., and Prussak-Wieckowska, T. (1989). *In vitro* import of cytochrome *c* peroxidase into the intermembrane space: Release of the processed form by intact mitochondria, *J. Cell. Biol., 109:* 101–112.

Kell, C., Pennell, E., and Rosenkrantz, M. (1992). Yeast citrate sythase gene *CIT1* is activated by HAP2, 3, 4-dependent and independent mechanisms, *Yeast, 8:* S149.

Keng, T. (1992). HAP1 and ROX1 form a regulatory pathway in the repression of *HEM13* transcription in *Saccharomyces cerevisiae, Mol. Cell. Biol., 12:* 2616–2623.

Keng, T., and Guarente, L. (1987). Constitutive expression of the yeast *HEM1* gene is actually a composite of activation and repression, *Proc. Natl. Acad. Sci. USA, 84:* 9113–9117.

Keng, T., Alani, E., and Guarente, L. (1986). The nine amino-terminal residues of δ-aminolevulinate synthase direct β-galactosidase into the mitochondrial matrix, *Mol. Cell. Biol., 6:* 355–364.

Keng, T., Richard, C., and Larocque, R. (1992). Structure and regulation of yeast *HEM3,* the gene for porphobilinogen deaminase, *Mol. Gen. Genet., 234:* 233–243.

Keyhani, J., and Keyhani, E. (1978). Mevalonic acid as a precursor of the alkyl sidechain of heme **a** of cytochrome *c* oxidase in the yeast *Saccharomyces cerevisiae, FEBS Lett., 93:* 271–274.

Kim, K. S., and Guarente, L. (1989). Mutations that alter transcriptional activation but not DNA binding in the zinc finger of yeast activator HAP1, *Nature, 342:* 200–203.

Kim, K. S., Pfeifer, K., Powell, L., and Guarente, L. (1990). Internal deletions in the yeast transcriptional activator HAP1 have opposite effects at two sequence elements, *Proc. Natl. Acad. Sci. USA, 87:* 4524–4528.

Kurlandzka, A., and Rytka, J. (1985). Mutants of *Saccharomyces cerevisiae* partially defective in the last steps of the haem biosynthetic pathway: Isolation and genetical characterization, *J. Gen. Microbiol., 131:* 2909–2918.

Labbe, P. (1971). Synthèse du protohème par la levure *Saccharomyces cerevisiae.* I. Mise en évidence de différentes étapes de la synthèse du protohème chez la levure cultivée en aérobiose et en anaérobiose. Influence des conditions de cultures sur cette synthèse, *Biochimie, 53:* 1001–1014.

Labbe, P., Dechateaubodeau, G., and Labbe-Bois, R. (1972). Synthèse du protohème par la levure *Saccharomyces cerevisiae.* II. Influence exercée par le glucose sur l'adaptation respiratoire, *Biochimie, 54:* 513–528.

Labbe-Bois, R. (1990). The ferrochelatase from *Saccharomyces cerevisiae.* Sequence, disruption, and expression of its structural gene *HEM15, J. Biol. Chem., 265:* 7278–7283.

Labbe-Bois, R., and Labbe, P. (1990). Tetrapyrrole and heme biosynthesis in the yeast *Saccharomyces cerevisiae, Biosynthesis of Heme and Chlorophylls* (H. A. Dailey, ed.), McGraw-Hill, New York, pp. 235–285.

Labbe-Bois, R., Simon, M., Rytka, J., Litwinska, J., and Bilinski, T. (1980). Effect of 5-amino levulinic acid synthesis deficiency on expression of other enzymes of heme pathway in yeast. *Biochem. Biophys. Res. Comm., 95:* 1357–1363.

Labbe-Bois, R., and Volland, C. (1977). Changes in the activities of the protoheme-synthesizing system during the growth of yeast under different conditions, *Arch. Biochem. Biophys., 179:* 565–577.

Lalonde, B., Arcangioli, B., and Guarente, L. (1986). A single *Saccharomyces cerevisiae* upstream activation site (UAS1) has two distinct regions essential for its activity, *Mol. Cell. Biol., 6:* 4690–4696.

LaMarche, A. E. P., Abate, M. I., Chan, S. H. P., and Trumpower, B. L. (1992). Isolation and characterization of *COX12,* the nuclear gene for a previously unrecognized subunit of *Saccharomyces cerevisiae* cytochrome *c* oxidase, *J. Biol. Chem., 267:* 22473–22480.

Laz, T. M., Pietras, D. F., and Sherman, F. (1984). Differential regulation of the duplicated isocytochrome *c* genes in yeast, *Proc. Natl. Acad. Sci. USA, 81:* 4475–4479.

Lodi, T., and Guiard, B. (1991). Complex transcriptional regulation of the *Saccharomyces cerevisiae CYB2* gene encoding cytochrome b_2: CYP1 (HAP1) activator binds to the *CYB2* upstream activation site UAS1-B2, *Mol. Cell. Biol., 11:* 3762–3772.

London, I. M., Levin, D. H., Matts, R. L., Thomas, N. S. B., Petryshyn, R., and

Chen, J. J. (1987). Regulation of protein synthesis, *The Enzymes, Vol. 18* (P. D. Boyer and E. G. Krebs, eds.), Academic Press, New York, pp. 359–380.

Lowry, C. V., and Lieber, R. H. (1986). Negative regulation of the *Saccharomyces cerevisiae ANB1* gene by heme, as mediated by the *ROX1* gene product, *Mol. Cell. Biol., 6:* 4145–4148.

Lowry, C. V., and Zitomer, R. S. (1984). Oxygen regulation of anaerobic and aerobic genes mediated by a common factor in yeast, *Proc. Natl. Acad. Sci. USA, 81:* 6129–6133.

Lowry, C. V., and Zitomer, R. S. (1988). *ROX1* encodes a heme-induced repression factor regulating *ANB1* and *CYC7* of *Saccharomyces cerevisiae, Mol. Cell. Biol., 8:* 4651–4658.

Lowry, C. V., Cerdan, M. E., and Zitomer, R. S. (1990). A hypoxic consensus operator and a constitutive activation region regulate the *ANB1* gene of *Saccharomyces cerevisiae, Mol. Cell. Biol., 10:* 5921–5926.

Ma, J., and Ptashne, M. (1987). Deletion analysis of GAL4 defines two transcriptional activating segments, *Cell, 48:* 847–853.

MacKay, V. L. (1983). Cloning of yeast *STE* genes in 2 μm vectors, *Methods Enzymol., 101:* 325–343.

Mahler, H. R., and Lin, C. C. (1974). The derepression of δ-aminolevulinate synthetase in yeast, *Biochem. Biophys. Res. Comm., 61:* 963–970.

Mahler, H. R., and Lin, D.-C. (1978). Molecular events during the release of δ-aminolevulinate dehydratase from catabolite repression, *J. Bacteriol., 135:* 54–61.

Malamud, D. R., Borralho, L. M., Panek, A. D., and Mattoon, J. R. (1979). Modulation of cytochrome biosynthesis in yeast by antimetabolite action of levulinic acid, *J. Bacteriol., 138:* 799–804.

Matner, R. R., and Sherman, F. (1982). Differential accumulation of two apo-isocytochromes *c* in processing mutants of yeast, *J. Biol. Chem., 257:* 9811–9821.

Marykwas, D. L., and Fox, T. D. (1989). Control of the *Saccharomyces cerevisiae* regulatory gene *PET494:* Transcriptional repression by glucose and translational induction by oxygen, *Mol. Cell. Biol., 9:* 484–491.

Mehta, K. D., and Smith, M. (1989). Identification of an upstream repressor site controlling the expression of an anaerobic gene (*ANB1*) in *Saccharomyces cerevisiae, J. Biol. Chem., 264:* 8670–8675.

Mehta, K. D., Leung, D., Lefebvre, L., and Smith, M. (1990). The *ANB1* locus of *Saccharomyces cerevisiae* encodes the protein synthesis intiation factor eIF-4D, *J. Biol. Chem., 265:* 8802–8807.

Miyake, S., and Sugimura, T. (1968). Coproporphyrinogenase in a respiration-deficient mutant of yeast lacking all cytochromes and accumulating coproporphyrin, *J. Bacteriol., 96:* 1997–2003.

Montgomery, D. L., Leung, D. W., Smith, M., Shalit, P., Faye, G., and Hall, B. (1980). Isolation and sequence of the gene for iso-2-cytochrome *c* in *S. cerevisiae, Proc. Natl. Acad. Sci. USA, 77:* 541–745.

Morris, N. R. (1976). Mitotic mutants of *Aspergillus nidulans, Genet. Res., 26:* 237–254.

Myers, A. M., Crivellone, M. D., Koerner, T. J., and Tzagaloff, A. (1987). Characterization of the yeast *HEM2* gene and transcriptional regulation of *COX5* and *COR1* by heme, *J. Biol. Chem., 262:* 16822–16829.

Nicholson, D. W., and Neupert, W. (1989). Import of cytochrome *c* into mitochondria: Reduction of heme, mediated by NADH and flavin nucleotides, is obligatory for its covalent linkage to apocytochrome *c, Proc. Natl. Acad. Sci. USA, 86:* 4340–4344.

Nicholson, D. W., Koehler, H., and Neupert, W. (1987). Import of cytochrome *c* into mitochondria. Cytochrome *c* heme lyase, *Eur. J. Biochem., 164:* 147–157.

Nicholson, D. W., Hergersberg, C., and Neupert, W. (1988). Role of cytochrome *c* lyase in the import of cytochrome *c* into mitochondria, *J. Biol. Chem., 263:* 19034–19042.

Nobrega, F. G., and Tzagoloff, A. (1980). Assembly of the mitochondrial membrane system. DNA sequence and organization of the cytochrome *b* gene in *Saccharomyces cerevisiae* D273-10B, *J. Biol. Chem., 255:* 9828–9837.

Olesen, J. T., and Guarente, L. (1990). The HAP2 subunit of yeast CCAAT transcriptional activator contains adjacent domains for subunit association and DNA recognition: Model for the HAP2/3/4 complex, *Genes Dev., 4:* 1714–1729.

Olesen, J., Hahn, S., and Guarente, L. (1987). Yeast HAP2 and HAP3 activators both bind to the *CYC1* upstream activation site, UAS2, in an interdependent manner, *Cell, 51:* 953–961.

Olesen, J. T., Fikes, J. D., and Guarente, L. (1991). The *Schizosaccharomyces pombe* homolog of *Saccharomyces cerevisiae* HAP2 reveals selective and stringent conservation of the small essential core protein domain, *Mol. Cell. Biol., 11:* 611–619.

Olson, R. E., and Rudney, H. (1983). Biosynthesis of ubiquinone, *Vitam. Horm.* (New York), *40:* 1–43.

Pathak, V. K., Schindler, D., and Hershey, J. W. B. (1988). Generation of a mutant form of protein synthesis initiation factor eIF-2 lacking the site of phosphorylation by eIF-2 kinases, *Mol. Cell. Biol., 8:* 993–995.

Pfeifer, K., Arcangioli, B., and Guarente, L. (1987a). Yeast *HAP1* activator competes with the factor RC2 for binding to the upstream activation site UAS1 of the *CYC1* gene, *Cell, 49:* 9–18.

Pfeifer, K., Prezant, T., and Guarente, L. (1987b). Yeast HAP1 activator binds to two upstream activation sites of different sequence, *Cell, 49:* 19–27.

Pfeifer, K., Kim, K.-S., Kogan, S., and Guarente, L. (1989). Functional dissection and sequence of yeast HAP1 activator, *Cell, 56:* 291–301.

Pinkham, J. L., and Guarente, L. (1985). Cloning and molecular analysis of the *HAP2* locus: A global regulator of respiratory genes in *Saccharomyces cerevisiae, Mol. Cell. Biol., 5:* 3410–3416.

Pinkham, J. L., Olesen, J. T., and Guarente, L. (1987). Sequence and nuclear localization of the *Saccharomyces cerevisiae* HAP2 protein, a transcriptional activator, *Mol. Cell. Biol., 7:* 578–585.

Poulson, R. (1976). The regulation of heme synthesis, *Ann. Clin. Res., 8;* 56–63.

Poulson, R., and Polglase, W. J. (1974). Site of glucose repression of heme biosynthesis, *FEBS Lett., 40:* 258–260.

Prezant, T., Pfeifer, K., and Guarente, L. (1987). Organization of the regulatory region of the yeast *CYC7* gene: Multiple factors are involved in regulation, *Mol. Cell. Biol., 7:* 3252–3259.

Raitio, M., Jalli, T., and Saraste, M. (1987). Isolation and analysis of the genes for cytochrome *c* oxidase in *Paracoccus dinitrificans, EMBO J., 6:* 2825–2833.

Repetto, B., and Tzagoloff, A. (1989). Structure and regulation of *KGD1,* the structural gene for yeast α-ketoglutarate dehydrogenase, *Mol. Cell. Biol., 9:* 2695–2705.

Repetto, B., and Tzagoloff, A. (1990). Structure and regulation of *KGD2,* the structural gene for yeast dihydrolipoyl transsuccinylase, *Mol. Cell. Biol., 10:* 4221–4232.

Rhee, S.-K., Icho, R., and Wickner, R. B. (1989). Structure and nuclear localization signal of the SKI3 antiviral protein of *Saccharomyces cerevisiae, Yeast, 5:* 149–158.

Rosenblum-Vos, L. S., Rhodes, L., Evangelista, C. C., Jr., Boayke, K. A., and Zitomer, R. S. (1991). The *ROX3* gene encodes an essential nuclear protein involved in *CYC7* gene expression in *Saccharomyces cerevisiae, Mol. Cell. Biol., 11:* 5639–5647.

Rothstein, J. R., and Sherman, F. (1980). Genes affecting the expression of cytochrome *c* in yeast: Genetic mapping and genetic interactions, *Genetics, 94:* 871–899.

Roussou, I., Thieros, G., and Hauge, B. M. (1988). Transcriptional-translational regulatory circuit in *Saccharomyces cerevisiae* which involves the GCN4 transcriptional activator and the GCN2 protein kinase, *Mol. Cell. Biol., 8:* 2132–2139.

Rytka, J., Bilinski, T., and Labbe-Bois, R. (1984). Modified uroporphyrinogen decarboxylase activity in a yeast mutant which mimics porphyria cutanea tarda, *Biochem. J., 218:* 405–413.

Sadler, I., Suda, K., Shatz, G., Kaudewitz, F., and Haid, A. (1984). Sequencing of the nuclear gene for the yeast cytochrome c_1 precursor reveals an unusually complex amino-terminal presequence, *EMBO J., 3:* 2137–2143.

Salemme, R. (1977). Structure and function of cytochrome *c, Ann. Rev. Biochem., 46:* 299–329.

Saltzgaber-Müller, J., and Schatz, G. (1978). Heme is necessary for the accumulation and assembly of cytochrome *c* oxidase subunits in *Saccharomyces cerevisiae, J. Biol. Chem., 253:* 305–310.

Schneider, J. C. (1989). "Mechanism of coordinate induction of cytochrome genes in *Saccharomyces cerevisiae,*" Ph.D. thesis, Massachusetts Institute of Technology.

Schneider, J. C., and Guarente, L. (1991). Regulation of yeast *CYT1* gene encoding cytochrome c_1 by HAP1 and HAP2/3/4, *Mol. Cell. Biol., 11:* 4934–4942.

Schnier, J., Schwelberger, H. G., Smit-McBride, Z., Kang, H. A., and Hershey, J. W. B. (1991). Translation initiation factor 5A and its hypusine modification are essential for cell viability in the yeast *Saccharomyces cerevisiae, Mol. Cell. Biol., 11:* 3105–3114.

Schultz, J., Marshall-Carlson, L., and Carlson, M. (1990). The N-terminal TPR region is the functional domain of SSN6, a nuclear phosphoprotein of *Saccharomyces cerevisiae, Mol. Cell. Biol., 10:* 4744–4756.

Shapleigh, J. P., Hosler, J. P., Tecklenburg, M. M., Kim, Y., Babcock, G. T., Gennis, R. B., and Ferguson-Miller, S. (1992). Definition of the catalytic site of cytochrome *c* oxidase: Specific ligands of heme *a* and the heme a_3-Cu_B center, *Proc. Natl. Acad. Sci. USA, 89:* 4786–4790.

Sikorski, R. S., Boguski, M. S., Goebl, M., and Hieter, P. (1990). A repeating amino acid motif in *CDC23* defines a family of proteins and a new relationship among genes required for mitosis and RNA synthesis, *Cell, 60:* 307–317.

Struhl, K. (1985). Naturally occurring poly(dA-dT) sequences are upstream promoter elements for constitutive transcription in yeast, *Proc. Natl. Acad. Sci. USA, 82:* 8419–8423.

Taanman, J.-W., and Capaldi, R. A. (1992). Purification of yeast cytochrome *c* oxidase with a subunit composition resembling the mammalian enzyme, *J. Biol. Chem., 267:* 22481–22485.

Tamura, Y., Yosida, Y., Sato, R., and Kumaoka, H. (1976). Fatty acid desaturase system of yeast microsomes. Involvement of cytochrome b_5–containing electron-transport chain, *Arch. Biochem. Biophys., 175:* 284–294.

Thorsness, M., Schafer, W., D'Ari, L., and Rine, J. (1989). Positive and negative transcriptional conrol by heme of genes encoding 3-hydroxy-3-methylglutaryl coenzyme A reductase in *Saccharomyces cerevisiae, Mol. Cell. Biol., 9:* 5702–5712.

Thrash-Bingham, C., and Fangman, W. L. (1989). A yeast mutation that stabilizes a plasmid bearing a mutated *ARS1* element, *Mol. Cell. Biol., 9:* 809–816.

Trawick, J. D., Wright, R. M., and Poyton, R. O. (1989). Transcription of yeast *COX6,* the gene for cytochrome *c* oxidase Subunit VI, is dependent on heme and on the *HAP2* gene, *J. Biol. Chem., 264:* 7005–7008.

Trawick, J. D., Kraut, N., Simon, F. R., and Poyton, R. O. (1992). Regulation of yeast *COX6* by the general transcription factor ABF1 and separate HAP2- and heme-responsive elements, *Mol. Cell. Biol., 12:* 2302–2314.

Trueblood, C. E., and Poyton, R. O. (1988). Identification of *REO1,* a gene involved in negative regulation of *COX5b* and *ANB1* in aerobically grown *Saccharomyces cerevisiae, Genetics, 120:* 671–680.

Trueblood, C. E., Wright, R. M., and Poyton, R. O. (1988). Differential regulation of the two genes encoding *Saccharomyces cerevisiae* cytochrome *c* oxidase subunit V by heme and the *HAP2* and *REO1* genes, *Mol. Cell. Biol., 8:* 4537–4540.

Trumbly, R. J. (1986). Isolation of *Saccharomyces cerevisiae* mutants constitutive for invertase synthesis, *J. Bacteriol., 166:* 1123–1127.

Trumbly, R. J. (1988). Cloning and characterization of the *CYC8* gene mediating glucose repression in yeast, *Gene, 73:* 97–111.

Turcotte, B., and Guarente, L. (1992). HAP1 positive control mutants specific for one of two binding sites, *Genes Dev., 6:* 2001–2009.

Turi, T. G., and Loper, J. C. (1992). Multiple regulatory elements control the *Saccharomyces cerevisiae* cytochrome P450, lanosterol 14α-demethylase (*ERG11*), *J. Biol. Chem., 267:* 2046–2056.

Turi, T. G., Kalb, V. F., and Loper, J. C. (1991). Cytochrome P450 lanosterol 14 α-demethlyase (*ERG11*) and manganese superoxide dismutase (*SOD1*) are adjacent genes in *Saccharomyces cerevisiae, Yeast, 7:* 627–630.

Urban-Grimal, D., and Labbe-Bois, R. (1981). Genetic and biochemical characterization of mutants of *Saccharomyces cerevisiae* blocked in six different steps of heme biosynthesis, *Mol. Gen. Genet., 183:* 85–92.

Urban-Grimal, D., Volland, C., Garnier, T., Dehoux, P., and Labbe-Bois, R. (1986). The nucleotide sequence of the *HEM1* gene and evidence for a precursor form of the mitochondrial 5-aminolevulinate synthase in *Saccharomyces cerevisiae, Eur. J. Biochem., 156:* 511–519.

Verdière, J., and Petrochillo, E. (1979). *CYP3-15:* An up-promoter mutation at the iso 2-cytochrome c structural gene locus and its interaction with other independent regulatory mutations of cytochrome c synthesis in *Saccharomyces cerevisiae, Molec. Gen. Genet., 175:* 209–216.

Verdière, J., Creusot, F., Guarente, L., and Slonimski, P. P. (1986). The overproducing *CYP1* and the underproducing *hap1* mutations are alleles of the same gene which regulates in trans the expression of the structural genes encoding isocytochromes c, *Curr. Genet., 10:* 339–342.

Verdière, J., Gaisne, M., Guiard, B., Defranoux, N., and Slonimski, P. P. (1988). CYP1 (HAP1) regulator of oxygen-dependent gene expression in yeast. II. Missense mutation suggests alternative Zn fingers as discriminating agents of gene control, *J. Mol. Biol., 204:* 277–282.

Verdière, J., Gaisne, M., and Labbe-Bois, R. (1991). *CYP1* (*HAP1*) is a determinant effector of alternative expression of heme-dependent transcription in yeast, *Mol. Gen. Genet., 228:* 300–306.

Waterland, R. A., Basu, A., Chance, B., and Poyton, R. O. (1991). The isoforms of yeast cytochrome *c* oxidase subunit V alter the *in vivo* kinetic properties of the holoenzyme, *J. Biol. Chem., 266:* 4180–4186.

Wickner, R. B. (1974). Mutants of *Saccharomyces cerevisiae* that incorporate deoxythymidine-5′-monophosphate into deoxyribonucleic acid *in vivo, J. Bacteriol., 117:* 252–260.

Williams, F. E., and Trumbly, R. J. (1990). Characterization of *TUP1,* a mediator of glucose repression in *Saccharomyces cerevisiae, Mol. Cell. Biol., 10:* 6500–6511.

Williams, F. E., Varanasi, U., and Trumbly, R. J. (1991). The CYC8 and TUP1 proteins involved in glucose repression in *Saccharomyces cerevisiae* are associated in a protein complex, *Mol. Cell. Biol., 11:* 3307–3316.

Winkler, H., Adam, G., Mattes, E., Schanz, M., Hartig, A., and Ruis, H. (1988). Coordinate control of synthesis of mitochondrial and non-mitochondrial hemoproteins: A binding site for the HAP1 (CYP1) protein in the UAS region of the yeast catalase T gene (*CTT1*), *EMBO J., 7:* 1799–1804.

Wright, R. M., and Poyton, R. O. (1990). Release of two *Saccharomyces cerevisiae* cytochrome genes, *COX6* and *CYC1,* from glucose repression requires the *SNF1* and *SSN6* gene products, *Mol. Cell. Biol., 10:* 1297–1300.

Yoshimoto, A., and Sato, R. (1970). Studies on yeast sulfite reductase. III. Further characterization, *Biochim. Biophys. Acta, 220:* 190–205.

Zagorec, M., and Labbe-Bois, R. (1986). Negative control of yeast coproporphyrinogen oxidase synthesis by heme and oxygen, *J. Biol. Chem., 263:* 9718–9724.

Zagorec, M., Buhler, J.-M. Treich, I., Keng, T., Guarente, L., and Labbe-Bois, R. (1988). Isolation, sequence, and regulation by oxygen of the yeast *HEM13* gene coding for coproporphyrinogen oxidase, *J. Biol. Chem., 263:* 9718–9724.

Zhang, M., Rosenblum-Vos, L. S., Lowry, C. V., Boayke, K. A., and Zitomer, R. S. (1991). A yeast protein with homology to the β-subunit of G proteins is involved in control of heme-regulated and catabolite-repressed genes, *Gene, 97:* 153–161.

Zitomer, R. S., and Lowry, C. V. (1992). Regulation of gene expression by oxygen in *Saccharomyces cerevisiae, Microbiol. Rev., 56:* 1–11.

Zitomer, R. S., Montgomery, D. L., Nichols, D. L., and Hall, B. D. (1979). Transcriptional regulation of the yeast cytochrome *c* gene, *Proc. Natl. Acad. Sci. USA, 76:* 3627–3631.

Zitomer, R. S., Sellers, J. W., McCarter, D. W., Hastings, G. A., Wick, P., and Lowry, C. V. (1987). Elements involved in oxygen regulation of the *Saccharomyces cerevisiae CYC7* gene, *Mol. Cell. Biol., 7:* 2212–2220.

Index